Nondestructive Characterization of Materials IV

Nondestructive Characterization of Materials IV

Edited by

Clayton O. Ruud
Pennsylvania State University
University Park, Pennsylvania

Jean F. Bussière
Industrial Materials Institute
National Research Council of Canada
Boucherville, Quebec, Canada

and

Robert E. Green, Jr.
The Johns Hopkins University
Baltimore, Maryland

PLENUM PRESS • NEW YORK AND LONDON

Library of Congress Cataloging-in-Publication Data

International Symposium on Nondestructive Characterization of Materials (4th : 1990 : Annapolis, Md.)
Nondestructive characterization of materials IV / edited by Clayton O. Ruud, Jean F. Bussière, and Robert E. Green, Jr.
p. cm.
"Proceedings of the Fourth International Symposium on Nondestructive Characterization of Materials, held June 11-14, 1990, in Annapolis, Maryland"--T.p. verso.
Includes bibliographical references and index.
ISBN 0-306-44047-4
1. Non-destructive testing--Congresses. 2. Materials--Testing--Congresses. I. Ruud, C. O. II. Bussière, Jean F. III. Green, Robert E. (Robert Edward), 1932- . IV. Title.
TA417.2.I56 1990
620.1'127--dc20 91-29691
CIP

Proceedings of the Fourth International Symposium on Nondestructive Characterization of Materials, held June 11-14, 1990, in Annapolis, Maryland

A Division of Plenum Publishing Corporation
233 Spring Street, New York, N.Y. 10013

Printed in the United States of America

PREFACE

There is a great deal of interest in extending nondestructive technologies beyond the location and identification of cracks and voids. Specifically there is growing interest in the application of nondestructive evaluation (NDE) to the measurement of physical and mechanical properties of materials. The measurement of materials properties is often referred to as materials characterization; thus nondestructive techniques applied to characterization become nondestructive characterization (NDC).

There are a number of meetings, proceedings and journals focused upon nondestructive technologies and the detection and identification of cracks and voids. However, the series of symposia, of which these proceedings represent the fourth, are the only meetings uniquely focused upon nondestructive characterization. Moreover, these symposia are especially concerned with stimulating communication between the materials, mechanical and manufacturing engineer and the NDE technology oriented engineer and scientist. These symposia recognize that it is the welding of these areas of expertise that is necessary for practical development and application of NDC technology to measurements of components for in-service life time and sensor technology for intelligent processing of materials.

These proceedings are from the fourth international symposia and are edited by C.O. Ruud, J. F. Bussiere and R.E. Green, Jr. The dates, places, etc of the symposia held to date area as follows:

TITLE:	Symposia on Nondestructive Methods for Material Property Determination
DATES:	April 6-8, 1983
PLACE:	Hershey, PA, USA
CHAIRPERSONS:	C.O. Ruud and R.E. Green, Jr.

TITLE:	Second International Symposia on the Nondestructive Characterization of Materials
DATES:	July 21-23, 1986
PLACE:	Montreal, Quebec, CANADA
CHAIRPERSONS:	J.F. Bussiere, R.E. Green, Jr., J.P. Mouchalin and C.O. Ruud

TITLE:	Third International Symposium on Nondestructive Characterization of Materials
DATES:	October 3-6, 1988
PLACE:	Saarbrucken, Germany
CHAIRPERSONS:	P. Holler, R.E. Green, Jr. and C.O. Ruud

TITLE:	Fourth International Symposium on the Nondestructive Characterization of Materials
DATES:	June 11-14, 1990
PLACE:	Annapolis, MD, USA
CHAIRPERSONS:	R.E. Green, Jr. and C.O. Ruud

The title, editors, etc of the proceedings of these symposia are as follows:

TITLE:	Nondestructive Methods for Material Property Determination
EDITORS:	C.O. Ruud and R.E. Green, Jr.
PUBLISHER:	Plenum Press, New York, NY, USA
YEAR PUBLISHED:	1984

TITLE:	Nondestructive Characterization of Materials II
EDITORS:	J.F. Bussiere, R.E. Green, Jr., J.P. Mouchalin and C.O. Ruud
PUBLISHER:	Plenum Press, New York, NY, USA
YEAR PUBLISHED:	1987

TITLE:	Nondestructive Characterization of Materials
EDITORS:	P. Holler, V. Hauk, G. Dobmann, C.O. Ruud and R.E. Green, Jr.
PUBLISHER:	Springler-Verlag, Berlin
YEAR PUBLISHED:	1989

TITLE:	Nondestructive Characterization of Materials IV
EDITOR:	C.O. Ruud, J.F. Bussiere and R.E. Green, Jr.
PUBLISHER:	Plenum Press, New York, NY, USA
YEAR PUBLISHED:	1991

The fifth International Symposium on Nondestructive Characterization of Materials was held May 27-30, 1991 in Karuizawa, Japan and the proceedings are in preparation at this printing.

CONTENTS

COATINGS AND THIN FILMS

WOOD, PAPER, CONCRETE, AND CERAMICS

MATERIAL CHARACTERIZATION AND MODELING

MICROSTRUCTURE AND PLASTIC DEFORMATION OF METALS

COMPOSITES

RESIDUAL STRESSES AND FRACTURE MECHANISMS

TEXTURE AND PROPERTY ANISOTROPY

ACCURATE STRUCTURAL CHARACTERIZATION OF ZrN COATINGS AND EPITAXIAL LAYERS BY X-RAY DIFFRACTION USING THE DOSOPHATEX SYSTEM

R. Y. Fillit*, A. J. Perry+, J. Pol Dodelet△, G. Perrier°, and R. Philippe°

*ENS des Mines de Saint-Etienne 42100 (France)
+GTE Valenite Corporation Troy MI48084 (USA)
△INRS Energie Varennes (Canada)
°IUT Saint-Etienne 42100 (France)

INTRODUCTION

Thin film characterization by X-Ray methods is often difficult because of the problems of preferred crystal orientations and residual stresses. For example, it is well known that coatings made by physical vapor deposition are very strongly textured so that measurements of residual stress by classical X-Ray methods are very inaccurate. Furthermore, the characterization of epitaxial thin layers of semi-conductors requires very accurate determination of crystal orientations and interfacial misfits.

A new diffractometer[1] has been developed to form a multipurpose 4-circle assembly (α, 2θ, ϕ, ψ), the different rotational positions of which are selected by computer driven high precision and high speed stepping motors (Fig. 1). Depending on the type of angular scanning selected, as well as the data processing and acquisition software, the following analysis may be rapidly performed on our system :

* Identification of textured phases,
* Volume fraction determination,
* Texture analysis by pole figures and Orientation Distribution function (ODF),
* Single crystal, or epitaxial layer orientations,
* Residual stress measurements (by psi setup),
* Grazing X-ray analysis of ultra-thin layers.

In fact these analyses complement each other and enable us to undertake any type of complex research. Here we shall present a few applications.

A- PHASE IDENTIFICATION

This is a basic analysis for materials characterization, which must precede all other analyses and must be perfect : all the peaks {hkl} of all the existing phases of the sample must be detected. Usually the well-known disturbing effects of texture lead to less accurate results.

In standard analyses of a single crystal of {111} oriented silicon, only the {111} peak and its multiples are detected (Fig. 2).

Nondestructive Characterization of Materials IV
Edited by C.O. Ruud *et al.*, Plenum Press, New York, 1991

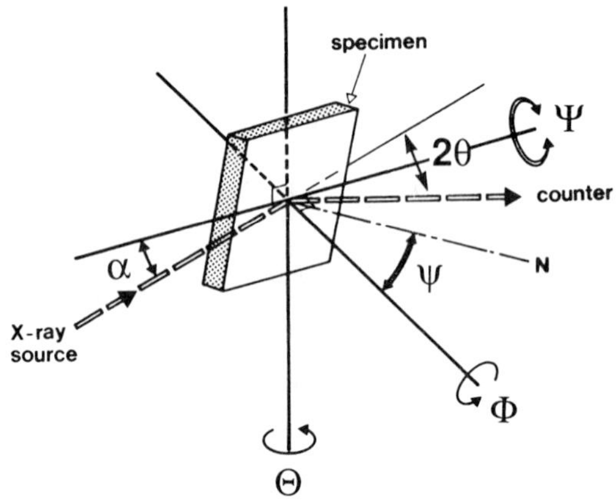

Fig. 1. The 4 circle assembly (α, 2θ, ϕ, Ψ) of the Dosophatex system.

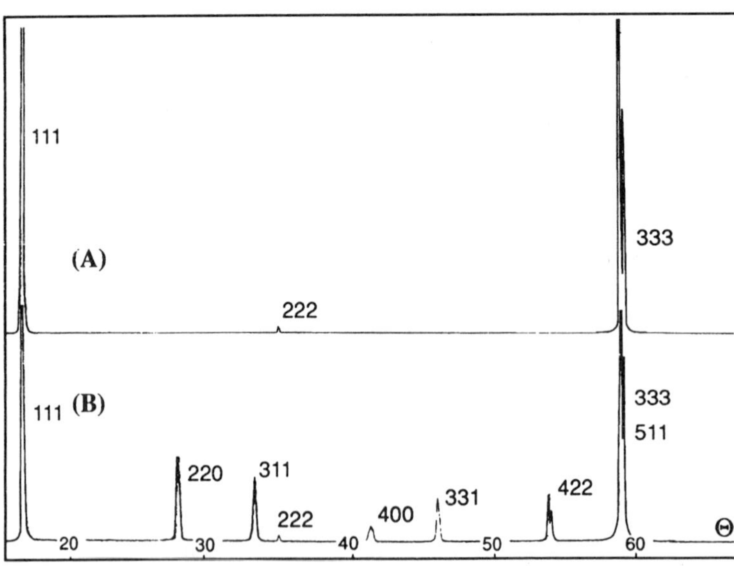

Fig. 2. θ-2θ diagrams of a single crystal of Silicon oriented (111)
(A) : Standard analysis.
(B) : Dosophatex analysis which performs integration over the range ϕ (0°-360°) and Ψ (± 60°).

On the other hand, using our system, due to a rapid sample rotation and oscillation, it is rendered isotropic with respect to the analysis : all the {hkl} peaks appear.

One of the main advantages of our system consists in the analysis of new phases[2].

B- QUANTITATIVE ANALYSIS

Determining the volume fraction of the phases of a multi-phased material is fundamental for knowledge of its strength properties or resistance to corrosion for example. With standard analyses, the effects of texture, which are almost always present, render quantitative phase analysis very difficult and rarely performed. But, with our system quantitative analysis is a frequent application.

Dosophatex enables one to solve problems as complex as determining the volume fraction of monoclinic and tetragonal zirconia in zirconia-toughened alumina ceramics[3]. This quantitative analysis is fundamental to understand toughening mechanisms. Dosophatex analysis is a general method capable of performing volume fraction determination of α, γ and carbides phases in high-speed steels where 6 to 7 phases are associated, while producing results to within 1 % ! Of course phase structure must be perfectly known in order to calculate theoretical intensities in connection with geometry assembly and for the wavelength used.

C - STRESS DETERMINATION

We shall present the possibilities our system offers by studying ZrN film deposited to a thickness of 5 μm onto stainless steel then annealed at 900° 1 hour[4].

In order to avoid the problem of an evolution in the composition with respect to the depth analysis in the deposit, we selected a deposit offering the best possible homogeneity, by means of preliminary Auger and electron probes[5,6] analysis.

The phase analysis was performed,using Cobalt Kα radiation, in order to select all the ZrN peaks {hkl} which did not overlap those of the austenitic steel substrate (Fig.3).

Moreover, for deformation measurements, we selected Chromium Kα radiation to limit as much as possible the analysis depth. We observe linear strain versus $\sin^2\phi$ plots (sspp) on each plane considered (Fig.4) showing a strong biaxial compressive stress on the surface of the deposit. To calculate the stress, we used a Young's modulus, of 460,000 MPa, and a Poisson's ratio 0.2. Also, since the X-ray elastic anisotropy was not known, we attempted to determine it directly : the measurement must give the same stress value, regardless of the {hkl} peak considered, if we enter the correct X-ray elastic anisotropy into the calculation[7,8].

The compressive stresses for the different planes considered are calculated by varying the X-ray elastic anisotropy (A_{XC}) (Fig.5). The curves roughly intersect at the same point, which according to the least square error method, gives us for A_{XC} the value of 0.83. By taking this A_{XC} value, we find that the surface compressive stress is σ = -7,000 MPa.

On the other hand, with more penetrating Copper Kα radiation passing entirely through the deposit, the deformation curves are no longer linear (Fig.6) and demonstrate that shear stresses exist[9]. This is directly linked to the stress gradient through the deposit, which may be explained by the different cooling rates between the substrate and the deposit.

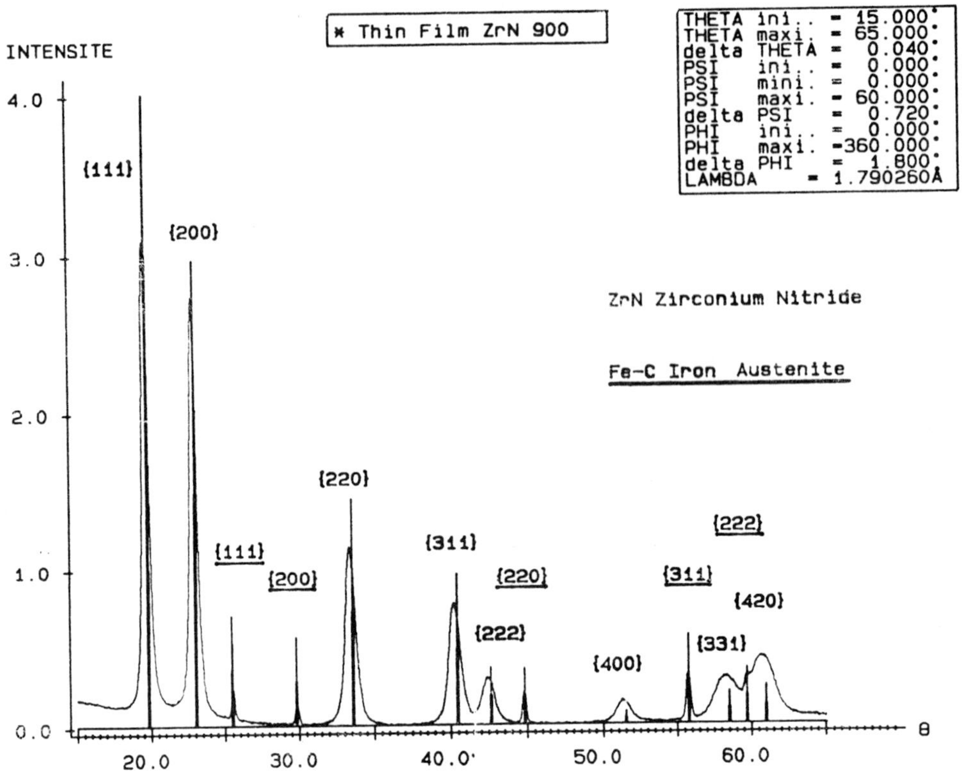

Fig. 3. XRD diagram with CoK_α for ZrN sample using Dosophatex system.

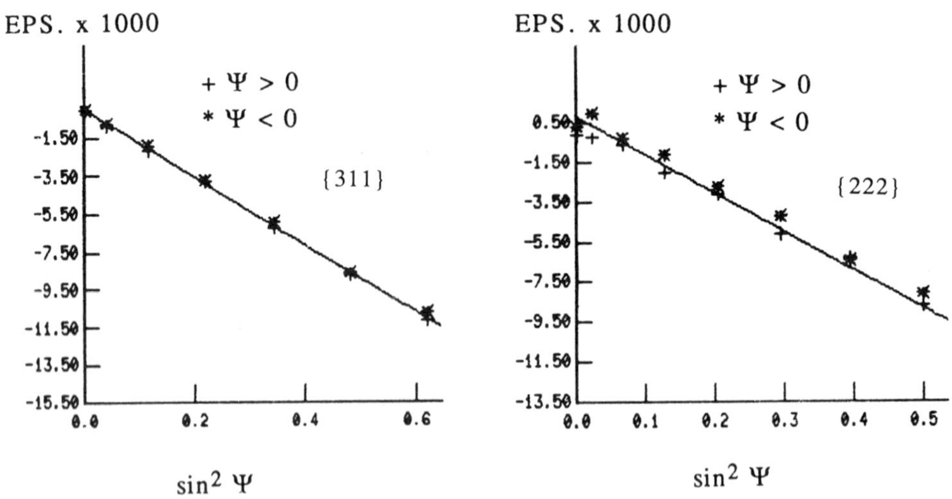

Fig. 4. Plots of residual strain as a function of $\sin^2 \Psi$ from {311} and {222} diffraction peaks of ZrN deposit as determined with $K_\alpha Cr$.

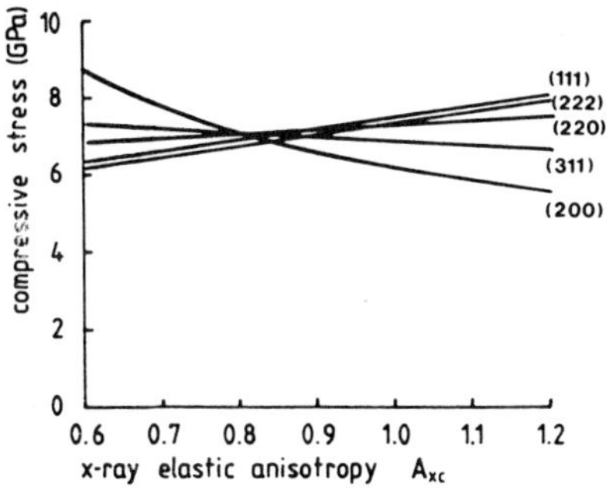

Fig. 5. Residual stress as a function of the ration $A_{xc} = \frac{S_2(h00)}{S_2(hhh)}$

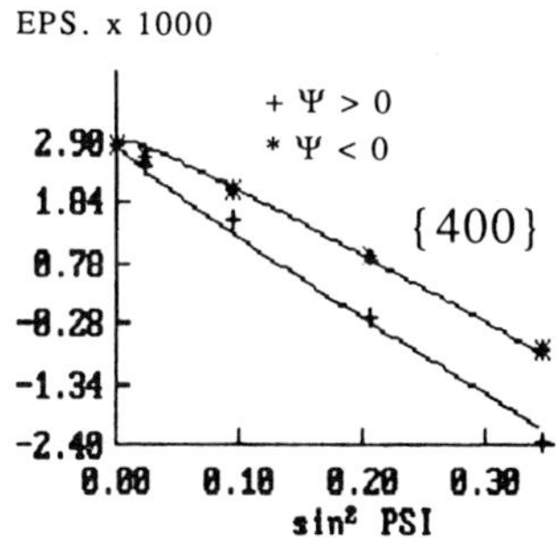

Fig. 6. sspp from {400} CuK_α diffraction peak of the ZrN film

D- TEXTURE ANALYSIS

Until now, we have considered textures with respect to their negative aspects on the accuracy of results, and we have presented a way of minimizing these negative effects.

However, it is equally essential to perfectly determine these textures in order, for example, to calculate the behavior of materials that are submitted to mechanical deformations.

We shall consider an extreme case of analysing epitaxial layers for the manufacture of photo-voltaïc cells[10].
Two types of layers were analysed :

(1) ZnSe layers on (100) oriented GaAs single crystal substrates. The purpose was to develop epitaxy via their isomorphism, which differs only by 0.3 %. Note, however, that ZnSe also crystallizes to a metastable hexagonal phase[11]

(2) GaAs layers on Ge single crystal substrate oriented (100). The isomorphism in this case is identical to within 0.07 %.

These layers were developed according to the close space vapor transport method (CSVT)[12]. This method is simple, low cost and offers the advantage of a high yield capable of reaching 90 %. We select two ZnSe layers through their morphology using SEM

(1) layer #1 was a non-bonding deposit and spontaneous debonding occured ; the interface between the layer and the substrat was smooth.

(2) layer #2 deposit was a highly bonding deposit.

Now, it is important to verify epitaxial relations, if they exist. In general, epitaxy is checked by means of Rocking Curves, which is a very high accuracy method, but which is used only to check simple epitaxial relations[13,14].

For more complex cases, with possibility of multiple epitaxial phases, the pole figure still remains the most efficient means for determining crystallographic orientation relations between the substrate and the deposit. For this operation, we developed an original method using high-resolution ultra-rapid pole figures (up to $\psi = 80°$) for which we performed the

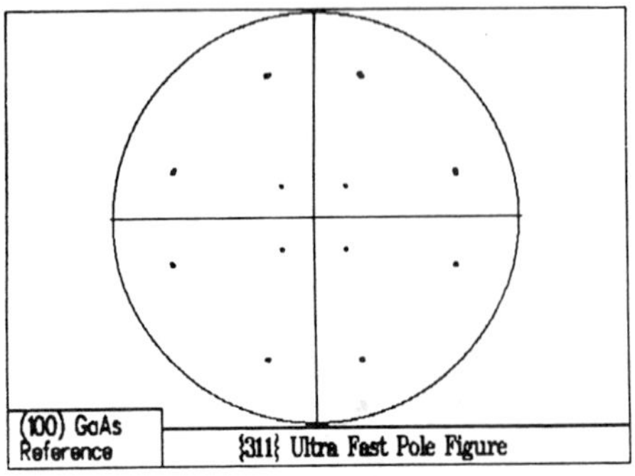

Fig. 7. High resolution {311} pole figure of a reference GaAs (100) single crystal

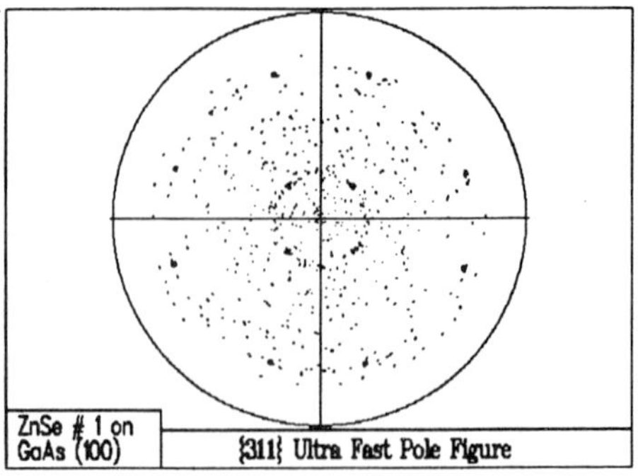

Fig. 8. High resolution {311} figure of layer ZnSe #1

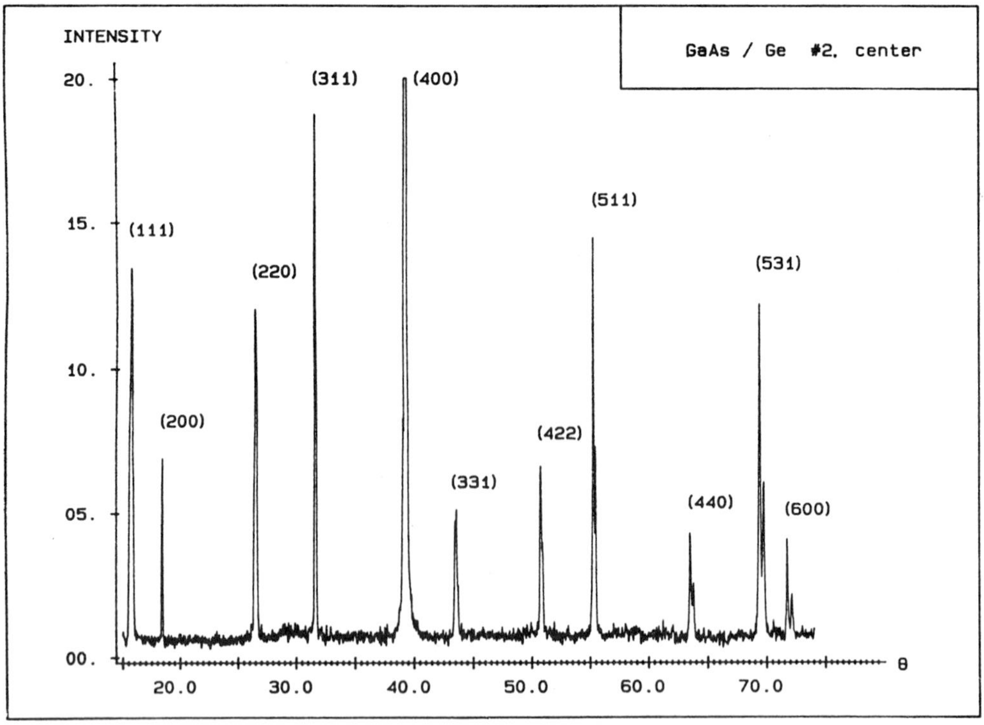

Fig. 9. XRD Dosophatex diagram of layer GaAs # 2

acquisition of 10,000,000 measurement points in only a few minutes, which is much better than standard pole figures that perform 2,000 to 3,000 measurement points in approximately 1 hour.

We selected the {311} pole figure for its high multiplicity (24) and also because the associated θ angle is small and therefore the depth of X-ray penetration is restricted to the layer only.

We analyse, as a reference, a (100) GaAs single crystal. The {311} pole figure (Fig.7), with these 12 poles, is that of an ideal single crystal having a {001} <010> orientation.

The {311} pole figure of the poor-quality Deposit #1 shows (Fig.8) in addition to epitaxial poles, a great amount of randomly disoriented grains. On the other hand, the {311} pole figure of Deposit #2 confirm a very good epitaxy (like Fig.7).

The deposits of GaAs epitaxial layers on Ge (100) also showed different qualities[15]. Let us examine the results obtained with two samples : layer #1 had a dull gray aspect ; layer #2, was perfectly mirror like.

The Dosophatex θ,2θ diagram of layer #2 (Fig.9), shows a single-phased GaAs deposit. Its {311} pole figure (like Fig.7) shows a perfect epitaxy with Ge.

In the {311} pole figure of the dull grey deposit #1 (Fig.10) we have detected, in addition to the 12 epitaxial poles, a considerable amount of poles perfectly and symmetrically distributed and whose intensities are characteristic of their position. This pole figure proves the existence of several simultaneous epitaxies. In order to interpret this result, we used complementary pole figures, {111} and {220}, selected also because their small θ angles avoided the simultaneous analysis of the Ge substrate. In the {111} pole figure (Fig.11) we observe, in addition to the 4 standard epitaxial poles, the abnormal epitaxial poles. The same applies for the {220} pole figure.

A complete interpretation of these figures demonstrates the existence of the following epitaxy components :

a) the standard component {001} <010>

b) 3 abnormal components $\{\bar{1}22\}$ $\langle 22\bar{1}\rangle$
$\{\bar{3}35\}$ $\langle 16\bar{3}\rangle$
$\{\bar{1}48\}$ $\langle 84\bar{1}\rangle$

These "abnormal" components have the property that a group of planes : one {110} plane and two {311} planes (or several of these groups at the same time) are in the same position as in the standard epitaxty.
It is clear that, a standard Rocking Curves analysis would not have established these components so easily.

A complete analysis of these results will be published[16]

CONCLUSION

We have demonstrated that the Dosophatex system is designed to be a multi-purpose apparatus, capable of undertaking complementary analyses : it enables one to undertake not only standard analyses, but extremely complex fundamental or applied researches.

Moreover, the system is distinguished by the wide angular ranges of α (0°, 90°), 2θ (0°, 170°), ϕ (0°,360°), ψ (± 90°) : c onsequently, it is possible to undertake grazing X-ray analyses in order to study ultra-thin layers and interfaces. We are presently testing this apparatus, in which Dosophatex equipment is linked to a Siemens rotating anode.

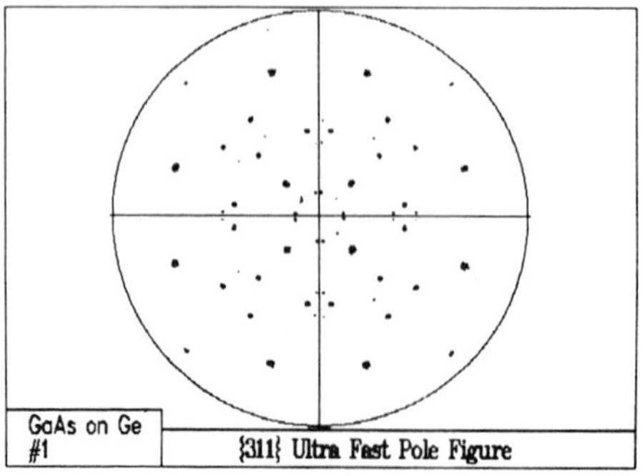

Fig. 10. High resolution {311} pole figure of GaAs layer #1

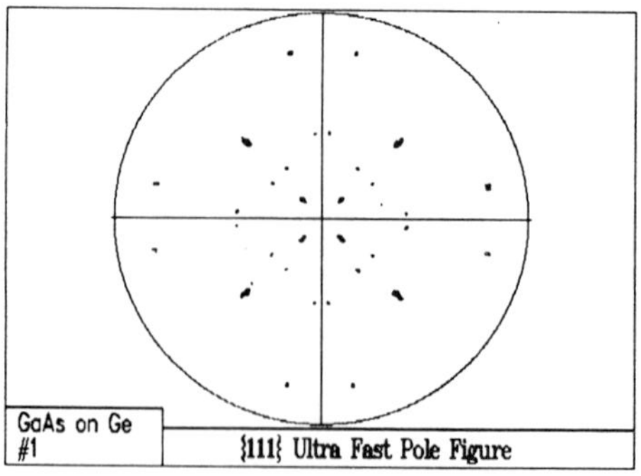

Fig. 11. High resolution {111} figure of GaAs layer #1

REFERENCES

1. R.Y.Fillit, H.Bruyas and F.Patay. Europatent 0161992.
2. J.P.Terrat, P.Fort, R.Y.Fillit, T.Naito and M.Hitomi. I.T.C.90 Nagoya.
3. R.Y.Fillit, P.Homerin, J.Schafer, H.Bruyas and F.Thévenot. J. Mater. Sci., 22 (1987) 3566.
4. R.Y.Fillit,A.J.Perry and S.Strandberg. Thin Solid Film, in press.
5. J.B.Brosse, R.Y.Fillit and M.Biscondi. Scripta Met. 15 (1981) 619.
6. A.J.Perry, C.Strandberg, W.D.Sproul, S.Hofmann, C.Ernsberger, J.Nickerson and L.Chollet. Thin Solid Films 153 (1987) 169.
7. H.Dolle J. Appl. Cryst. 12 (1979) 489.
8. M.Barral, J.L.Lebrun, J.M.Sprauel and G.Maeder. Met. Trans. 18A (1987) 1229.
9. D.S.Rickerby, A.M.Jones and B.A.Bellamy. Surface Coating Tech.36 (1988) 661.
10. B.A.Lombos,M.Bartkowski, M.Buchbinder, D.Cote and J.P.Dodelet. 5th Photovoltaic Solar Conference, W.Palz and F.Fittipaldi Editors, D.Reidel Co. (1983).
11. H. Hartmann. Kristall und Technik 5 (4) (1970) 527.
12. G.Perrier, R.Philippe and J.P.Dodelet. J. Mater. Res.3 (5) (1988) 1031.
13. I.C.Bassignana, A.-J.SpringThorpe and C.-C.Tan. Inst. Phys. Conf., Ser. No 91 (3) (1987) 267.
14. Nobuo Itoh and Keiichi Okamoto. J. App. Phys. 63(5) (1988) 1486.
15. D.Cossement, N.Guelton, E.Koskiahde, R.Y.Fillit, R.Saint.Jacques and J.P.Dodelet. 5th Canadian Semiconductor Technology. Conference Ottawa Canada august 14-16, 1990.
16. R.Y.Fillit, J.P.Dodelet and D.Cossement. J. Appl. Cryst. in press.

NONDESTRUCTIVE CHARACTERIZATION OF THE STRUCTURE AND MECHANICAL PROPERTIES OF METALLIC SUPERLATTICE THIN FILMS

T.E. Schlesinger and R.C. Cammarata

Department of Materials Science and Engineering
The Johns Hopkins University
Baltimore, MD 21218

C. Kim, S.B. Qadri, and A.S. Edelstein

Naval Research Laboratory
Washington, D.C. 20375

ABSTRACT

Superlattice thin films of Cu-Ni (50 atomic percent Cu-50 atomic percent Ni) with bilayer repeat lengths between 1.6 and 19.4 nm were produced by ion beam sputtering. The films were characterized by x-ray diffractometry, from which the bilayer repeat length and the average lattice parameter perpendicular to the plane of the film were determined. The average lattice parameter displayed a small increase as the bilayer repeat length was reduced. Films were also investigated using low load indentation methods. Such methods produce submicron deep indents, and leave most of the film unaffected, and as such can be considered nondestructive characterization techniques. A small enhancement above the rule of mixtures value of the effective elastic modulus of the films was observed (independent of the bilayer repeat length); however, a large enhancement in the hardness was measured for small bilayer repeat lengths. This hardness enhancement was attributed to an interface hardening effect. This behavior suggests that metallic superlattices may be have applications as high strength coatings or thin films.

SUPERLATTICE PREPARATION AND X-RAY CHARACTERIZATION

Superlattice thin films are composed of two different, alternately layered materials. The thickness of layers can be quite small, on the order of 1 nm. Superlattice thin films have several technologically important applications, such as strained-layer semiconductor devices, and as x-ray mirrors and gratings. Metallic superlattices have displayed interesting magnetic, superconducting, and mechanical properties that make them potentially important materials technologically.

Nondestructive Characterization of Materials IV
Edited by C.O. Ruud *et al.*, Plenum Press, New York, 1991

In this work, superlattice thin films composed of an equal amount of Cu and Ni were produced by a dual target ion beam sputtering system, and deposited onto Corning 0211 cover glass slides held at room temperature; details can be found elsewhere[1]. The microstructure of the films was investigated by θ-2θ x-ray diffractometry using a Rigaku rotating anode system. Typical scans are presented in Figure 1. All the films displayed a large degree of (111) in-plane texture. Also, the presence of several satellite peaks (at least in the films with the larger bilayer repeat lengths) indicates that the interfaces were relatively sharp. The bilayer repeat length λ of a film can be calculated from the angular positions of the central and satellite peaks using the equation

$$\lambda = n\Lambda / 2\,\Delta\theta\cos\theta,$$

where n is the order of the satellite peak, Λ is the x-ray wavelength, θ is the Bragg angle of the central peak, and $\Delta\theta$ is the angular separation between the central and nth order satellite peak[1]. The average lattice parameter a, perpendicular to the plane of the film, was also calculated from the Bragg peaks. It was found that the lattice parameter slightly increased as the bilayer repeat length decreased. As an illustration, for λ = 2.5 nm, a = 0.3595 nm, and for λ = 14.1 nm, a = 0.3581 nm. This general trend of a lattice expansion perpendicular to the plane of the film as the bilayer repeat length is decreased is consistent with the behavior reported for other metallic superlattices[2] and can be explained as due to interfacial effects[2,3].

MECHANICAL PROPERTIES CHARACTERIZATION

A series of films (each with a total thickness of between 1.1 to 1.3 μm) with bilayer repeat lengths between 1.6 and 16 nm, as well as homogeneous Cu and homogeneous Ni films, were investigated by low load indentation techniques. One method involved the use of a nanoindenter, which uses a three sided (Berkovitch) diamond tip indenter. Force versus depth curves were generated during the loading and unloading of the indenter. The total depth of the indentation was 50 nm. Hardness values were determined from the loading curves and elastic modulus values were determined from the unloading curves. The analysis used to determine these values has been given by Doerner and Nix[4]. The hardness of the same films were also measured using an Anton Paar MHT4 Ultramicrohardness Tester. In contrast to the nanoindentation method, which monitors the load for a given depth, the hardness was determined using the ultramicrohardness tester by measuring the indent size (and therefore its depth) for a given load. The Knoop indent geometry was chosen because it produces relatively shallow indents, and the load applied was 0.75 g. Using these parameters, the depth of indentation never exceeded more than about 10% of the film thickness, ensuring that the mechanical properties of the film and not of the substrate were being measured.

The low load indentation methods discussed above are nondestructive in the sense that only a small region (area of a few square microns, and a depth of less than 100 nm) is

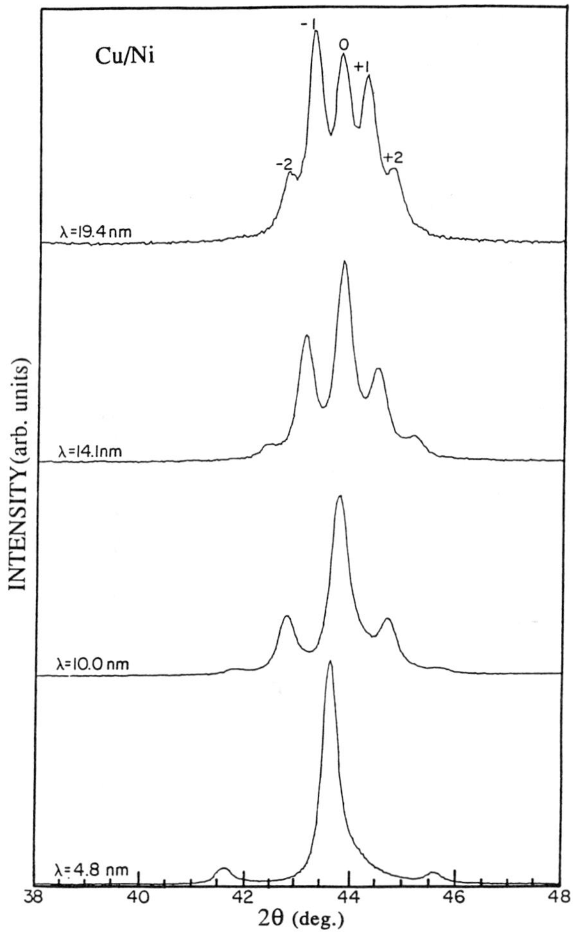

Figure 1. X-ray diffraction scans for Cu-Ni superlattices.

sampled, and most of the film is unaffected. These techniques are currently used as a nondestructive method for characterizing substrate-film adhesion, mechanical integrity of microelectronic structures, etc.

The relative elastic moduli of the superlattice films as a function of bilayer repeat length are shown in Figure 2. A relative modulus value of 1 represents the rule of mixtures value calculated using the elastic measurements obtained from the homogeneous films. Within the accuracy of the measurements, all of the films displayed about a 6% increase over the rule of mixtures values. Previous measurements of the elastic moduli of Cu-Ni multilayered thin films revealed that for bilayer repeat lengths near 1.5 to 2 nm, enhancements by a factor of two or more (the so-called supermodulus effect[5,6]) have been reported, though these results are still a subject of some controversy[7-9]. Clearly, a supermodulus effect was not observed in this study.

The relative hardness values as measured by the nanoindenter are given in Figure 3. As before, a value of 1 represents the rule of mixture value. All of the films displayed at least a 20% enhancement over the rule of mixtures, and the film with the 1.6 nm repeat length gave a 55% increase. This is close to the hardness of the homogeneous nickel film (which was about 2.4 times greater than that of the copper film). The enhanced hardness, and its dependence on the bilayer repeat length for the range 1.6 to 12 nm can be understood as due to an interfacial hardening mechanism[10]. These results suggest that multilayered thin films may have uses in application where increased mechanical strength is needed; e.g. as strong coatings or as VLSI interconnect materials that experience large stresses due to thermal cycling, etc. The enhanced hardness of the film with λ = 16 nm is more difficult to explain; it may be a manifestation of a large density of interfacial misfit dislocations that act as pinning sites for mobile dislocations.

The hardness of the superlattice films as measured by the ultramicrohardness tester is shown in Figure 4. As can be seen, the results are very similar to those presented in Figure 3. However, some of the films displayed a slight decrease relative to the rule of mixtures value. The reasons for this behavior are not clear.

SUMMARY

Nondestructive characterization of the microstructure and the mechanical properties of superlattice thin films have been presented. The microstructure of the films was investigated by x-ray diffractometry, and the mechanical properties were studied using low load indentation techniques. These microindentation techniques, although plastically deforming the sample during testing, create such shallow and small area indentations that the sample is virtually unharmed. Along with the capability of investigating films of submicron thickness, low load indentation methods, when used in conjunction with other nondestructive techniques such as x-ray diffraction, allow a comprehensive evaluation of the

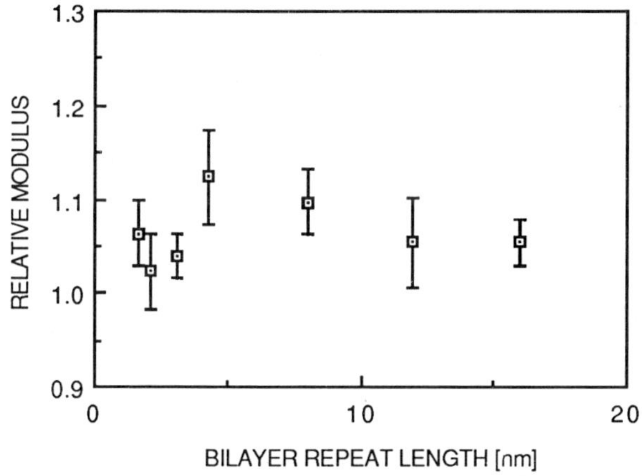

Figure 2. Relative moduli of Cu-Ni films as measured by the nanoindenter (indent depth = 50 nm).

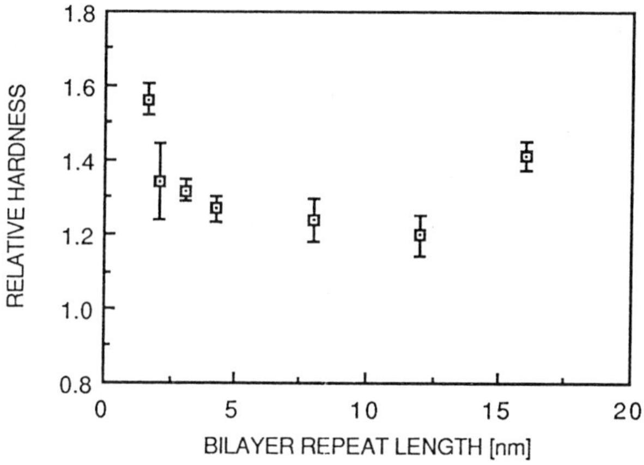

Figure 3. Relative hardness of Cu-Ni films as measured by the nanoindenter (indenter depth = 50 nm).

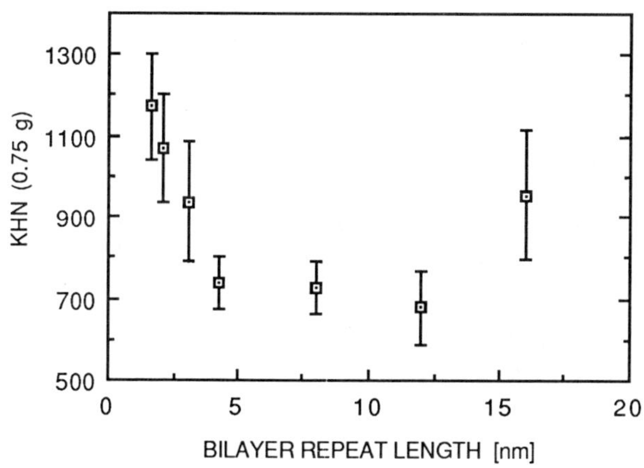

Figure 4. Knoop hardness of Cu-Ni films as measured by the low load indenter (rule of mixtures value = 825).

mechanical integrity and underlying structure of thin films and devices.

Superlattice Cu-Ni thin films displayed enhanced hardness behavior when the layer repeat length became small. This hardening was attributed to an interfacial hardening effect. Such materials may be useful in applications that require high strength thin films or coatings.

ACKNOWLEDGEMENTS

The authors would like to thank M. Ferber and D. Sammons for their assistance in performing the nanoindentation measurements, which were performed at the High Temperature Materials Laboratory of the Oak Ridge National Laboratory. This work was funded in part by the Department of Energy contract DE-AC05-76OR00033, and support from the Office of Naval Research is also gratefully acknowledged. One of us (TES) was supported by the Center for Nondestructive Evaluation at The Johns Hopkins University.

REFERENCES

1. C. Kim, S. B. Qadri, M. Twigg, and A. S. Edelstein, Synthesis and structural properties of ion-beam sputtered Cu-Ni multilayers, J. Vac. Sci. Technol. (in press).
2. B. M. Clemens and G. L. Eesley, _"Relationship between interfacial strain and the elastic response of multilayered metal films," Phys. Rev. Lett. 61:2356 (1988).
3. R. C. Cammarata and K. Sieradzki, "Effects of surface stress on the elastic moduli of thin films and superlattices," Phys. Rev. Lett. 62:2005 (1990).
4. M. F. Doerner and W. D. Nix, "A method for interpreting the data from depth-sensing indentation instruments," J. Mater. Res. 1:601 (1986).
5. T. Tsakalakos and J. E. Hilliard, "Elastic modulus in composition-modulated copper-nickel foils," J. Appl. Phys. 54:734 (1983).
6. D. Baral, J. B. Ketterson, and J. E. Hilliard, "Mechanical and thermoelectric behavior of composition modulated foils," in "Modulated Structure Materials," T. Tsakalakos, ed., Martinus Nijhoff, Dordrecht (1984).
7. B. S. Berry and W. C. Pritchett, "Effect of interdiffusion on the elasticity and internal friction of compositionally modulated copper-nickel thin films," Thin Solid Films 33:19 (1976).
8. H. Itozaki, Ph.D. Thesis, Northwestern University (1982).
9. A. Moreau, J.B. Ketterson, and J. Mattson, "New measurements of the elastic properties of composition modulated Cu-Ni thin films," Appl. Phys. Lett. 56:1959 (1990).
10. R. C. Cammarata, T. E. Schlesinger, C. Kim, S. B. Qadri, and A. S. Edelstein, "Nanoindentation study of the mechanical properties of copper-nickel multilayered thin films," Appl. Phys. Lett. 56:1864 (1990).

X-RAY STRESS STUDIES OF ALUMINUM METALLIZATIONS ON SILICON SUBSTRATE

M.A. Korhonen*, L.S. Suominen**, and Che-Yu Li

Department of Materials Science and Engineering
Cornell University, Ithaca, New York 14853

* On leave from Helsinki University of Technology
02150, Espoo, Finland

** Mexpert Instrument Technology
Otaniemi Science Park, 02150 Espoo, Finland

INTRODUCTION

Thin and narrow aluminum alloy metallizations represent the most important interconnects used in integrated circuit technology at the microchip level. Recently, stress induced voiding and notching of these films has become a serious reliability concern[1-3]. Moreover, since there is a steady trend towards larger packaging densities, the industry standard for the line width approaching now to the submicron scale, the failure risk of the interconnects, inversely proportional to the cube of the line width in case of grain boundary cavitation[3], is likely to increase significantly.

The largest stresses in these metallizations usually result from the thermal expansion mismatch between the silicon substrate and the thin film. A typical heat treatment used in manufacturing these chips, involves an anneal at about 400°C, during which the existing stresses largely relax. Then, upon cooling, very high tensile stresses develop in the aluminum metallization because of the large difference in the thermal expansion coefficient between aluminum and silicon. For continuous thin films deposited on silicon wafers the stress is equi-biaxial; the magnitude of the stress can be determined by wafer curvature techniques[4,5] or by x-ray analysis[6-13]. Wafer curvature techniques have been applied for continuous films to monitor the stress build-up and subsequent relaxation following a temperature change[4]. However, by x-rays it is possible to record the whole triaxial stress state, arising in real sized passivated aluminum lines[11-13]. Moreover, because the stress relaxes with time even at RT, it is necessary to record the stress as much in real time as possible. This paper reports preliminary studies in applying a fast solid state x-ray camera to record the stress relaxation in aluminum metallizations at RT after heat treatment at a higher temperature. Additionally, some previous stress studies on aluminum metallizations obtained by conventional x-ray diffractometry are described, and discussed with the present findings.

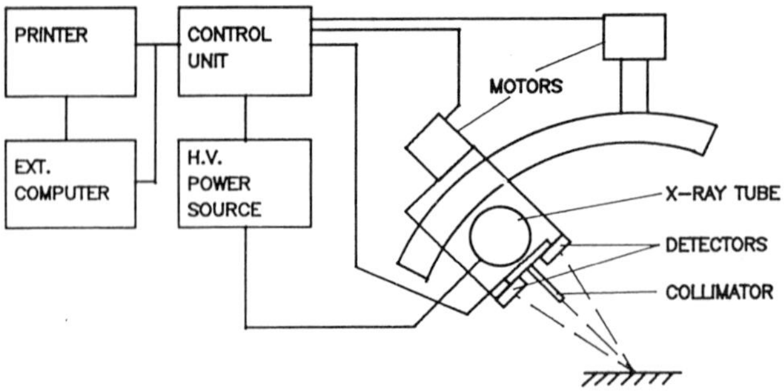

Fig. 1. General layout of the camera system

SOLID STATE X-RAY CAMERA

In the solid state camera, detailed elsewhere[14], the film has been replaced by a photosensitive microchip, Fig. 1. The microchips that can be used for direct detection of x-rays, without any fluorescent coating, include linear and matrix charge coupled devices (CCD's) and photodiode arrays (PDA's). In the present application we used a PDA consisting of 512 pixels per 12.5 mm; a CCD application has been described previously[15]. To increase the sensitivity to low level radiation, the pixels are exceptionally wide, 2.5 mm, in relation to their mutual spacing, about 25 μm. Two PDA's, positioned symmetrically around the collimator in analogy with the conventional back-reflection flat film camera technique[16], are driven by a custom made driver board, interfaced to a microcomputer, Fig. 1. To facilitate the measurement, stepper motor driven settings for the inclination angle, and for the sample to PDA board distance are provided. In order to gain as large intensity as possible, the sample to PDA board distance is set to the practical minimum of 50 mm, while the collimator size is selected to be the largest possible without excessively compromising the resolution. The camera construction described is very versatile in that the exposures may be taken either according the ω- or ψ-geometries of a conventional stress diffractometer (consult e.g. the recent textbook by Noyan and Cohen[17] for x-ray measurement techniques) or according to corresponding modifications for the camera technique[14].

STRESS MEASUREMENT OF ALUMINUM THIN FILMS

An earlier x-ray study revealed that the aluminum films under investigation had a very strong [111]-fiber texture, with the fiber axis perpendicular to the substrate[8,9]. These highly textured films must be

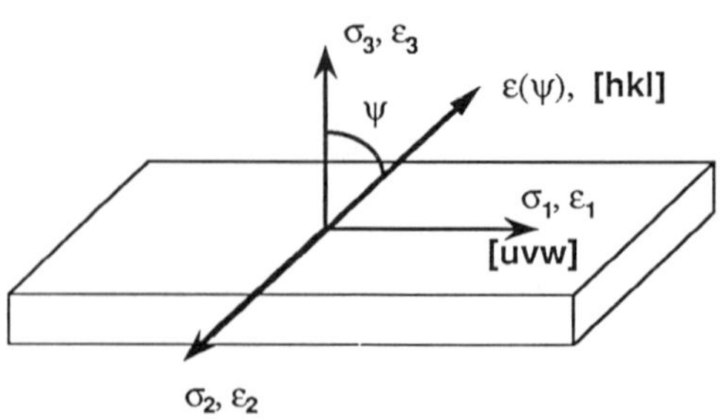

Fig. 2. The principal stress system used

analysed by the single crystal techniques, applying the single crystal elastic compliances s_{11}, s_{12}, s_{44} and $s_0 = s_{11}-s_{12}-(s_{44}/2)$. The principal stress system used in the analysis is depicted in Fig. 2: σ_1 points into the direction of the aluminum lines, σ_2 lies in the width direction and σ_3 is normal to the substrate. The strain in any crystallite into ψ-direction, whose projection lies along σ_1, becomes

$$\epsilon(\psi) = 2(s_{12}+s_0/3)(\sigma_1+\sigma_2+\sigma_3)+s_0(G-1/3)(\sigma_1-\sigma_2)$$

$$+(s_{44}/2)[\sigma_3+(\sigma_1-\sigma_3)]\sin^2\psi. \qquad (1)$$

as detailed elsewhere[8,13]. Here G is an orientation factor defined as

$$G = h^2u^2+k^2v^2+l^2w^2, \qquad (2)$$

in the terms of the directions [hkl] and [uvw] of the reflection vector and of the principal stress σ_1, respectively, in the crystal coordinate system, see Fig. 2. For 111-reflection, the orientation factor readily reduces to 1/3, and for a general reflection it can be shown to be

$$G = 1/3 + \sin^2\psi/6 + \sqrt{3}\ uvw\ \sin(2\psi) \qquad (3)$$

in case the projected strain lies along σ_1-direction, as shown in Fig. 2. Because principal stress directions are known by the symmetry, and eitherone of the in-plane stresses can be taken as σ_1, this limitation poses no problems for the present case.

According to Reuss hypothesis[18] the stress state in each grain is simply the macroscopic stress, and thus the above equations yield directly the connection between the x-ray strains and the stress state. On the other hand, according to Voigt hypothesis[19] the strain into a given direction in each grain equals to the mechanical average. For the present fiber texture the averaging should be made around the fiber axis [111], resulting in

$$G = 1/3 + \sin^2\psi/6. \qquad (4)$$

The real mechanical behavior of crystals usually lies between these two extremes as first noted by Neerfeld[20]. For reflection vectors with projections [uvw] of the form $[110]/\sqrt{2}$ no errors due to elastic anisotropy should arise. For other directions [uvw] the anisotropy effect remains small when large inclinations ψ are used for the measurement. As an example, for (111)-planes with $\psi = 70.5°$, the error in σ_1 remains smaller than 2% when $\sigma_2 = \sigma_3 = 0$. Further still, the anisotropy error goes to zero when σ_2 approaches σ_1. Thus, when only the ideal pole orientations of fiber textured aluminum are studied, elastic anisotropy can be accounted for in a way that no complications should arise in the stress analysis. A more comprehensive treatment of the anisotropic effects in connection to [111]-fiber textured aluminum films can be found elsewhere[13].

The stresses in aluminum thin films can be quite large[4,7-13] while the aluminum lattice is relatively soft, which means that the common differential formulas for the strain[16,17] can lead to significant errors in stress. In the present study the strains were derived directly from the Bragg law as

$$\epsilon = \sin\theta^*/\sin\theta - 1, \qquad (5)$$

where θ is the Bragg angle and the asterisk refers to the stress free state.

EXPERIMENT

Materials and sample preparation

Metallizations of 99.999% pure aluminum were deposited on cleaned, (100)-silicon wafers which had been oxidized at 1100°C prior to film deposition. A continuous, 300 nm thick, film was deposited on wafers in a cryo-pumped electron beam evaporator. Sets of 1.5 and 5 μm wide, 300 nm thick, lines were prepared by the conventional lift-off technique on several one centimeter square sections on their own wafers. After fabrication the wafers were diced into samples of one centimeter squares. The base pressure during all depositions was approximately 2.0E-7 torr, and the evaporation rate for all depositions was 1 nm/s. Texture studies revealed that aluminum metallizations had a very strong [111]-fiber texture[8,9]. After deposition one set of continuos and 5μm line samples were heat treated at 400°C and an other set in 450°C for one hour in an inert atmosphere. 1.5 μm line samples were passivated with 400 nm layer of Si_xN at 300°C, after which they were heat treated at 400°C for an hour in an inert atmosphere. After the heat treatment, the samples were allowed to cool in an inert atmosphere to approximately 50°C in 6-15 minutes, after which the stress analysis was started.

X-ray stress measurement

Stresses were measured with Cr-Kα radiation yielding a 111-reflection at an angle $2\theta \approx 157°$. Because of the strong texture, the reflections could only be studied at inclinations $\psi = 0.0°$ and 70.5°. During the initial fast stages of stress relaxation, it was possible to record the strain at $\psi = 0°$, only. However, the use of a single exposure, only, presupposes that the stress free value $2\Theta^*$ is known. Also, this value is needed for the determination of the whole triaxial stress state. The stress free $2\Theta^*$ was determined at a later stage of stress relaxation in the manner described previously[8]. Elastic compliance values of s_{11} = 1.57E-5/MPa, s_{12} = -0.568E-5/MPa and s_{44} = 3.51E-5/MPa, as given by Segmueller and Murakami[21] were used.

The initial stages of stress relaxation were studied in a set of continuous and 5 μm line samples, heat treated at 400°C, by the solid state x-ray camera. The irradiated area was about 5 mm^2. Because in these samples $\sigma_3 = 0$, eqn. (1) yields the average in-plane stress $(\sigma_1+\sigma_2)/2$ from a single, perpendicular exposure with $\psi = 0$. The peak shifts were recorded by the cross correlation method[22]. Also the half-value width of the peaks were determined. Reasonably stable stress readings were achieved in less than one minute. The precision of a single stress reading was estimated to be about 5 MPa as based on the scatter of a large number of repeated stress determinations about the mean.

In previous studies, reported elsewhere[10,13] in detail, a conventional ω-diffractometer was used for the stress measurement; the irradiated area was typically about 50 mm^2. For unpassivated continuous film and 5 μm lines, with $\sigma_3 = 0$ and $\sigma_1 \simeq \sigma_2$, two strains at $\psi = 0$ and 70.5° were measured. For 1.5 μm passivated lines, one perpendicular and two inclined strains, with projections lying in σ_1- and σ_2-directions, see Fig. 2, were determined, in order to be able to find the triaxial stress state. A more general technique for the case when the principal stress directions are not known beforehand has been described by Ruud et al.[23]. In our studies conducted at longer times (> 2 hours) after the heat treatment, the peaks were covered by .1 degree steps at 60 points, and the peak shifts were determined by the cross correlation method. The measurement times for a peak ranged from .5 to 4 hours depending on the inclination ψ and the reflecting power of the sample. The standard deviation of the stress readings obtained at the slow stages of relaxation was estimated to be 2 - 4 MPa as detailed elsewhere[8].

Despite the fact that cross correlation was proposed several years ago for line shift determination in stress measurement[22], the method has been applied only occasionally. The inherent advantages of the cross correlation method (most importantly its unsensitivity to changes in line width and its self-smoothing property), although well-known to mathematicians and physicists, appear to have gone largely unregocnized by practicing x-ray stress analysts.

In connection to thin film stress studies, the cross correlation method appears to offer marked benefits. By definition, the reflecting volume of thin films is much smaller than that of bulk samples which results in much weaker intensity, and spotty Debye-rings because of the limited number of grains able to reflect. However, consideration of the whole line shape, with the smoothing property of cross correlation, largely offsets these drawbacks.

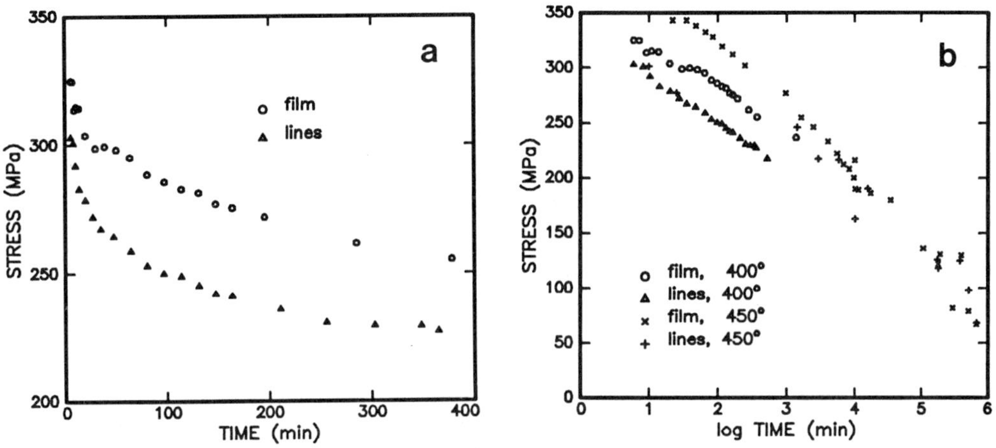

Fig. 3. (a) stress relaxation after an anneal at 400°C
(b) comparison to previous results[10] after 450°C anneal

Often, in thin film studies diffraction lines of various width have to be compared. In one exposure method the measured line profile is compared to the profile of a stress-free state, where the line width is likely to be significantly smaller. The same situation applies when the triaxial state of stress needs to be determined. The differing degree of overlapping of $K\alpha$-diffraction lines can give rise to significant artificial line shifts, when using the parabola fit for peak location, for example. Even in comparing exposures from the same sample, taken at different inclinations ψ, artificial line shifts may arise, either due to geometrical errors (mainly focusing), or because of directional microstresses. According to our experience, the angle dependent line broadening can be quite significant in thin aluminum films, and leads to artificial line shifts up to .2°. Apart from this systematic shift, the repeatability of cross correlation readings turned out to be better than that of parabola fit. Even when intensity profiles from some aluminum samples were quite ill-defined, cross correlation generally produced repeatable readings within some hundredths of a degree, while the peaks determined from parabola fit fluctuated to much larger extent. However, no systematic study, besides the one described earlier[14], was conducted to determine the quantitative differences in accuracy between the two methods of peak location.

RESULTS AND DISCUSSION

Stress relaxation of unpassivated continuous films and 5 μm lines at RT, as analysed by the x-ray camera, is displayed in Fig. 3 a. The first reading was obtained about 6 minutes after taking the samples from the furnace. Exposure times were about 30 s. Four first stress readings are displayed as such, while the rest of the points displayed are averages of at least half a dozen single readings. The maximum, tensile biaxial stress, $\sigma_1 = \sigma_2$, for the continuous film was 325 MPa and the average stress, $(\sigma_1+\sigma_2)/2$, for the 5 μm lines was 305 MPa. Because the stress in the width direction must be zero at the edge of the line, the slight difference between the stress readings can readily be explained. Because of the edge effect we anticipate 5 - 10 % lower average stress in the width direction. Also, as a result of local relief of σ_2 close to edges, the in-plane shear stress increases, and the enhanced plastic deformation is likely to result also in a local, partial relief of σ_1 component of the stress. We found the equi-biaxial stress of the continuous film to relax to about 260 MPa in 4 hours at room temperature.

These fast stress readings complement very satisfactorily our earlier measurements obtained by an x-ray diffractometer study on similar samples, the only difference being that the diffractometer samples were annealed at 450°C, instead of 400°C used for the samples analysed by the x-ray camera. The thermal stress in continuous films can be calculated from

$$\sigma = 3\ \Delta\alpha\ \Delta T/(2s_{11}+4s_{12}+s_{44}/2), \qquad (6)$$

where ΔT is the difference between the annealing and room temperature, and $\Delta\alpha = 20$ ppm/°C is the difference in thermal expansion coefficients between aluminum and silicon. According to this formula the thermal stress at room temperature after anneals at 400°C and 450°C would be about 860 MPa and 970 MPa, respectively. In practice, however, these high stresses never arise because the film deforms plastically during cooling. Thus the highest observable stresses at RT should reflect rather the RT yield strength than the fictitious thermal stress predicted by eqn. (6). Accordingly, in otherwise identical samples we can expect nearly equal stresses at RT, even if the annealing temperatures differ somewhat. Fig. 3 b depicts the relaxation of samples heat treated at 450°C, up to one year, together with the present results. The stresses of continuos films and lines, respectively, agree within 20 MPa, thus confirming our expectation that RT stresses are determined mainly by the (time dependent) yield strength at RT; the remaining small difference in stress readings is likely to reflect the different amount of work hardening during cooling. These results demonstrate that a

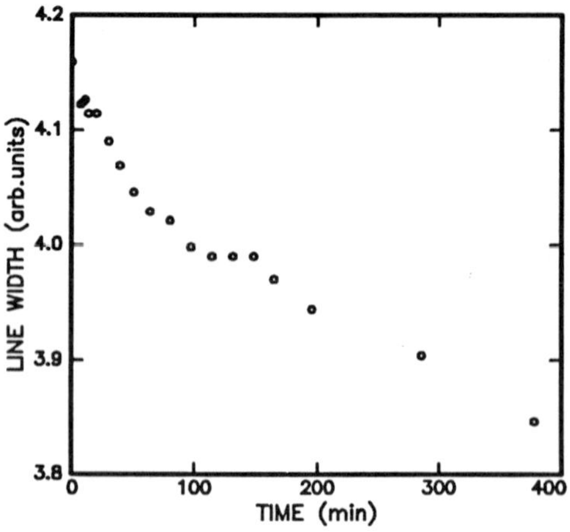

Fig. 4. Decrease of x-ray line width at RT after an anneal at 400°C

considerable portion of the thermal stresses relax already upon cooling. We estimate the maximum equi-biaxial stress at RT to be 330 - 350 MPa in continuous films; the same equi-biaxial stress appears also at the center of wide enough lines. This maximum stress relaxes about 15 % in four hours in pure 300 nm thick aluminum metallizations, and after two weeks the stress is about 200 MPa. Further, stress relaxation at a decreasing pace continues over several months at room temperature. After one year the stress level is still about 100 MPa, and Fig. 3 b suggests that stress relaxation is still continuing.

It is very interesting that also the microstresses appear to relax in the same manner as the macrostresses, as displayed by the x-ray line widths, Fig. 4. This observation suggests that the deformation is not localized to the proximity of the interface, for example, but the mechanism responsible for stress relaxation is a bulk process. Previously determined activation volume, about $(4\ A)^3$ is suggestive of diffusion as the rate controlling activation step[10].

We are now working with the application of the solid state camera technique to study the triaxial stress state arising in passivated lines. In a triaxial stress study one perpendicular and two inclined exposures at 70.5° need to be taken, and the inclined exposures need to be much longer than the perpendicular exposure to compensate for the lower intensity. Nevertheless, we expect to be able to measure triaxial stress states in aluminum thin films in a matter of some minutes in the near future.

Fig. 5 a depicts results of a diffractometer study of longer term relaxation of passivated and heat treated 1.5 μm aluminum lines[13]. The first point corresponds to about 4 hours from taking the sample from furnace, and thus serves as a good example why a faster means of stress measurement is desirable. Nevertheless, even the longer term behavior shows some very interesting features. First, $\sigma_1 \simeq \sigma_2$ even if the lines are relatively narrow. The maximum in-plane stresses are about 350 Mpa after four hours at RT. The original stresses in reaching the room temperature have most probably been much higher. Second, the third stress component σ_3 appears to stay at the neighbourhood of 200 MPa over extended times, while the in-plane components relax much faster. Consequently, there is a stage where the lines are in a hydrostatic tension. Because of voiding, as also experimentally verified[9],

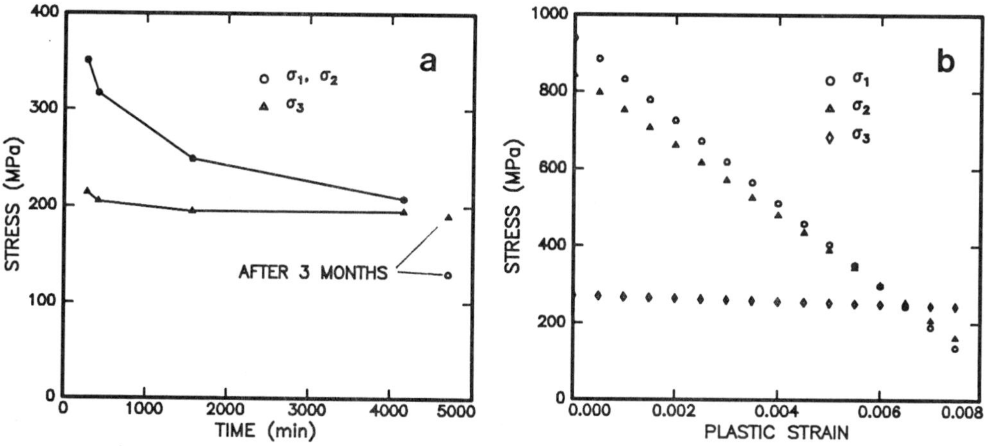

Fig. 5. Stress relaxation of passivated lines at RT after an anneal at 400°C, (a) experimental results, (b) simulation

the in-plane stress continues to relax so that with time σ_3 becomes the largest stress. This behavior is confirmed by recent results by Flinn[12].

It is possible estimate that the triaxial stress state arising in passivated lines by considering aluminum lines as Eshelby inclusions[24] in an infinite silicon matrix. For ellipsoidal inclusions, including plate-like and cylindrical inclusions as special cases, the stress state is uniform, and the stresses can readily be given in an analytic form as a function of the thermal and plastic strains[17,24]. Of course, the present aluminum lines are rather rectangular than elliptic in cross section; however, it is possible to replace a rectangular cylinder by an "equivalent" elliptic cylinder, so that the average stresses are the same, and then treat the "equivalent" cylinder as an Eshelby inclusion. Neither the 400 nm thick passivation layer may appear to be representative of the infinite matrix as modelled in Eshelby theory. However, an exercise by a simple shrink fit model, as described e.g. by Timoshenko[25], shows that the stresses build up very fast with increasing passivation layer thickness. As an example we consider an aluminum bar of diameter h shrink fitted to a hollow silicon cylinder of wall thickness t. At a ratio t/h = 1, the normal stresses are only 10% less than in case t = infinite. The same basic fact is displayed also by recent FEM simulations of passivated aluminum metallizations by Sauter and Nix[26]. They report, for an example, that the average stress of lines with thickness to width ratio of 1/3, increases only 30 Mpa, or about 5%, when the ratio t/h increases from 1.5 to 2. As a conclusion, we expect that modelling of aluminum lines as cylindrical Eshelby inclusions in an infinite matrix, gives a reasonable upper bound for the stresses we can expect to arise during cooling and subsequent stress relaxation. The modelling details are given elsewhere[13].

Instead, here we state only a fundamental result derivable from Eshelby analysis[17,24]. The relation between the thermal strains e^T_{ij}, plastic strains e^P_{ij}, and the stresses σ_{ij} in the aluminum line embedded in an infite silicon matrix can be represented in the form

$$e^T_{ij} - e^P_{ij} = T_{ijkl}\,\sigma_{kl} \,, \qquad (7)$$

where T_{ijkl} is a constant tensor. It is given explicitly in the above cited study[13], and depends only on elastic constants of aluminum and silicon, and on the axial ratios of the cross section of the aluminum line. Because the thermal strains at RT are known ($\approx$.0075), we see that the above equation actually connects the plastic strains and residual stresses. Now, we can deduct from the close equivalence of the lateral stresses, Fig. 5 a, that also the corresponding plastic strains must become close to each other, i.e., $e^P_1 \simeq e^P_2$. By inserting the measured stresses, Fig. 5 a, to eqn. (7), we found that the volume is not conserved during the stress relaxation[13]; in fact, the result suggested that the normal strain e^P_3 decreases only at half the pace needed to conserve the volume:

$$\dot{e}^P_1 = \dot{e}^P_2 = - \dot{e}^P_3 \,. \qquad (8)$$

Evidently, the volume change must be connected to void growth[10,13]. Now we can use the Eshelby theory, eqn. (7), complemented with the condition of nonconservation of the volume, eqn. (8), to simulate the change of the stress state with the lateral plastic strain, Fig. 5 b. It is remarkable that this model predicts all the general features of the experimentally found stress relaxation behavior of the passivated aluminum lines, shown in Fig. 5 a.

SUMMARY

In the present x-ray study of pure aluminum thin films, nearly real time stress measurement was demonstrated by using a solid state x-ray camera. Cross correlation method was found to yield repeatable stress values even if the x-ray lines were diffuse and of low intensity. The maximum biaxial stress at RT for 300 nm thick, continous films and 5 μm lines (central parts) after heat treatments at 400 - 450°C was 325 - 340 Mpa; the stress relaxed about 15% in 4 hours, to 200 MPa in two weeks, and after 1 year the stress level still was on the order of 100 MPa. In passivated 300 nm thick, 1.5 μm wide lines the stress state was triaxial: four hours after the heat treatment at 400°C the lateral stresses were $\sigma_1 = \sigma_2 = 350$ MPa while the normal stress was 215 MPa. The stresses relaxed in three days to a state of hydrostatic tension of about 200 MPa. The relaxation of lateral stress components continued even after this stage by a mechanism connected to void growth, so that with time σ_3 became the largest stress component. The general features of stress relaxation in passivated lines were well in accord with a model based on Eshelby theory of inclusions.

ACKNOWLEDGEMENTS

The authors would like to thank Bill LaFontaine for supplying the samples used in this study. The fast x-ray stress analyses were made by Mexstress camera, kindly provided for this study by Mexpert Instrument Technology. Portions of this work were supported by an IBM Shared University Project, by the National Science Foundation through the Materials Science Center, by the National Nanofabrication Facility at Cornell, and by the Academy of Finland.

REFERENCES

1. K. Hinode, N. Owada, T. Nishida, and K. Mukai, J. Vac. Sci. Technol. B5:518 (1987)
2. J.W. McPherson and C.F. Dunn, J. Vac. Sci. Technol. B5:1321 (1987)
3. Che-Yu Li, R.D. Black, and W.R. LaFontaine, Appl. Phys. Lett. 53:31 (1988)
4. P.A. Flinn, D.S. Gardner and W.D. Nix, IEEE Trans. ED-34:689 (1987)
5. M.F. Doerner, D.S. Gardner and W.D. Nix, J. Mater. Res. 1:845 (1986)
6. T.J. Louzon and T.H. Spencer, Sol. St. Technol. 7:25 (1975)
7. C.J. Shute, J.B. Cohen, and D.A. Jeannotte, Proc. MRS Conf., Boston, Nov. 28 - Dec. 3, 1988
8. M.A. Korhonen and Christine A. Paszkiet, Scripta Met. 23:1449 (1989)
9. Christine A. Paszkiet, M.A. Korhonen and Che-Yu Li, Proc. MRS Conf., San Francisco, Apr. 16-20, 1990
10. M.A. Korhonen, Christine A. Paszkiet, R.D. Black, and Che-Yu Li, to appear in Scripta Metallurgica
11. A. Tezaki, T. Mineta, H. Egawa, and T. Noguchi, Proc. Conf. IEEE/IRPS, 1990, p. 221
12. P.A. Flinn and C. Cheng, J. Appl. Phys. 67:2927 (1990)
13. M.A. Korhonen, R.D. Black, and Che-Yu Li, submitted to J. Appl. Phys.
14. M.A. Korhonen, V.K. Lindroos and L.S. Suominen, Adv. X-ray Anal. 32:407 (1989)
15. K. Simomaa, V. Kelha, T. Tuomi, M.A. Korhonen, and L.S. Suominen, Proc. Conf. Image Detection and Quality, Jul. 16-18, 1988, Paris, 203
16. B.D. Cullity, "Elements of X-ray Diffraction", Addison-Wesley, first ed., 1956
17. L.C. Noyan and J.B. Cohen, "Residual Stress: Measurement by Diffraction and Interpretation", Springer-Verlag, 1987
18. A. Reuss, Z. Angew. Math. Mech. 9:49 (1929)
19. W. Voigt, "Lehrbuch der Kristallphysik", Teubner, Leipzig, 1928

20. H. Neerfeld, Mitt. Kaiser Wilhelm Inst. fur Eisenforschg. 24:61 (1942)
21. A. Segmueller and M. Murakami, Treatises on Materials Science and Technology 27:143 (1988)
22. K. Toenshoff, E. Brinksmeier, and H.H. Noelke, Z. Metallknde 72:349 (1981)
23. C.O. Ruud and P.C. Chen, Exp. Mech. 25:245 (1985)
24. J.D. Eshelby, Proc. Roy. Soc. A241:561 (1957)
25. S. Timoshenko, "Strength of Materials", Van Nostrand, 1956, 210
26. Anne I. Sauter and W.D. Nix, Proc. MRS Conf., San Francisco, Apr. 16-20, 1990

MICROMECHANICAL ANALYSIS OF THE TiO_2/Si INTERFACE USING THE SCANNING ACOUSTIC MICROSCOPE

M. H. McCord, J. C. Duke and S. B. Desu

Virginia Polytechnic Institute
and State University
Blacksburg, VA 24061

INTRODUCTION

With the increasing use of thin film devices, a need has arisen for a better understanding of the mechanical properties of these film/substrate systems. Of particular importance, in many cases, is the adhesion between substrate and film. Several methods are commonly used to test the mechanical behavior of these interfacial systems, but for the most part, they are not quantitative, lack repeatability, and result in destruction of the sample [1]. In cases where the sample is rare or difficult to prepare, a nondestructive method would be highly desirable.

Since its introduction in 1974 [2], the scanning acoustic microscope (SAM) has proven to be a very versatile and useful tool for making nondestructive material characterizations. High spatial resolution and subsurface imaging capabilities suit it well to thin film applications. As a qualitative tool, the SAM can be used as a defect detection system capable of identifying debonds, poor bonds, or anomalous areas in the film itself [1,3,4,5]. The SAM can even be used to measure film thicknesses. In the non-imaging mode, the SAM becomes a quantitative microprobe capable of measuring acoustic properties precisely and rapidly "at a point." Measurements of sample anisotropy are also possible using specially prepared, line focused lenses [6,7,8].

PRINCIPLES OF OPERATION

The SAM operates according to a well known pulse-echo technique and, therefore, requires only a brief review here. The SAM uses ultrasonic energy to illuminate a specimen and, hence, is sensitive to changes in the elastic properties of the target material. This allows observation of phenomena that is not possible with an optical system.

In most instruments, ultrasonic waves are generated by a piezoelectric transducer which is deposited onto one end of a sapphire rod. The other end of this rod is ground to a conical shape with a carefully controlled spherical or cylindrical cavity fabricated into it (Fig. 1). This concavity serves to focus the acoustic waves both upon leaving the lens assembly and upon returning from the sample [9,10].

This incident acoustic wave energy mechanically excites the material of the specimen and can be used as a method of generating stresses within the specimen [10]. In the case of the acoustic microscope, the area being excited is small due to the focusing of the acoustic waves and the lens design. By scanning the lens over the sample at a constant height, and gating the return signal by time of arrival, a map of reflection intensities which can be used to locate variations in the elastic properties of the sample. By gating on a later arriving reflection, subsurface images are produced.

The lens of the acoustic microscope can also be used to excite surface acoustic waves (SAWs). These waves are generated by moving the lens into a "defocused" position (Fig. 1), which serves to allow waves generated at the periphery of the lens to strike the sample at an oblique angle that can, depending on the aperture angle of the lens and the properties of the sample/couplant interface, approach the critical "Rayleigh" angle (θ_{LSAW}). The waves thus generated are bound to the surface of the sample and tend to have the general characteristics of Rayleigh waves, although other types can also be present.

The SAWs generated in this manner "leak" energy back to the lens/transducer and, therefore, can be used as a tool for improving image contrast. In the case of grain boundaries and surface cracks, the LSAW (Leak Surface Acoustic Wave) will interact with the side of the feature (to a depth of approximately one wavelength) rather than just the exposed surface gap [9]. The subsequent reflections can interfere with incoming waves to produce rings around the object which are easily observed. In addition, the perturbation of the LSAW as it encounters non-surfacing features within the shallow subsurface will add to the returning energy and have an effect on the image.

The LSAW can also interfere with the direct reflected wave [12] in a way that is both troublesome and useful. If the lens is scanned in the z (vertical) direction only, and the transducer output is monitored as a function of z, the resulting curve shows a distinctive regular oscillation of period Δz [13] (Fig. 2). In trying to interpret acoustic images, the V(z) curve can result in contrast reversals when the lens to sample distance falls into one of the minima.

The V(z) relationship itself is the heart of the non-imaging mode of SAM operation. In the case of a homogeneous, isotropic material, the V(z) curve would, at all points, be the same. It was later shown that Δz is related to the velocity of the LSAW (V_{LSAW}) by the relation

$$V_{LSAW} = \frac{V_w}{(1 - (1 - \frac{V_w}{2f\,\Delta z})^{2})^{1/2}}$$

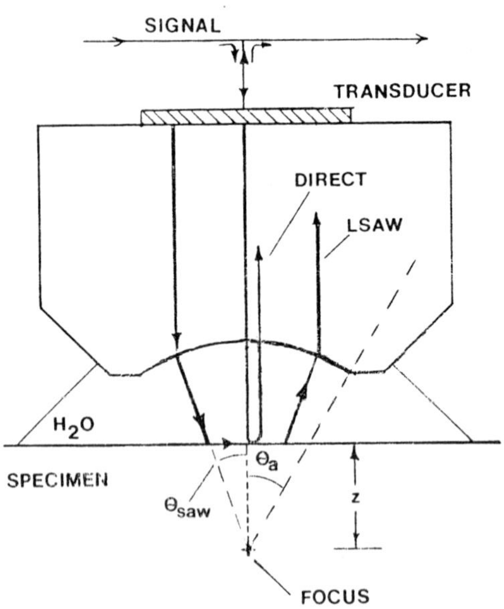

Fig. 1. Ray diagram of a typical acoustic microscope lens in a defocused position.

where V_w is the longitudinal wave velocity in water and f is the operating frequency of the transducer [13]. Each LSAW mode can be identified if its Δz can be isolated. In the case of a highly deformed V(z) curve, this is not a trivial matter.

Due to the spherical shape of the lens cavity, a circular focal spot is produced which allows LSAWs to propagate in all directions equally whereupon the transducer response will be an average of the signals received. If the material in question is anisotropic, the LSAW velocity will be dependent on the direction of propagation and will not be properly determined by the spherical (or Point Focus Beam, PFB) lens configuration. To properly measure anisotropy, a line-focus lens (LFB) is used (Fig. 3). The advantage here is the LSAWs can only propagate perpendicular to the lens axis. If V_{LSAW} is measured for various angular orientations, the anisotropy becomes apparent [14].

The interpretation of the V(z) curve can be difficult, particularly if more than one mode of LSAW is present and the individual oscillations tend to overlap. A Fourier transform relationship exists between the position z (in planes parallel to the sample surface) and the component of the wave number, k, that lies in the direction of wave propagation. This relationship provides a very useful tool that helps in interpreting the V(z) curve. In fact, the FFT procedure makes it possible to calculate V_{LSAW} to within 0.1% if the measurements are done carefully [7].

APPLICATION TO THE TiO_2 INTERFACE

The acoustic microscope system used in this investigation is a commercially produced unit manufactured by Olympus Corp. (Model UH3). It operates over a frequency range of 30 MHz. to 1 GHz. at discrete frequencies of 30, 50, 100, 200, 400, 800 MHz. and 1 GHz. The imaging lenses include a variety of aperture angles and focal lengths to allow as much flexibility as possible. In addition, 200 and 400 MHz. LFB lenses are available for non-imaging applications.

Most of the operations of the UH-3 are automated to some extent, with V(z) curve generation, sample leveling, and image formation in particular being simplified. Support software allows image enhancement, image overlays, and FFT processing of V(z) curves.

The LFB lenses used here have aperture angles of 120 deg. an focal lengths of 0.310 mm. The length of the beam axis is approximately 1 mm. This configuration gives the V(z) measurement system an optimal range of 2000 to 6000 m/s [7].

The samples used in this study are single crystal silicon wafers, cut parallel to the <100> crystallographic plane that have been polished flat, and coated with approximately 2500, 5000, and 10,000 angstroms of

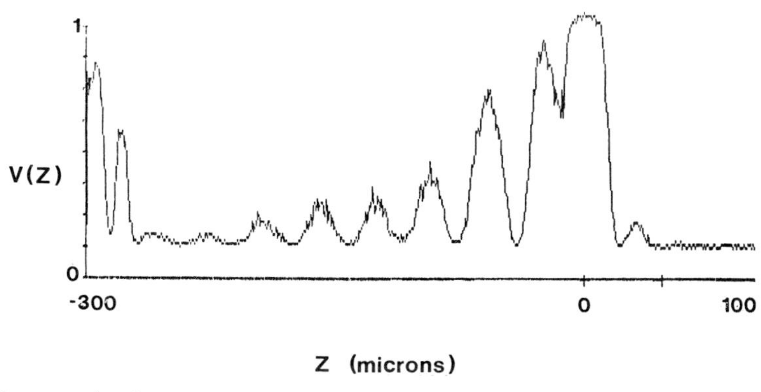

Fig. 2. Typical V(z) curve for a Silicon wafer coated with 10,000 Angstroms of TiO_2.

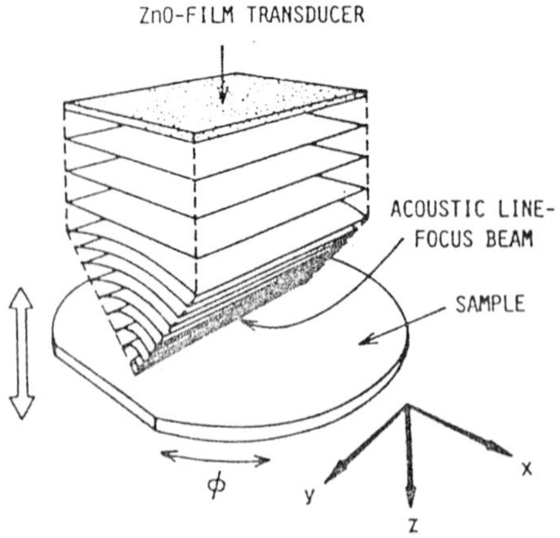

Fig. 3. Diagram of a LFB lens showing 0 deg. orientation of specimens. (From Kushibiki and Chubachi, 1985).

TiO_2 by a low pressure chemical vapor deposition process. An uncoated wafer was also obtained for determination of substrate properties. Each sample was inspected optically and acoustically, and suitable test areas were selected for further evaluation.

A series of 400 MHz. V(z) curves were developed for each specimen in order to assess the material properties. The LFB lenses were used, and the orientation of the specimen was varied in 10 deg. increments from 0 to 90 deg. All measurements were made without disturbing the lens in order to avoid changing either the axis orientation or the impedance characteristics of the transducer connection. The 0 deg. sample orientation was taken parallel to the <010> plane. Specimen level was checked by equalizing travel time across the X- and Y-scan directions. Measurements were performed at 51.2 C with distilled water as a coupling fluid.

A Vickers microhardness tester was used in an attempt to produce a debond in an area well removed from the original test area. The 10,000, 5000 and, and uncoated wafers were indented at 100, 200 and 300 grams. The indentation marks were compared optically and acoustically for evidence of debonding. The resulting analysis showed evidence of fractures in the substrate rather than debonding at the interface. This would tend to imply good bonding between film and substrate in the test area. Hardness values derived from the indentations showed evidence of elastic rebound due to the thickness of the wafers and were considered to be unreliable.

DISCUSSION

Imaging Examples

The imaging capability of the SAM provides a good way to qualitatively characterize material properties. Debonds, inclusions, and cracks are of particular interest to the thin film researcher. To the SAM, these types of defects appear as local changes in the reflectivity of the specimen and can be recognized by the corresponding change in image contrast. As an illustration, we have included images 4a and b. Figure 4a is an optical image of the 200 g indent on the 10,000 angstrom thick wafer. Note, that the bright part of the image is due to cracking in the substrate and not to debonding of the coating. Some inhomogeneity of the coating is apparent in this image, but the source is not readily apparent.

Figure 4b is a 400 MHz image of the same area, at approximately the same magnification. The lens is defocused 4.5 microns to make use of the LSAWs in the imaging process. Note here, the interference fringes around the indent. These are the result of LSAWs reflecting off of the side of the indent and the radial cracks and can be used to calculate the wavelength of the incident wave. Also note the increased

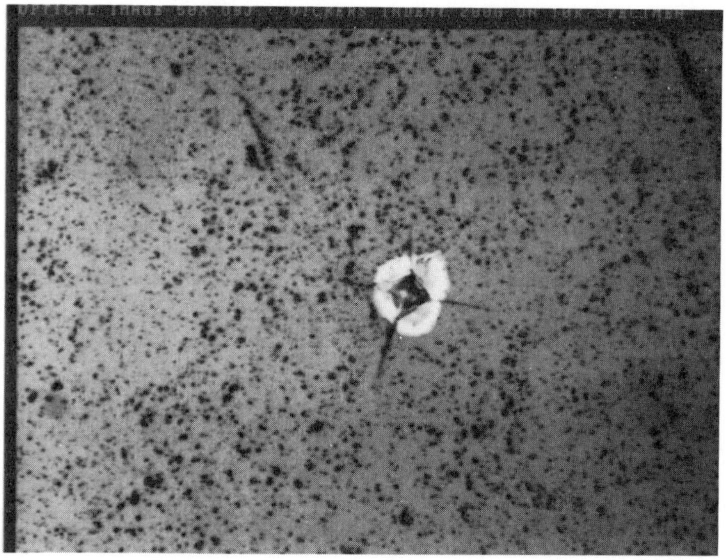

Fig. 4a. Optical image of Vickers indent of 10,000 Angstrom thick specimen. (50x objective).

Fig. 4b. 400 MHz. acoustic image of the same indent as Figure 4a.

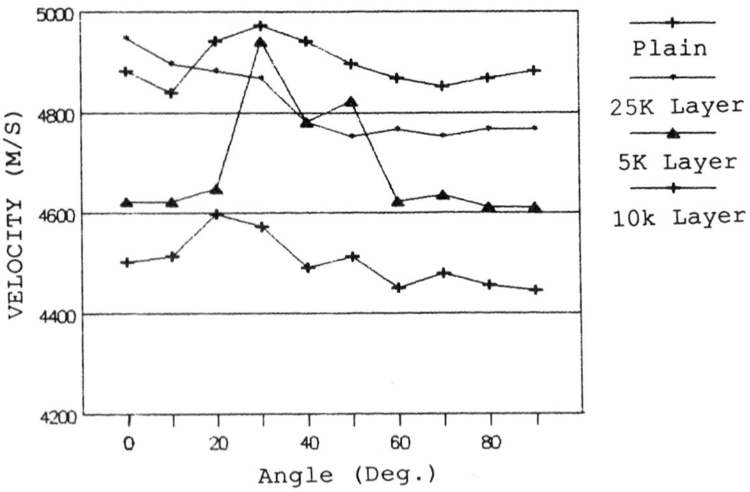

Fig. 5. Plot of Saw velocity vs. sample orientation for the various coating thicknesses.

contrast visible on the upper radial crack. The bifurcation of this crack is much more apparent in this image than in the optical image.

Non-Imaging Examples

The non-imaging mode of SAM operation provides several applications that are of interest to the thin film researcher. The first of these is the ability of detect anisotropy in the film or substrate. Figure 5 is a plot of LSAW velocity against wafer orientation relative to the beam axis. The values are plotted for each of the film thicknesses. In each case, several LSAW modes were present, and the predominant mode changed in several cases. The values plotted in figure 5 are velocities for the dominant mode. The LSAW velocities for single crystal silicon reported in other works range from 4624 to 5000 M/S [8,6,15]. Values plotted fall within this range and tend to average toward 4900 M/S for the uncoated wafer.

The data for the uncoated wafer in figure 5 shows a clear velocity anomaly in the region of 10 to 50 degrees. The data for the other (coated) wafers tends to mimic this trend. This would imply that either the coating is relatively isotropic, or that the coating is not thick enough to alter the overall propagation qualities of the substrate. The latter option is rejected because of the obvious velocity change with coating thickness seen in figure 5.

The coating thicknesses used in this study are considerably less than the wavelength of the energy used to examine the samples. This ensures that the surface waves that propagate in the near surface layer will interact with the coating/substrate interface. In doing so, any change in the interface condition should become apparent as a variation in the shape of the V(z) curve. If the absolute thicknesses of the coatings were measured, then the data in figure 5 could be used to measure the thickness of other coatings of the same type.

CONCLUSION

It has been the goal of this project to illustrate the application of the acoustic microscope to the "thin film" problem and to demonstrate a few of the useful analyses that can be performed with it. The applications available are many, and at the moment, many questions remain unanswered. Continuation of the present work into SAW attenuation and characterization of the LSAW modes is planned.

The SAM fills a niche where few other instruments can be used. By itself, it has some limitations that cannot be denied, including:

specimen size, surface condition, and flatness. Traditionally, limitations in one area have been the building blocks for discoveries in others. By using the SAM in conjunction with other techniques, much can be learned and new techniques can be developed that will overcome the present deficiencies.

REFERENCES

1. R. C. Addison, M. G. Somekh and G. D. Briggs, "Techniques for the Characterization of Film Adhesion," in: I.E.E.E., 1986, Symposium on Ultrasonics, Proceedings, B. R. McAvoy, ed., I.E.E.E., New York, (1986).
2. R. A. Lemmons and C. F. Quate, "A Scanning Acoustic Microscope," in: Proceedings, 1973, I.E.E.E. Ultrasonics Symposium, J. DeKlerk, ed., I.E.E.E., New York, (1974).
3. J. Kushibiki and N. Chubachi, "Application of LFB Acoustic Microscope to Film Thickness Measurement," Elect. Lett., V 23, No. 12, (1987).
4. R. C. Addison, M. W. Kendig and S. J. Jeanjaque, "Measurement of Cathodic Disbonding Using a Scanning Acoustic Microscope," in: I.E.E.E. 1988 Symposium on Ultrasonics, Proceedings, B. R. McAvoy, ed., I.E.E.E., New York, (1988).
5. R. C. Bray, C. F. Quate, J. Calhoun and R. Koch, "Film Adhesion Studies Using the Acoustic Microscope," Thin Solid Films, 74, (1980).
6. G. M. Crean, M. G. Somekh, A. Golanski and J. C. Oberlin, "The Influence of Thin Film Microstructure on SAW Velocity," in: I.E.E.E. 1987 Symposium on Ultrasonics, Proceedings, B. R. McAvoy, ed., I.E.E.E., New York, (1987).
7. J. Kushibiki and N. Chubachi, "Material Characterization by Line-Focus-Beam Acoustic Microscope," I.E.E.E. Trans. Son. and Ultrasonics, SU-32, No. 2, (1985).
8. J. Kushibiki, K. Horii and N. Chubachi, "Leaky SAW Velocity on the Water/Silicon Boundary Measured by Acoustic LFB," Elect. Lett., Vol. 18, No. 17, (1982).
9. G. A. D. Briggs, "An Introduction to Scanning Acoustic Microscopy," Oxford University Press, New York, (1985).
10. G. S. Kino, "Fundamentals of Scanning Systems," in: Scanned Image Microscopy," E. A. Ash, ed., Academic, New York, (1980).
11. S. I. Rokhlin, "Adhesive Joint Characterization by Ultrasonic Surface and Interface Waves," in: Adhesive Joints: Formation Characteristics and Testing," K. L. Mittal, ed., Plenum, New York, (1984).
12. W. Parmon and H. L. Bertoni, "Ray Interpretation of the Material Signature in the Acoustic Microscope," Elect. Lett., Vol. 15, (1979).
13. R. D. Weglein, "A Model for Predicting Acoustic Material Signature," Appl. Phys Lett., Vol. 34, (1981).
14. J. Kushibiki, A. Ohkubo and N. Chubachi, "Anisotropy Detection in Sapphire by Acoustic Microscope Using Line Focus Beam," Elect. Lett., Vol. 17, (1981).
15. G. M. Crean and A. Waintal, "Rapid Nondestructive Thickness Measurement of Opaque Thin Films on Anisotropic Substrates," Elect. Lett., Vol. 22, No. 1, (1986).

REAL TIME SPECTROSCOPIC ELLIPSOMETRY FOR NON-INVASIVE CHARACTERIZATION OF THIN FILM GROWTH AND ETCHING

R. W. Collins, Ilsin An, Yue Cong, A.R. Heyd, H.S. Witham, R. Messier, and K. Vedam

The Pennsylvania State University
Materials Research Laboratory
University Park, Pennsylvania, 16802

INTRODUCTION

Determination of material properties in real time during preparation or processing is an important problem in materials research. Many technologically relevant thin film materials are fabricated in adverse environments such as plasmas or reactive gases. In these situations, an optical probe is among the few non-invasive methods of real time characterization.

With the development of automatic optical ellipsometers for rapid data acquisition, ellipsometry has been employed to characterize thin films during preparation and processing. Until recently, however, all such real time studies were performed at a single wavelength.[1] Nevertheless, the measurements were effective in determining film thickness vs. time and rudimentary information on deviations from atomic layer-by-layer growth owing to nucleation, interfacial reactions, void density evolution, and surface roughness. Unfortunately, the optical functions of the material, ie. the index of refraction and extinction coefficient or dielectric function, are determined only at the single wavelength.

Although spectroscopic ellipsometry (SE) provides more complete information on material properties,[2] the time required to scan a monochromator over its spectral range, has limited this method to static situations. Two recent developments, however, have led to the first real time SE studies of thin film growth. In the first, a rapid-scanning interference filter monochromator and a single channel detector were used to study electrochemical deposition of thin films.[3] In the second, a prism spectrometer and rapid scanning photodiode array detector were used to characterize the vacuum deposition of thin films.[4] With these instruments, detailed information including bulk film thickness, void density, composition and band gap, interface layer thickness and composition, and surface roughness and damage can be deduced from real time measurements during materials preparation and processing.

In this article, we review recent real time SE studies to characterize film growth and etching in vacuum. First, the unique multichannel spectroscopic ellipsometry experiment will be described, and then two different applications of the technique will be presented: (1) ion beam sputter-etching of GaAs/AlGaAs heterostructures, fabricated by molecular beam epitaxy, and (2) ion beam deposition of hydrogenated (diamond-like) amorphous carbon (a-C:H).

EXPERIMENTAL DETAILS

Instrumentation

A schematic of the multichannel ellipsometer is shown in Figure 1. In the experiment, linearly polarized white light is reflected from a dynamic surface, and after dispersal of the light, the change in polarization state upon reflection is deduced as a function of wavelength. The sequence of optical elements is the following: Xe light source, collimator, rotating polarizer, sample in process chamber, fixed analyzer, prism spectrometer, and optical multichannel analyzer (OMA). The photodiode array detector, the heart of the OMA, consists of 1024 pixels. For purposes of ellipsometry, the output signals of 8 neighboring pixels are "grouped", providing 128 total spectral points from 1.2 eV to 4.7 eV.

For the ellipsometer of Fig. 1, each pixel-group signal obeys the following equation:[5]

$$I_P(t) = I_0\{1 + \alpha'\cos 2\omega t + \beta'\sin 2\omega t\}. \quad (1)$$

ω is the angular frequency of the rotating polarizer, and $\{\alpha',\beta'\}$ are the measured 2ω Fourier coefficients (normalized to the dc coefficient) which depend on pixel group (or wavelength).

Because each pixel group in the array integrates the impinging irradiance over the exposure time (ie. the time between read-out scans), a Hadamard approach is used to evaluate $\{\alpha',\beta'\}$ for each pixel group. In this scheme, the array is scanned four times per half-mechanical cycle of the polarizer, triggered by an optical encoder. The resulting four spectra, S_i, i=1-4, obtained in the four exposures, can be used to reconstruct the waveform for each pixel group:

$$\alpha' = (\pi/2)(S_1 - S_2 - S_3 + S_4)/(S_1 + S_2 + S_3 + S_4) \quad (2a)$$

$$\beta' = (\pi/2)(S_1 + S_2 - S_3 - S_4)/(S_1 + S_2 + S_3 + S_4). \quad (2b)$$

The measured coefficients are related to those in theoretical equations through a rotation:

$$\alpha = \alpha'\cos 2\phi + \beta'\sin 2\phi \quad (3a)$$

$$\beta = -\alpha'\sin 2\phi + \beta'\cos 2\phi. \quad (3b)$$

where ϕ is a phase angle, determined in calibration, which accounts for the fact that, when the first exposure begins (S_1), the polarizer transmission axis is not necessarily aligned in the plane of incidence.[5] Hence ϕ depends on pixel group owing to the finite group readout time.

The ellipsometry angles, $\{\psi,\Delta\}$, can now be determined, defined by $\rho \equiv r_p/r_s = \tan\psi\exp(i\Delta)$, where r_p and r_s are the Fresnel reflection coefficients for p- and s-polarized light:[1]

$$\tan\psi = [(1 + \alpha)/(1 - \alpha)]^{1/2} |\tan(A - A_S)| \quad (4a)$$

$$\cos\Delta = \beta/(1 - \alpha^2)^{1/2}. \quad (4b)$$

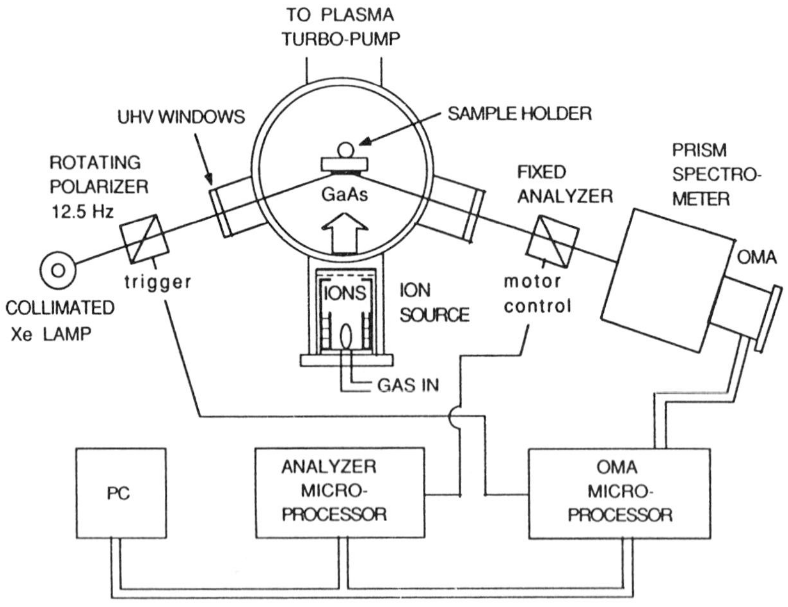

Fig. 1 Real time spectroscopic ellipsometer and vacuum etching/deposition system used for the studies of GaAs/AlGaAs etching and diamond-like carbon deposition.

Here A is the scale reading on the fixed analyzer, and A_S is a scale correction, also determined in calibration. To summarize then, the measured Fourier coefficients vs. pixel group number, $\{\alpha',\beta'\}$ and polarizer phase and analyzer angle calibration data, ϕ and A_S, are used along with a photon energy calibration of the spectrometer/OMA to deduce $\{\psi(h\nu),\Delta(h\nu)\}$.

For the studies below, $\omega/2\pi$ =12.5 Hz, meaning that a single $\{\psi(h\nu),\Delta(h\nu)\}$ pair of spectra could be collected in 40 ms, or in one-half rotational period. To improve precision, independent raw data, S_i, i=1-4, from 40 rotations were accumulated and averaged for an acquisition time of ~3 s. With averaging, monolayer sensitivity to film thickness could be obtained over the spectral range from 1.6 eV to 4.0 eV. Outside this range, the light throughput of the optical system is too low. The precision in $\{\psi,\Delta\}$ on a Au film was (0.01°,0.02°) at 2.0 eV. The complete data cycle time, ie. the time from the acquisition of one time-averaged $\{\psi(h\nu),\Delta(h\nu)\}$ to the next was ~16 s. This included time for background correction, data transfer, and real time graphical output. The background correction consisted of an identical 3 s acquisition but with a shutter blocking the incident beam. For the processes studied here, background correction was needed owing to the light emitted by the ion source filament.

Film growth and etching were performed in a stainless steel chamber with a base pressure of 10^{-8} Torr, which was mounted at the ellipsometer axis. The sample mount was attached to a precision manipulator for accurate alignment under vacuum. Window ports in the chamber walls were oriented for optical access to the sample surface at a 70° angle of incidence. The windows were fused silica and stress-compensated to eliminate birefringence.

Data Analysis

The ellipsometry angles, $\{\psi(h\nu),\Delta(h\nu)\}$, can be converted to a complex quantity called the pseudo-dielectric function through the following equation:

$$<\varepsilon_1>+i<\varepsilon_2> = \sin^2\theta\{1 + [(\rho-1)/(\rho+1)]^2\tan^2\theta\} \qquad (5)$$

where θ is the angle of incidence. The pseudo-dielectric function becomes an actual dielectric function when the sample is opaque and presents an atomically abrupt interface to the ambient. When more than one interface is present, as in most cases below, the pseudo-dielectric function is a function of the thickness(es) and dielectric functions of the components of the structure.

The goal of SE data analysis is to invert the problem and deduce layer thicknesses vs. time and layer properties from the time evolution of the pseudo-dielectric function during film growth or etching. The problem is straightforward when the sample is composed of materials, or of microscopic composites of materials, having known dielectric functions. In this case, all unknowns in the problem are photon energy independent: layer thicknesses and composite volume fractions. Thus, each pseudo-dielectric function can be interpreted independently by linear regression analysis (LRA) in order to deduce the best fit unknown parameters and their confidence limits.[2] LRA is performed by testing out various plausible models for the film structure, and choosing the one that provides the lowest unbiased estimator of the mean square deviation while maintaining sufficiently low confidence limits to justify all details of the model. A consistency check is that the volume fractions and thicknesses obtained by analyzing successive pseudo-dielectric functions vs. time follow physically realistic trends. In the LRA, the Bruggeman effective medium approximation is used to calculate the dielectric functions of the composites from the volume fractions and dielectric functions of the components. This overall approach has been used to study sputter-etching of GaAs/AlGaAs heterostructures.

If one of the materials in the structure has an unknown dielectric function, data analysis becomes more difficult, and the approach to be used is a function of the problem at hand. Because the optical properties of a-C:H are very sensitive to the preparation, which can lead to a wide range of possible structures for the C:H system,[6] dielectric function data for use in the LRA do not exist. New procedures developed for the a-C:H problem are outlined below.

RESULTS AND DISCUSSION

Etching of GaAs/AlGaAs Heterostructures

Figure 2(a) depicts the evolution of the pseudo-dielectric function from 1.5 to 4.0 eV during sputter-etching of a molecular beam epitaxy (MBE) $GaAs/Al_{0.23}Ga_{0.77}As/GaAs$ heterostructure. Such structures are being used increasingly as components of novel electronic devices. The intended thicknesses of the top-most GaAs and underlying $Al_{0.23}Ga_{0.77}As$ were 300 and 500 Å, respectively. The alloy composition was determined by source calibration in the MBE apparatus. Etching was performed using an Ar^+ beam accelerated through 200 V, having a fluence of $5x10^{15}$ $cm^{-2}s^{-1}$. Although this is not a technologically important process, it provides a simple model system to demonstrate the capabilities of SE for in-process characterization. Figure 2(a) includes ~75 spectra which were collected over 20 minutes. Each spectrum consists of ~80 points, giving a total of ~12K data values in the entire Figure.

The dielectric functions of GaAs and $Al_{1-x}Ga_xAs$, for selected x values, have been measured,[7] and an interpolation scheme provides the dielectric function of $Al_{1-x}Ga_xAs$ for any value of x. In addition, the dielectric functions of a-GaAs and GaAs oxide, components of the damaged layer on the heterostructure, are also available.[8,9] As a result, the data of Fig. 2(a) can be analyzed by the LRA. Figures 3 and 4 show the best fit structural parameters for the model, depicted in Figure 2(b), that leads to the lowest overall unbiased estimator of the mean square deviation. The free parameters in this model include: (1) the top-most GaAs thickness, (2) the underlying $Al_{0.23}Ga_{0.77}As$ thickness, (3) the thickness of the surface damage layer, modeled as a (GaAs)/(GaAs oxide) mixture, and (4) the damage layer composition.

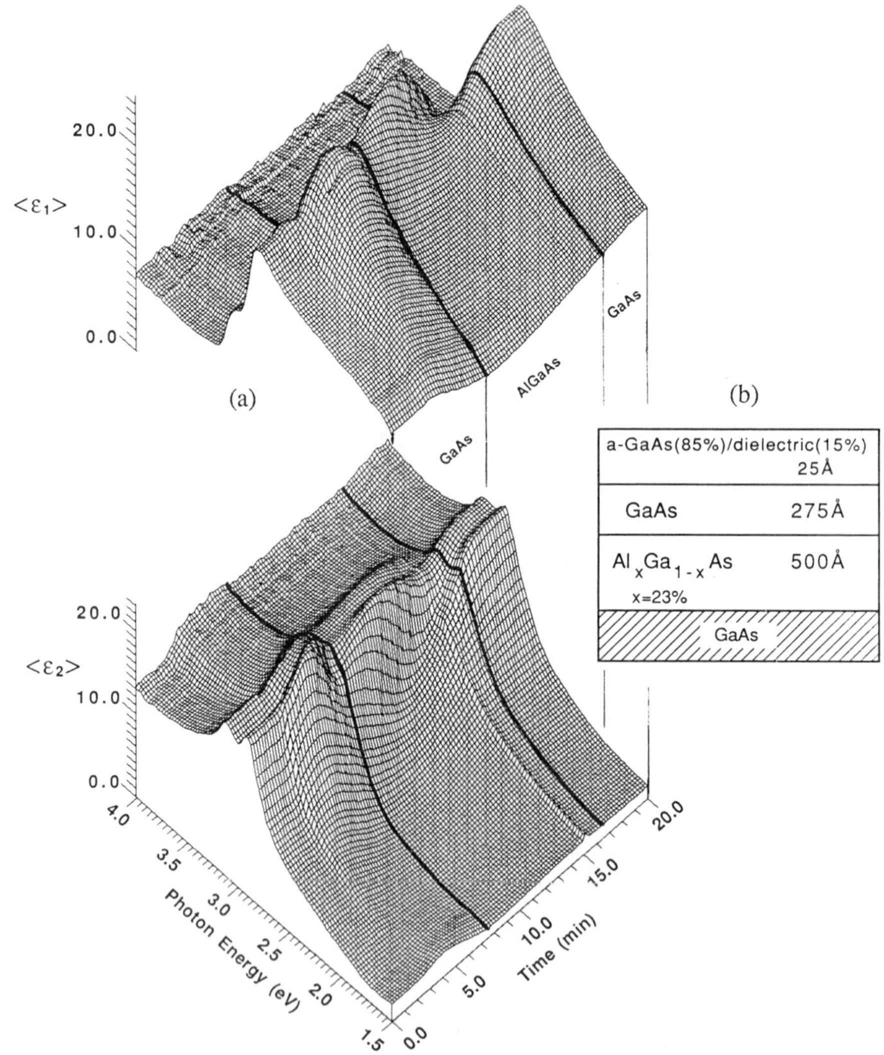

Fig. 2 (a) Pseudo-dielectric function obtained during Ar^+ sputter-etching of an GaAs/-$Al_{0.23}Ga_{0.77}As$/GaAs heterostructure. The lines on the plot indicate the times when the interfaces were crossed. (b) Model for the heterostructure in the initial stages of etching.

In a study of bulk GaAs etching, the surface temperature could also be determined in the LRA. For the more complex heterostructure this was not attempted, and the rise in surface temperature (estimated to be from 25°C to ~80-90°C) that occurs during etching at 200 V is neglected. This appears to lead to a 5 Å overestimation of the damage thickness of Figure 4.

The following behavior was observed after Ar^+ beam ignition at 1.6 min. First, 2 min of etching was required before the oxide component in the surface layer (originating from the 10 Å initial native oxide) was completely removed. In this time, the damage layer changed from an a-GaAs/oxide mixture to pure a-GaAs, and its thickness increased to ~30 Å. Furthermore, ~80 Å of underlying GaAs was removed before the oxide disappeared. This provides a measure of the ability of Ar^+ to mix the near-surface region and bury oxygen. From 2.5 to 7 min, the 300 Å top-most GaAs was sputter-etched at a constant rate of 66 Å/min. The data were not analyzed in the brief interval near 7 min when the GaAs/$Al_{0.23}Ga_{0.77}As$ interface was crossed as it was expected that the damaged layer then is a complex mixture of materials. From 7 to 17 min, the 500 Å $Al_{0.23}Ga_{0.77}As$ layer was removed at a constant rate of 52 Å/min, showing clearly that the alloy has a lower sputter-etch rate. If linear extrapolations of both etch rates are made near 7 min, it is found that the $Al_{0.23}Ga_{0.77}As$ thickness has decreased by ~20 Å at the 0 Å GaAs thickness. It is this material that forms the 30 Å damaged layer. The damaged layer on the alloy is also simulated as an a-GaAs/oxide mixture, owing to the fact that dielectric functions for a-AlGaAs alloys are not available. The apparent 0.18 volume fraction oxide in the damaged layer for the $Al_{0.23}Ga_{0.77}As$ probably simulates the effect of the Al content. It is also possible that the more reactive Al in the surface damage layer segregates and then oxidizes from residual contaminants in the vacuum. Finally, the data after 17 min show again that ~80 Å of GaAs base material must be removed before the surface damage composition returns to pure a-GaAs.

Through this example, the effectiveness of real time SE as a probe of semiconductor etching has been demonstrated. Instantaneous bulk thicknesses, etch rates, and damage layer thickness and composition can be deduced from the data. Although the example represents a physical sputtering process, the potential is great for studying reactive etching in which etch rates and surface quality are important figures of merit. Real time SE will provide even more precise thicknesses during MBE growth because the damaged layer, whose presence tends to increase the overall confidence limits, will be not be required in the modeling.

Growth of Diamond-like Carbon

Figure 5 shows the evolution of the pseudo-dielectric function during a-C:H growth by ion beam deposition onto an unheated Ta substrate. The Ta was prepared as a thin film on c-Si by dc magnetron sputtering using conditions that led to a high bulk density and smooth surface. Just prior to a-C:H deposition, the Ta native oxide was removed by Ar^+ bombardment under real time control. To prepare the a-C:H, pure CH_4 was introduced into the ion source which was directed toward the substrate at a 30° angle of incidence. The ion beam voltage and current were 200 V and 7 mA. Figure 5 incudes ~55 spectra collected during the 14 min deposition.

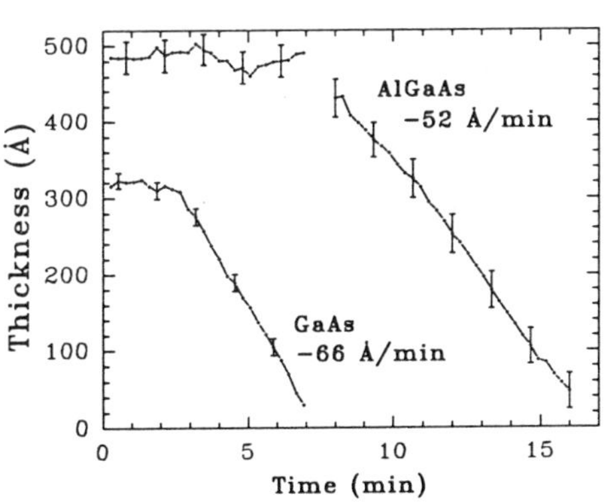

Fig. 3 Time dependence of the top GaAs and underlying $Al_{0.23}Ga_{0.77}As$ thickness values during Ar^+ sputter-etching.

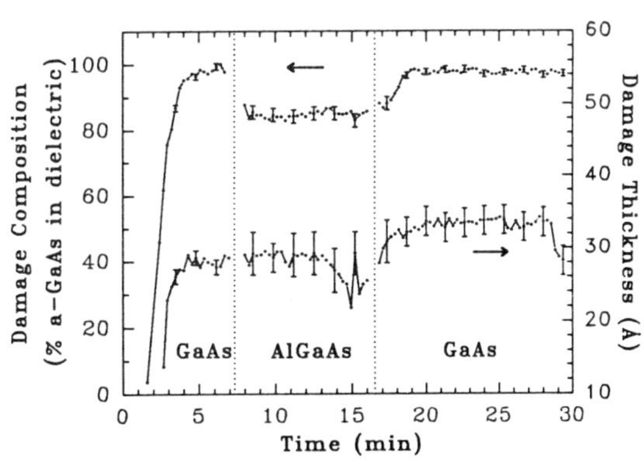

Fig. 4 Time dependence of the surface damage thickness and composition on GaAs/$Al_{0.23}Ga_{0.77}As$ during Ar^+ sputter-etching.

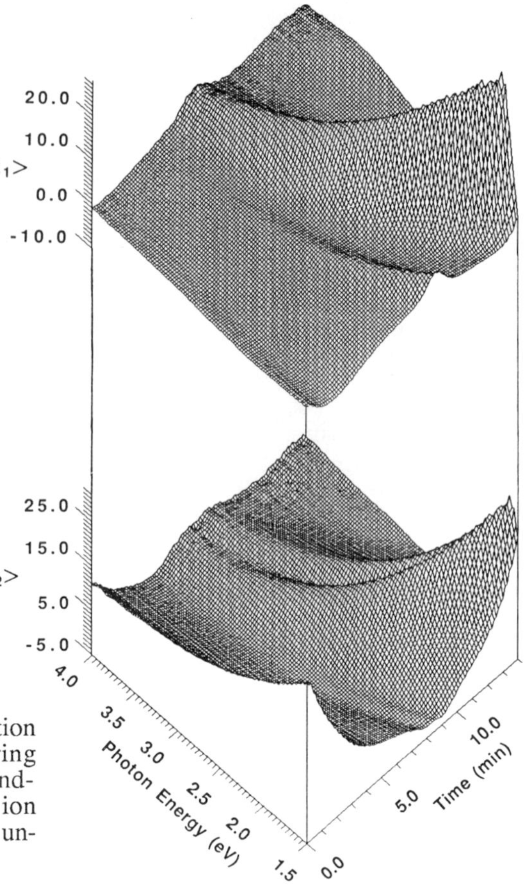

Fig. 5 Pseudo-dielectric function spectra obtained during the growth of diamond-like a-C:H by direct ion beam deposition on an unheated Ta substrate.

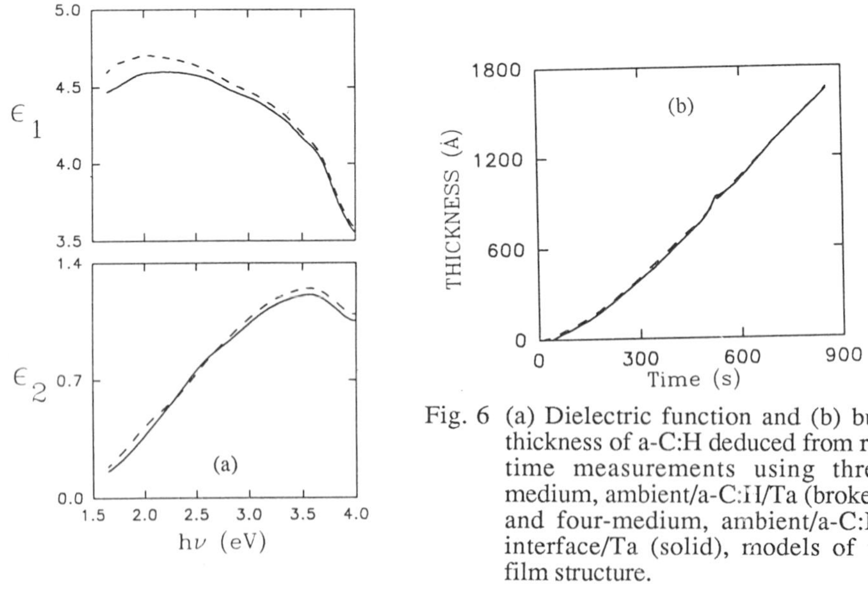

Fig. 6 (a) Dielectric function and (b) bulk thickness of a-C:H deduced from real time measurements using three-medium, ambient/a-C:H/Ta (broken), and four-medium, ambient/a-C:H/-interface/Ta (solid), models of the film structure.

For precleaning, care was taken to optimize the processing parameters for removal of the Ta oxide without causing significant damage to or roughening of the resulting surface. Thus, at t=0 in Fig. 5 the substrate is assumed to form a single abrupt interface. As a result, the pseudo-dielectric function is taken to be the actual dielectric function of Ta. In the initial data analysis, a three-medium model (ambient/a-C:H/Ta) is assumed.

The data analysis steps required to deduce the thicknesses and the dielectric function of the a-C:H will be enumerated. (1) One pseudo-dielectric function, mid-way through the deposition is chosen, and the thickness at that time is guessed. (2) Then the only unknowns in the analysis of that pair of spectra are the real and imaginary parts of the a-C:H dielectric function, which can be obtained by numerical inversion, ie. $\{<\varepsilon_1(h\nu)>,<\varepsilon_2(h\nu)>\} \rightarrow \{\varepsilon_1(h\nu),\varepsilon_2(h\nu)\}|_{d'}$. The latter notation represents the a-C:H dielectric function evaluated using the guessed thickness, d'. (3) This dielectric function is used in a LRA of the full data set of Fig. 5, with the thickness as the deduced parameter. The sum of all unbiased estimators for the best fits to the ~55 spectra are used to assess the quality of the overall fit. (4) The entire procedure, starting with step (1), is repeated with a new d'. The correct d' is that which minimizes the sum of the unbiased estimators. The a-C:H dielectric function and the best fit thickness vs. time using the correct guess are then also correct, at least within the first approximation of the three medium model. Figures 6(a) and (b) (broken lines) show the final results for the deposition of Fig. 5.

To assess the three-medium model, single wavelength $\{<\varepsilon_1>,<\varepsilon_2>\}$ trajectories (with time as a parameter) are compared. Figure 7 shows results at 2.5 eV for the first 3 min of growth. In this Figure, the model (broken line) deviates significantly from the data (points) although at later times the agreement is better. This suggests, in agreement with earlier single wavelength studies,[10] that the material in the early growth stages is different than in the bulk. To obtain more information on the interface, a fit to a four-medium model, (ambient/a-C:H/interface/Ta), was performed. The optical properties of the interface layer were simulated using materials or material mixtures having known dielectric functions: single crystal and polycrystalline TaC and physical mixtures of TaC/a-C:H and Ta/a-C:H. The approximate three-medium a-C:H dielectric function in Fig. 6(a) was used for the mixtures. With known dielectric functions, the LRA was used to analyze the early growth stage. The solution which minimizes the overall unbiased estimator uses a 0.7/0.3 mixture of a-C:H/Ta, probably simulating the optical response of a C-rich carbide. The best fit thicknesses in Figure 8 are reasonable, showing that once the carbide is buried by the a-C:H, its thickness remains constant at ~18 Å. The solid line in Fig. 7 for the four-medium model is in excellent agreement with the data.

Finally, the previous analysis applied to deduce the a-C:H dielectric function is repeated using a four-medium model including the known interface layer. Although the procedure will not be outlined here, the results appear in Figs. 6(a) and (b) (solid lines). Figure 6(a) shows that if the interface layer is neglected, 5% errors in the a-C:H dielectric function are possible.

In summary, real time SE can be used to determine the optical functions of diamond-like carbon thin films. This capability is important considering the sensitivity of the optical

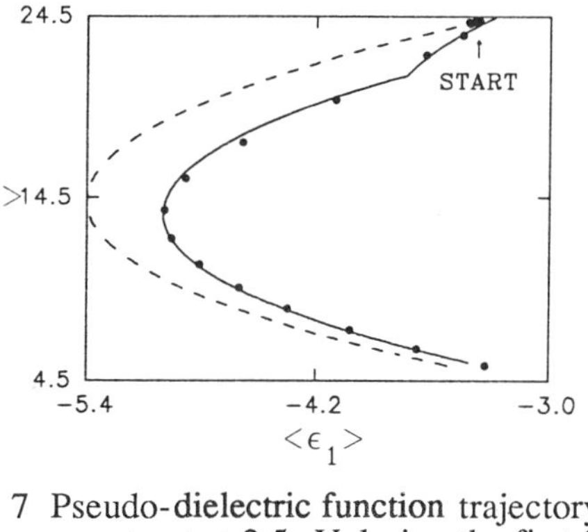

Fig. 7 Pseudo-dielectric function trajectory swept out at 2.5 eV during the first 3 min of a-C:H growth on Ta. The broken and solid lines represent fits for three- and four-medium models, respectively.

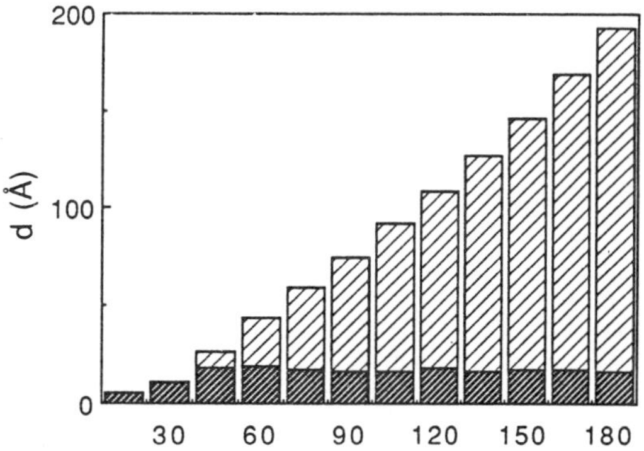

Fig. 8 Time dependence of the bulk a-C:H and carbide interface layer thickness obtained in the first 3 min of the deposition of Fig. 6.

functions to the deposition conditions. It is also important because the major goal in deposition is a film with optical and mechanical properties comparable to those of diamond.[6] Real time SE also provides the bulk a-C:H and interfacial carbide thicknesses as a function of time. Information on the optical properties of the carbide are also obtained, and this is expected to be important in correlation with the mechanical stability of the films.

CONCLUSIONS

Applications of newly-developed ellipsometric techniques for in-process, non-destructive characterization of vacuum deposition and etching of thin film structures have been described. The key development is the use of a rotating polarizer ellipsometer with multichannel detection so that $\{\psi(h\nu),\Delta(h\nu)\}$ spectra from 1.6 to 4.0 eV can be collected in 40 ms. The new capabilities represent a breakthrough in the determination of thickness, deposition rate, and optical properties (ie. dielectric functions, $\{\varepsilon_1(h\nu),\varepsilon_2(h\nu)\}$) of materials from real time data. For in-process characterization of semiconductor structures, which are composed of layers of materials having known optical properties, linear regression analysis can be applied to each spectrum independently to determine the time evolution of layer thicknesses and compositions. For in-process characterization of thin films whose optical properties are not known in advance, the full data set, $\{\psi(h\nu),\Delta(h\nu)\}$ vs. time, can be analyzed to determine the bulk dielectric function and thickness as a function of time. In the particular example discussed here, vapor deposition of diamond-like amorphous carbon, deviations from perfect film growth could be understood by incorporating an interfacial carbide layer into the model. The optical properties and thickness as a function of time for the carbide completed a comprehensive optical analysis of the growth process. We anticipate that such techniques can address a wide range of problems in film growth, etching, and materials modification, particularly processes for which a detailed understanding of surface and interface phenomena are required.

ACKNOWLEDGMENTS

The authors would like to thank Prof. D.L. Miller, Dr. S.S. Bose, and Prof. L.J. Pilione for indispensible collaborations. Support for this work was provided by an AT&T Foundation Grant, by the Office of Naval Research (with funding from the Strategic Defense Initiative Organization's Office of Innovative Science and Technology) under Contract No. N00014-86-K-0443. RC, IA, and KV received support from the National Science Foundation (Grant No. DMR-8901031). RC, KV, HW, and RM received support from The Air Force Office of Scientific Research (Contract No. AFOSR-87-0343). AH was supported by the US Army CECOM, Ft. Monmouth, NJ.

REFERENCES

1. J.B. Theeten, Surf.Sci. 96, 275 (1980).
2. D.E. Aspnes, Proc. Soc. Photo-Opt. Instrum. Eng. 276, 188 (1981).
3. R.H. Muller and J.C. Farmer, Rev. Sci. Instrum. 55, 371 (1984).
4. Y. Cong, I. An, R.W. Collins, K. Vedam, H.S. Witham, L.J. Pilione, and R. Messier, Proc. 8th Int. Conf. on Thin Films, San Diego CA, April 2-6 1990, Thin Solid Films (in press).
5. D.E. Aspnes and A.A. Studna, Appl. Opt. 14, 220 (1975).
6. J.C. Angus, P. Koidl, and S. Domitz, in: Plasma Deposited Thin Films, edited by J. Mort and F. Jansen, (CRC, Boca Raton FL, 1986), p. 89.
7. D.E. Aspnes, S.M. Kelso, R.A. Logan, and R. Bhat, J. Appl. Phys. 60, 754 (1986).
8. D.E. Aspnes, G.P. Schwartz, G.J. Gualtieri, A.A. Studna, and B. Schwartz, J. Electrochem. Soc. 128, 590 (1981).
9. A. Gheorghiu and M.-L. Theye, Philos. Mag. B 44, 285 (1981).
10. R.W. Collins, J. Vac. Sci. Technol. A 7, 1378 (1989).

REAL-TIME THICKNESS MEASUREMENT OF METAL OXIDE FILM GROWTH

R. T. Allemeier and J. W. Wagner

The Johns Hopkins University
102 Maryland Hall
34th and Charles Streets
Baltimore, Maryland 21218

ABSTRACT

The use of Dynamic Imaging Microellipsometry (DIM) for mapping and measuring oxide film thickness during growth has been demonstrated. The DIM system facilitates the analysis of oxide film characteristics by determining the polarization properties of a beam of light reflected from an oxide film/substrate system. While the principles of DIM are much the same as those for conventional ellipsometry there are significant differences. These differences, most notably, the use of a CCD video camera as a detector coupled with image processing capabilities, permit the production of spatially resolved ellipsometric "images" at high repetition rates. Using the high spatial and temporal resolution of the DIM system, a sequence of ellipsometric images have been generated showing the growth of cuprite (Cu_2O) on copper. This data clearly demonstrates the capability of DIM for measuring dynamic changes in oxide film thickness.

INTRODUCTION

DIM is an extension of a traditional optical technique, known generally as ellipsometry, which is used for the determination of physical properties of thin films (i.e. thickness and refractive index). Ellipsometry derives its name from the fact that the state of a polarization ellipse of reflected light is measured and used to determine the properties of the film in question. Upon interacting with the material, the state of polarization of the light will be changed by the different Fresnel coefficients of the material. The characteristics of importance for the reflected or transmitted beam are the changes in relative amplitudes of P-component of polarization (component parallel to plane of incidence) and the S-component of polarization (component perpendicular to the plane of incidence), in addition to the relative phase differences between these two components.

Nondestructive Characterization of Materials IV
Edited by C.O. Ruud *et al.*, Plenum Press, New York, 1991

These characteristics of the reflected light may be analyzed and used to determine the physical properties of the film through an iterative solution[1] of the ellipsometric equations.

Prior to the advent of DIM[2], ellipsometric measurements of high spatial resolution combined with good temporal resolution were not possible. The dramatic improvement in these qualities can best be demonstrated by comparing the DIM technique with the results of a method where the incident beam is focused to a small spot size (on the order of 25μm) and then scanned over the surface[3]. In order to produce the spatial resolution of DIM over an area equivalent to 60,000 pixels squared this method would require approximately 66 hours while the DIM system can produce similar data in 45 seconds, indicating that the DIM method is fully 3 and 1/2 orders of magnitude faster.

PRINCIPLES OF DYNAMIC IMAGING MICROELLIPSOMETRY

A block diagram of the DIM system is shown in Figure 1. The DIM system consists primarily of a quasi-monochromatic light source, polarizer, half-wave plate, compensator (quarter-wave plate), analyzer (polarizer), CCD video camera, and image processing system. The primary difference between the DIM system and traditional ellipsometers is that a traditional null type ellipsometer uses a single element photodetector such as a photomultiplier tube to detect output intensities. A minimum intensity output is obtained by adjusting the azimuth angles of the polarizer and analyzer to determine the Δ and ψ values[4] for the film.

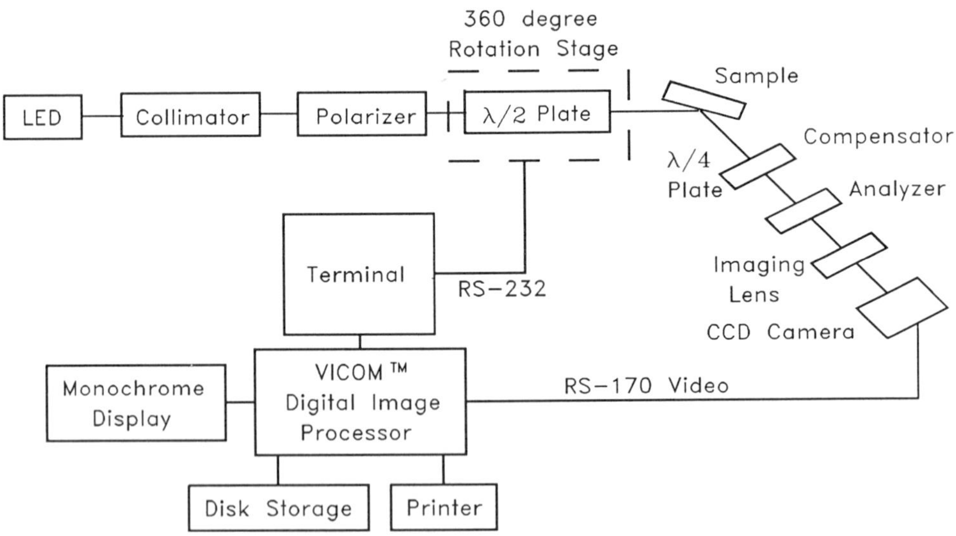

Figure 1. Block diagram of the Dynamic Imaging Microellipsometer.

Alternatively, DIM uses a CCD video camera as the detector. Since the CCD is essentially composed of a large number of individual detectors of small area, packaged in an array, it yields ellipsograms which are composed of details which are a function of the CCD element (pixel) size and of the imaging optics magnification and resolution. As mentioned previously, this detection scheme makes possible dynamic studies of passive film growth and decay since the DIM system can produce results of high spatial quality at rates which are improved by more than 3 orders of magnitude compared to those of null ellipsometers.

Using a video camera as the detector, data from this system can be collected and processed by a digital image processor. The approach established by Cohn[5] is to use a standard PSCA (polarizer, specimen, compensator, analyzer) configuration where the polarizer is followed by a half-waveplate, mounted in a computer controlled rotation stage, which permits rotation of the incident polarizations. Images are captured at incident polarization angles of $-45°, 0°, 45°$, and $90°$ and used in equations 1 and 2 to determine the Δ and ψ values for the material in question.

$$\Delta = \cos^{-1}\left(\frac{I(45°) - I(-45°)}{2\sqrt{I(0°)I(90°)}}\right) - \tan^{-1}\left(\frac{tan2(A-C)}{sin2C}\right) \qquad (1)$$

$$\psi = \tan^{-1}\left(\sqrt{\frac{1 - cos2Ccos2(A-C)}{1 + cos2Ccos2(A-C)}}\sqrt{\frac{I(0°)}{I(90°)}}\right) \qquad (2)$$

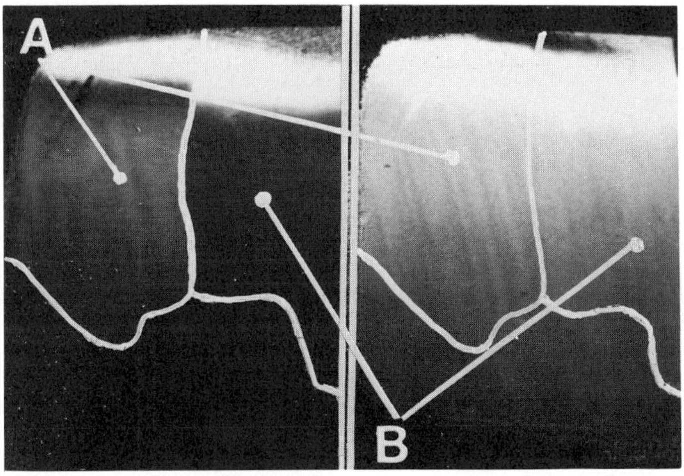

Figure 2. Δ,ψ (Δ left half, ψ right half) image for the copper sample showing Grains A and B.

Constants A and C are the azimuth angles of the analyzer and compensator respectively. The Δ and ψ values are then used to solve for film thickness and refractive index using an iterative numerical approach previously mentioned.

In practice a polished sample is placed in the ellipsometer and data is collected under computer control at each of the four polarization settings. After collecting the fourth image an ellipsogram such as the image in Figure 2 is computed using equations 1 and 2. This ellipsogram consists of grey levels corresponding to the Δ values on the left half and the ψ values on the right. The ellipsogram is then stored on the computer hard disk for future analysis.

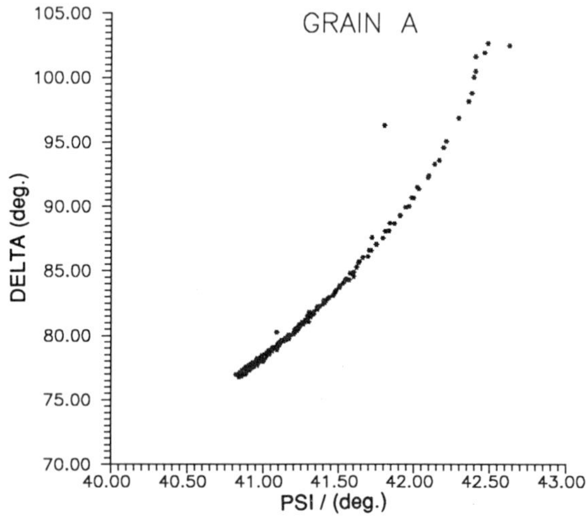

Figure 3. Δ versus ψ for Grain A, $(\bar{1}\bar{2}2)$ orientation.

EXPERIMENTAL PARAMETERS

To verify the ability of the DIM system to measure dynamic changes in oxide film thickness, a copper sample was chosen for the experiment. This sample was polycrystalline having relatively large grains on the order of 5mm in diameter. Since it is necessary for the material under study to be a good specular reflector the sample was both mechanically and chemically polished. Subsequent x-ray backscatter measurements used to determine orientation of two of the grains showed no evidence of plasticly deformed material on the sample surface.

In order to produce a desireable cuprite growth rate, a temperature controlled heating element was affixed to the sample. The sample temperature was monitored using a thermocouple throughout the data collection interval of 15 hours. Data collection was initiated simultaneously with the application of heat, with the sampling period fixed at 5 minutes. Thermal equilibrium at 100°C was reached prior to the collection of the second data set and was constant throughout the data collection. In addition, there had been some initial oxide growth on the sample surface prior to the collection of the first data set.

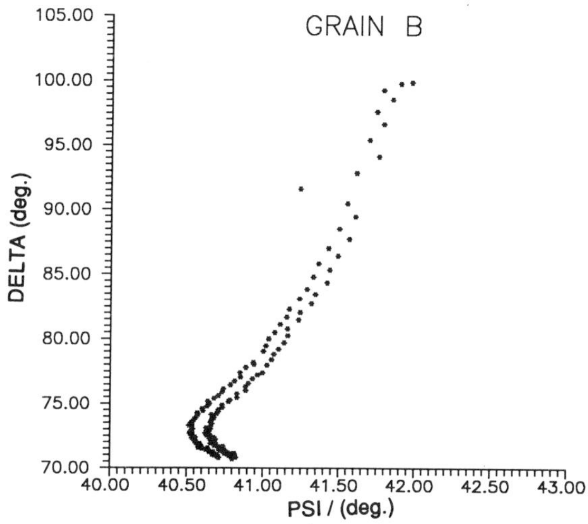

Figure 4. Δ versus ψ for Grain B, undetermined orientation

EXPERIMENTAL RESULTS

Shown in Figure 2 is a Δ,ψ map (ellipsogram) of the sample. The cuprite growth rates of a 20 by 20 pixel area of two grains were followed as indicated by the outlined areas in the figure. X-ray backscatter data was collected on each of these grains revealing that grain A (left grain in figure 2) has an orientation of $(\bar{1}\bar{2}2)$ while grain B has a more complex and as yet undetermined indices.

Figures 3 and 4 show Δ versus ψ for grains A and B respectively. The increasing density of points as Δ and ψ become smaller is indicative of slower film growth rates over time. In Figure 4 the apparent oscillation from point to point is caused by quadrature error in the half-waveplate.

Thickness versus time plots for grains A and B are shown in Figure 5. In order to arrive at these values for the thickness a numerical solution of the ellipsometric equations was used. Since there are three unknowns (thickness, refractive index, and absorption) but only two equations, the program iterates on thickness and refractive index for fixed absorption coefficients. In practice, a range of absorption coefficients is selected with a specified step size, so that it is possible to obtain multiple solutions for a given Δ,ψ pair, one solution for each absorption coefficient. However, even with a reasonably large range of possible absorption coefficients (0.1 to 0.25) most Δ,ψ pairs have only one solution. A value found in the literature[6] for the refractive index of cuprite at our wavelength (650.0 nm) is approximately 2.75. In the case of more than one solution this was the criteria used to determine the desired solution.

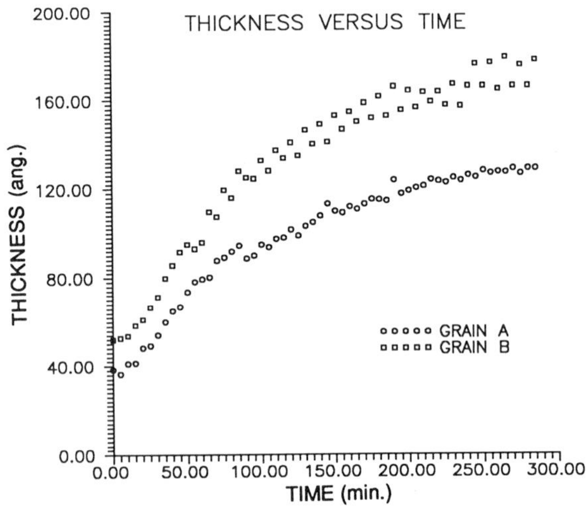

Figure 5. Thickness versus time for Grains A and B of the copper sample.

The data of Figure 5 are for the first 285 minutes of data collection. Data in this figure indicate that the cuprite growth rates for different grain orientations are different. This type of growth rate anisotropy has been previously documented[7,8] under various conditions.

CONCLUSIONS

The DIM system has been shown to be capable of measuring dynamic changes in oxide film thickness in-situ. In particular, cuprite growth rates on copper have been measured for two different grain orientations. This work demonstrates that the system should be capable of use in studies of passive film dynamics such as that prior to pitting in aluminum. This latter application will require the unique imaging capabilities of the DIM system since one cannot predict exact locations of film thinning which may occur prior to pitting.

ACKNOWLEDGEMENTS

The authors would like to thank Alcoa for providing the funding for this research. In addition, special thanks to Kirsten Green for her heroic efforts in determining crystallographic orientations of the copper grains studied. Finally, thanks to Chris Strienz for his work on the third generation DIM.

REFERENCES

1. F. L. McCrackin, A Fortran Program for Analysis of Ellipsometric Measurements, NBS Technical Note 479 April 1969
2. R. F. Cohn, J. W. Wagner, and J. Kruger, Dynamic Imaging Microellipsometry: Proof Concept Test Results, J. Electrochem. Soc. v135, n4:1033-1034 (1988)
3. Gaertner Scientific Corp., "Auto Gain Ellipsometers", Bulletin EE, pg. 13
4. R. M. A. Azzam and N. M. Bashara, Theory and Analysis of Measurements in Ellipsometry, in: "Ellipsometry and Polarized Light," North-Holland, Amsterdam (1987)
5. R. F. Cohn and J. W. Wagner, Nondestructive Mapping of Surface Film Parameters with Dynamic Imaging Microellipsometry, in: "Review of Progress in Quantitative NDE," 8A, Donald O. Thompson and Dale Chimenti eds., Plenum Press, New York (1989)
6. Harold Wieder and A. W. Czanderna, Optical Properties of Copper Oxide Films, J. App. Phys. v37, n1:184-187 (1966)
7. T. N. Rhodin Jr., Low Temperature Oxidation of Copper. II. Reaction Rate Anisoptropy, J. Am. Chem. v73, (July, 1951)
8. Jerome Kruger, The Oxide Films on Copper Single Crystal Surfaces in Water II. Rate of Growth at Room Temperature, J. Electrochem. Soc. v108, n6 (June, 1961)

NONDESTRUCTIVE EVALUATION OF ADHESIVE STRENGTH OF CVD POLYCRYSTALLINE DIAMOND COATINGS DEPOSITED ON SIALON SUBSTRATES

J. C. Conway, Jr., P. H. Cohen, B. R. Tittmann,
W. R. Drawl, M. Meloncelli, D. F. Poeth, and R. Kropiewnicki

The Pennsylvania State University
University Park, PA 16802

1. INTRODUCTION

In this report we demonstrate the use of optical and acoustic microscopy to nondestructively examine chemically vapour deposited (CVD) diamond coatings with simulated machining damage. During machining, a coated tool insert is subjected to both normal and shearing stresses in the process zone. Machining tests conducted early in this investigation revealed that these stresses resulted in spalling of the CVD diamond coating on the tool flank and rake face resulting in substantially increased wear. Further, it was discovered that the adhesive strength of the coating was extremely sensitive to many of the coating parameters such as roughness of the substrate, coating thickness, percent methane, coating temperature and pressure, etc. A decision was therefore made to utilize a simpler test to evaluate the adhesive strength of a particular coating and thus demonstrate a nondestructive evaluation method in order to optimize the parameters of the coating test matrix prior to conducting machining tests.

Damage to coatings and thin films can be simulated by a variety of procedures including the scratch test (1-3), indentation techniques (4-5) and laser spallation (6). Most of these techniques are qualitative in nature and are used primarily to optimize coating parameters for a given substrate. In the case of this investigation, an indentation technique was selected in which coated SIALON (a ceramic consisting of silicon, aluminum, oxygen and nitrogen) machining tools were indented with a Brale indenter over a load range of from 15 to 150 Kg. Lateral cracking occurred between the coating and substrate resulting in a significant weakening of the adhesion of the coating to the substrate. Both optical microscopy with polarized light and scanning acoustic microscopy were employed to nondestructively characterize the adhesive strength of the coating.

2. SAMPLE PREPARATION

The following sections describe the deposition and damage simulation procedures.

2.1 Deposition Process

Diamond depositions utilized a Toshiba tube type microwave system. The system operates at a frequency of 2.45 GHz with a maximum output power of 1500 W. The microwave cavity surrounds a 1.5 inch quartz tube, which contains the specimen and susceptor. The susceptor is boron nitride (BN) and it is supported by a quartz rod. A view port allows for observation of the specimen surface during the deposition process. The view port also permits measurement of the surface temperature by optical pyrometry.

A mixture of 1 to 5 percent ultra-high-purity (UHP) methane in hydrogen gas flows through the tube and over the substrate. The microwave energy couples with the gas mixture creating a plasma. Both the plasma and the microwave energy combine to heat the substrate. Substrate temperatures ranged between 900 C and 1000 C. Gas pressure ranged between 70 and 90 torr. The total flow rate was held constant at 100 standard cubic centimeters-per-minute (s.c.c.m.).

In an attempt to obtain uniform films, the surfaces of all the substrates were polished with a mixture of methanol and .5 micron diamond powder. All substrates should be considered seeded with diamond. The SIALON substrates are 12.7 mm square and 6.3 mm thick. The substrates were then placed in the reactor. The reactor was evacuated and back filled with hydrogen gas. This step was repeated several times to insure the presence of only hydrogen. A plasma was ignited at 10 torr; it was maintained while the pressure was increased to the desired level. Once the desired pressure was reached, the temperature was then adjusted by varying the input power to the magnetron. After a stable pressure and temperature were obtained, the methane was introduced to the plasma. These conditions were maintained until the desired deposition time was reached, after which the methane was discontinued and the magnetron was turned off. Hydrogen gas was allowed to pass over the substrate until the substrate reached room temperature. Upon cooling, the substrate was removed for evaluation.

2.2 Growth Parameters

There are several variables which are expected to significantly alter the adhesive strength of CVD polycrystalline diamond as deposited on a SIALON substrate. These variables include deposition time or coating thickness, surface preparation, the methane concentration of the chamber atmosphere, temperature of deposition, and pressure of deposition. For the work presented in this paper a two percent methane atmosphere was used. Deposition was performed at a temperature of 1000°C and a pressure of 90 torr. Deposition time and surface preparation were varied as shown in Table 1. The ground surface is an "as received" lapped surface while the polished specimen has a 1 micron finish.

Table 1. Experimental test matrix for indentation testing with 2% methane, 1000°C, 90 torr atmosphere

Variable	Low	High
Deposition Time	5 hrs.	6 hrs.
Surface Preparation	polished	ground

2.3 Damage Simulation

The machining damage was simulated by performing indentations made with a diamond Brale indenter over a load range of from 15 to 150 kg. During the unloading half cycle, indentation induced delamination occurred at the interface between the diamond coating and substrate (4). Residual stresses induced during deposition of the diamond film are assumed to further enhance cracking between the film and substrate through buckling of the film above the crack and by providing an additional driving force during buckling. Delamination results in a post-indentation, axially symmetric weakening of the film surrounding the indentation site.

3. EXPERIMENTAL NDE TECHNIQUE

The measurement of the debond between the film and the substrate was performed by two independent techniques: optical microscopy for transparent coatings and scanning acoustic microscopy (SAM) on opaque coatings. The two techniques were compared side-by-side on the same films to establish a common reference base.

3.1 Optical Microscopy

In the case of optical microscopy an interference contrast method was adopted, which accentuated the vertical relief of the coating surrounding the indent. Specifically, with the use of a Leitz microscope polarized light was reflected from the back surface of the delaminated coating and viewed through an analyzer. This technique, due to reflectivity variations as a function of coating back surface gradient, accentuated the vertical relief of the coating surrounding the indent. It was discovered that increases in methane content produced more opaque coatings and corresponding loss of back surface reflectivity. It was for this reason that two approaches for measuring the delamination, optical microscopy and SAM were utilized.

3.2 Acoustic Microscopy

Scanning acoustic microscopy was carried with the Olympus UH3 which provides reflection images in the frequency range from 30 MHz to 1000 MHz. The use of a high numerical aperture transducer allows the generation of surface acoustic waves as well as the bulk waves encountered in simple C-scans with focused transducers. In this application the SAM transducer was defocused to image the vertical relief resulting from the damage. The observed lack of interference of surface and bulk waves induced at 400 MHz in the film in the region surrounding the indent gave a striking image of the zone of weak adhesion. The resulting image was very similar when the SAM was used to focus on the interface between the substrate and the film, so as to reveal the extent to which interface waves could travel at that interface. In either case, the area of the nearly circular zone of weak adhesion was measured and plotted as a function of indentation load.

4. RESULTS

Prior to evaluating the results from the test matrix presented in section 3.4, the impact of indenting a diamond coating with a diamond indentor had to be evaluated. In addition, the results (i.e. debond areas) generated by optical and scanning acoustic microscopy had to be compared to obtain independent verification of the debond areas. These results are discussed in sections 4.1 and 4.2 respectively. Results on coating reproducibility and spatial uniformity is discussed in sections 4.3 and 4.4 while sections 4.5 and 4.6 discuss the effects of surface preparation and coating thickness on adhesive strength, respectively.

4.1 Damage Simulation on Samples

A high quality diamond Brale indentor was used for all indentation testing. Scanning electron microscopy revealed that the indentation of CVD diamond coatings did inflict progressive damage on the indentor, as shown in Figure 1. Experimentation was then performed on ground SIALON (KYON 2000) specimens coated at 1000°C, with a pressure of 90 torr, in a two percent methane atmosphere for five hours. These specimens were indented under a load of 60 kg and the area of debond measured. When plotted as a function of sequence, no patterns were found (see Figure 2). In addition, the slope of the best fit line cannot be statistically shown to be different than zero. Therefore, although there is undeniably indentor damage, it does not significantly influence the debond area measured.

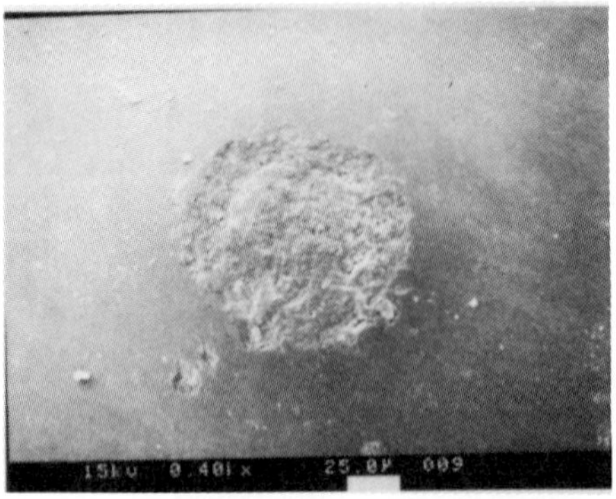

Figure 1. Damage at tip of Brale indenter after 19 indents.

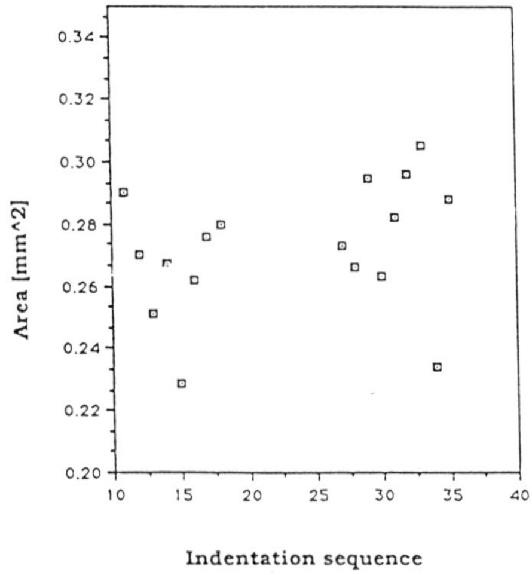

Figure 2. Area of delamination as a function of indentation sequence.

4.2 Comparison of Optical and Scanning Acoustic Microscopy Results

The measurement of debond area is a difficult task. While optical microscopy is an effective and quick technique it will not be suitable under certain conditions (such as high methane content atmospheres) where the coating becomes more opaque. It is therefore necessary to find an alternate method and correlate it to optical microscopy as well as to independently verify the optical results. Scanning acoustic microscopy was chosen since it is not affected by the opaqueness of the diamond film.

Figures 3a and 3b show the debond area using scanning acoustic and optical microscopy respectively for the same 100 kg indent performed on a ground specimen under the following conditions:

- atmosphere: 2% methane
- pressure: 90 torr
- temperature: 1000°C
- thickness (approx.): 2.5 µm

The star-shaped pattern is revealed by both techniques. Figure 4 compares the areas measured by both methods over a load range from 30 to 150 kg. As expected, the debond area is linearly correlated with indentation load over the range tested with correlation coefficients of 0.994 and 0.995, respectively, for the optical and SAM techniques. There is no statistical difference between the optical and SAM techniques.

The results presented in Figure 4 are the best results obtained. The difference between the optical and SAM techniques is larger under certain techniques depending on the surface topography. Continued refinement of SAM procedures and the use of a 1 GHz lens should give excellent correlation under all conditions.

4.3 Coating Reproducibility

In order to be commercially viable a coating must be reproducible giving each application similar, predictable mechanical properties. The experiment performed to

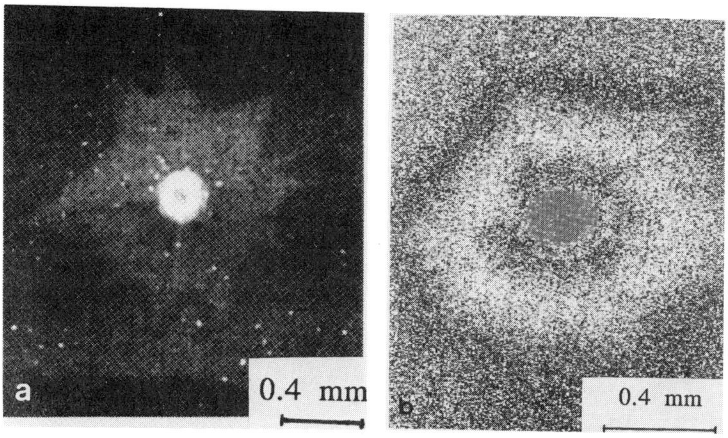

Figure 3. Area of debond as measured by (a) optical microscopy and (b) scanning acoustic microscopy.

explore the affect of simulated machining damage by using an indentor also served to determine the reproducibility of the mechanical properties of the film since more than one specimen was utilized.

While Figure 2 shows the result of indentation sequence, the data points fall into two clusters. Each cluster represents a series of identical indents performed on like specimens with the coating applied on different days. Therefore, the zero slope of the best fit line (see section 4.1) can also be interpreted as showing no difference in debond area (as measured by optical microscopy) between the two specimens. The mean area of debond for indentations 11 - 19 on the first specimen is 0.266 mm^2 with a standard deviation of 0.018 mm^2. The second specimen (indentations 26 - 35) has a mean area of debond of 0.277 mm^2 with a standard deviation of 0.023 mm^2. There is no statistical difference between the debond areas of the two specimens.

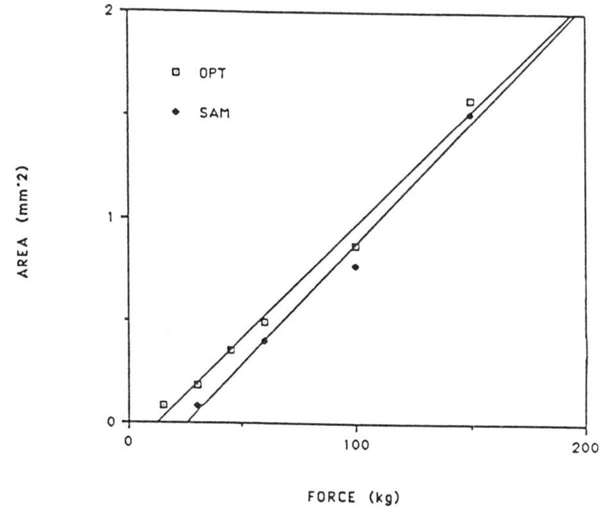

Figure 4. Debond area measured by acoustic and optical microscopy as a function of indentor load.

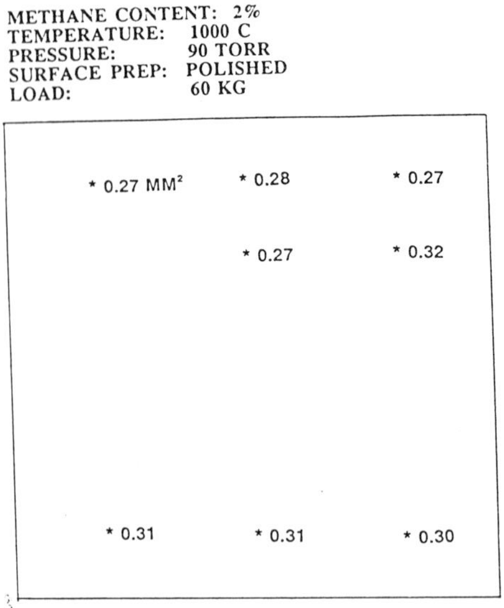

Figure 5. Spacial uniformity results for a ground specimen.

4.4 Spacial Uniformity of Coating

In order to perform adequately in many applications, especially when coating cutting tools, the coating must exhibit uniformity in properties within a given specimen. To determine spatial uniformity eight 60 kg indents were made on each specimen and the patterns observed. Two specimens coated in a two percent methane atmosphere, at 1000°C, under a pressure of 90 torr were used -- one ground and one polished.

Figures 5 and 6 depict the location of testing and debond area (as measured by optical microscopy) at each location for the ground and polished substrates, respectively. The ground specimen showed no identifiable pattern. The mean area of debond was

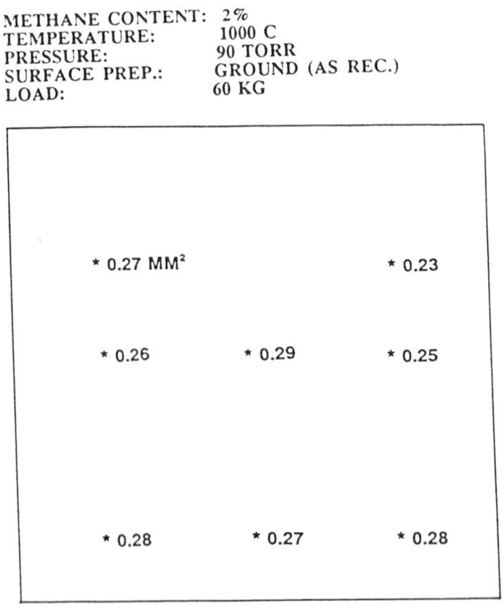

Figure 6. Spacial uniformity results for a polished specimen.

0.291 mm^2 with a range of 0.05 mm^2. The polished specimen had a mean debond area of 0.266 mm^2 and range of 0.06 mm^2. Although the mean areas of debond are different neither specimen exhibits large or patterned variation.

4.5 Effect of Surface Preparation

The type of surface (ground versus polished in this case) can have a significant impact on the adhesive strength due to differences in nucleation of the diamond. In the ground specimen the film is grown on the as-received, lapped surface. The polished specimen is obtained from a ground specimen which is polished to a 1 μm finish.

Figure 7 shows the results from both ground and polished specimens coated in a two percent methane atmosphere, at 1000°C and 90 torr indented in the range from 15 to 60 kg. The load and debond area are highly linearly correlated with a correlation coefficient of 0.929 for the polished specimen and 0.957 for the ground specimen. At all loads the ground specimen yielded smaller debond areas than the polished specimen.

4.6 Effect of Deposition Time

In an attempt to determine the influence of deposition time, two specimens were prepared in a two percent methane atmosphere, at 1000°C and 90 torr pressure. One was coated for five hours and a second for six hours. The debond area is plotted as a function of indentation load in Figure 8. For each deposition time the debond area is again highly

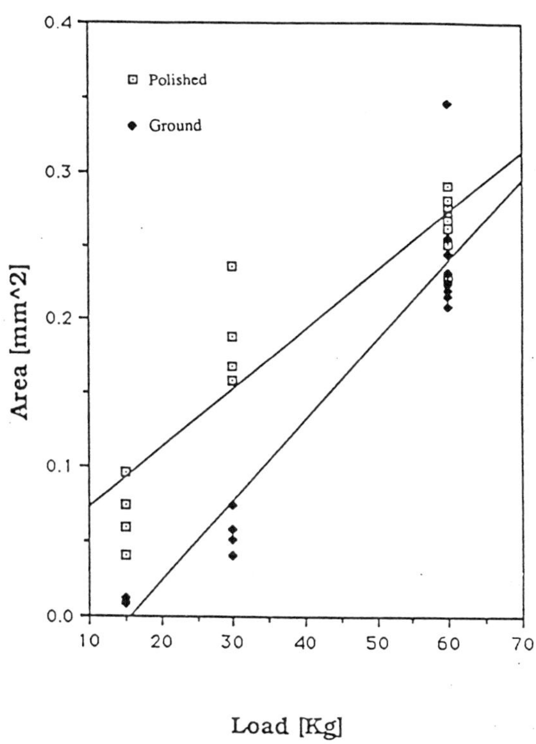

Figure 7. Area of debond as a function of indenter load for ground and polished specimens.

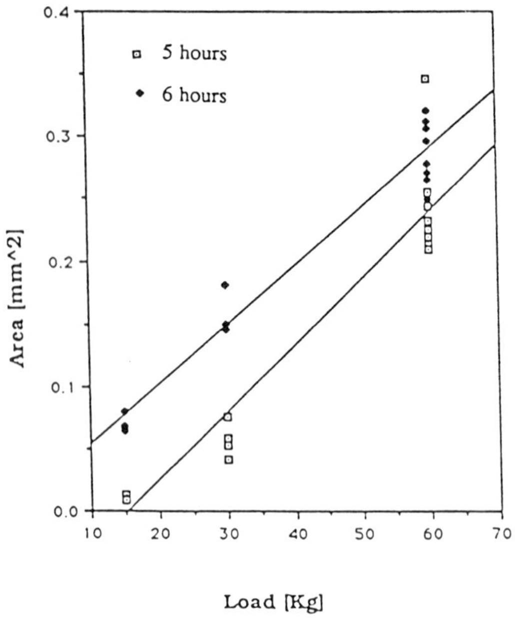

Figure 8. Area of debond as a function of indenter load for 5 and 6 hour deposition times.

linearly correlated to the load with correlation coefficients of 0.957 for the 5 hours and 0.979 for the 6 hours coating. At all loads, the thinner film had a smaller debond area indicating superior adhesive strength of the film to the SIALON substrate.

5. CONCLUSIONS

The results presented here are still preliminary but lead to the following tentative conclusions: The two NDE techniques, optical and acoustic microscopy, are complementary in that they give comparable results. The optical technique is very useful and rapid for the inspection of optically transparent films. For the film growth under conditions involving high methane content, the films are opaque and then scanning acoustic microscopy is the only reliable technique. In the comparison tests we found good correspondence (at the conditions tested) of measured debond area between optical and scanning acoustical microscopy.

In the sample preparation effort we found that the CVD diamond coating reproducibly yielded statistically equivalent debond areas between specimens. We also found good spatial uniformity of adhesive strength within each specimen. Under similar loading conditions, ground specimens yielded smaller debond areas than the polished specimens. Thin films yielded smaller debond areas than the thicker films.

Future work will include studies of the influence of methane content, pressure and temperature on adhesion and the exploration of other techniques to induce debonding such as laser spallation.

ACKNOWLEDGEMENTS

The authors thank the generous support of Thin Film and Diamond Consortium of the Materials Research Laboratory at Penn State for their support of this work.

REFERENCES

1. A. J. Perry, "Scratch Adhesion Testing of Hard Coatings," Thin Solid Films, 107, pp. 167-180 (1983).
2. A. J. Perry, "Scratch Adhesion Testing: A Critique," Surface Engineering, 2, pp. 183-185 (1986).
3. P. A. Steinmann, Y. Tardy and H. E. Hinterman, "Adhesion Testing by the Scratch Test Method: The Influence of Intrinsic and Extrinsic Parameters on the Critical Load," Thin Solid Films, 154, pp. 333-349 (1987).
4. D. B. Marshall and A. G. Evans, "Measurement of Adherence of Residually Stressed Thin Films by Indentation: 1. Mechanics of Interface Delamination," J. Appl. Phys, 56, pp. 2632-2638 (1984).
5. Prem C. Jindal, Dennis T. Quinto and George J. Wolfe, "Adhesion Measurements of Chemically Vapor Deposited and Physically Vapor Deposited Hard Coatings on WC-Co Substrates," Thin Solid Films, 154, pp. 361-375 (1987).
6. J. L. Vossen, "Measurements of Film-Substrate Bond Strength by Laser Spallation," ASTM STP 640, pp. 122-133 (1978).

NONDESTRUCTIVE EVALUATION OF WOOD—PAST, PRESENT, AND FUTURE

Robert J. Ross

USDA Forest Service
Forest Products Laboratory[1]
Madison, Wisconsin USA

Roy F. Pellerin

Washington State University
Pullman, Washington

INTRODUCTION

Nondestructive evaluation (NDE) of materials is, by definition, the science of identifying the physical and mechanical properties of a piece of material without altering its end-use capabilities. Such evaluations rely upon nondestructive testing (NDT) techniques or "tools" to provide accurate information pertaining to the properties and performance of the material in question.

During the past 30 years, forest products researchers and the forest products industry have developed and used NDT tools for a wide range of applications—from the grading of structural lumber to the in-place evaluation of the mechanical properties of individual members in wood structures. The purpose of this paper is to provide a brief overview of the evolution of NDT tools for the forest products industry by reviewing fundamental concepts and pioneering research efforts related to NDT of wood, presenting two current industrial applications of NDT, and examining current research activity that will serve as a basis for future applications.

FUNDAMENTAL CONCEPTS AND PIONEERING RESEARCH EFFORTS

Nondestructive testing techniques for wood differ greatly from those for homogeneous, isotropic materials such as metals, glass, plastics, and ceramics. In such nonwood-based materials, whose mechanical properties are known and tightly controlled by manufacturing processes, NDT techniques are used to detect the presence of discontinuities, voids, or inclusions. Because wood is a biological material, these irregularities occur naturally and, further, may occur because of agencies of degradation in the environment. Hence, NDT techniques for wood are used to measure how natural and environmentally induced irregularities interact in a wood member to determine its mechanical properties.

---

[1] The Forest Products Laboratory is maintained in cooperation with the University of Wisconsin. This article was written and prepared by U.S. Government employees on official time, and it is therefore in the public domain and not subject to copyright.

Armed with this concept, early forest products researchers vigorously examined several techniques for grading structural lumber and evaluating the quality of laminated materials (Bell et al. 1950, Galiginaitis et al. 1954, James 1959, Jayne 1955, Jayne 1959, Hoyle 1961, McKean and Hoyle 1962, Senft et al. 1962). Out of these pioneering efforts evolved a hypothesis, founded on fundamental material properties, for establishing relationships between NDT parameters and static mechanical properties of wood products.

The fundamental hypothesis for NDT of wood materials was first presented by Jayne (1959). He proposed that the energy storage and dissipation properties of wood materials, which can be measured nondestructively by using a variety of static and dynamic techniques, are controlled by the same mechanisms that determine the mechanical behavior of such materials. As a consequence, useful mathematical relationships between these properties and elastic and strength behavior should be attainable through statistical regression analysis methods.

To elaborate on Jayne's hypothesis, consider how the microscopic structure of clear, straight-grained wood affects mechanical behavior and energy storage and dissipation properties. Clear wood is a composite material composed of many tube-like cells cemented together. At the microscopic level, energy storage properties are controlled by orientation of the cells and their structural composition, factors that contribute to elasticity and strength. Such properties are observed at frequency of oscillation in vibration or speed of sound transmission. Energy dissipation properties, conversely, are controlled by internal friction characteristics to which bonding behavior between constituents contributes significantly. Rate of decay of free vibration or acoustic wave attenuation measurements are frequently used to observe energy dissipation properties.

Early laboratory studies to verify Jayne's hypothesis were conducted with clear wood and lumber products. Jayne (1959) was successful in demonstrating a relationship between energy storage and dissipation properties, measured using forced transverse vibration techniques, and the static bending properties of small, clear wood specimens. His technique was based on forcing a bending member to produce transverse oscillations. Pellerin (1965) verified the hypothesis using free transverse vibration techniques and dimension lumber. Pellerin's effort differed from that of Jayne in that the specimens were allowed to oscillate in response to an initial deflection. No forcing function was applied. Kaiserlik and Pellerin (1977) furthered the hypothesis by using stress wave techniques (longitudinal oscillation) to evaluate the tensile strength of a small sample of clear lumber containing varying degrees of slope of grain.

CURRENT INDUSTRIAL APPLICATIONS

Machine stress rating (MSR) of lumber and ultrasonic veneer grading for laminated veneer lumber manufacture are two currently utilized industrial applications of NDT that developed from early research efforts.

Machine stress rating, as currently practiced in North America, couples visual sorting criteria with nondestructive measurements of the stiffness of a piece of lumber to assign it to an established grade (Galligan et al. 1977). Machines measure the modulus of elasticity (MOE) of individual specimens that are passed through endwise. They accomplish this by measuring the bending deflection resulting from a known load or by measuring the load required to accomplish a given amount of deflection. Special features of some machines include sensitivity to low-point elasticity; that is, the ability to detect sections of lumber that may have MOE values much lower than the average of the piece. Design stresses are determined from machine MOE values, using a regression between MOE and strength.

Laminated veneer lumber manufacturing facilities use stress wave NDT techniques to sort incoming veneer into strength classes prior to processing the veneer into finished products. Veneers are assigned to strength categories, which are established through empirical relationships between stress wave velocity and strength, based on the velocity at which an ultrasonically induced stress wave travels through the wood (Sharp 1985).

CURRENT RESEARCH ACTIVITY

Considerable research activity has recently focused on applying of NDT concepts to wood-based composites, decay detection, and in-place evaluation of wood structures. In addition, current research is aimed at enhancing current MSR techniques and developing low-cost MSR grading systems.

Wood-Based Composites

Successful verification of Jayne's (1959) hypothesis using stress wave techniques on wood-based composites has been shown by Pellerin and Morschauser (1974), Ross and Pellerin (1984, 1989), and Vogt (1985). Pellerin and Morschauser (1974) showed that stress wave speed, a measure of energy storage properties, could be used to predict the flexural behavior of underlayment-grade particleboard. Ross and Pellerin (1984, 1989) revealed that wave attenuation, a measure of energy dissipation properties, was sensitive to bonding characteristics and is a valuable NDT parameter that contributes significantly to the prediction of the tensile and flexural mechanical behavior of wood-based particle composites. Vogt (1985) furthered application of the hypothesis to wood-based fiber composites. In another study, Vogt (1986) also found a strong relationship between internal bond and stress wave parameters of particle and fiber composites.

Decay Detection

Verification of Jayne's hypothesis with wood subjected to different levels of deterioration by decay fungi, which have a detrimental effect on the mechanical properties of wood and are commonly found in wood structures, has been limited to studies that have employed only energy storage parameters. Wang et al. (1980), for example, found that the frequency of oscillation of small, eastern pine, sapwood cantilever bending specimens was significantly affected by the presence of decay. Pellerin et al. (1985) showed that stress wave speed could be used successfully to monitor the degradation of small, clear wood specimens exposed to brown-rot fungi. They showed a strong correlative relationship between stress wave speed and the compressive strength parallel to the grain of exposed wood. Rutherford (1987) showed similar results. He also revealed that MOE perpendicular to the grain, measured using stress wave techniques, was significantly affected by degradation from brown-rot decay and could be used to detect incipient decay. Chudnoff et al. (1984) reported similar results from experiments that utilized an ultrasonic measurement system and several hardwood and softwood species. Patton–Mallory and DeGroot (1989) reported similar results from a fundamental study dealing with the application of acousto-ultrasonic techniques. Their results also showed that energy loss parameters may provide useful additional information pertaining to early strength loss from incipient decay resulting from the presence of brown-rot fungi.

Acoustic emission techniques have also been investigated for use in decay detection. Utilizing a small sample of clear, white fir specimens infected with brown-rot fungi. Beall and Wilcox (1986) showed a relationship between selected acoustic emission parameters and radial compressive strength.

In-Place Evaluation

Several organizations have published results of their efforts on the use of these concepts for in-place evaluation of wood structures. Pellerin (1989) summarized his results and those of others (Browne and Kuchar 1985, Hoyle and Pellerin 1978, Hoyle and Rutherford 1987, Neal 1985) on successful use of stress wave methods for in-place detection of decay in wood structures. All studies used energy storage parameters from stress wave, time-of-flight-type measurement systems. Structures evaluated successfully included bridges, footbridges, grandstands for spectator activities, piers, laminated arches in school buildings, and the world's largest wood structure—TRESTLE.

Anthony and Bodig (1989) reported on the use of sonic spectral analysis techniques they developed and used for the inspection of wood structures. Murphy et al. (1987) and Dunlop (1983) reported on the use of acoustic and vibration techniques for evaluating wood poles. The technique developed by Murphy et al. (1987) involved measuring the vibrational response of a pole after it is tapped by a rubber mallet. Dunlop (1983) utilized an electronic system, sweeping through a selected range of excitation frequencies, to develop an acoustic signature of a pole. Resonant frequencies and their bandwidths were examined for use as NDT parameters.

MSR Techniques and Low-Cost Systems

Research using noncontact scanning technology with conventional MSR systems is yielding encouraging results. This research has been aimed at providing more accurate estimates of the strength of lumber products and promises to significantly enhance the use of lumber in engineered applications (Bechtel and Allen 1987, Cramer and McDonald 1989).

Considerable effort is also being devoted to developing tools for low-cost MSR grading systems (Geske et al. 1989). This effort couples relatively inexpensive personal computer technologies and transverse vibration NDT techniques by innovative software programming. Tools developed from this effort promise to yield MSR grading systems at costs considerably lower than the costs of those currently used. This will make MSR lumber grading available to a wider range of wood products manufacturers and users.

SUMMARY

Pioneering research efforts and lessons learned by using current state-of-the-art nondestructive evaluation (NDE) will serve as the basis for future NDE activities with wood.

Technologies will be needed to enhance existing machine stress rating lumber grading for evaluating the performance of wood-based composites and the properties of the wood resource before it is processed into finished products. Low-cost systems will also be needed for the in-place evaluation of wood structures. Opportunities will continually arise as the acceptance of NDE for wood grows and the use of wood as an engineered material expands.

REFERENCES

1. Anthony, R. W., and Bodig, J. 1989, Nondestructive evaluation of timber structures for reliable performance. Proceedings of the Second Pacific Timber Engineering Conference. Aug. 28–31, 1989. Auckland, New Zealand.
2. Beall, F. C., and Wilcox, W. W., 1986, Relationship of acoustic emission to mass loss from decay during radial compressive testing. Unpublished research note, Weyerhaeuser Company, Tacoma, WA.

3. Bechtel, F. K., and Allen, J. R., 1987, Methods of implementing grain angle measurements in the machine stress rating process. Proceedings of the Sixth Symposium on Nondestructive Testing of Wood. Sept. 14–16, 1987. Washington State University, Pullman, WA. pp. 303–353.

4. Bell, E. R., Peck, E. C., and Krueger, N. T., 1950, Youngs modulus of wood determined by a dynamic method. Report 1775. U.S. Department of Agriculture, Forest Service, Forest Products Laboratory, Madison, WI.

5. Browne, C. M., and Kuchar, W. E., 1985, Determination of material properties for structural evaluation of TRESTLE. Proceedings of the Fifth International Nondestructive Testing of Wood Symposium. Sept. 9–11, 1985. Washington State University, Pullman, WA.

6. Chudnoff, M., Eslyn, W. E., and McKeever, D. B., 1984, Decay in mine timbers: Part III, Species-independent stress grading. Forest Prod. J., 34(3):43–50.

7. Cramer, S. M., and McDonald, K. A., 1989, Predicting lumber tensile stiffness and strength with local grain angle measurements and failure analysis. Wood and Fiber Sci., 21(4): 393–410.

8. Dunlop, J. I., 1983, Testing of poles by acoustic resonance. Wood Sci. Technol., 17:31–38.

9. Galiginaitis, S. V., Bell, E. R., Fine, A. M., Auer, G., and Roy, J., 1954, Nondestructive testing of wood laminates. Final Report, Office of Naval Research, University of Louisville, Institute of Industrial Research.

10. Galligan, W. L., Snodgrass, D. V., and Crow, G. W., 1977, Machine stress rating: practical concerns for lumber producers. FPL-GTR-7, U.S. Department of Agriculture, Forest Service, Forest Products Laboratory, Madison, WI.

11. Geske, E. A., Ross, R. J., Larson, G., and Murphy, J. F., 1989, PC-Based E measurement. Proceedings of Seventh Nondestructive Testing of Wood Symposium. Sept. 27–29, 1989. Washington State University, Pullman, WA.

12. Hoyle, R. J., 1961, A nondestructive test for stiffness of structural lumber. Forest Prod. J., 11(6):251–254.

13. Hoyle, R. J., and Pellerin, R. F., 1978, Stress wave inspection of a wood structure. Proceedings of Fourth Nondestructive Testing of Wood Symposium. Aug. 28–30, 1978. Washington State University, Pullman, WA.

14. Hoyle, R. J., and Rutherford, P. S., 1987, Stress wave inspection of bridge timbers and decking. Final Report for Research Project Y-3400. Department of Civil and Environmental Engineering, Washington State University, Pullman, WA.

15. James, W. L., 1959, A method for rapid measurement of the rate of decay of free vibrations. Bulletin 2154, U.S. Department of Agriculture, Forest Service, Forest Products Laboratory, Madison, WI.

16. Jayne, B. A., 1955, A nondestructive test of glue bond quality. Forest Prod. J., 5(5): 294–301.

17. Jayne, B. A., 1959, Vibrational properties of wood as indices of quality. Forest Prod. J., 9(11):413–416.

18. Kaiserlik, J. H., and Pellerin, R. F., 1977, Stress wave attenuation as an indicator of lumber strength. Forest Prod. J., 27(6):39–43.

19. McKean, H. B., and Hoyle, R. J., 1962, Stress grading method for dimension lumber. ASTM Special Tech. Pub. No. 353. American Society of Testing and Materials, Philadelphia, PA.

20. Murphy, M. W., Franklin, D. E., and Palylyk, R. A., 1987, A nondestructive testing technique for wood poles. British Columbia Hydro Research and Development, Surrey, British Columbia.

21. Neal, D. W., 1985, Establishment of elastic properties for in-place timber structures. Proceedings of Fifth Nondestructive Testing of Wood Symposium. Sept. 9–11, 1985. Washington State University, Pullman, WA.

22. Patton–Mallory, M., and DeGroot, R. C., 1989, Acousto-ultrasonics for evaluating decayed wood products. Proceedings of the Second Pacific Timber Engineering Conference. Aug. 28–31, 1989. Auckland, New Zealand.

23. Pellerin, R. F., 1965, A vibrational approach to nondestructive testing of structural lumber. Forest Prod. J., 15(3):93–101.

24. Pellerin, R. F., 1989, Inspection of wood structures for decay using stress waves. Proceedings of the Second Pacific Timber Engineering Conference. Aug. 28–31, 1989. Auckland, New Zealand.

25. Pellerin, R. F., DeGroot, R. C., and Esenther, G. R., 1985, Nondestructive stress wave measurements of decay and termite attack in experimental wood units. Proceedings of Fifth Nondestructive Testing of Wood Symposium. Sept. 9–11, 1985. Washington State University, Pullman, WA.

26. Pellerin, R. F., and Morschauser, C. R., 1974. Nondestructive testing of particleboard. Proceedings of Seventh International Particleboard Symposium. March 1973. Washington State University, Pullman, WA.

27. Ross, R. J., 1984, Stress wave speed and attenuation as predictors of the tensile and flexural properties of wood-based particle composites. Ph.D. Dissertation, Washington State University, Pullman, WA.

28. Ross, R. J., and Pellerin, R. F., 1988, NDE of wood-based composites with longitudinal stress waves. Forest Prod. J., 38(5):39–45.

29. Rutherford, P. S., 1987, Nondestructive stress wave measurement of incipient decay in Douglas–fir. M.S. Thesis, Washington State University, Pullman, WA.

30. Senft, J. F., Suddarth, S. K., and Angleton, R. D., 1962, A new approach to stress grading of lumber. Forest Prod. J., 12(4):183–186.

31. Sharp, D. J., 1985, Nondestructive testing techniques for manufacturing LVL and predicting performance. Proceedings of Fifth Nondestructive Testing of Wood Symposium. Sept. 9–11, 1985. Washington State University, Pullman, WA.

32. Vogt, J. J., 1985, Evaluation of the tensile and flexural properties and internal bond of medium density fiberboard using stress wave speed and attenuation. M.S. Thesis, Washington State University, Pullman, WA.

33. Vogt, J. J., 1986, Longitudinal stress waves as predictors of internal bond strength. Proceedings of Twentieth International Particleboard/Composite Materials Symposium. March 1986. Washington State University, Pullman, WA.

34. Wang, S. C., Suchsland, O., and Hart, J. H., 1980, Dynamic test for evaluating decay in wood. Forest Prod. J., 30(7):35–37.

THE HISTORICAL DEVELOPMENT OF NDT OF WOOD

William L. Galligan

Frank Lumber Company
Drawer 79
Mill City, OR 97360

Roy F. Pellerin and Brian K. Brashaw

Wood Materials & Engineering Lab
Washington State University
Pullman, WA 99164-1806

INTRODUCTION

Wood is the source of structural fiber for use in many different ways. It is a renewable resource, plus, many wood products are recyclable. In the current controversy over the forest environment, it is easy to lose sight of the fact that the worldwide environment includes critical wood fiber usage for shelter, for communication media, and for hygienic products. Since its beginning, research on nondestructive testing of wood has had the goal of increasing efficiency in the use of the wood resource. Today, this goal is more important than ever.

The purpose of this paper is to provide an overview of the major elements of nondestructive testing of wood and also in tracing the developments of wood over the last 30 years. Critical future needs are noted and a short bibliography that contains the proceedings from the Symposia on Nondestructive Testing of Wood as well as several workshops are provided.

THE NATURE OF WOOD

Wood is a material which, aside from trimming here and there and drying, often is used exactly as it has been cut from the tree. No smelting, no catalytic polymerization, no alloying, no fundamental alteration of nature's wood fiber is required to produce the basic structural form of wood. Wood in this natural form has been used as a structural product for thousands of years.

What are the features of wood that bring about its high strength-to-weight ratios, yet also produce anisotropy and variability? Both the macro and also the microstructure of wood provide relatively easy answers. The familiar cross-section of a tree (Figure 1) illustrates a few of the reasons for the variability. Starting at the center, or pith, the tree grows outward, adding a "growth ring" each year. This growth ring usually is composed of two types of cells: spring or earlywood (formed in the period of rapid growth in the spring) and summer or latewood (formed in the period of slow growth, usually in the summer). These two types of cells have different densities and properties and the ratio of earlywood to latewood often differs year by year. Further, the quality of the wood formed near the pith will be less than that found some 10 to 20 rings further out. Also shown in the figure is the cross section of a

Edited by C.O. Ruud *et al.*, Plenum Press, New York, 1991

limb that originated at the pith and will appear as a knot when lumber is sawn from the log. This knot can be a significant source of weakness in a product.

The basic structural element of wood is the fiber. A bundle of these fibers is shown in Figure 2 to illustrate the microstructure of a coniferous wood. The majority of the fibers are long and narrow, generally aligned longitudinally with the tree trunk. The high tensile and compressive strength of wood results from this alignment, while the transverse properties, as could be predicted, are much lower. The tubes (fibers) are held together by an adhesive (lignin), completing nature's composite.

The wood in a tree has a generally circular macrostructure, while at a micro level the structural orientation emphasizes the longitudinal fibers, it is from these fibers that anisotropy results. Further, it is virtually impossible to reduce this complex structure to a product (through cutting, peeling, or pulping) without disturbing nature's efficiency. As a consequence, much of the wood that nondestructive testing (NDT) addresses deals with the heterogeneity, including knots and similar features, as well as the anisotropy as it affects the performance of the final product.

DEVELOPMENT OF NDT OF WOOD

Nondestructive Measurement of Wood Characteristics

As noted, wood is heterogeneous, containing knots and other characteristics that are considered attributes in some products, defects in others. One thrust of NDT in wood, as in other materials, is the detection and quantification of these characteristics. An important example of this is in producing "cut stock" or wood for the furniture industry. Commonly, knots are allowed in only very limited size and frequency. Acceptable wood must be obtained by identifying the knots and cutting an acceptable piece that excludes the knots. In other processes, distinctive patterns of grain and knots are desirable and must be identified for subsequent processing.

Fig. 1. Cross section of a tree to illustrate the form of the annual growth rings and the presence of a knot

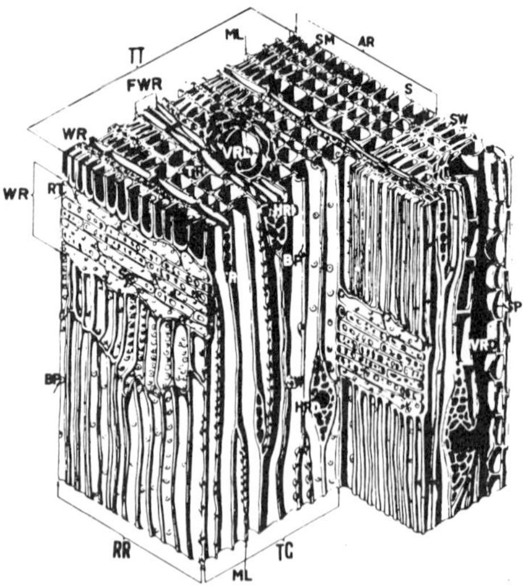

Fig. 2. A drawing of the microscopic structure of a coniferous tree

Much of the same basic technology used for other materials is applied also in wood NDT. Differences are found usually in the resonant or transmission frequencies, the energy levels of radiation, and the complicating interactions with moisture content and anisotropy. An example that can be considered is radiation. Infrared has been used to look at grain patterns in order to grade high-priced veneer; it currently also serves as the basis of a moisture meter system. Light in the visible range has been used to detect knots and other defects in the grading of softwood veneer. New technology has extended research and application in this field to automated positioning of wood pieces for sawing. Radiation in the ultraviolet spectrum is used to study the microstructure of wood. X-rays have been used commercially to find decay in power poles and other products. Gamma rays and neutrons have been teamed up in moisture detectors that can separate the density of the wood from the moisture measurement.

Measuring the moisture content of wood is an example of a common NDT challenge found in the testing of wood. The only methods to directly measure the moisture content of wood are destructive. Consequently, inferential methods are used that depend on NDT techniques. The most common of the methods is electrical conductivity because the conductivity is proportional to moisture content and independent of density. Species differences are important, however. Other electrical characteristics, such as dielectric properties, are used in important commercial moisture monitoring systems. All of these techniques are nondestructive; all require calibration to predict moisture content.

Test methods employing ultrasonics, microwaves, stress waves, and transverse vibration are all used in wood NDT. Ultrasonics have been used commercially to find toredo attack in marine pilings by underwater inspection, to detect decay in power poles in service, to quantify the complicated grain patterns in wood for subsequent machining, to measure the modulus of elasticity in full-size products, and to study bad glue bonds. The microstructure of the wood dictates that the frequencies for propagation are relatively low, usually in the range of 500 to 1000 Khz for effective transmission. Coupling can be a problem in the industrial environment if rapid testing is required because of the irregular surface of some wood products and because of a reluctance to use couplants that may be absorbed. Transmission parallel-to-the-grain is much greater than across-the-grain (modulus of elasticity ratio of 20 to 1), which can be either an advantage or disadvantage.

To date, the application of microwaves has been quite limited. A microwave signal responds to many characteristics of wood, but interactions sometimes limit the use. An example is the response to both specific gravity and moisture. More research on the application of microwaves to wood is needed.

Stress waves in wood are compression waves induced by a shock pulse or by an impact. Research on stress waves has been extensive at Washington State University in Pullman, WA, and at the U. S. Forest Products Laboratory in Madison, WI (Appendix). The principal application, to date, is for the calculation of the modulus of elasticity through measuring the transit time of the stress wave and the density of the member. This has been applied successfully to sawn lumber, to logs and poles, and to reconstituted particle panel products. Commercial devices are available in differing configurations that permit testing sizes that range from small specimens on a laboratory bench to large laminated beams in-place. However, often the motivation has included the ability to infer a strength (destructive) property from the modulus measurement. This aspect will be discussed in more detail later in the paper.

One of the most promising recent applications of stress wave systems is identification of decay and the prediction of the effect of the decay on strength. This has long been an area of intended interest, but one complicated by the nature of both wood and decay. Now, research is underway (Appendix) on use of the "small specimen" stress wave system to monitor controlled decay tests. Recent field studies (Appendix) have demonstrated that the stress wave system can be used to map the presence of decay in such different products as large laminated beams and power poles.

When stress wave research was initiated in the late 1950s at Washington State University, the intent was to also use the decay of the propagating wave as an additional indicator of the strength properties of the piece. For years this goal has eluded researchers. Recent advances in instrumentation have given a boost to this research, however, and interest has been renewed. Again, since strength properties of full-size products are being predicted, an inferential technique is needed.

Transverse vibration has been a successful NDT method for some wood products for approximately 25 years. Early applications were portable testing systems that could be transported to a plant or a customer site to measure the stiffness of lumber. For example, lumber screened and categorized by stiffness using transverse vibrations played a major role in research that spurred the evolution of new designs for the glued laminated beam industry. Industrial application began in the early 1970s with a custom installation (Appendix) that sorted structural members into quality classes for the construction of cooling towers. Portable systems that are now available commercially can be interfaced directly to a personal computer, permitting automatic recording of the modulus of elasticity, density, and parameters of the envelope of vibration decay.

The major commercial application of stiffness-measuring equipment in wood NDT is in grading lumber. In most developed countries that either produce or consume large quantities of lumber, "grading machines" are used to measure the stiffness of lumber at speeds up to 1200 ft/min. Figure 3 is a schematic of the Continuous Lumber Tester or CLT, the most common grading machine in North America. Lumber passes in a flatwise orientation through the machine as the force to bend the piece is measured. The stiffness of the piece dictates the grade sort to be applied. This and other similar devices resulted from research in the United States, England, South Africa, and Australia over the last 35 years. About 50 machines now operate in the United States and Canada. While the lumber stiffness measured by these machine grading systems is an important property for design, another important feature of these systems is to predict the lumber strength properties.

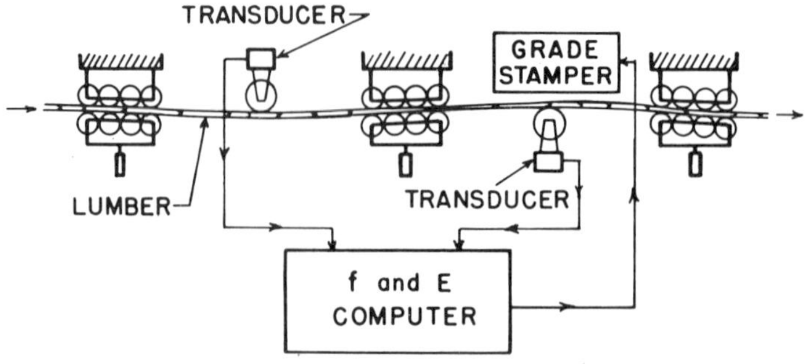

Fig. 3. Schematic drawing of the CLT machine for grading structural lumber

Estimating the Strength Properties of Wood Products

A major industrial use of wood is as a structural element, not only in familiar house construction, but also in commercial and industrial structures like roof systems and bridges. As with any engineering material, the design properties describe axial, bending, or crushing strength performance. For many years, these properties were established for wood on the basis of small specimen tests. With the advance of technology and expanded use of wood in engineered structures, more emphasis now is placed on tests of full-size products. These test results are the basis of performance models. Nondestructive observations made on each piece are the predictors employed in the models. This process is necessary because the combination of anisotropy and heterogeneity results in individual pieces having significant departures from the idealized performance of small wood specimen-based models.

Correlation between the nondestructive predictors and actual strength performance has R^2 values in the range of 0.5 to 0.7 for highly variable materials and coefficients of variation of strength of 25 to 40%. Some of the newer, proprietary, reconstituted products have higher correlations and less variability--often a result of using NDT methods in selection of the raw materials as well as the homogenization of the reconstitution process.

In solid sawn lumber, the principal strength predicators have been modulus of elasticity and knot size. Improvement of the resulting correlations are needed if efficiency of resource use is to improve. Structural panel properties are predicted with a combination of elements; principal among these is transit time of a stress wave. Use of wave decay parameters is the subject of some current research (Appendix).

Correlation models based primarily on the modulus of elasticity appear to have reached their optimum efficiency. New models or new model parameter input is vital for efficient sorting and grading of wood products.

WOOD NDT RESEARCH NEEDS

Research can lead to significant improvements in modeling for short term strength prediction and for identification of specific wood characteristics. In addition, however, there are related needs that can be addressed with NDT technology. These include so-called "environmental" aspects of wood performance and more use of NDT to enhance wood processing.

Improvements in Strength Prediction

To achieve more efficient strength prediction, it will be necessary to depart from the empiricism of the current models and devote more effort to phenomenological models. The mechanical wood properties that control failure of the wood structure at the "fiber" level are primarily tension perpendicular to grain and shear--neither of which is directly incorporated in the empirical strength models.

Elements of Structural Integrity in End-Use

Much effort has been expended on prediction of the short-term strength performance of wood products; however, wood exhibits long-term creep and duration-of-load (failure) characteristics as well. Both of these are considered in contemporary wood design. There are no NDT measuring or sorting criteria to enhance the design process by reducing the conservatism required to accommodate the variability of the wood. Identification of "fringe" performers, in the sense of duration-of-load, for example, would result in more efficient models of long term performance.

Strength degradation is an area of wood structural research that has had emphasis placed on it only in recent years. In the end-use environment, it is desirable to be able to monitor strength loss (if any) as it occurs in critical elements. This degradation may be caused by biological agents, by chemicals, or by excessive heat. The goal of such research is to predict performance of full-size members in the hostile environment. It is important such research be approached on an interdisciplinary basis.

Selection Criteria for Processing

The wood industry needs to place more emphasis on screening materials prior to processing. While common for some products, in other situations the technology and economics have not come together for efficient application. Potential areas for research and development include "treatability" and degrade.

Adequate preservative treatment of wood products is a critical issue; the importance will increase. Present standards for the selection of materials for treatment are not sufficiently sensitive to actual treatment performance. If products can be screened for "treatability," the efficiency of the treating process (chemicals and effectiveness) will increase markedly and quality control issues within the process will be reduced.

The term "degrade" encompasses the degradation that occurs under process conditions. A significant portion of this degrade can be caused by the lack of appropriate selection or sorting prior to processing. In wood drying, for example, methods for separating easy and hard-to-dry lumber are primitive. Although a great deal of effort is being placed on such techniques as whole log CAT scanning (Appendix), at the current time, and also on a practical level, the sorting of logs for the most efficient subsequent use is based principally on surface analysis.

CONCLUSION

Wood NDT has progressed significantly in the areas of prediction of short-term mechanical properties and in the identification of many characteristics. Many nondestructive techniques are employed, some in production environments, others as diagnostic tools. More emphasis is needed to improve the efficiency of models for predicting structural performance. Long-term performance prediction and monitoring methods require research. Wood processing systems will improve if more NDT technology can be applied.

The current environmental issues that affect the wood resources are likely to remain. Improved efficiency in wood-product processing and in structural application will become more critical if wood, as a renewable material, is to continue its contribution as an important natural resource.

APPENDIX

The overview nature of this paper suggests the best system of referencing to be a chronological list of the proceedings series from some symposia and workshops on nondestructive testing of wood prepared by Washington State University and the USDA Forest Products Laboratory. In the years since inception in 1963, these series have been devoted exclusively to wood, in all product forms, and as commercial products as well as a medium of basic research. NDT research and application from many countries are represented. The references listed below start with the earliest date and work towards the most recent date.

REFERENCES

Forest Products Research Laboratory, 1964, Proceedings of the Symposium on Nondestructive Testing of Wood, U. S. Forest Service Research Note, FPL-040, Forest Products Laboratory, Madison, WI.

Forest Products Research Laboratory, 1965, Proceedings of the Symposium on Needs for Nondestructive Testing in the Forest Products Industries, U. S. Forest Service Research Note, FPL-080, Forest Products Laboratory, Madison, WI.

Washington State University, 1965, Proceedings of the Second Symposium on the Nondestructive Testing of Wood, William L. Galligan, Editor, Wood Technology Section, Washington State University, Pullman, WA.

Washington State University, 1970, Proceedings of the Third Washington State University Short Course on Nondestructive Testing of Wood, Commercial Machine Stress-Rating for Profit, Robert J. Hoyle, Jr., Editor, Wood Technology Section, Washington State University, Pullman, WA.

Washington State University, 1978, Proceedings of the 4th Nondestructive Testing of Wood Symposium, Wood Technology Section, Washington State University, Pullman, WA.

Washington State University, Proceedings of the Fifth Nondestructive Testing of Wood Symposium, Engineering Publications, College of Engineering and Architecture, Washington State University, Pullman, WA.

Washington State University, 1987, Proceedings of the Sixth Nondestructive Testing of Wood Symposium, Engineering Publications, Washington State University, Pullman, WA.

Washington State University, 1990, Proceedings of the Seventh International Nondestructive Testing of Wood Symposium, Conferences and Institutes, Washington State University, Pullman, WA.

NDE APPLIED TO THE CONSERVATION OF WOODEN PANEL PAINTINGS

A. Murray,† R. E. Green,† M. F. Mecklenburg,* and C.M. Fortunko°

† Department of Materials Science and Engineering and the Center for Nondestructive Evaluation, The Johns Hopkins University, Baltimore, MD, 21218, USA
* The Conservation Analytical Laboratory, The Smithsonian Institution, Washington, D.C., 20560, USA
° National Institute of Standards and Technology, Boulder, CO 80303, USA

Partially supported by the Conservation Analytical Laboratory The Smithsonian Institution

Abstract

Various nondestructive methods for determining the condition of wooden panel paintings have been investigated. Methods are being developed to detect voids, hidden cracks, and fine fractures that are often a source of premature failure in such objects. Computer analysis shows that cracks in wood promote severe stress concentrations which are aggravated by the mechanical constraints imposed on the wood by the construction of a panel painting.[1] The panel's structural condition needs to be mapped in a nondestructive fashion in order to predict how these objects will mechanically respond over time or in different environments. Decisions can then be made as to whether an object can be safely removed from storage, if it can withstand conservation treatment, if it can be continuously displayed, or if it can be safely shipped to special exhibitions. We have demonstrated the NDE capabilities of different techniques by examining standard panels of white oak (*Quercus* sp.), tulip poplar, hard maple (*Acer* sp.), true mahogany (*Swietenia* sp.), black cherry, and western fir (*Abies* sp.). The panels contain voids between 0.15 and 0.6 cm in diameter and cracks smaller than 0.02 cm. The techniques for investigation included x-ray radiography, xeroradiography, and ultrasonics. Results show that both x-ray radiography and xeroradiography techniques can easily find voids at least as small as 0.15 cm. Xeroradiography has benefits over x-ray radiography because its edge enhancement property enables flaws to be displayed prominently. At certain angles, cracks less than 0.02 cm were easily seen. The results of using these two techniques on a panel painting are discussed. Ultrasound techniques using an all-air coupled system and a hybrid air-coupled/dry-coupled system have shown promising preliminary results for finding voids and cracks.

Introduction

Wood has been a support for paintings through the ages. There are a variety of cuts, and species of wood possible depending upon the epoch, the country of

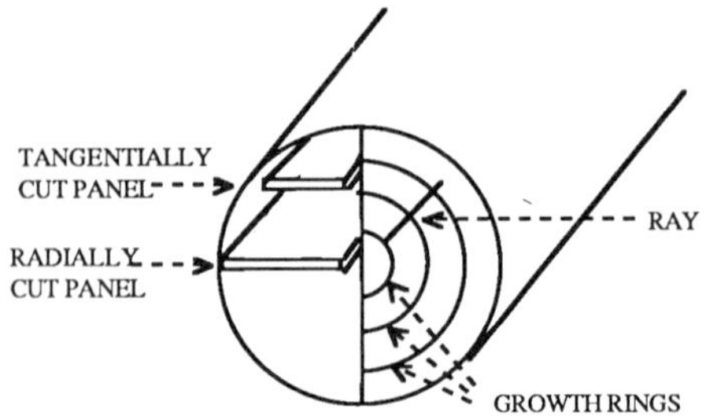

Figure 1 Cross section of wood

origin, and the workshop in which the panel painting was made.[2] The possible layers that make up a panel painting include: the wooden support, the glue layer, the ground or gesso layer, the underdrawing, the imprimatur, the underpaintings, the paint layer, the glazing, the varnish, and any retouchings.[3] Cennini's 1437 treatise explains the craft in great detail.[4] The panels are usually cut either radially, which is more uniform in its dimensional response to relative humidity changes, or tangentially (Figure 1). Radially cut panels are usually between 0.7 and 1.5 cm thick, while tangentially cut panels are 4.0 to 6.0 cm thick.

Although many of the panels still exist today, there has been considerable concern with their deterioration for centuries.[5,6] Panels are dimensionally less stable with changes in relative humidity, especially when they have been thinned down and restrained during "conservation treatments". Studies are underway at various institutions including the Conservation Analytical Laboratory at the Smithsonian Institution, the Canadian Conservation Institute (CCI) in Ottawa, the National Gallery of Art in Washington, D.C., and the Tate Gallery in London, to examine the risk caused by incorrect travelling or storage conditions.

The structure of wood is well-known.[7] The anomalies that exist naturally and that may evolve need to be studied to understand results from different NDE techniques. Different natural inhomogeneities in wood include seasonal growth rings, knots, and ray cells, which are sheets of cells that grow in a radial direction along the longitudinal radial axis of the tree (Figure 1). Voids are often created by insects, including the common furniture beetle, (*Anobium punctatum* in the anobiidae family), the death-watch beetle (*Xestobium rufovillosum*), the powder post beetle (Lyctidae), termites, and ants.[8,9] The tunnels can be as small as 1 mm and can lead to a completely riddled panel. The panel is then weakened, allowing cracks and breaks to occur more easily. Cracks also appear in panel paintings because of changes in the relative humidity that cause dimensional changes.[6,7,10,11] When restrained, wood compresses with water adsorption, crushing wood cells. Wood develops tensile stress and cracks with desorption. The structure of wood leads to severe anisotropic behaviour as the dimensional changes are greater in the radial and tangential directions than in the longitudinal direction. Both the adsorption and desorption of water can, therefore, lead to warping and splitting of the wood. The expansion and contraction of the wood can also cause the ground and paint layers to crack, cleave and buckle if they are not flexible.

Nondestructive Testing Techniques

The wood industry has used a variety of nondestructive techniques including: visual tests, ultrasonics, stress-wave measurements, vibration tests, x-ray radiography, computer aided tomography (CAT), collimated photon scattering, magnetic resonance imaging (MRI), acousto-ultrasonics (AU), acoustic emission (AE), microwaves, pulsed electric current, and even the skills of sniffer dogs to discover rot in buildings. Many industrial x-ray studies have

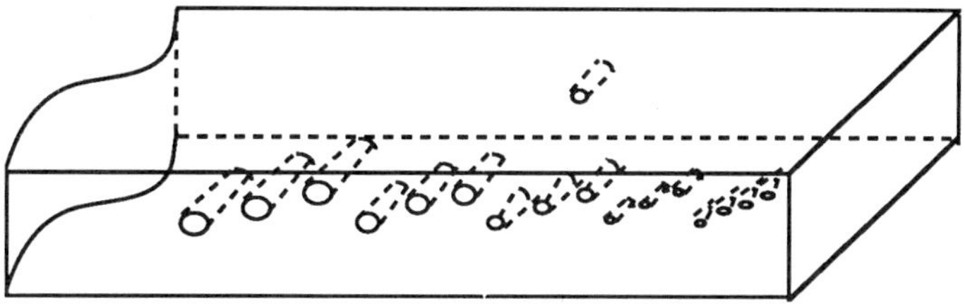

Figure 2 Angled view of wooden panel with voids

been concerned with wood-boring insects. Ultrasonics has been compared to other methods to determine quality, strength, elastic constants, and degree of deterioration of wood. The theory, experiments, and viabilities have been discussed in many articles and proceedings.[12,13]

The museum conservation field has used many different nondestructive techniques on many materials. A few examples include: x-ray radiography,[14] which has been used extensively; xeroradiography,[15,16] on ceramics, porcelain, Egyptian faience, and underdrawings of paintings; ultrasound,[17-19] on murals, masonry, and wetted archaeological woods; infrared thermography,[20,21] to find deterioration in wooden panel paintings; x-ray computer tomography;[22] visible light; ultraviolet radiation; and other techniques. The first four techniques, x-ray radiation, xeroradiography, ultrasound with various coupling techniques, and infrared thermography were used in this study. Infrared thermography had only limited success and, therefore, will not be discussed further. In previous published studies, x-ray radiography and xeroradiography have not been compared while trying to map out the voids and cracks found in panel paintings; some work in this area has been performed at CCI. Air-coupled and dry-coupled ultrasonic systems have not been used on art objects, including wooden panel paintings.

Experimental Procedure

Samples

The samples used throughout the present experiments were six panels of wood, each with dimensions 40 cm x 23.5 cm x 3.5 cm. The six species of wood, with their densities were: white oak (*Quercus* sp.) (720 kg/m^3),[23] tulip poplar (460 kg/m^3),[23] hard maple (*Acer* sp.) (660 kg/m^3),[23] true mahogany (*Swietenia* sp.) (545 kg/m^3),[24] black cherry (530 kg/m^3),[23] and western fir (*Abies* sp.) (410 kg/m^3).[23] The panels were approximately at 12% moisture content.

In order to simulate voids and flaws in the panels, holes of different diameters (0.6, 0.4, 0.35, 0.2, and 0.15 cm) were drilled into the sides of the panel at different depths from the surface, as shown in Figure 2. The holes closest to the surface were between 0.2 and 0.5 cm from the surface, the next closest were 1.0 cm, and the furthest away were between 1.5 and 2.0 cm from the surface. On each side, a solitary 0.35 cm diameter hole was drilled 1 cm from the surface. The cracks in the sample panels occurred naturally from swings in relative humidity. These cracks of different sizes and lengths occurred in the white oak, hard maple, true mahogany, and western fir panels.

An oak panel painting, probably made around the turn of this century, (Figure 3a) was x-ray radiographed and xeroradiographed (Figures 3b and 3c). The width and length of the panel were 20 x 30 cm^2. The thickness was 0.7 mm at the edges, and it tapered to 1.2 cm in the centre. This panel had a number of cracks along the grain, at the top and bottom edges of the panel (less than 0.1 cm) and some voids (around 0.1 cm in diameter). As this painting is part of a

Figure 3 Panel painting: (a) photograph; (b) x-ray radiograph; and (c) xeroradiograph.

research collection at the National Gallery of Art in Washington, D.C., strain gauges and metal wires had been attached to the back of the painting for another set of experiments. These are visible on the radiographs. The painting shows a winter scene of a house surrounded by trees with eight figures at the bottom. The pigment had been laid directly onto the wooden oak panel. A white pigment was used a great deal, for example in the areas surrounding the figures, the house, and the trees. This allowed the picture to be seen on the radiographs.

Techniques

Each panel was x-ray radiographed. The hard maple panel is shown as an example (Figure 4a). A Philips' MG 320 Constant Potential X-Ray Unit was used. This included a 300 kV double focus tube and a tube current of up to 10 mA. A small focal spot of 1.5 mm was used. The distance from the tube to the film was 125 cm. The film (Kodak M industrial large) and a 1 mm sheet of lead, which lay under the film, were placed in a vacuumed envelope. The settings used were 80 kV, 2 mA, for 3 minutes. Each wooden panel was xeroradiographed. A positive and negative image of the hard maple panel are shown (Figure 4b and 4c). A XEROX 125 System was used. The plate could only image half of each panel with some overlap, therefore two images needed to be taken for each panel. This caused the horizontal line seen in the xeroradiographic image. Both positive and negative images were taken. For the positive images, the settings were 42 kVp, 300 mA, for 0.4 seconds. For the negative images, the settings were 44 kVp, 300 mA, for 0.3 seconds. The panel painting was x-ray radiographed at the National Gallery of Art, in Washington, D.C., using the settings 25 kV, 50 mA, for 12 seconds. The xeroradiographs of the painting were taken using the same equipment as for the wooden panels and the settings 45 kVp, 10 mA, and 0.2 seconds. Both radiographs were taken from the front of the painting (Figures 3b and 3c).

The first ultrasonic set-up used an all-air coupled configuration (Figure 5). The transducers could be positioned anywhere over the tulip poplar panel. A unipolar, burst pulser (3 Ω impedance, 450 V) generated the signal which was then sent through an impedance-matching network (high-pass inverter). Both air transducers were 0.5 MHz, with 1-inch diameters face-plates and 2-inch focal lengths. They were both 0.8 inches away from the sample and were therefore defocussed on the sample. They were also off-set by 0.5 inches along the horizontal axis. One transducer sent the signal through the sample and the

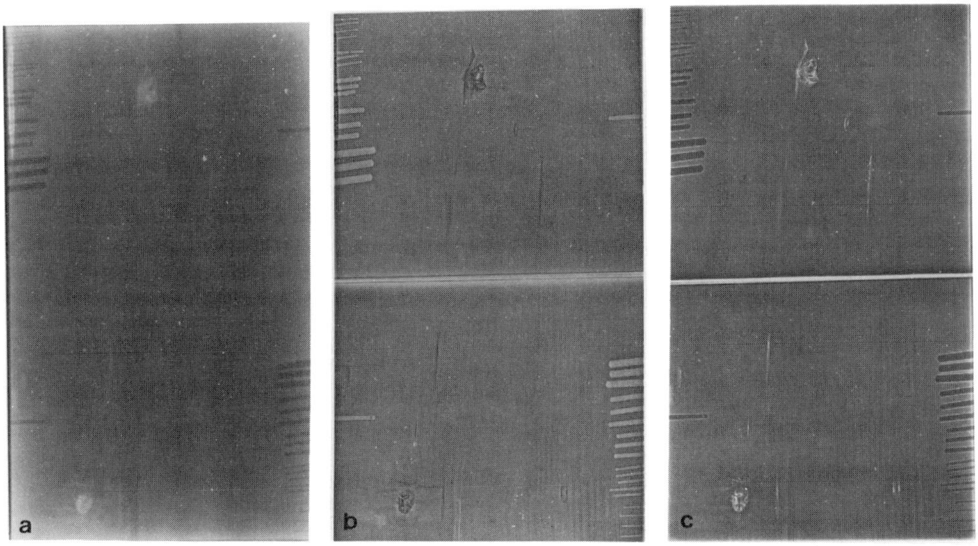

Figure 4 Panel of hard maple: (a) x-ray radiograph; (b) positive xeroradiograph; and (c) negative xeroradiograph.

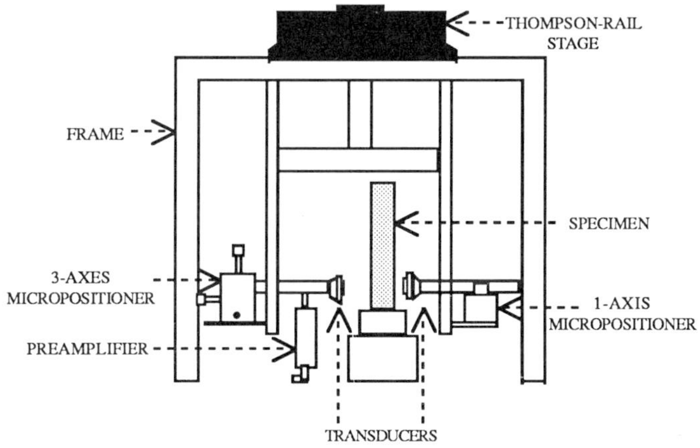

Figure 5 Experimental configuration of all-air coupled ultrasound.

other received it after it traversed through the sample. The signal was then sent through a very low-noise bipolar amplifier, two precision attenuators (12 x 10 dB and 12 x 1 dB), a band-pass filter, and a linear detector. The transducer-transducer spacing was fixed by a "U-shaped" holder. The second set-up was a hybrid air-coupled/dry-coupled through-transmission configuration on the white oak panel. The set-up was identical to the previous one, however, the signal was transmitted through an air-coupled transducer (0.5 MHz, 1-inch diameter, 2-inch focal length, and 0.9 inches from the surface) and was received by a dry-coupled transducer (ULTRAN WD 50-1, 113065, 0.5 MHz, and 0.375-inch diameter). Scissor-like tongs were used to hold the transducers.

Experimental Results

The results showed that all voids were clearly seen with both radiographic techniques. The xeroradiographic technique displayed cracks in the maple panel at its bottom edge, which were smaller than 0.02 cm and which the x-ray radiographic technique did not show as well (Figure 4). Other features include mineralization pockets (in the central region and along the top left hand side), which are anatomical features of hard maple, and knots, including two large ones, one near the top edge in the centre and one in the bottom left hand corner of the panel. The xeroradiographic images of the panel paintings showed cracks (less than 0.1 cm), voids (around 0.1 cm in diameter), vessel lines in the wooden support, and the white pigment (Figure 3). There appears to be two types of cracks, both less than 1 mm; one at the top edge with an increase in density (a light line) and one at the bottom edge of the painting with a loss of material (a dark line). The cracks with an increase in density were probably in the wood prior to painting and then filled up with the paint. The other cracks, which showed a decrease in density, were probably formed after the painting of the panel. The voids, approximately 0.1 cm in diameter, were filled with pigment and, therefore, now show an increase in density. The x-ray radiographs do not show the cracks as clearly. The strain gauges and wires appear on the left hand side of the image, in the centre above the roof of the painting, and on the lower right hand side above the figures.

When using the all-air coupled ultrasonic system, various sections of the poplar panel were analysed. Pieces of lead tape were placed onto the sample to give a reference for the analysis. The signal is greater over areas where the beam is not skewed by the grain of the wood. Distinct decreases in signal occur over the solitary holes and over the pieces of lead tape, whereas gradual decreases correspond to changes in grain. Over areas with many voids of different diameters and depths, the signal is smaller where there are larger diameter voids and less signal is permitted through, while the signal is larger where the voids are smaller and more signal is allowed through.

The all-air coupled ultrasonic system worked for the lower density of tulip poplar, but not for the higher density woods such as white oak. Only the higher density woods had cracks and therefore to evaluate these types of voids, a different system, the hybrid air-coupled/dry-coupled system, had to be used. The results (Figure 6) show a definite signal decreases around 1600-2200 mil and around 3200-3600 mil, corresponding to cracks. There is also a strong correlation between signal size and the direction of the grain. There is a larger signal over areas where the grain is perpendicular to the edge of the sample. The opposite is also true. Any analysis using ultrasound to discover flaws will have to consider grain orientation, which causes beam skewing, and any other artifacts within the wood. Resolution may be adversely affected and other information about the sample may get hidden because of speckle caused by the inhomogeneities in the wood.

Conclusions

Both techniques, the x-ray radiography and xeroradiography, detect the voids with the smallest diameter (0.15 cm), at all depths from the surface; the edge enhancement in the xeroradiograph technique does, however, make the voids more visible. The xeroradiographic results show the smallest cracks in the panels (less than 0.02 cm) which were not always clear in the x-ray radiograph. The other flaws, including knots, grain, and cutting defects, were more easily discernible with xeroradiography. This technique was able to show flaws in actual panel paintings and differentiate between the flaws that occurred before and after painting the wooden panel. Either a positive or a negative xeroradiographic image may be more useful according to what the individual researcher is used to and what the artifacts being looked for are. It is, therefore, valuable to take both images. X-ray radiography remains a more easily accessible technique, but clearly not as useful as xeroradiography.

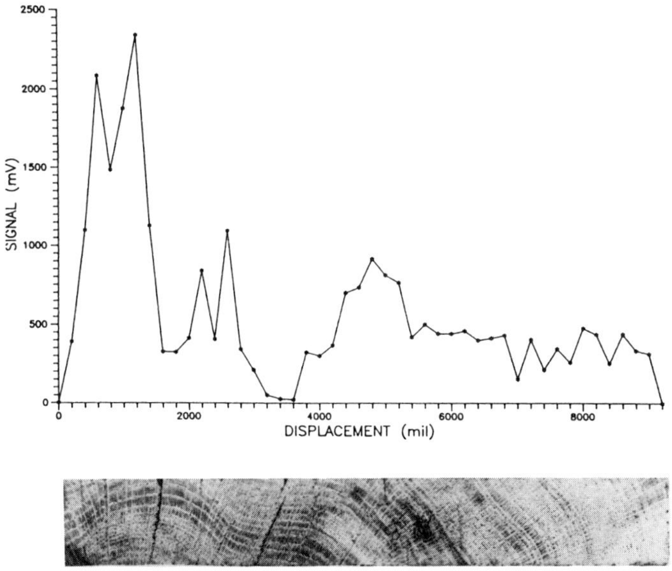

Figure 6 Results from hybrid air-coupled/dry-coupled through-transmission ultrasonic technique performed on white oak.

The two ultrasonic systems have shown promising preliminary results in showing cracks that cannot be detected with x-ray radiography and xeroradiography. The conventional coupling problem has been overcome using these systems. The all-air coupled system is useful on lower density wood such as tulip poplar and distinguished voids as small as 0.15 cm in diameter. The hybrid air-coupled/dry coupled technique was able to map out the cracks of higher density woods, such as white oak. The system was able to show cracks of 0.1 cm.

Acknowledgements

The authors would like to thank the following people: Donald C. Williams, Senior Furniture Conservator, and Melvin Wachoviak, Furniture Conservator, both at the Conservation Analytical Laboratory, the Smithsonian Institution, for providing the samples and for their instructive discussions; Mervin Richard, Conservator at the National Gallery in Washington D.C., for allowing me to have xeroradiographs, x-ray radiographs, and ordinary pictures of the panel paintings in the research collection; Eric Jensen, a summer coop student at the NIST in Boulder, for his help with the ultrasound data collection; Wilma Savage at the Clinical Radiologists, in Silver Hill, Maryland, for her work and interest in taking the xeroradiographs of the samples; Melanie Feather for her help with the x-ray radiography; and Ian Wainwright, David Grattan, and others at the Canadian Conservation Institute for our discussions on their work with xeroradiography and other techniques.

References

1. Mecklenburg, M.F., private communication, current research work ongoing at the Conservation Analytical Laboratory, the Smithsonian Institution.
2. Van Schoute, R., and H. Verougstraete-Marcq, "Painting technique: Supports and frames" *Journal of the European Study Group on Physical, Chemical and Mathematical Techniques Applied to Archaeology (PACT)*, edited by Van Schoute, R., and H. Verougstraete-Marcq, 13, 1986, p. 19.
3. Mairinger, F., and M. Schreiner, "Analysis of supports, grounds and pigments," *PACT*, 13, 1986, pp. 172-173.

4. Cennini, C. d'A, *Il Libro del l'Arte*, Translated by D.V. Thompson, *The Craftsman's Handbook*, Dover Pulications, (New York, 1960).
5. Samet, W.H., "An Evaluation of Wood Panel Painting Conservation at the Walters Art Gallery: 1934 to 1980," *The American Institute for Conservation of Historic and Artistic Works (AIC) Preprints*, 1988, pp. 161-175.
6. Buck, R.D., *The Behaviour of Wood and the Treatment of Panel Paintings,* The Upper Midwest Conservation Association (Minneapolis, 1978).
7. Panshin, A.J., and C. de Zeeuw, *Textbook of Wood Technology, Volume 1*, Third Edition, McGraw-Hill, (U.S.A., 1970).
8. Rowell, R.M., and R.J. Barbour, *Archaeological Wood Properties, Chemistry, and Preservation*, Advances in Chemistry Series 225, American Chemical Society (ACS), (Washington D.C., 1990), p. 86.
9. Panshin and de Zeeuw, pp. 362-382.
10. Gibson L.F. and M.F. Ashby, *Cellular Solids Structure and Properties*, Pergamon Press (New York, 1988), pp. 278-315.
11. Williams, D.C., and M.A. Williams, "Treatments for "dry" wooden objects," paper given at the *Symposium on Archaeological Wood: Properties, Chemistry, and Preservation, 196th ACS National Meeting, September 25-30, 1988.*
12. Pellerin, R.F., and K.A. McDonald, *Proceedings: Sixth Nondestructive Testing of Wood Symposium, September 14-16, 1987, Washington State University, Pullman, Washington,* Engineering Publications, and Conferences and Institutes, Washington State University.
13. Murray, A., "The Nondestructive Evaluation of Wooden Art Objects," M.S.E. Thesis, 1990, The Johns Hopkins University.
14. Gilardoni, A., R.A. Orsini, and S. Taccani, *X-Rays in Art*, Gilardoni S.p.A., Mandello Lario (Como) (Italy, 1977).
15. Ellingson, W.A., P.D. Vandiver, T.K. Robinson, and J.J. Lobick, "Radiographic Imaging Technologies for Archaeological Ceramics," *Materials Research Society Symposium Proceedings*, 123, 25-32, 1988.
16. Magliano, P., B. Boesmi, "Xeroradiography for Paintings on Canvas and Wood," *Studies in Consservation,* 33 (1988) pp.41-47.
17. Asmus, J.F., and S.T. Pomeroy, "Ultrasonic Mapping of Detachments in Mural Paintings by Vasari," *AIC Preprints*, 1978, pp.12-20.
18. Berra, M., L. Binda, G. Baronio, A. Fatticcioni, "Ultrasonic Pulse Transmission: A Proposal To Evaluate The Efficiency of Masonry Strengthening By Grouting," *2nd International Conference on "Nondestructive Testing, microanalytical methods and environment evaluation for study and conservation of works of art,* I/10.1-19.
19. Miura, S., "Ultrasonic Characteristic of Waterlogged-wood," *Science for Conservation*, (5), 1976, pp. 14-18.
20. Miller, B.F., "The Feasibility of Using Thermography To Detect Subsurface Void in Painted Wooden Panels," M.A. Thesis, 1976, Oberlin College.
21. Asperen de Boer, J.R.J. van, "Infrared Reflectography: A Method for the Examination of Paintings," *Applied Optics*, VII, (9), September, 1968, pp. 1711-1714.
22. Taguchi, E., I. Nagasawa, S. Yabuuchi, and M. Taguchi, "Investigation of a Wooden Sculpture Using X-ray Computed Tomography," *Scientific Papers on Japanese Antiques and Art Crafts*, (29), 1984, pp. 43-50.
23. Ten Wolde, A., McNatt, J.D., and L. Krahn, "Thermal Properties of Wood and Wood Panel Products for Use in Buildings," Forest Produts Laboratory, Madison,Wisconsin, Semptember 1988, pp.17-18.
24. Weast, R.C., *CRC Handbook of Chemistry and Physics*, 63rd edition, CRC Press, Inc., (Boca Raton, Florida, 1982-1983), pp. E-5 and E-2.

FAR-INFRARED TESTING OF DIELECTRIC SHEET MATERIALS

Bernard Drouin and Richard Gagnon

College F.X. Garneau
1660 Blvd de l'Entente
Québec, PQ, Canada, G1S 4S3

INTRODUCTION

Quality control in fibre based dielectric sheet materials like paper and non woven fabric is hampered by complex manufacturing processes and usually very high production rates. Far-infrared (FIR) waves are interesting for non destructive characterization of such materials: their low scattering and moderate penetration allow accurate high resolution measurement of basis weight, fibre orientation anisotropy and dominant alignment angle[1]. In most applications, sheet materials need to have uniform mechanical properties in order to achieve dimensional stability and problem free handling by end users' equipment[2,3]. In the case of newsprint paper for instance, small scale variations in basis weight will affect readability[4]. Similar scale variations in fibre orientation anisotropy or angle of alignment will result in cockling, bagging, waving or curling of the sheet during the printing process[5]. This, in turn, causes local loss of register and poor quality colour rendering, due to the multiple superimposed prints required. The herein described technique should alleviate this problem by giving the producers a way of testing their product before shipment.

After going through the basics of this characterization method, typical results involving actual production sheet materials will be presented, followed by an in depth discussion of the proposed system.

BASIC FAR-INFRARED EXPERIMENT

The above mentioned characteristics of the sheet material sample are obtained by measuring FIR transmission. As shown in the schematic presentation of the experimental set up (Fig. 1), the first operation on the FIR beam is mechanical chopping. Most FIR detectors, pyroelectric or Golay type, are sensitive to intensity variations only, consequently, chopping is mandatory, and also welcome because it allows synchronous lock-in detection thus improving the precision of the measurement.

Nondestructive Characterization of Materials IV
Edited by C.O. Ruud *et al.*, Plenum Press, New York, 1991

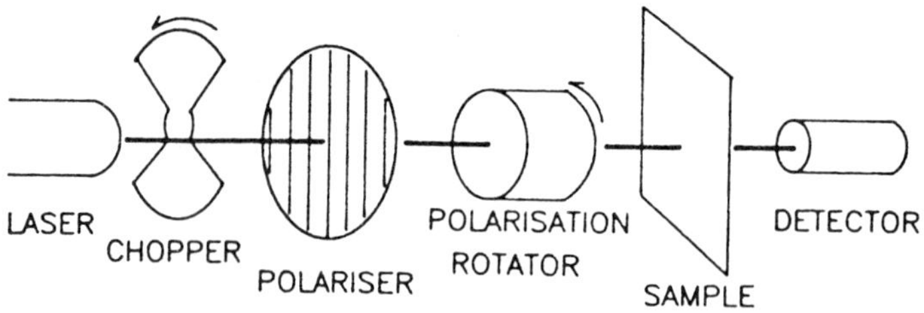

Fig 1. Experimental set-up

Next comes the polarizer: determination of fibre orientation anisotropy and alignment requires a linearly polarised beam. Then, with the polarisation rotator, the experimenter can vary the angle between the beam plane of polarisation and a reference direction on the fixed sample. Measuring FIR transmission as a function of this angle yields the kind of data presented on Fig. 2.

Mean FIR transmission through most of the tested samples is an exponential function of basis weight, hence, once a sample of known mass and area has been characterised, basis weight determination for any sample of the same composition is straightforward. As illustrated on Fig. 2, FIR transmission through these samples varies with the polarisation angle, a typical dichroic material behaviour; we have established in earlier work[6], that the amplitude of this variation is related to fibre orientation anisotropy in the sample. Furthermore, the phase of this sinusoidal variation is an almost direct measurement of the dominant fibre orientation angle with respect to the reference direction, often referred to as the fibre alignment angle. This simple experimental procedure allows simultaneous determination of basis weight, anisotropy and alignment. Spatial resolution associated with this measurement depends on the beam size which can be easily modified with suitable optical components.

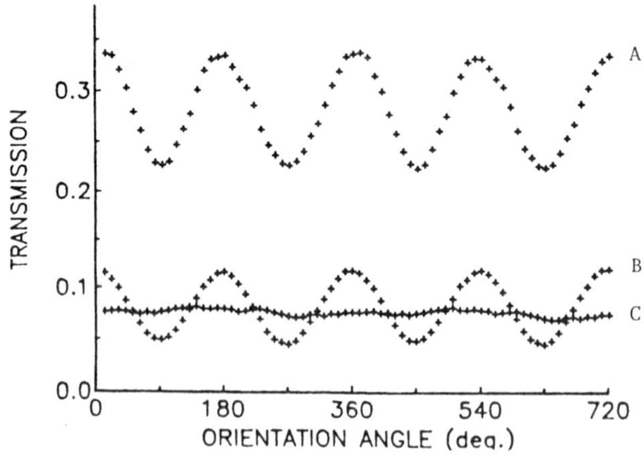

Fig. 2. Transmission as a function of the polarisation angle for A: a single polyester sample, B: two parallel tread polyester samples and C: two crossed tread polyester samples.

RESULTS

We have chosen to illustrate the performance of this technique with typical one dimensional high resolution basis weight data from two samples, and two dimensional fibre orientation and alignment data obtained at medium resolution with synthetic fibre materials. The next figures (Fig. 3 A and B) show the measured variations of basis weight as a function of position using a half millimetre beam size for A, 50 gram per square metre (gsm) standard newsprint, and B, 80 micron thick mylar.

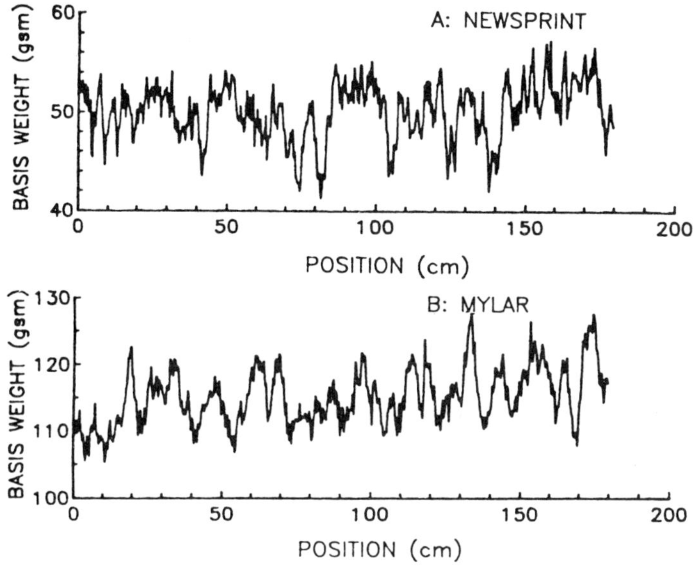

Fig.3. FIR measured basis weight variations as a function of position for two different samples.

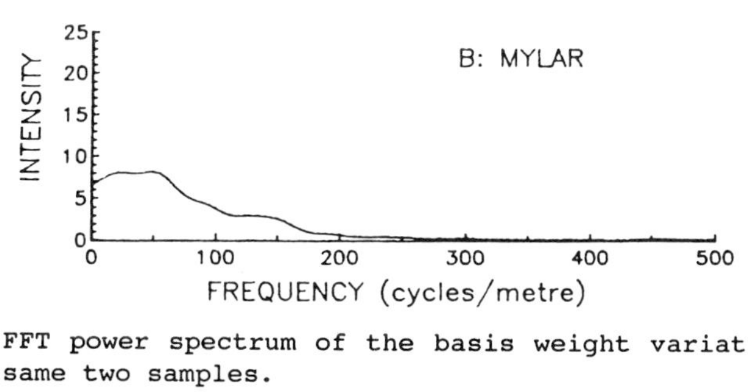

Fig. 4. FFT power spectrum of the basis weight variations for the same two samples.

On Fig. 3, we have chosen a scale that exagerates the basis weight variations. Normalised FFT power spectra (Fig. 4) reveal that mylar is relatively more uniform than newsprint. All this data and analysis demonstrate the performance of the FIR technique in precision basis weight measurement, at high spatial resolution.

Fibre orientation anisotropy is determined using the type of transmission data shown on fig. 2; a dimensionless coefficient is calculated from the average transmission and the modulation amplitude of the transmission curve[6]. This coefficient also called FIR anisotropy ratio is equal to 1.00 when the sample is perfectly isotropic. In fig. 5A and B, we present the results of two dimensional anisotropy measurements obtained with nylon fabric and non woven polyester felt. As expected, anisotropy is smaller and more uniform in the woven nylon material.

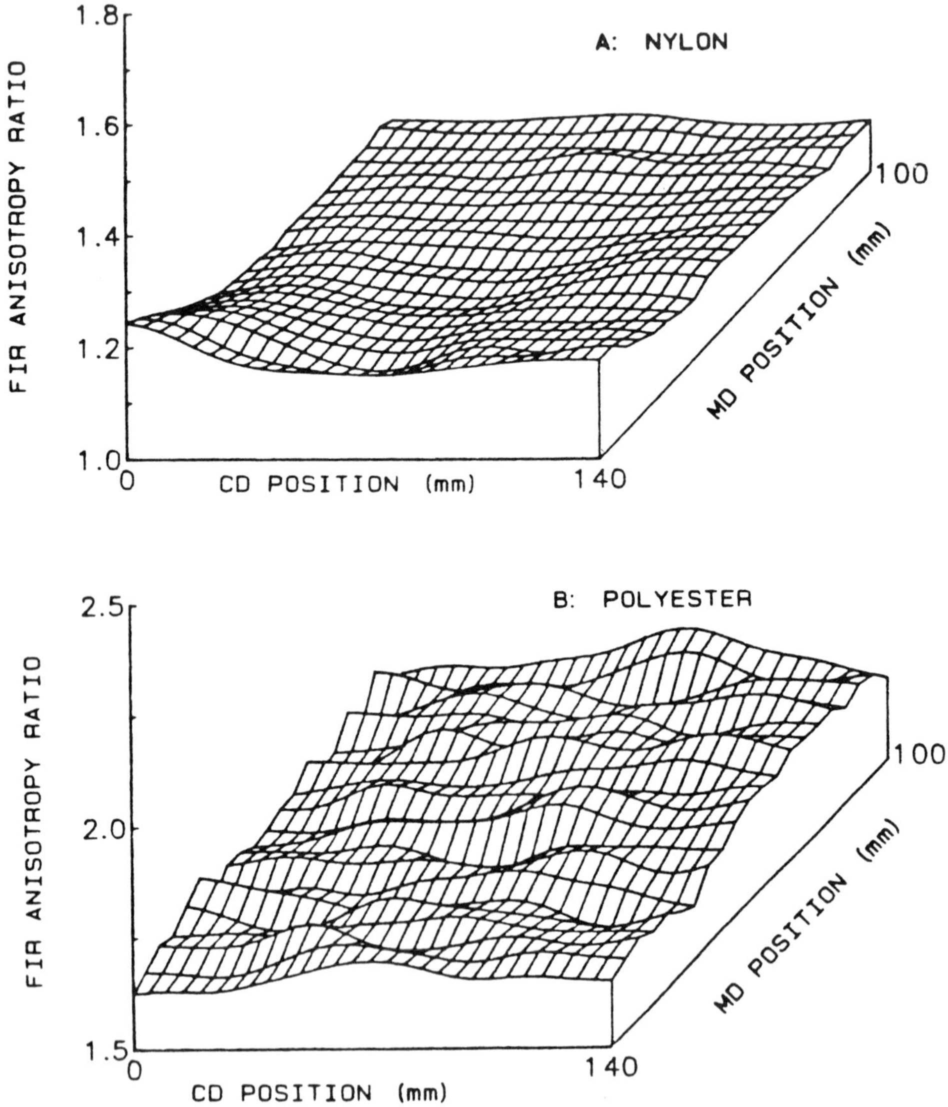

Fig. 5. FIR anisotropy ratio map for a 10 cm x 14 cm area of A, a nylon fabric sample, and B, a non woven polyester felt sample.

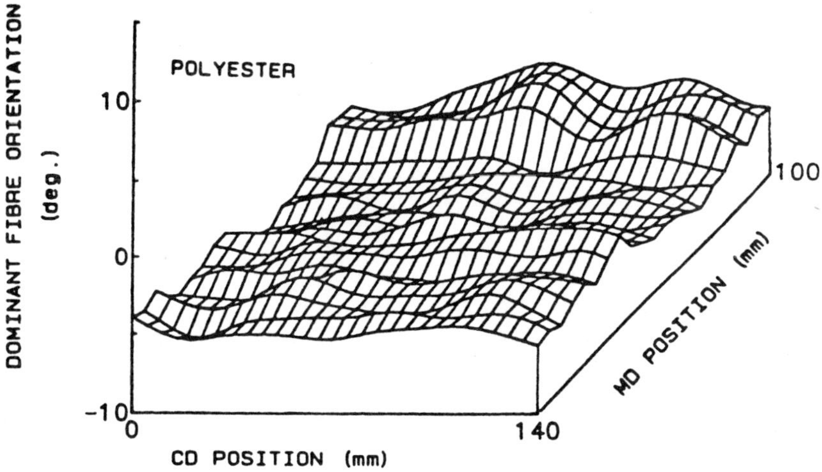

Fig. 6. Dominant fibre orientation map for the polyester sample

Dominant fibre orientation or alignment angle for the same area of the same polyester sample is presented in Figure 6. This angle has a strong effect on the local tensile strength of the material, although in this case, its variations are limited to a few degrees. All those results demonstrate the ability of this FIR technique to perform various tests on many different dielectric sheet materials.

DISCUSSION

Among the non contact, non destructive characterization methods for dielectric sheet materials, FIR has many advantages over its main rival, microwaves. First, FIR transmission measurement is more sensitive to weight, anisotropy and alignment variations because far infrared waves are more absorbed by this type of substance. This characteristic also helps to reduce the occurrence of annoying standing waves often experienced by microwave based systems. Because of the shorter wavelength, a FIR beam can be focalised on a smaller area thus allowing higher spatial resolution than microwave[7]. On the other hand, despite shorter wavelength, scattering by the fibres remains low. This is shown on Figure 7, where transmission from a source at 50 cm from the detector is almost constant while the sample (non woven polyester) is moved along the beam.

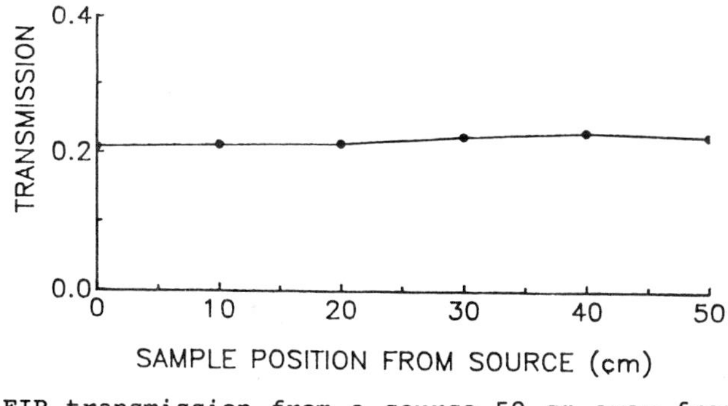

Fig. 7. FIR transmission from a source 50 cm away from the detector, as a function of the position of a non woven polyester sample.

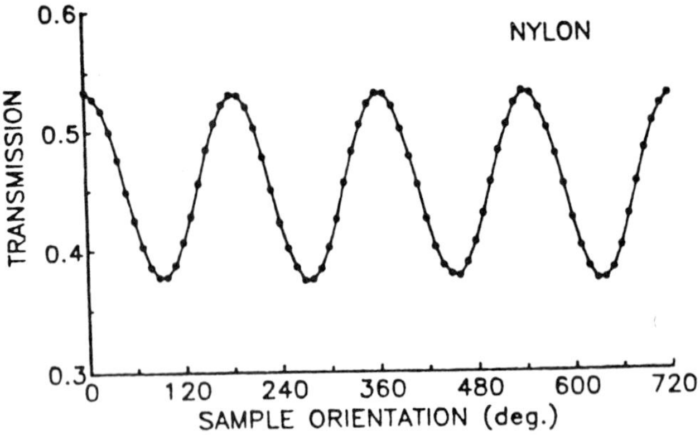

Fig. 8. FIR transmission through a nylon fabric sample, as a function of polarisation angle, using a filtered thermal source.

All the results presented in this article were obtained with an optically pumped FIR laser. This type of source is very useful while experimenting the technique, but its cost, bulk, complexity and troublesome operation hinder its use in an industrial environment. To overcome this obstacle, we are now developing a much cheaper, smaller and simpler source based on properly filtered black body radiation. Preliminary results are promising (Figure 8), despite band pass filters needing considerable further improvement.

Another part of the experimental system is limiting its performance, the polarisation rotator. It comprises three mirrors, their alignment is critical, and rotating the whole device is time consuming. We plan to use a "no moving part" approach by tapping three unpolarised beams from the thermal source and separately polarising each beam in a different direction, 0, 45 and 90 degrees. The signals from three detectors, one for each beam, should allow much faster determination of the three wanted parameters, average transmission, maximum transmission and phase angle. We have started preliminary experimentation of this type of set up and, for all the tested samples, precision and reproducibility turned out to be equal or better than with the previous laser and polarisation rotator system.

CONCLUSION

The measuring methods developed in the course of this project are suitable for basis weight, fibre orientation anisotropy and fibre alignment angle determination in dielectric sheet materials. These methods offer a unique combination of features: high spatial resolution, non contact and non destructive operation, and adequate speed, precision and reproducibility. This research has brought the FIR characterization technique closer to application, for quality control, in an extensive range of widespread industrial products.

ACKNOWLEDGEMENT

The authors want to acknowledge funding for this research by the Quebec government Fonds FCAR (AR-0044).

REFERENCES

1. R. Gagnon, B. Drouin, P. Bernard and M. Bergeron, FAR INFRARED NON DESTRUCTIVE TESTING OF SHEET MATERIALS, TAPPI 1989 annual meeting, proceedings, p 73-78, New-York, March 1989.

2. J. Koskimies, LEANING BEHAVIOUR OF FORM PAPER STACK, Preprints CPPA-TS 72nd annual meeting, p A123-A129, Montreal, January 1986.

3. A.G. Hoyle, NONWOVEN APPLIED IN ELECTRICAL MARKETS, TAPPI Journal, vol 72, no 8, p 83-87, August 1989.

4. S.I. Shapiro, ON LINE FORMATION SENSORS: SURVEY OF CURRENT USE. TAPPI Journal, vol 72, no 9, p 281-283, September 1989.

5. D.F. Rutland, DIMENSIONAL STABILITY AND CURL OF PAPER, Preprints CPPA-TS 71st annual meeting, p A277-A282, Montreal, January 1985.

6. R. Boulay, B. Drouin, R. Gagnon, METHOD FOR MEASURING THE FIBRE ORIENTATION ANISOTROPY IN A FIBROUS STRUCTURE, U.S. Patent # 4 654 529, October 1986.

7. C.C. Habeger and G.A. Baum, THE USE OF MICROWAVE ATTENUATION AS A MEASURE OF FIBER ORIENTATION ANISOTROPY, TAPPI Journal, vol 70, no 2, p 109-113, February 1987.

Measurements of Attenuation and Dispersion in Multi–Phase Materials Using Ultrasonic Point–Source/Point–Receiver Technique

Young H. Kim†, Sekyung Lee†, and H. Chul Kim‡

† Nondestructive Evaluation Lab, Korea Standards Research Institute
P.O.Box 3, Taedok Science Town, Taejon, Korea
‡ Department of Physic, Korea Advanced Institute of Science and Technology
Chungryang P.O.Box 150, Seoul, Korea

ABSTRACT

The attenuation and dispersion of stress waves in multi–phase materials such as granite, paste, mortar and concrete were evaluated by the point–source/point–receiver technique. Glass capillary breaking and a calibrated pin–type transducer were employed as a point source and point receiver, respectively. The frequency dependence of the attenuation and dispersion in the multi–phase materials was obtained by a single measurement and Fourier analysis. Mortar and paste were found to be less attenuative and dispersive than granite and concrete.

INTRODUCTION

Condition assessment of concrete structures such as buildings, bridges and nuclear containment structures is essential since the performance of such structures affects public safety. Structures exposed to severe weather and other adverse conditions such as chemical attacks undergo gradual degradation. The rate of degradation is dependent largely upon the severity of the environmental conditions and the quality of the structure. Nondestructive testing methods need to be developed so that the condition of existing structures can be assessed accurately and reliably to estimate the structural integrity for repair and retrofit. Concrete hardens with increase in age, and most nondestructive methods for concrete are used for determining the strength of hardened concrete. Hardness tests such as indentation[1], penetration[2], rebound number[3] and pull–out[4] are used for strength estimation, however, there is little apparent theoretical relationship between the bulk strength of concrete and these test results; thus they are not regarded as substitutes for standard destructive compression tests, but as methods for determining the uniformity of concrete in structures and comparing one concrete against another. The alternatives are elastic wave methods such as, ultrasonic pulse velocity[5], resonant frequency[6] and impact resonant[7] methods. These methods involve measurements of the transient time of stress waves passing through the concrete and the resonant frequencies of vibration of specimens in calculating the moduli of elasticity and rigidity, measuring degradation during durability studies and detecting flaws in the structures.

Elastic waves in concrete are subjected to scattering due to the large size of the aggregate in concrete, and this scattering leads to attenuation and velocity change. However, concretes were regarded as homogeneous and isotropic medium in previous work. Recently, differences of vibration modes between mortar and concrete block were studied by impact resonance tests[8], however,the reasons for the differences has not been clarified. In the present work, elastic wave properties, attenuation and dispersion in the multi–phase materials, such as concrete are studied to clarify elastic wave propagation in these materials. Point–source/point–receiver technique[9] (PS/PR) is employed to measure attenuation and dispersion of elastic waves in the multi–phase materials. Glass capillary and pin–type transducer are used as ideal elastic wave point–source and ideal point–receiver, respectively.

THEORETICAL BACKGROUND

As shown in Fig.1, elastic waves generated by a source propagate through a medium, reach a sensor and convert to electrical signals. The observed signal, V, is expressed by the convolution of the source function, S, and the system transfer function, T. The transfer function, T, is composed of the coupling between the source and medium, C_1, Green's function of the medium, G, the coupling between the medium and sensor, C_2, receiving characteristics of the sensor, R, and the transfer function of the electric system, E. This can be represented as

$$V = E \otimes R \otimes C_2 \otimes G \otimes C_1 \otimes S. \tag{1}$$

where "$\otimes$" denotes the convolution integral. In this relation, C_1 could be regarded as the source and C_2 and E could be regarded as the receiving characteristics. Thus the observed signal, V, can be written as,

$$V = R \otimes G \otimes S. \tag{2}$$

When the observed signal, V is given and two functions in the right handed side of (2) are known, the remaining term could be obtained by the deconvolution procedure. That is, the receiver function R could be obtained from known G and S (sensor calibration), the source function S from R and G (source characterization), and the Green's function G from R and S (materials characterization). Eqn.(2) involves area integrals if source and/or detector are of some size, however, the area integrals will be omitted when point source and point receiver are considered in similar to the seismic problem. The computed Green's function[10] of a point source and a point receiver in an ideal, homogeneous, elastically isotropic and non–attenuative material provides the basis by which the signals measured in a real material to be identified, processed and interpreted, and thus characterize real material in terms of its testing configuration and structural features.

The Fourier transform of a given signal $f(t)$ is represented in terms of magnitude V and phase ϕ which are functions of frequency as

$$\mathfrak{F}\{f(t)\} = V(\omega)\, e^{i\phi(\omega)}. \tag{3}$$

The phase function ϕ for P–wave arrival of the velocity signal is given by[11]

$$\phi = kL + A/kL \tag{4}$$

where k is the wave number, L is the source/receiver separation and A is correction factor specified by

$$A = 2 - 8(c_s/c_p)^2 + 16(c_s/c_p)^3 \tag{5}$$

where c_s and c_p are shear and longitudinal wave speeds, respectively. The amplitude corresponding to the first shear wave arrival can be processed similarly. The frequency dependence of a particular wave arrival is determined by properly windowing that portion of the waveform and by transforming this windowed signal to obtain the phase function. With this phase function (4) can be solved numerically to obtain a solution for the dispersion relation of the wave in the material. The phase velocity, v_p, for the material can be evaluated as a function of frequency from the following equation:

$$v_p(\omega) = \omega/k \qquad (6)$$

Since the computed waveform corresponds to the propagation in a non–attenuative material, the attenuation of either the P– or the S–wave amplitude in a real material can be determined by comparison of the measured and computed waveform amplitudes of the corresponding wave arrivals. The frequency dependence of the attenuation, $\alpha(\omega)$ can be determined by processing the windowed signal amplitude in the frequency domain to form its magnitude spectrum $V(\omega)$ and evaluating

$$\alpha(\omega) = 20\ log_{10}[V(\omega)/V_r(\omega)] \qquad (7)$$

where $V_r(\omega)$ is the magnitude spectrum of the computed wave velocity amplitude.

The error in this procedures may be caused by the time shift, magnitude scaling and vertical offset of the input signals. The vertical offset of signals affects only the DC components of the frequency spectrum, and thus, can be ignored. The magnitude scaling affects only the magnitudes of the frequency spectrum which are independent of the frequency, and thus, can also be ignored because attenuation obtained by this procedure is a relative one. When the received signal is an ideal impulse function at $t=\tau$, ie, $\delta(t-\tau)$, then the Fourier transform of this signal is $exp(i\omega\tau)$ and the phase function is $\omega\tau$. For plane waves, the second term on the right hand side of eqn.(4) vanishes, and becomes $kL = \omega\tau$. Thus, the resulting phase velocity is $v_p = L/\tau$, independent of the frequency, and it means a simple distance/transient time. Therefore, use of Fourier analysis to obtain phase velocity becomes meaningless for ideal cases, and obtained velocities are neither absolute nor precise. However, when the received signal is dispersive, the Fourier analysis has some relative meaning of the wave velocity as a function of frequency.

EXPERIMENTS

The system configuration for the materials characterization is illustrated in the Fig.2. Breakage of glass capillaries made of fused quartz round tubing with 0.4 mm OD and 0.3 mm ID was identified by universal test machine and laser interferometer and used as an ideal step force to generate elastic waves. A calibrated pin–type piezoelectric transducer with contact diameter of 1.3 mm was used as a point–receiver to detect velocity signals.

Circular plates of multi–phase materials, granite, paste, mortar and concrete, and steel as a reference, were characterized by the point–source/point–receiver technique. The compositions of cement:water:sand:coarse aggregate in weight for paste, mortar and concrete plate are 3:1:0:0, 2:1:4:0 and 2:1:4:6, respectively. The maximum

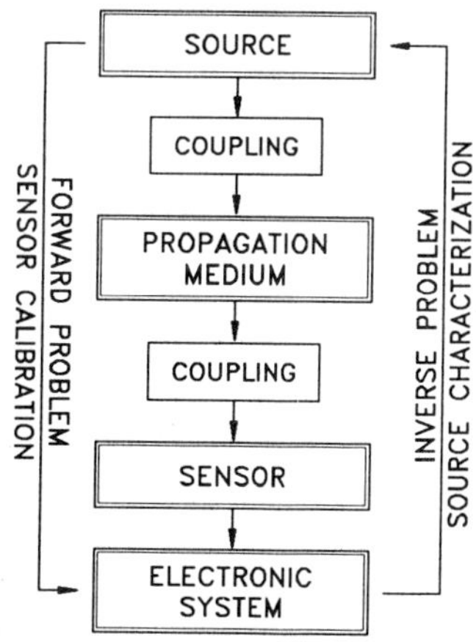

Fig.1 Diagram of stress wave observation

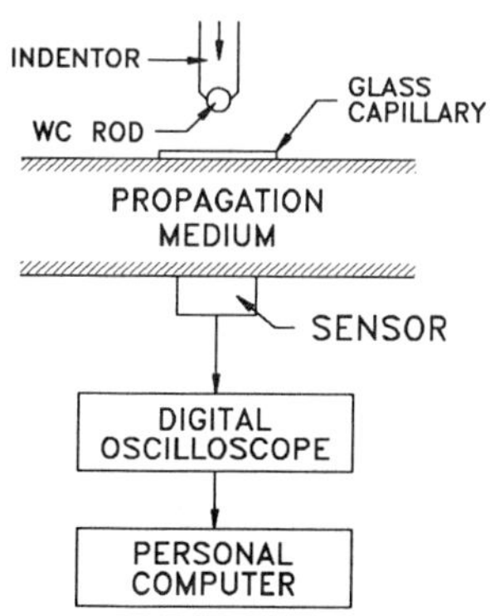

Fig.2 Diagram of the PS/PR test system.

size of aggregate in concrete is 25 mm. The P– and S–wave speeds of these plates are measured by ultrasonic techniques using KB 0.5L and Panametrics V151 0.5 MHz ultrasonic transducers, respectively. The Young's modulus, shear modulus and Poisson's ratios are calculated from the measured wave speeds and densities, and are listed in Table 1.

Detected waveforms at the epicenter of plates due to glass capillary breaking are stored in the digital oscilloscope, transferred to the personal computer and analyzed with Fast Fourier Transform to obtain attenuations and dispersions in plates.

Table 1 Measured materials parameters of the plates

Materials	Steel	Granite	Paste	Mortar	Concrete
Diameter(cm)	24.57	25.02	25.13	25.13	25.13
Thickness(cm)	4.99	5.32	5.85	6.05	5.72
Density(g/cm^3)	7.82	2.62	2.08	2.13	2.36
Wave speeds ($mm/\mu s$)					
– Longitudinal	5.97	3.99	3.99	4.13	4.32
– Shear	3.28	2.50	2.26	2.43	2.45
Poisson's ratio	0.28	0.18	0.26	0.24	0.26
Modulus ($x10^{10}N/m^2$)					
– Young's	21.63	3.87	2.69	3.10	3.59
– Shear	8.43	1.64	1.06	1.26	1.42

RESULTS AND DISCUSSIONS

The glass capillary was broken using universal test machine (Instron), and the force applied was approximately 12 N. Displacement at the epicenter of the 15.8 mm thick steel plate due to glass capillary breaking measured by the laser interferometer is shown in Fig.3 with the computed displacement for unit step force[12]. In this figure, the measured displacement of 1S wave arrival is 3.5 nm. Comparing computed displacement 0.3 nm/N, applied force is calculated by 11.7 N, which is very similar to the measured one by universal test machine. Therefore, the glass capillary breaking is assumed as a step force of 12 N magnitude. The output signal of pin–type transducer at the epicenter for the glass capillary breaking is shown in Fig.4 with computed velocity for unit step force[13]. The sensor output signal is in agreement with theoretical velocity, therefore the pin–type transducer can be considered as a velocity sensitive transducer.

The Green's functions of each plate obtained with measured parameters and velocity signals for the glass capillary breakage for each plate are shown in Fig.5.

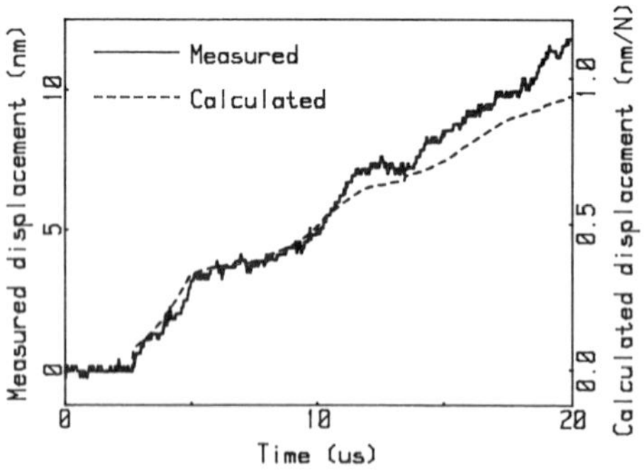

Fig.3 Computed and measured displacements using laser interferometer

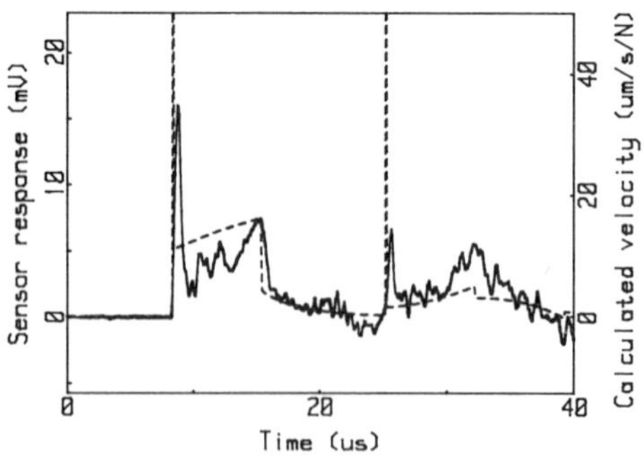

Fig.4 Pin–type transducer response and theoretical velocity at epicenter.

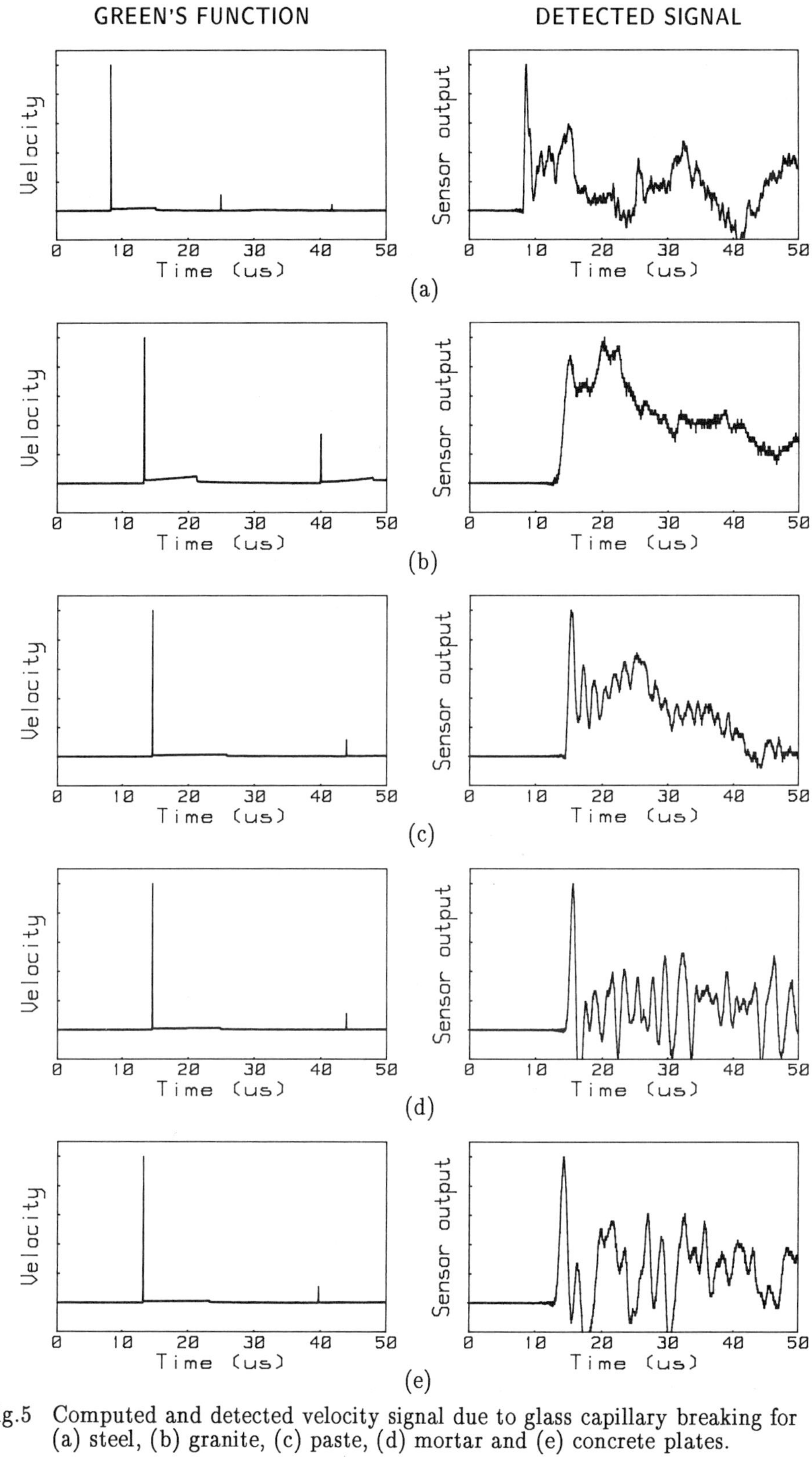

Fig.5 Computed and detected velocity signal due to glass capillary breaking for (a) steel, (b) granite, (c) paste, (d) mortar and (e) concrete plates.

Even though Green's function of each plates are similar, the obtained signals are different from each other. The differences are considered to originate from the different attenuations and dispersion of P– and S–waves in the materials. For the steel plate, the S–wave arrival could be identified, but not in other plates. Therefore, the P–wave arrival is considered only in this analysis. The procedure to obtain the attenuation and dispersion relation of the steel plate, as a typical case, is shown in the Fig.6 where (a) is received signal, (b) windowed signal, (c) and (d) are magnitude and phase spectrum of windowed signal, respectively. In (c), the solid line is for real materials and the dashed line is for ideal material. The scales for ideal and real materials are arbitrary, so the resulting attenuation obtained in this spectrum is a relative one and is shown in (e). The dispersion of P–wave driven from the dispersion relation (6) is shown in (f).

The attenuation for unit thickness and dispersion values of the plates are shown in Fig.7. From this figure, the frequency ranges within –6dB/cm of steel, granite, paste, mortar and concrete are found as 1.88, 0.96, 1.27, 1.86 and 0.68 MHz, respectively. Steel has relatively low attenuation than other materials as expected. Mortar, paste, granite and concrete are in sequence of attenuation, however, paste, mortar, granite and concrete are in sequence of constitution size. In general, higher attenuation is attributed to larger scatterers in materials, however, mortar and paste are changed in those orders and this substitution is considered to be due to the air bubbles in paste.

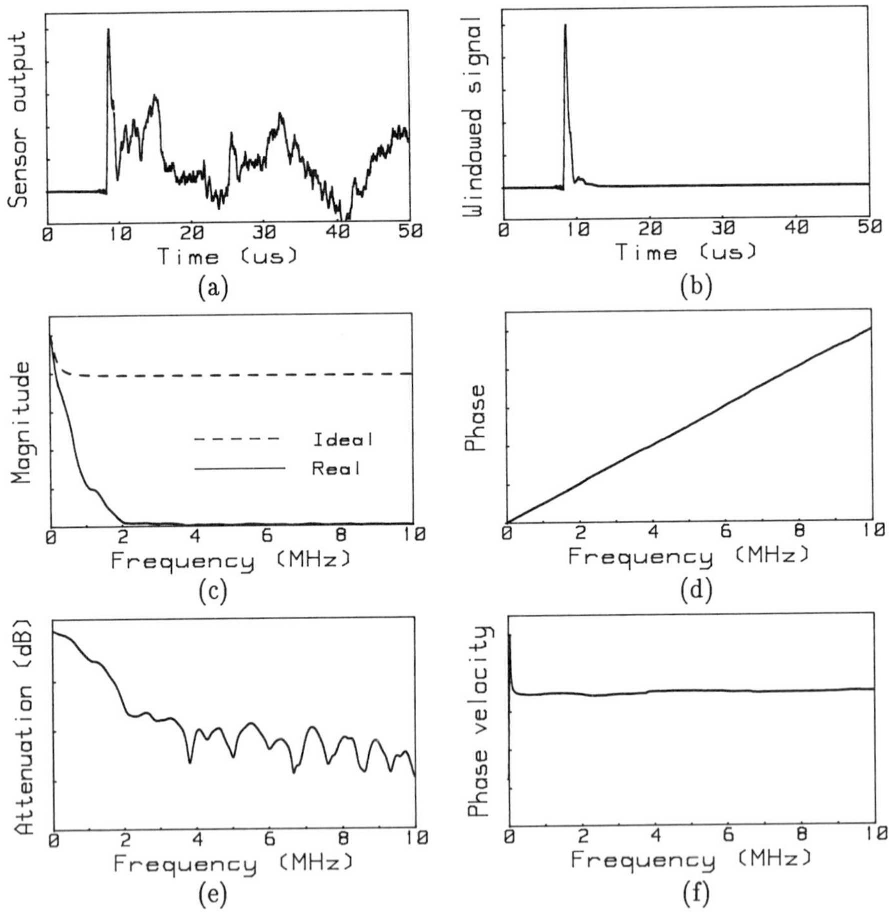

Fig.6 Procedure to obtain attenuation and dispersion by Fourier analysis. (a) Velocity signal, (b) Windowed signal, (c) magnitude and (d) phase spectrum, (e) attenuation and (f) dispersion of phase velocity

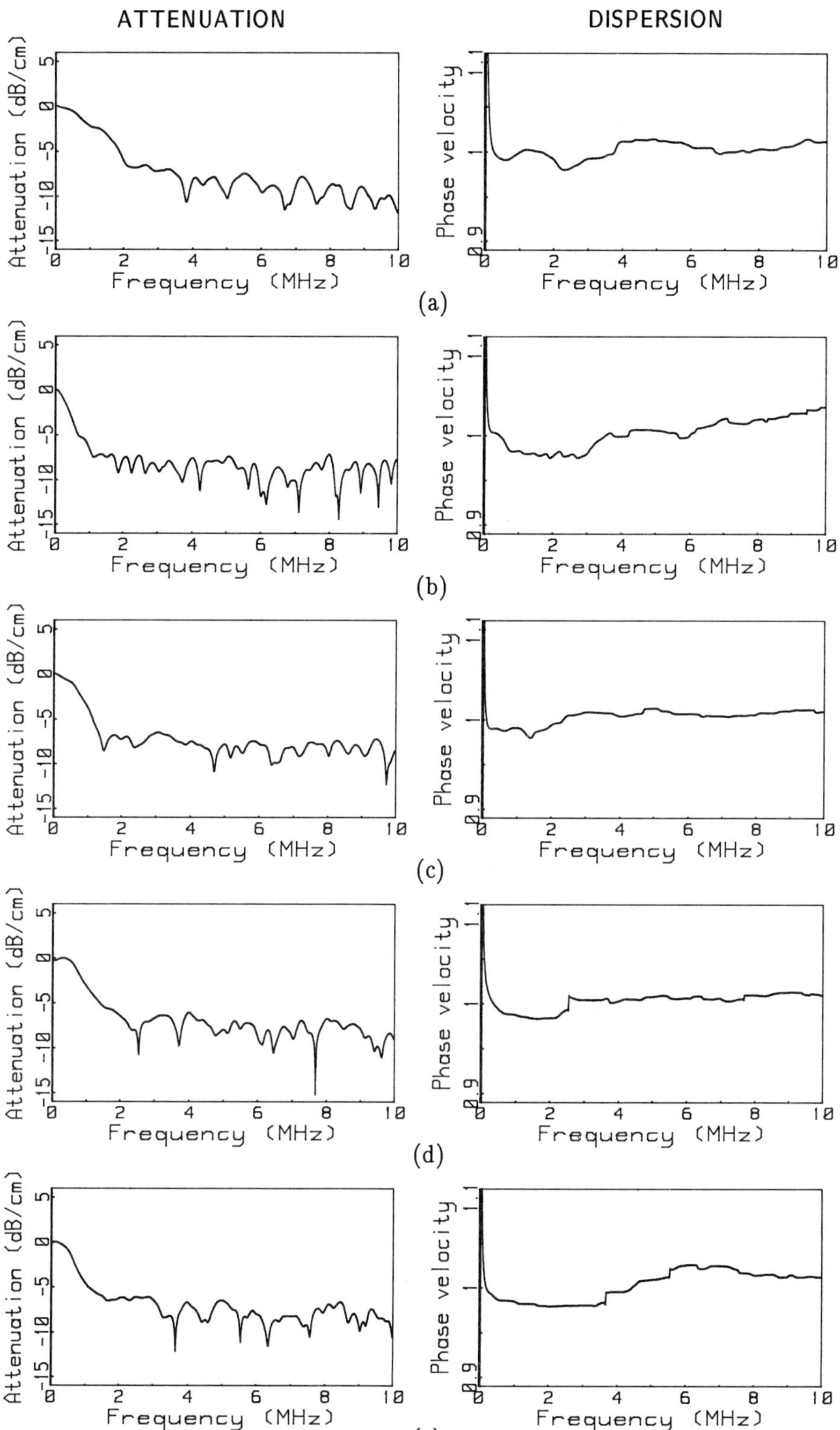

Fig.7 Attenuation and dispersion of phase velocity obtained by Fourier analysis (a) steel, (b) granite, (c) paste, (d) mortar and (e) concrete plates.

The normalized velocities in the dispersion relations of steel, granite, paste, mortar and concrete are 5.72, 3.58, 3.88, 3.94 and 4.42 mm/μs, respectively. These values agree with the values in Table 1 within an accuracy of 5%. The mean deviation of the phase velocity from the normalizing velocity is introduced to define the degree of dispersion. Those values for steel, granite, paste, mortar and concrete are 1.26, 1.81, 1.20, 1.26 and 1.97 %, respectively. Therefore granite and concrete are dispersive media compared to the other materials. The commonness of granite and concrete is that the sizes of constituents are larger than the others and multi compositions are involved. Therefore, the dispersions of granite and concrete are considered to arise from the material properties and scattering. The obtained elastic properties of plates are summarized in Table 2.

Table 2 Obtained elastic properties of the plates

Materials	Steel	Granite	Paste	Mortar	Concrete
Attenuation (MHz)	1.88	0.96	1.27	1.86	0.68
Phase velocity (mm/μs)	5.72	3.58	3.88	3.94	4.42
Dispersion (%)	1.26	1.81	1.20	1.26	1.97

CONCLUSIONS

The ultrasonic point–source/point–receiver technique is applied to measure attenuation and dispersion of elastic waves in the multi–phase materials. Attenuation and dispersion of elastic waves in materials are measured by single experiments with Fourier analysis. Mortar and paste were found to be less attenuative and dispersive than granite and concrete.

REFERENCES

1. Ball impact test for normal concrete, DIN 4240, No.6, 311–319 (1966)
2. Test for penetration resistance of hardened concrete, ASTM C 803–75T
3. J.Kolek, An appreciation of Schmidt rebound hammer, Mag. Concr. (London), **10**(28):27 (1958)
4. B.Tremper, The measurement of concrete strength by embedded pull–out bars, Proceedings, ASTM, **44**:880 (1944)
5. Measurement of the velocity of ultrasonic pulses in concrete, BS4408
6. Standard method of test for fundamental transverse, longitudinal and torsional frequencies of concrete specimens, ASTM C 215–60
7. J.Koo, S.Lee, Y.Kim and H.S.Lew, Quantitative spectral analysis of the flaw detection in concrete, "Proceedings of 12th world conference on NDT", 1250, Amsterdam (1989)
8. Y.H.Kim, S.Lee and H.C.Kim, Quantitative vibration mode analysis of mortar and concrete blocks containing flaws, NDT Int'l, **23**:67 (1990)
9. W.Sachse and K.Y.Kim, Quantitative acoustic emission and failure mechanics of composite materials, Ultrasonics, **25**:195 (1987)
10. N.N.Hsu, Dynamic Green's function of an infinite plate – a computer program, Report NBS IR85–3234, National Bureau of Standards (1985)
11. Y.H.Pao and W.Sachse, On the determination of phase and group velocities of dispersive waves in solids, J.Appl.Phys., **48**:4320 (1978)
12. Y.H.Kim, D.H.Lee and O.Y.Kwon, Absolute measurements of dynamic micro–displacements by laser interferometry(in Korean), Kor.Appl.Phys., **3**:461 (1990)
13. Y.H.Kim, D.J.Yoon, S.Lee and H.C.Kim, Characterization of acoustic emission sensors using transient stress waves(in Korean), Kor.Appl.Phys., **3**:374 (1990)

REAL-TIME MONITORING OF CERAMIC SINTERING WITH LASER ULTRASONICS

K. L. Telschow

Idaho National Engineering Laboratory
EG&G Idaho, Inc. Idaho Falls, ID. 83415-2209

SUMMARY

A Fabry-Perot Doppler interferometer that directly measures elastic waves from rough, optically diffuse, surfaces of materials has been constructed at the Idaho National Engineering Laboratory (INEL). This apparatus, coupled with a pulsed laser for elastic wave generation, is currently being used for real-time monitoring of the sintering of zinc oxide. From the ultrasonic propagation measurements, actual densities and shrinkage values can be calculated as the material sinters; the algorithm requires previous knowledge of the ultrasonic velocity dependence on density and assumptions concerning the method of shrinkage in the material.

INTRODUCTION

The properties of ceramic components are often determined by residual porosity in the material. Increased production efficiency and improvements in the consistency of properties of ceramics could be achieved if the porosity could be monitored during the sintering stage. In a material with porosity, the effective compliance is increased due to the presence of voids, with an attendant decrease in the elastic modulus and the velocity of ultrasonic wave propagation with increasing porosity. Therefore, ultrasonic measurements can be used to monitor the sintering process; however, to be most effective, a noncontacting technique is needed. This paper describes the successful application of laser ultrasonic techniques, which require only optical access to the sample surface, to the monitoring of sintering of zinc oxide at temperatures from 850 to 1050°C. The basic measurement process has been previously described.[1,2] This paper focusses on some of the complications of ultrasonic measurements at high temperatures in ceramics and on an algorithm that allows real-time determination of the sample density and shrinkage during the sintering process.

CERAMIC MATERIAL

The material used was commercial zinc oxide (Fisher Z52-500) with a very small particle size of around 1 μm. The green state samples were pressed into disks 25 mm in diameter and 8 to 10 mm thick with a theoretical density of approximately 47%. They were particularly soft with roughly the

Nondestructive Characterization of Materials IV
Edited by C.O. Ruud *et al.*, Plenum Press, New York, 1991

consistency of chalk. Sintering was done in a tube furnace at temperatures from 850 to 1050°C; the ultrasonic measurements were performed in a through transmission geometry.

ULTRASONIC LASER GENERATION AND DETECTION

A pulsed Nd:YAG laser with a 10 ns pulse width and pulse energies of 100 to 150 mJ was used as the ultrasound source. The laser beam was focused to a diameter of approximately 3 to 4 mm. The generation mechanism for ultrasound in ceramics generally includes both thermoelastic and ablation sources. Most ceramics exhibit a significant optical penetration depth, which alters the thermoelastic generation mechanism and can be used to advantage. Thermoelastic generation alone at the surface produces elastic wave radiation that is largely directed away from the surface normal, whereas ablation produces its largest wave motion in the direction of the surface normal.[3] In ceramics that exhibit a sizeable optical penetration depth, however, the thermoelastic mechanism itself produces a significant ultrasonic wave in the direction of the surface normal because the optical penetration extends the thermoelastic expansion into the material. This subsurface expansion produces a precursor waveform normal to the surface and a significant normal surface motion.[3-5]

Fig. 1 shows the ultrasonic displacement waveforms recorded with a capacitive detector from a sintered 6.3 mm thick sample of zinc oxide. The source was on the top of the sample and the detector on the bottom. To illustrate that this material absorbs green light more than infrared light,

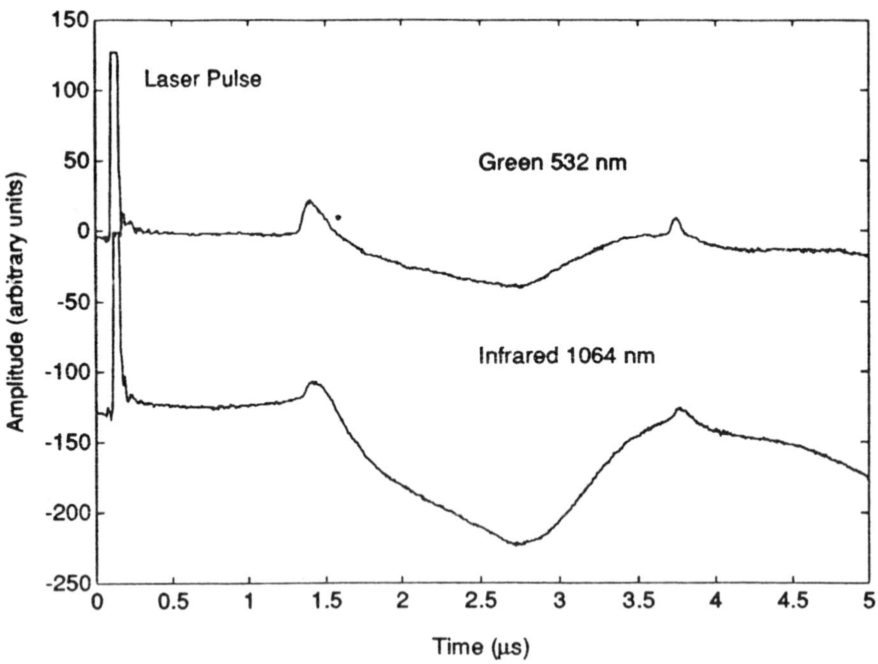

Fig. 1. Pulsed laser generated ultrasonic waveforms recorded with a capacitive detector through a fully sintered zinc oxide sample using the frequency doubled pulse at 532 nm, 40 mJ/pulse and the infrared pulse at 1064 nm, 145 mJ/pulse.

more pulse energy was needed with the infrared than with the green. This wavelength dependence of optical properties needs to be considered carefully when adapting a particular laser source to a given material.

The initial spike displacement seen in the waveforms of Fig. 1 is the precursor waveform due primarily to optical penetration into the sample and was used for the waveform time of flight measurements. It was observed that the optical properties of the zinc oxide change with temperature, which can markedly alter both the generation and detection of ultrasonic motion at the surface. Fig. 2 shows a time series of waveforms, again detected with the capacitive detector, as a function of the number of laser pulses absorbed in the material. The waveform at t=0 s is that recorded after the first laser pulse. Subsequent waveforms show the cumulative effect of the laser pulses. The repetition rate is 10 Hz, so the waveform at t=1.1 s is after 11 pulses, etc. Clearly, the precursor waveform grows significantly during the first 32 laser pulses, after which it remains constant. This is because the optical absorption coefficient for zinc oxide increases with temperature, which was confirmed by direct reflection measurements from the sample surface at different temperatures (Fig. 3). The increased absorption at high temperatures was both a help and hindrance. It helped by increasing the efficiency of the ultrasonic generation process, but it hindered the optical detection process since it reduced the light scattered back to the detector. Quite different effects could pertain to other ceramics.

The laser beam was absorbed to a depth of about a millimeter at 532 nm and produced a strong longitudinal precursor waveform due to the optical penetration and some ablation of material from the surface, as was evident from minor pitting and discoloration. However, the sintering measurements could be completed without ablation of the ceramic of more than about 25 to 50 μm. The green state samples were more susceptible to ablation than those sintered to a density greater than 50%.

A confocal Fabry-Perot interferometer, modeled after that described by Monchalin and Heron,[6] was used to detect the ultrasonic waves on the surface of the samples. The detector is sensitive to the Doppler shift of light reflected from the sample surface, which is proportional to the velocity of the sample surface. This is confirmed by the comparison in Fig. 4, which shows the same ultrasonic waveform recorded with the capacitive displacement detector and the optical detector. The optical detector waveform is essentially the time derivative of the capacitive detector waveform. This response is maintained for frequencies up to approximately 8 MHz for this interferometer. Note that the major distinguishing feature in the optical detector waveform is that part due to the precursor, which can be generated by optical penetration, ablation or by using special coatings.

An argon ion continuous laser (1.0 W maximum at 514 nm) was used to illuminate the sample surface. This unit is sensitive to the frequency of the reflected light and not its phase, which allows it to collect light from a relatively large surface area (about 4 mm^2) including many speckles. This gives it the sensitivity needed for detection from rough and/or optically diffuse material surfaces, as encountered in ceramic materials in general and "green" state ceramic materials in particular. No surface damage resulted from using the detection laser at powers up to 900 mW, which were necessary at high temperatures due to the increased sample absorption noted earlier.

The zinc oxide samples used for the sintering study exhibited nearly complete diffuse reflection from their surfaces, even after sintering. The optical collection of scattered light from the sample, the determining factor for the signal to noise ratio, was limited by the opening of the furnace tube.

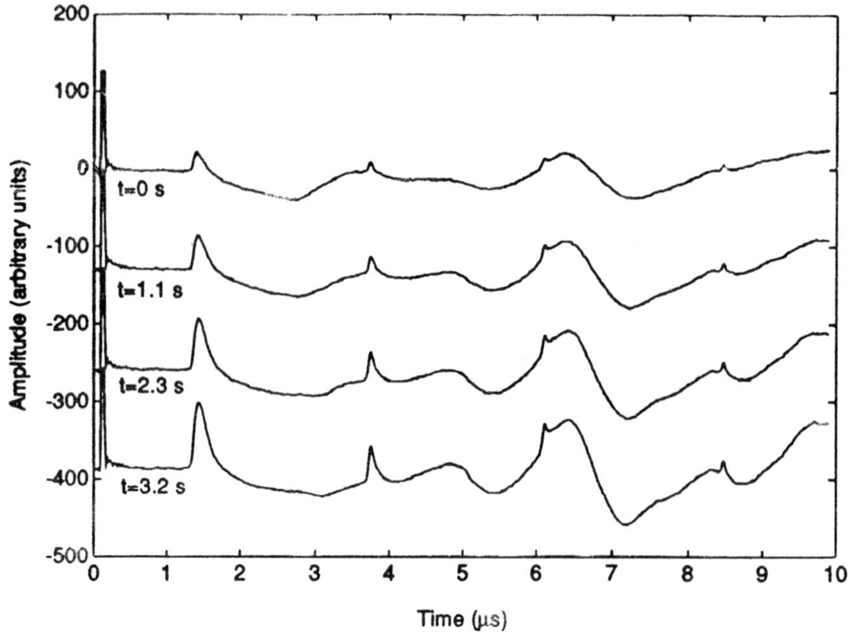

Fig. 2. Time sequence of the displacement waveforms recorded as in Fig. 1 showing the cumulative effect of the laser pulses. The repetition rate is 10 Hz.

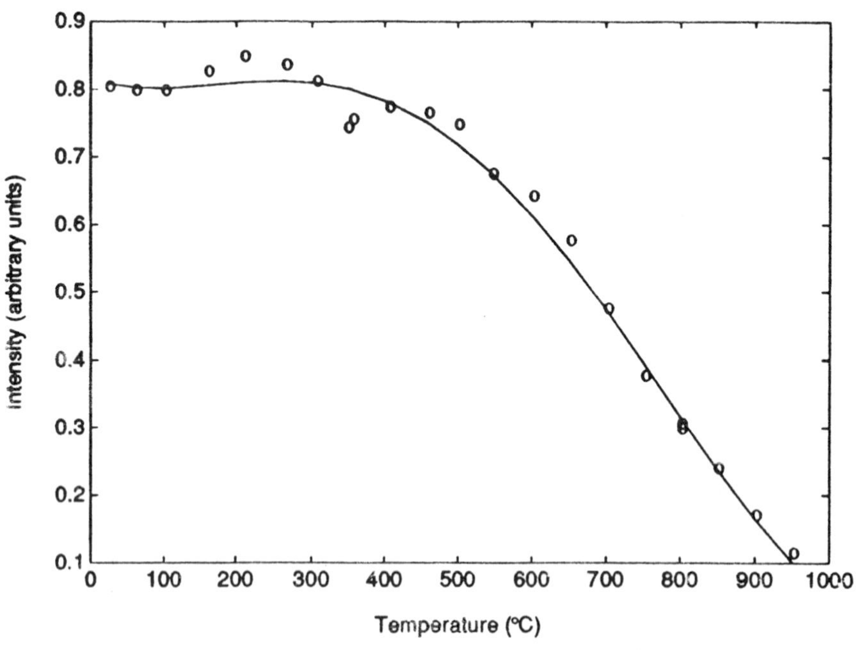

Fig. 3. Reflected optical intensity at 514 nm from a fully sintered sample of zinc oxide as a function of temperature.

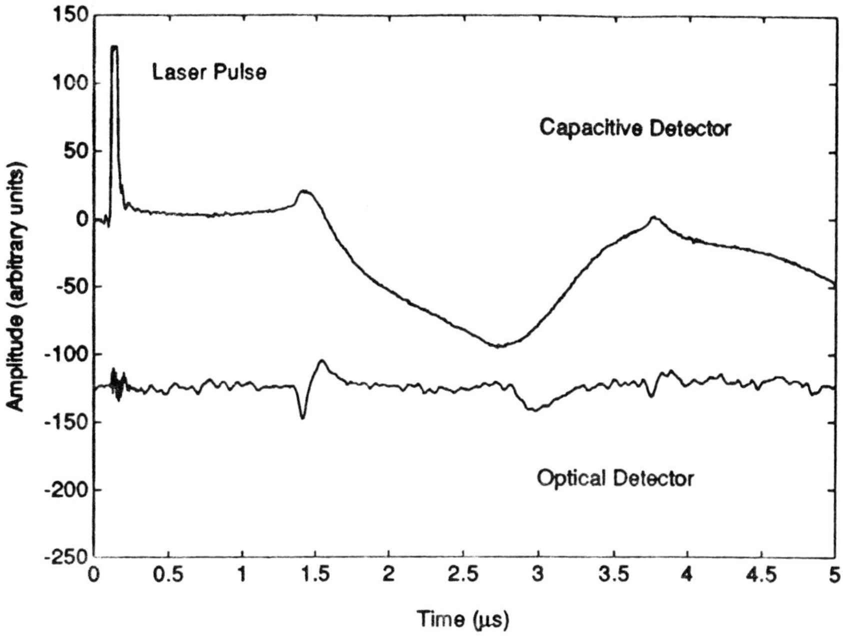

Fig. 4. Comparison of the ultrasonic waveform recorded with the capacitive detector showing surface displacement and the optical detector showing surface velocity for a fully sintered zinc oxide sample with pulsed laser excitation.

REAL-TIME SINTERING MEASUREMENTS AND THE RELATIVE DENSITY ALGORITHM

The procedure for the sintering measurements was to insert a sample into the furnace and preheat it at 700°C for about 1 h. This step was necessary as it was found that initially a large change in elastic constant occurred with very little change in average density, believed to be due to the powder particles beginning to join together. The measurements were of the time of flight of longitudinal ultrasonic waves propagating through the sample, as it sinters. In order to determine the sample average density from these measurements, it is necessary to know the sample thickness and the dependence of the ultrasonic velocity on density and temperature. In addition, there are most likely density and temperature gradients in the material as it sinters, since it is heated from the outside. There is no way to determine these gradients with the through transmission measurements reported here, so only average sample densities are considered in what follows. There is a connection between the sample thickness and density since many ceramics shrink uniformly during sintering if compacted uniformly with little residual stress. It was confirmed that the zinc oxide experiences uniform shrinkage, i.e. the sample diameter shrinks in the same proportion as its thickness. In addition, no mass loss occurred to within a precision of 4%. Densities are reported in percent of the theoretical density (5.6 g/cm^3).

The temperature dependence of the ultrasonic velocity in the zinc oxide compacts during sintering is unknown. For fully sintered samples the longitudinal wave velocity was found to be essentially temperature independent (to within 5%) over the range of temperatures from ambient to 1100°C. Therefore no correction for temperature dependence has been considered for the data presented here.

During sintering, the ultrasonic time of flight decreases significantly as the material's density increases and its thickness decreases. To complete the density determination algorithm, the dependence of the ultrasonic wave velocity on density must be known for the material. Several samples of zinc oxide were prepared in the "green" state and then sintered at various temperatures and times to produce a set of partially sintered samples with relative densities from 52% to 97%. Longitudinal ultrasonic wave velocities were recorded for this sample set to provide basic information on the effects of sample density. Measurements were taken with a 5 MHz contact piezoelectric transducer, using a vacuum coupled polymer film[7] with gel couplant and pulse echo overlap, and the laser ultrasonic method. In the green state the measured velocities were low, around 0.7 mm/μs, and not very dependent on the sample density. However, the velocity rises abruptly after a small amount of sintering, up to 52% density, probably due to bonding between the powder particles in the initial sintering. This results in a large change in elastic constant for the sample but very little densification. Further sintering produces an approximately linear change in velocity with relative density up to values approaching that of the bulk material as full densification is achieved.

The assumptions of conservation of mass and uniform shrinkage yield the following relation between density and thickness (applicable before, during and after, sintering): $\rho_i L_i^3 = \rho L^3 = \rho_f L_f^3$ where (i,f) refer to (before, after) sintering. This, coupled with the longitudinal velocity dependence on porosity, $C(\rho)$, determined previously, yields a relation between density and ultrasonic wave time of flight of the following form: $\rho^{1/3}C(\rho)T = \rho_m^{1/3}C_m T_m$. Here (m) can refer to either before or after sintering values. This equation can be inverted to obtain the density as a function of time of flight through a given sample. The size of different samples is taken into account by the normalization values, determined either before or after the sample is sintered. The range of applicability for this algorithm is determined by the range over which the velocity as a function of density is reliably known. For the zinc oxide samples described here, only the data recorded for samples of density greater than 50% of theoretical density were used.

Fig. 5 shows typical results obtained for zinc oxide sample #49 with starting density around 47% and presintered by heat treating at 700°C for about 1 h to 50% density and 8.62 mm thickness. At the time indicated in the graph the temperature is raised quickly to a final temperature of 975°C and the longitudinal wave time of flight recorded as a function of time. Both the sample relative density and shrinkage can be calculated as a function of time using the above algorithm. These values are also shown in Fig. 5. Several samples with roughly the same starting density and size were monitored and sintered in this way to a variety of final temperatures ranging from 850 to 1050°C. Similar results to that shown in Fig. 5 were obtained in all cases; the speed at which the sample sintered and the final density reached were monotonic functions of the final sample temperature. The slope of the densification and shrinkage curves becomes small after times of about 1.5 to 2.0 h for all the samples tested. The final density of the sample after sintering was predictable from this procedure to a precision of about 2-3% of theoretical density, which is about the precision to which the various parameters used in the algorithm are known. That the final sample density seems to level off after some time, or at least continues to increase very slowly, indicates that the furnace temperature is the most important parameter in determining the final density for zinc oxide.

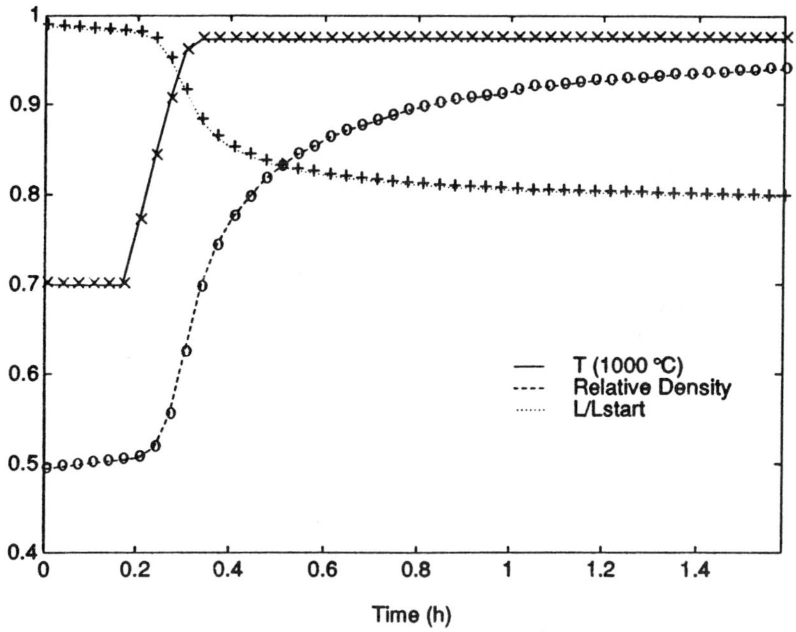

Fig. 5. Sintering plot for sample #49, with initial density of 50% and thickness 8.62 mm, showing the relative density, shrinkage ratio, and sintering temperature history with time.

CONCLUSIONS

These results show that laser ultrasonic techniques, which allow measurements on samples with rough and unpolished surfaces, can be used for noncontacting *in situ* measurements of elastic properties and, with an appropriate algorithm, sample density and shrinkage at high temperatures. The technique has been successfully applied to the sintering of zinc oxide. It can be concluded that the ultrasonic velocity provides a direct measure of the densification of the material during sintering. Several different factors in the sintering process are still unknown, such as the temperature uniformity of the material during heat up and the effects of powder size, compaction pressure, and composition. These factors can now be easily and quickly explored by using noncontacting laser ultrasonics to measure densification, which allows the direct monitoring of the sintering process as a function of both time and temperature. This technique should also be applicable to other materials and processing environments, such as heat treatment of metals.

ACKNOWLEDGMENT

The author thanks S. T. Schuetz, R. S. Wallace, and B. H. Park for preparing the zinc oxide samples and J. B. Walter, G. V. Garcia and C. Stander for assistance with the measurements. This work was supported by the Interior Department 's Bureau of Mines under Contract No. J0134035 through Department of Energy contract No. DE-AC07-76ID01570.

REFERENCES

1. K. L. Telschow, J. B. Walter, G. V. Garcia and D. C. Kunerth, "Process Monitoring using Optical Ultrasonic Wave Detection," Review of Quantitative Nondestructive Evaluation, Vol. 9, D. O. Thompson and D. C. Chimenti eds., Plenum Press, New York, (1990), to be published.
2. K. L. Telschow, J. B. Walter, and G. V. Garcia, "Laser Ultrasonic Monitoring of Ceramic Sintering," in preparation.
3. D. A. Hutchins, "Ultrasonic Generation by Pulsed Lasers," Physical Acoustics, vol.XVIII, W. P. Mason editor (Academic Press, N.Y.,1988), 21-123.
4. K. L. Telschow and R. J. Conant, "Optical Parameter Effects on Laser Generated Ultrasound for Microstructure Characterization," Nondestructive Materials Characterization, P. Holler, V. Hauk, G. Dobmann, C. O. Ruud and R. E. Green eds., (Springer-Verlag, N.Y., 1989), 799-806.
5. K. L. Telschow and R. J. Conant, "Optical and Thermal Parameter Effects on Laser Generated Ultrasound," To be published in J. Acoust. Soc. Am. (1990).
6. J.-P.Monchalin and R.Heron, "Laser Ultrasonic Generation and Detection with a Confocal Fabry-Perot Interferometer," Materials Evaluation, 44, 1231 (1986).
7. R. A. Roberts, "Ultrasonic NDE of Green-State Ceramics by Focused Through-Transmission, "Review of Quantitative Nondestructive Evaluation, Vol. 6B, D. O. Thompson and D. C. Chimenti eds., Plenum Press, New York, 1443 (1987).

MEASUREMENTS OF ELASTIC CONSTANTS OF METAL MATRIX CERAMIC COMPOSITES USING ULTRASONIC PLATE MODE ANTIRESONANCES

W. Wang and S. I. Rokhlin

The Ohio State University
Department of Welding Engineering
Columbus, Ohio 43210

INTRODUCTION

The concept and the technique of using ultrasonic antiresonance to determine elastic constants of thin-plate anisotropic media have recently been introduced by the authors [1]. The technique is based on measurement of obliquely incident ultrasonic beam zero-transmission angles and reconstruction from these angles of the composite elastic constants. Such a technique is particularly useful for measuring elastic constants of anisotropic plates and it has a unique capacity to measure in-plane elastic constants of thin anisotropic plates. So far this technique has been applied successfully to the determination of elastic constants of composite materials containing many rows of fibers [1-3].

This work focuses on measuring elastic constants of composites with only one row of fibers, whose diameter may be comparable to the sample thickness. The composite selected for this study is a 220 μm thick fiber-reinforced SiC/Al composite. The fibers (SiC) have a diameter of 65 μm and are embedded in the plane of the plate.

Based on the results of this measurement we shall analyze the possibility of considering composites with only one row of fibers as an effective anisotropic medium. This question is of major importance for laminated composites. Our aim is to investigate experimentally the validity of microstructure theories applicable to a bulk composite containing multiple rows of fibers for estimation of the effective elastic constants of a sample with only one row of fibers.

THEORY

Dilatational (symmetric), flexural (antisymmetric), and quasi-SH (shear) waves can be excited in a fluid coupled anistropic plate upon oblique incidence of ultrasonic waves. When the normal displacement at the back of the plate vanishes due to compensation of the symmetric and antisymmetric components of the displacement, a zero of the transmission (full isolation) occurs. We will call such a phenomenon an antiresonance.

Nondestructive Characterization of Materials IV
Edited by C.O. Ruud *et al.*, Plenum Press, New York, 1991

Theory is available [4] to calculate the transmission coefficient (T) of a plane ultrasonic wave through a fluid-loaded arbitrarily-oriented orthotropic plate. T is given by

$$T = \frac{i\,Y\,(S + A)}{(S + i\,Y)\,(A - i\,Y)}\,, \tag{1}$$

where S is the symmetric Lamb wave characteristic function for an orthotropic plate in vacuum, A the antisymmetric Lamb wave characteristic function, and Y expresses the effect of fluid loading and depends on the fluid parameters. It is apparent that zero transmission occurs when

$$S + A = 0. \tag{2}$$

As can be seen from equations (1) and (2) the vanishing of the transmission coefficient is not affected by fluid properties.

EXPERIMENTAL APPROACH

The experimental apparatus for the measurement uses a novel ultrasonic goniometer as shown in Figure 1. The goniometer is distinguished by its ability to make simultaneous measurements of both doubly transmitted (through the sample) and doubly reflected (from the sample) ultrasonic signals with the use of only one ultrasonic transducer. The sample is held by two rotation tables at the center of a cylindrical reflector. One rotation table rotates the sample around the central axis of the cylinder so that different incident angles can be used. The other rotation table has a hollow aperture in the center to allow wave transmission and it rotates

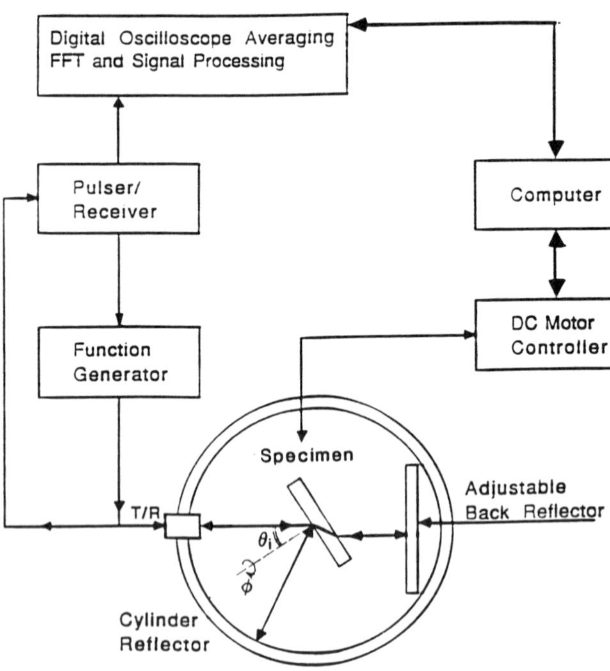

Fig. 1. Configuration of the experimental system.

the sample in the plane parallel to the sample surface. This makes it possible to make measurements as a function of the angle of deviation of the incident plane from the fiber direction.

The wave propagation is such that both incident and reflected ultrasonic beams propagate along cylinder radii. After being reflected from the cylinder wall, the ultrasonic signal returns to the sample, reflects a second time from its surface, and returns to the receiving transducer. The ultrasonic signal transmitted through the sample is reflected from an adjustable back reflector and returned to the transducer after the second through transmission. In this experiment only the transmission mode was used.

Ultrasonic tone-burst signals are generated by a function generator. The signals received are amplified by a Panametrics ultrasonic analyzer, digitized and averaged by a LeCroy 9400 125-MHz digital oscilloscope, and then fed to a computer through an IEEE bus as a function of the angle of rotation. For sample rotation around the cylinder axis a computer-controlled DC motor is used. Angle resolution and repeatability of 0.01° are achieved in our setup. The water tank is temperature stabilized at 29.8 ± 0.1°C.

The representative microstructure of the sample is shown in Figure 2. The volume fraction of fibers (SiC) is 4% and the diameter of the fibers is about 65 μm. The average sample thickness of the SiC/Al composite is 0.22 mm.

EXPERIMENTAL RESULTS AND RECONSTRUCTION OF ELASTIC CONSTANTS

Doubly-transmitted wave amplitudes are measured as a function of incident angle at various fiber orientations and ultrasonic frequencies. The amplitude coefficient of a doubly-transmitted signal is equal to a single energy transmission coefficient through the plate. Typical results of the measurements are shown in Figures 3 and 4. Figure 3 shows the data obtained at fiber orientation 0°, i.e., fibers parallel to the plane of incidence, and at ultrasonic frequency of 3 MHz. The dots are the experimental measurements and the dashed line is calculated from equation (1) using the reconstructed elastic constants. The reconstruction of elastic constants will be discussed below. A transmission minimum is observed at incident angle 16.5°, which corresponds to the transmission zero in the theory. Figure 4 shows the results obtained at fiber orientation 50° relative to the incident plane and ultrasonic frequency 3 MHz. As can be seen by comparing Figures 3 and 4, the zero-transmission angle increases as the fiber orientation angle increases.

In addition to the shift of the position of the angle, a second transmission zero is observed in the theoretical curve when the incident plane deviates from a plane of symmetry (e.g., 50° deviation from fiber orientation, as shown in Figure 4). The occurrence of the additional transmission zero is due to the generation of a quasi-SH wave, whose interaction with the symmetric and antisymmetric plate waves leads to the transmission zero. Experimentally, however, the second transmission was observed only as a very shallow minimum in the neighborhood of the theoretical second zero-transmission angle. The absence of the second transmission zero in the experiment may be attributed to ultrasonic attenuation due to scattering by fiber/matrix interfaces. The effect of scattering can be accounted for by introducing a frequency-dependent attenuation coefficient. An example of the calculated result with consideration of ultrasonic attenuation due to scattering is shown by the solid line in Figure 4. The second transmission zero disappears after the attenuation was taken into consideration.

a

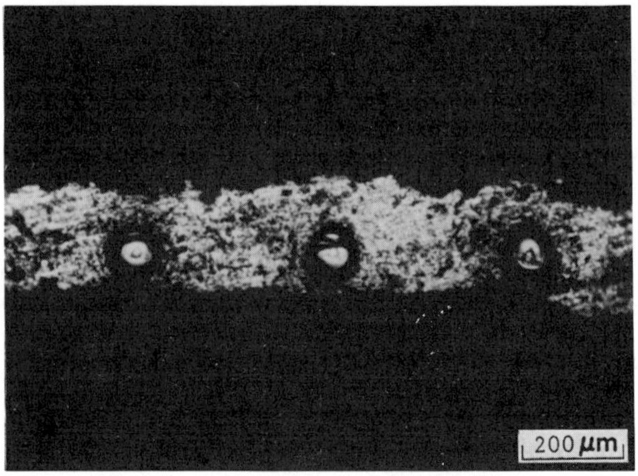

b

Fig. 2. Microstructure of the SiC/Al composite sample. (a) Top surface. (b) Cross section.

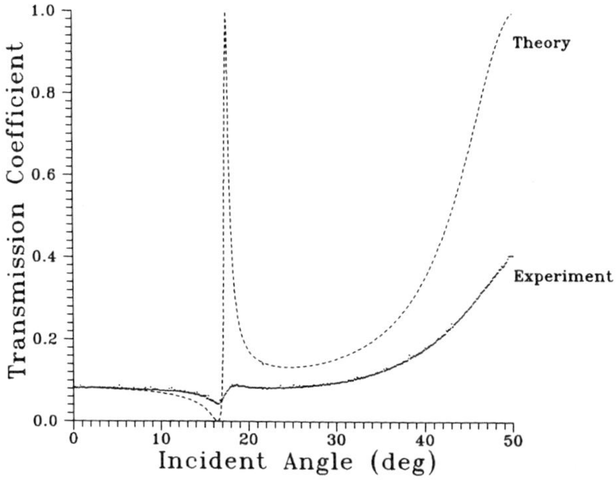

Fig. 3. Measured and calculated transmission coefficient as a function of incident angle at fiber orientation 0° and ultrasonic frequency 3 MHz. The dots represent experimental data and the dashed line calculations using the reconstructed elastic constants.

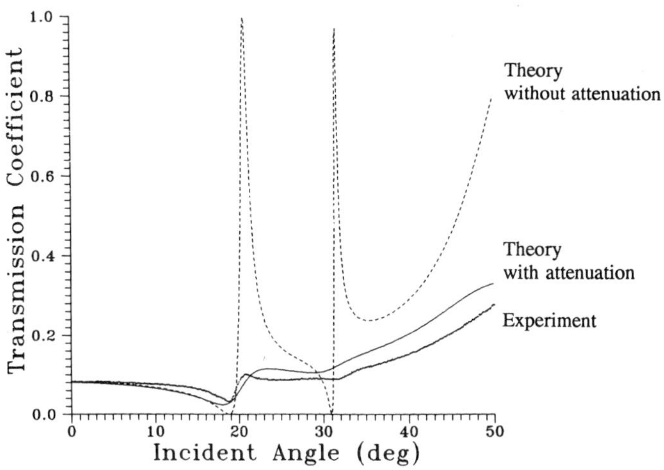

Fig. 4. Measured and calculated transmission coefficients as a function of incident angle at fiber orientation 50° and ultrasonic frequency 3 MHz. The dots represent experimental data and the solid and dashed lines, respectively, calculations using the reconstructed elastic constants with and without attenuation.

Repeating the measurements for a fiber orientation angle range from 0° to 90°, we obtain a polar diagram of the transmission zero. For example, the polar diagram obtained at ultrasonic frequency 3 MHz is shown in Figure 5. The zero-transmission angle increases monotonically from fiber orientation 0° to 90°.

Prior to the reconstruction process, a numerical study of the sensitivity of the angle dependence of the zero transmission on the elastic constants was conducted. Such a study is important since one may expect the elastic constants reconstructed from the experimental data to have meaningful precision only if the data are sensitive to elastic constant changes. Examples of the calculated sensitivities are shown in Figures 6 and 7. Figure 6 shows the dependence of the zero-transmission angle on the effective elastic constant C_{11}. Here the coordinate system is defined such that axis 1 is along the fiber direction, axis 2 is in-plane but transverse to the fiber direction, and axis 3 is normal to the plane of the plate. As shown in the figure, the zero-transmission angle increases significantly with a decrease in C_{11}, especially for fiber orientation near 0°. A similar effect was found for C_{22} at fiber orientation near 90°. Figure 7 shows the effect of C_{66}. A decrease of C_{66} increases the zero-transmission angle when the incident plane does not coincide with a plane of symmetry, e.g., a plane parallel or perpendicular to the fibers.

The behavior of the zero-transmission angle is quite sensitive to the value of the shear modulus in the fiber orientation range between 30 and 60 degrees. At zero degrees the behavior is determined by C_{11} and at 90° by C_{22}. So these values set the range of variation for the zero transmission angle, depending on the microstructural properties of the composite. The full range of variation in Fig. 7, for example, is approximately 4°. There the shear modulus C_{66} changes by 20% which causes the zero transmission curve to shift by about 0.8°, which is about 20% of the full scale (4°). This sensitivity indicates that the method has good potential for reconstruction of shear moduli from experimental data. (Recall that our angle measurement resolution is about 0.01°.)

To reconstruct the elastic constants from the experimental results, we use a nonlinear least square optimization in a multi-dimensional space, where the elastic constants are considered as free parameters and the zero-transmission angles are calculated to curve fit the experimental data. The effective elastic constants of the composite sample are the ones that give the best fit theoretical values to the experimental data. Such an approach can be carried out either in fitting the polar diagram at a given ultrasonic frequency, or in fitting the frequency dependence of the zero-transmission angle for a given fiber orientation.

The reconstruction was performed in several steps: first C_{11} and C_{13} were found from the data in the vicinity of zero degrees, then C_{22} and C_{23} from the data in the vicinity of 90°. Next C_{12} and C_{66} were found from the middle region of angles. The elastic constants found in this way have been used as initial guesses for multidimensional optimization of the fit over the full range of angle dependence of the experimental data. We leave until the next section discussion of C_{33}.

The effective elastic constants reconstructed from the measured zero-transmission angles are shown in Table 1. Using the set of reconstructed elastic constants we can calculate the transmission coefficient as a function of incident angle as well as the polar diagram of the zero-transmission angle then compare the results with those from the experiment. Examples of such a comparison are shown in Figures 3 to 5.

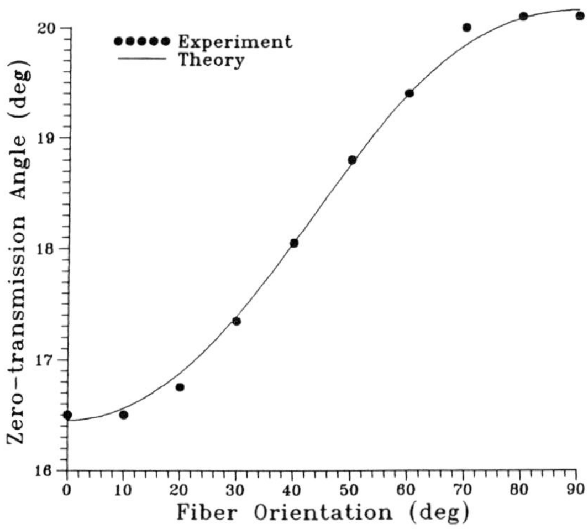

Fig. 5. Transmission zeros measured and calculated at ultrasonic frequency 3 MHz. The discrete points represent experimental data and the solid line calculations using the reconstructed elastic constants.

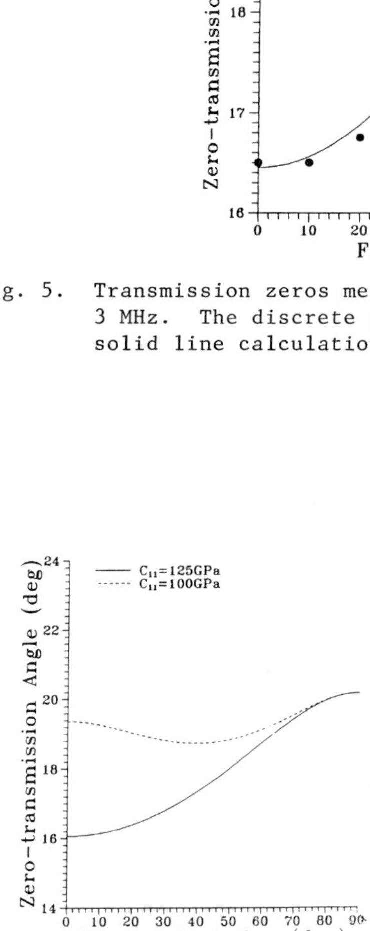

Fig. 6. Polar dependence of the zero transmission angle for two values of elastic constant C_{11}.

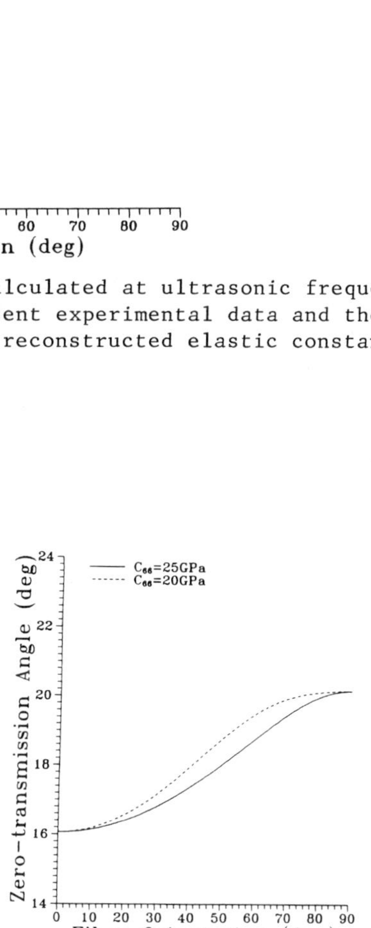

Fig. 7. Polar dependence of the zero transmission angle for two values of elastic constant C_{66}.

Table 1. Elastic constants of fiber reinforced SiC/Al composite. The volume fraction of fiber (SiC) is 4%. The unit for the elastic constants shown below is GPa.

	C_{11}	C_{22}	C_{66}	C_{12}	C_{23}	C_{13}
Experiment	121	108	20.9	65.0	69.6	61.5
Nielsen [6]	125	115	28.4	58.1	58.8	58.1
Christensen [7]	123	115	28.3	56.0	58.8	56.0

DISCUSSION

As previously discussed, a fundamental question for composite laminates is the validity of using effective elastic moduli of a composite with an infinite number of fiber rows to estimate the elastic moduli of composites with only one or a few rows of fibers. This question has been partially addressed previously [5], where the authors numerically calculated the Young's modulus in the direction transverse to the fiber direction. They demonstrated that the difference is about 10% and depends on the properties of the fibers and the matrix.

In this work we compare the measured set of effective elastic constants with those calculated from models for a bulk composite with an infinite number of fiber rows. Such a comparison is shown in Table 1, where the models of Nielsen [6] and Christensen [7] are used. As can be seen in the table, the measured and calculated effective elastic constants are generally in agreement. The agreement is especially good for C_{11}. The elastic constants estimated from the models, such as C_{22}, C_{12}, and C_{13} are within 12% of the experimental values. This deviation is close to that predicted [5] as discussed above. The difference for the shear modulus C_{66} is about 20%. It is reasonable to suggest that the effect of the free surface on the shear stress distribution in the 1-2 plane will play a greater role.

To measure the elastic constant C_{33} we performed an independent measurement using a normally incident ultrasonic beam and measuring the plate thickness resonance frequency. The resonance at thickness equal to a half wavelength occurred at 12.1 MHz. This gives a value of the elastic constant C_{33} of 76.6 GPa. This is significantly below the estimated value, 115 GPa. It is clear that such a measurement cannot be used for this composite, since as one can see from Fig. 2(b) the distance between fibers is greater than the plate thickness (therefore greater than half the wavelength) and acoustic measurement in this case does not give averaged properties.

Finally, one should realize that the one row of fibers is located in the central plane of a composite plane which is a neutral plane under flexural vibrations. For this reason, in the case of flexural vibrations the measured effective Young's modulus will be less influenced by the fibers than the average cross-sectional Young's modulus. In contrast, in dilatational vibration at low frequencies the measured effective elastic properties will depend on the average cross-sectional properties of the composite plate. In our method of measurement the zero-transmission angle depends mostly on the dilatation wave characteristic function since it varies sharply at zero transmission angle and determines the zero of equation (2) (see Ref. [1]). Therefore, using plate mode antiresonance the average properties across the plate can be accurately determined. In addition, since the plate mode antiresonance technique is insensitive to thickness variations [1], the accuracy of the determination of elastic constants is not affected by the sample surface roughness.

CONCLUSION

Interactions between symmetric, antisymmetric, and quasi-SH guided waves result in the occurrence of transmission zeros, which are not affected by fluid loading. The zero-transmission angles depend on the elastic constants of the materials. In this work, a set of elastic constants for a SiC/Al composite has been reconstructed from the measured zero transmission angles. It was found experimentally that micromechanical theories applicable for estimation of effective elastic constants of bulk composite materials can give very good results for estimation of C_{11} ('1' is the fiber direction). C_{22}, C_{12}, C_{23}, and C_{13} have been found to be estimated within 12%. The effect of free surface was most important for the shear modulus in the 1-2 plane, C_{66}. Since the distance between the fibers in this material is greater than the lamina thickness we were unable to recover C_{33} (axis 3 is perpendicular to the sample surface).

ACKNOWLEGMENTS

The SiC/Al composite sample was obtained from Alcoa Technical Center, Alcoa Center, PA 15060.

REFERENCES

1. W. Wang and S. I. Rokhlin, in Quantitative Nondestructive Evaluation, D. O. Thompson and D. E. Chimenti eds., Vol. 9B, Plenum Press, N.Y., 1629-1636 (1990).
2. S. I. Rokhlin, C. Y. Wu, and L. Wang, in Quantitative Nondestructive Evaluation, D. O. Thompson and D. E. Chimenti eds., Vol. 9B, Plenum Press, N.Y., 1403-1410 (1990).
3. S. I. Rokhlin and W. Wang, in Elastic Waves and Ultrasonic Nondestructive Evaluation, S. K. Datta et al. eds., Elsevier Science Publishing Co., N.Y., 249-254 (1990).
4. A. H. Nayfeh and D. E. Chimenti, J. Appl. Mech., 55, 863-870 (1988).
5. L. E. Hulbert and E. F. Rybicki, J. Composite Materials, 5, 164-175 (1971).
6. L. F. Nielsen, Material Science and Engineering, 52, 39-62 (1982).
7. R. M. Christensen, "Mechanics of Composite Materials", John Willy & Sons, 73-89 (1979).

After the conversion the visible light image is magnified by the macro lens for a 35mm still camera and then the image is focused on to the CCD detector which finally converts the light image to electric signals. The magnification factor is approximately 7. The CCD contains 576 x 394 pixels and the pixel size of the CCD is 23mm. However we had to reduce the pixel number down to 256 because of the limitation of memory size of the PC used. We have merged neighboring two pixels together to one. Therefore the actual pixel size at the object position 46 ∏ 7 or approximately 7mm. The CCD is operated in the accumulation mode by cooling down to -45°C for the reduction of the thermal noise. The accumulation capability provides the high signal to noise ratio that is indispensable to reconstruct cross sections in good quality. The image data are finally converted into 12 bits digital data by the camera and the digital data are stored in a magneto-optical disk memory through the PC. The prism mirror is inserted to keep the CCD detector off from the X-ray axis and to protect the detector against the hard X-ray irradiation. The CT measurement is controlled by the PC which conducts the acquisition of the image data at different angles by synchronizing the rotation of the object step by step. The tomographs are reconstructed from the stored data by running a standard filtered liner back-projection software[8] on the PC and the results are shown on its display.

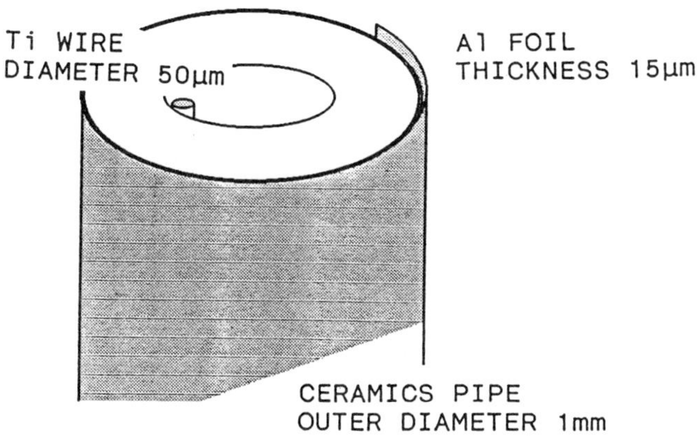

Figure 2. Schematic illustration of the test object.

3. RESULTS AND DISCUSSION

A porous ceramics pipe of 1mm in outer diameter, and about 0.5mm inside diameter, containing a titanium wire of 50μm in diameter and also wrapped twice by a piece of aluminum foil of 15μm thickness (Fig. 2) has been selected for the test object in order to examine the performance of the present system. An example of a tomograph obtained is shown in Fig. 3. The rotating anode was operated at the power of 40kV x 30mA for the molybdenum target. The shadow images were taken at 200 different angles over 180° by exposing 10s for each direction. The fluorescent screen No. 4, which was fabricated from zinc oxide by the standard sedimentation technique and has the average thickness of about 15μm, was used. The reconstructed image has fairly good quality. We can roughly distinguish the titanium wire, the ceramic pipe, and the peeled portion of the wrapped foil. However it is hard to see any pores in ceramics region although we observed many of them in it by a scanning electron microscope.

Then we tried to change the fluorescent screen to No. 3 which was also fabricated through the same process but has the thinner thickness of approximately 7μm. Figure 4 b) shows the result obtained for the same object in the same operating condition except the exposure time of 20s because of the decreased conversion efficiency of the screen. Apparently the spacial resolution is greatly improved. Some large pores are seen in the ceramics pipe as white dots and also some detail of the aluminum foil, i.e., whether the foil is wrapped twice or once and whether the foil is stuck tightly or loosely, can be observed. The smallest white dots correspond to one to three pixels which lead to the spacial resolution of roughly 10 to 20μm.

Those three tomographs shown in Fig. 4 a), b) and c), which were obtained simultaneously from the measurement during single CT rotation, show different cross sections at distinct vertical levels as indicated in Fig. 4 d). In those three pictures white dots are seen in different positions, and also the situations of wrapping are distance from each other. In order to achieve three-dimensional tomography we can merely increase reconstruction planes at every vertical level from the data acquired during single CT rotation. However it has taken approximately 20 minutes to reconstruct a cross section of 256 x 256 pixels. Thus, it is not practical to carry out all the reconstruction for three-dimensional tomography by using the present system. Some further improvements have to be made for the reconstruction software or for the computing power itself. The fast Fourier transformation (FFT) is the favorable technique for the reconstruction in the Fourier space instead of the standard filtered line back-projection in real space. Subsequently hardware specially designed for FFT processing may accelerate the computing speed much more.

As seen in Figs. 3, 4 a), b), and c) there were noisy rings around the center of CT rotation. All the shadow image data were calibrated by subtracting the dark level, which was taken without X-ray illumination, and by normalizing the white level, which was taken with x-rays but without the object. Thus the inhomogeneity of CCD sensitivity, that of X-ray illumination and that of conversion efficiency to the visible light was eliminated. Another possible origin of the ring noise is the slight nonlinearity which was reported recently.[9] However we could not find such nonlinearity in the lower intensity region where the present work was done. As another possibility the phosphor thickness of 7 to 15μm and the particle size of roughly 1μm were confirmed by observing them both by an optical

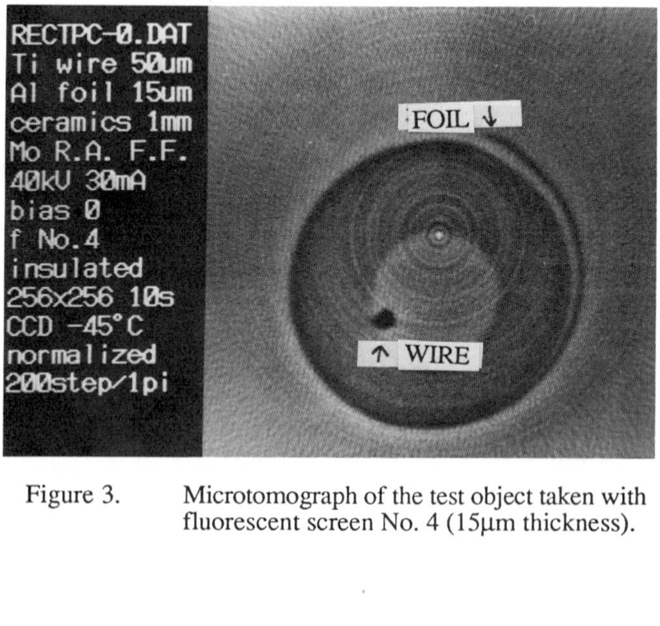

Figure 3. Microtomograph of the test object taken with fluorescent screen No. 4 (15μm thickness).

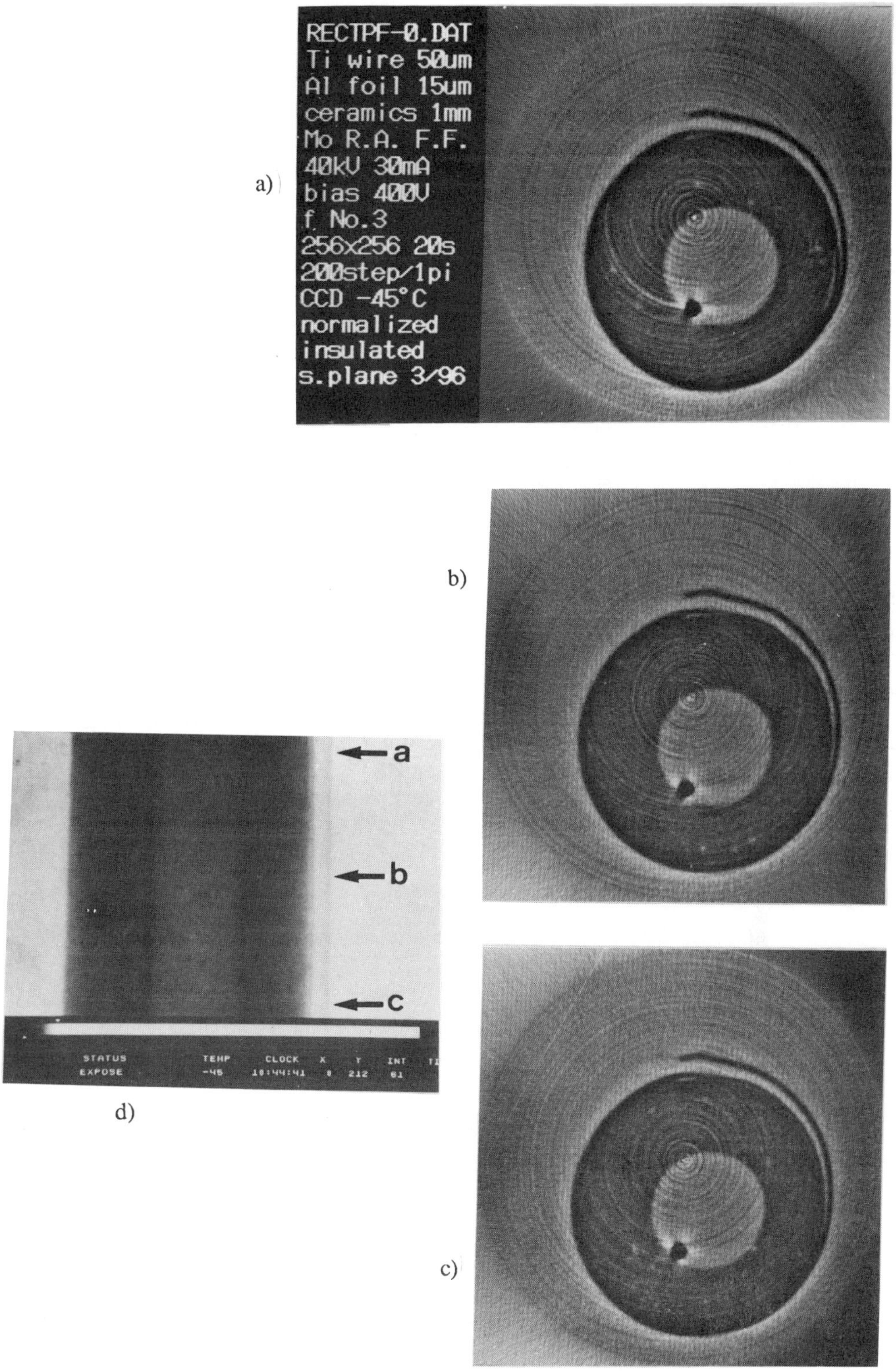

Figure 4. Microtomographs a), b), c) of different cross sections as shown in microradiograph d) of the test object. All were taken with the fluorescent screen No. 3 (7μm thickness).

microscope and by scanning electron microscope. This means that only 10 to 20 phosphor particles are piled up and that spiked defects, which can not be restored by the subtraction and normalization described above, may exist. Therefore some improvements have to be made for the fluorescent screens in their homogeneity and in efficiency by fabricating specific column structures which prevent the diffusion of visible light in the phosphor but still keep the thickness adequate.[10]

4. CONCLUSIONS

It has been demonstrated that the microtomography system described here provides high resolution of 10 to 20μm in cross section planes for relatively small objects of about 1mm. Also it provides the capability to take multi-sliced tomographs which enables three-dimensional tomography in practical uses. The limitating factor, which rules the lateral resolution, was found not to be the geometrical blurring in the X-ray optics but the diffusion of the visible light in the fluorescent screen. In order to improve the resolution thinner screens have to be fabricated by coating with finer phosphor grains of sub-micron size. However from the point of conversion efficiency the heavier fluorescent materials, which have larger stopping power for X-rays, or the fine column structures of certain thickness have to be developed.

ACKNOWLEDGEMENTS

The authors are grateful to Prof. J.S. Elliot of the London Hospital Medical College and Dr. W. Graeff of the HASYLAB for their discussions especially about the noise rings. Thanks are due to Dr. K. Sakurai for his support on the operation of the rotating target and for his advice.

REFERENCES

1. G.N. Hounsfield, Brit. J. Radiol. 46, 1016 (1973).
2. D.K. Bowen, L.S. Elliott, S.R. Stock and S.D. Dover, Proc. SPIE-Int. Soc. Opt. Eng. 94, 691 (1986).
3. J.H. Kinney, Q.C. Johnson, R.A. Saroyan, M.C. Nichols, U. Bonse, R. Nusshardt and R. Pahl, Rev. Sci. Instrum. 59, 196 (1988).
4. K. Sakamoto, Y. Suzuki, T. Hirano and K. Usami, Jpn. J. Appl. Phys. 27, 127 (1988).
5. B.P. Flannery, H.W. Deckman, W.G. Roberge and K.L. D'Amico, Science 237, 1439 (1987).
6. S. Aoki, M. Oshiba and Y. Kagoshima, Jpn. J. Appl. Phys. 27, 1784 (1988).
7. J.E. Bateman, G.R. Moss, P. Rockett and S. Webb, Nucl. Instrum Methods A273, 767 (1988).
8. L.A. Shepp and B.F. Logan, IEEE Trans. Nucl. Sci. NS-21, 21 (1974).
9. J.H. Kinney, Q.C. Johnson, R.A. Saroyan, M.C. Nichols, U. Bonse, R. Nusshardt and R. Pahl, Rev. Sci. Instrum 60, 2471 (1989).
10. H.W. Deckman, J.H. Dunsmuir, and S.M. Gruner, J. Vac. Sci. Technol. B7, 1832 (1989).

SENSOR DEVELOPMENT AND PROCESS CONTROL IN THE FIELD OF POLYMER COMPOUNDING

H.-G. Fritz and S. Ultsch

Institut fuer Kunststofftechnologie (IKT)
University of Stuttgart

INTRODUCTION

The growing number of engineering plastics and polymer blends as well as the increasing efficiency of modern compounding units make an on-line characterization of the resulting polymer quality necessary, whereby the trend is directed towards sensor systems fixed to processing units. This continuous characterization of the material is a point of considerable progress when compared with discontinuous sample analysis in the laboratory because of the delay-time intervals hereby involved. At the same time, a complete documentation of product quality is possible, which becomes essential under product liability aspects.

Since many relevant product characteristics cannot be measured directly, one must correlate them with processing and material parameters which can be detected using in- or on-line techniques. The heterogeneity of the quality characteristics to be assessed hereby necessitates the simultaneous use of two or more sensor systems.

As shown in our studies, a comprehensive characterization of polymer quality may be achieved using a combination of the following sensor equipment.

On-line Rheometer Systems for the on-line measurement of rheological material parameters to obtain information about changes in the macromecular structural parameters, the concentration of additives and/or the degree of dispersion.

On-line FTIR spectrometers for analysis of the polymer composition (copolymers, type and concentration of additives, residual monomer content) and for determining molecular structure modifications in reactive polymer compounding.

In-line density measurement systems for the monitoring of filler and pigment concentration.

On-line pressure filter test sensors which can be used together with an on-line rheometer to obtain information concerning the actual degree of dispersion.

Nondestructive Characterization of Materials IV
Edited by C.O. Ruud *et al.*, Plenum Press, New York, 1991

On-line image analysis systems for the assessment of contamination particles, content of gel particles ('fish-eyes') and the structure of fillers and reinforcing materials.

On-line color analysis systems for the detection of the colour value and the colour constancy.

This contribution will focus on FTIR-spectroscopy in combination with on-line rheometry and density measuring systems. Design, operating principles and potentials of these sensors will be described in the following for compounding problems as encountered in practice.

INFRARED SPECTROSCOPY ANALYSIS

FTIR spectroscopy allows the assessment of copolymer composition, the detection of type and content of additives (thermo-stabilizers, coupling molecules) and the control of synthesis processes. In addition, molecular structure modifications and changes in chemical bond ratios can be precisely detected[1-3]. An FTIR spectrometer recently introduced by Automatik Maschinenbau GmbH, Grossostheim, Germany, can be coupled to the compounding extruder and operated on-line (Fig. 1).

A bypass melt stream is fed into the measuring gap of the spectrometer by means of a gear pump. The distance between the IR-transparent windows made of zinc selenide can be manually adjusted between 30 and 2000 μm; values between 2000 and 300 μm have been proven as sufficient for measurement of the harmonics and values between 150 and 30 μm for measurement of strong absorption bands. A FTIR spectrometer was used which works under the refraction principle (→ refraction interferometer) in the middle infra-red range (wavelength 2,3 - 25 μm). The actual IR measurement, which takes about 4 s with a resolution of 3 cm^{-1}, is very fast when compared with delay and selfcleaning times of the measuring system. The efficiency of this sensor in combination with an on-line rheometer will be demonstrated using the graft polymerization of polyethylene with organosilanes. This process, which is a typical example for reactive polymer compounding, can be described as follows (Fig. 2).

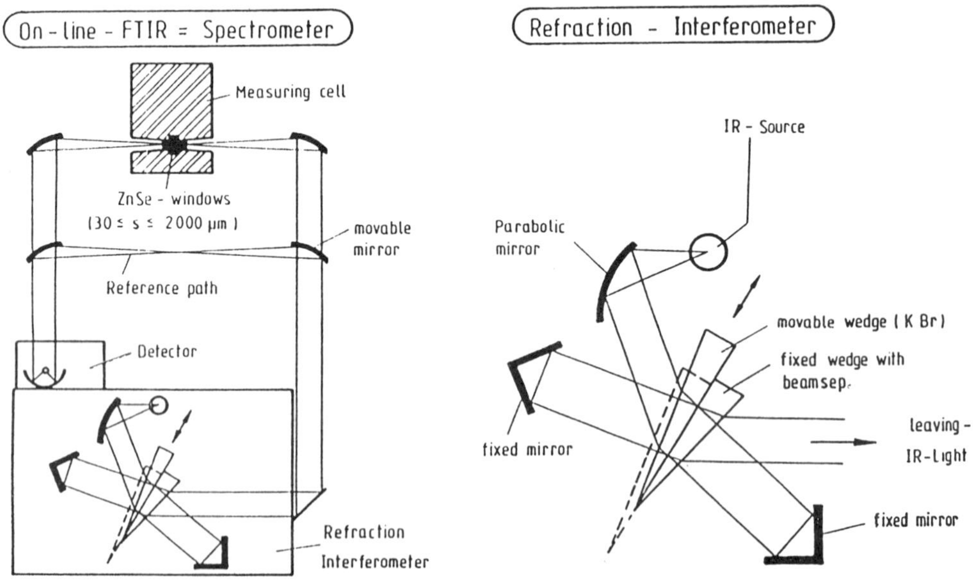

Fig. 1. Optical concept of an On-line-FTIR-Spectrometer

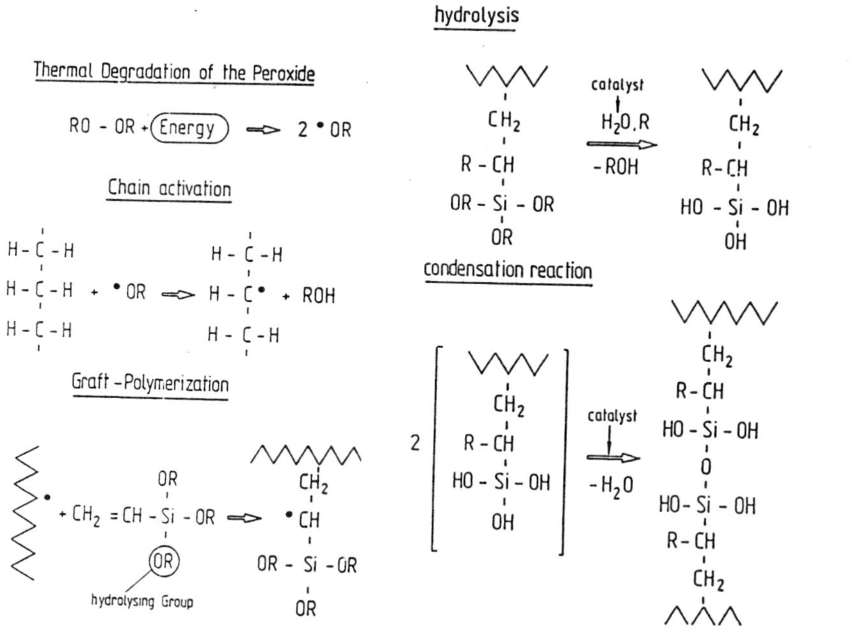

Fig. 2. Silane grafting and crosslinking of PE

By activation of PE-macromolecules with peroxide radicals, a grafting of vinyltrimethoxysilane (VTMOS) molecules onto the resultant chain radicals is performed. The coupling of these voluminous units is obtained by breaking-up of the vinyl double bond in the VTMOS molecule. In subsequent reactions the grafted hydrolizable groups are degraded to silanols under the action of water molecules, which exposed to a catalytically initiated condensation process, permits a linkage of the PE chains to a spatial network structure via Si-O-Si bridges. Only small amounts of additives are required for this grafting process. Peroxide additions of only $0{,}01 \leq c_p \leq 0{,}1$ % by weight and VTMOS additions of only $0{,}5 \leq c_s \leq 2.5$ % by weight are sufficient to yield crosslinking levels between 40 and 85 %. Modifications in macromolecular structure initiated by the grafting reaction and detectable by FTIR spectroscopy, cause significant changes in the rheological behaviour of PE as well as in the thermomechanical properties of the crosslinked polymer. The application of this sensor combination seems to be highly profitable because the delay-time between the grafting reaction and the fixing of the final product quality is more than six hours.

Fig. 3 shows the IR spectrograms taken of a PE-LLD melt to which VTMOS without peroxide has been added and thus only mixed physically into the polymer. A linear increase in the specific VTMOS bands is clearly detectable as the silane addition is increased: The most strongly pronounced IR peak being caused by the γ(Si)-O-C oscillation of the Si-O-C groups (1080 cm^{-1}). Smaller displacements result from the $\delta(O-)CH_3$ resonances (1190 cm^{-1}) and from the ($CH=CH_2$) inherent oscillations (965 and 1009 cm^{-1}) as well as from two further VTMOS group oscillations $\nu(Si-O_3)$. An on-line quantification of VTMOS proportions is possible without problems. Thereby it is advantageous to correlate the absorption levels of the VTMOS bands with the PE-peak at the wave number λ = 1375 cm^{-1}, which is not influenced by VTMOS addition.

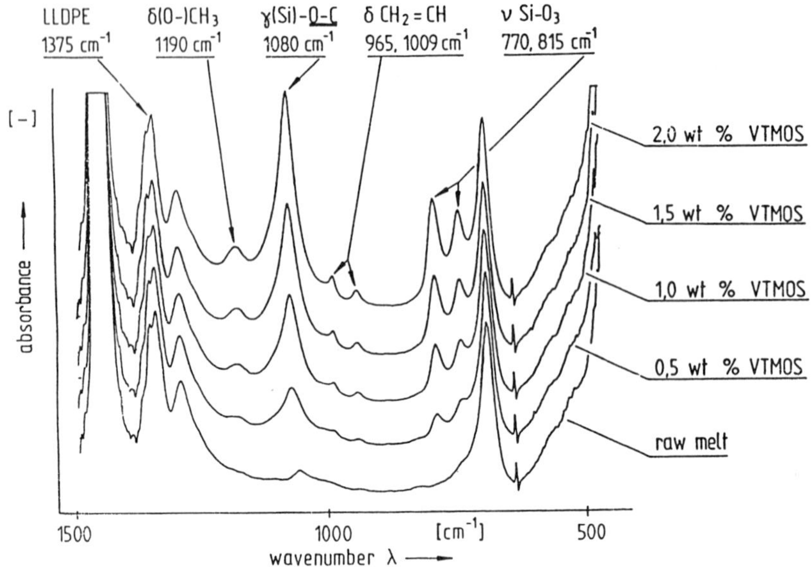

Fig. 3. IR spectra of different graft polymers: Variation of VTMOS content

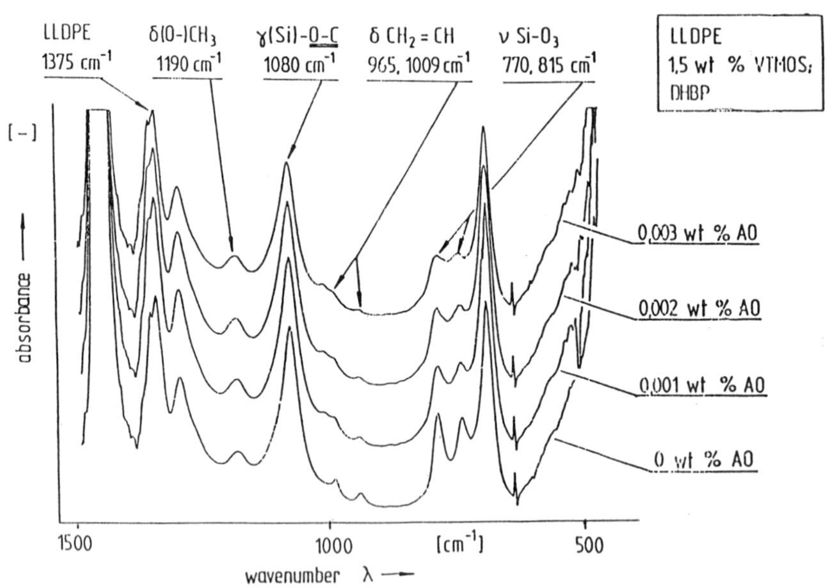

Fig. 4. IR spectra of different graft polymers: Variation of AO content

In a second step samples were prepared including a constant VTMOS content but variable peroxide additions (Fig. 4). The comparison of the individual spectrums reveals large absorption differences which are dependent on the peroxide content and thus on the grafting level, and which are particulary obvious for the $\nu(Si-O_3)$ oscillations at 770 and 815 cm^{-1} and the $\delta(CH = CH_2)$ bands. Whereas the oscillation bands of the vinyl coupling group ($CH_1 = CH_2$) disappear due to the dissolution of the double bond during the grafting process as could be expected, the VTMOS-linkage to the PE chains has a decreasing effect on the oscillations as larger as the distance between the relevant atomic groups and the coupling site is.

The extremely small peroxide contents can be detected by FTIR spectroscopy only under enormous problems. However, rheological data show in a significant manner the influence of variable peroxide additions on the grafting process. Fig. 5 shows the complex viscosity function $\eta^*(\omega)$, on-line measured with an oscillating Couette system, for three specimens prepared with a constant VTMOS content but with different additions of active oxygen (AO). The viscosity values increase as the AO content grows, resulting from linkage reactions. If activated chain segments are inadequately saturated by VTMOS, these reactions lead to chain elongation and partially branched molecule structures which exert a significant influence on the deformation behaviour.

However, when increasing the VTMOS addition at constant AO-content, the viscosity values are decreased (Fig. 6). An explanation for this effect may be seen in the fact that activated chain sites being increasingly saturated with growing VTMOS concentration by the grafting of organosilane molecules, with the result of a decreasing number of chain extentions and branchings. But due to the grafting of these molecules storage and loss moduli generally shift to higher values than determined for the modified polymer.

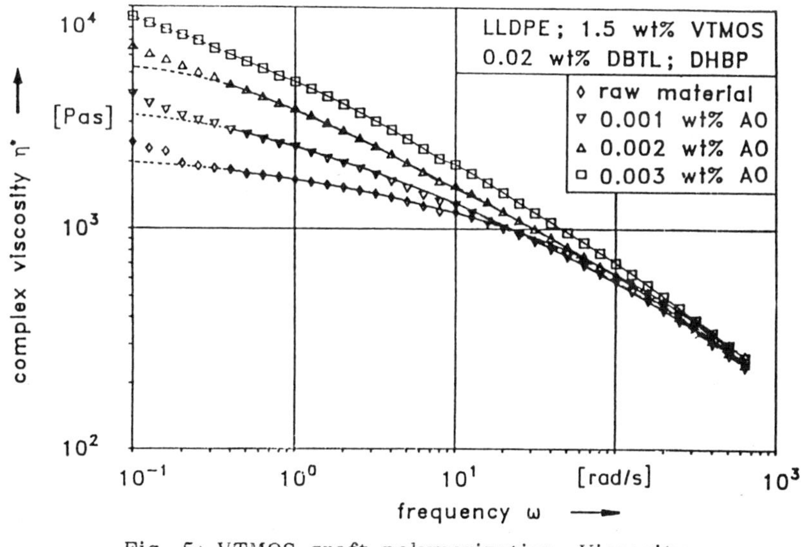

Fig. 5: VTMOS graft polymerization: Viscosity variation as function of AO content

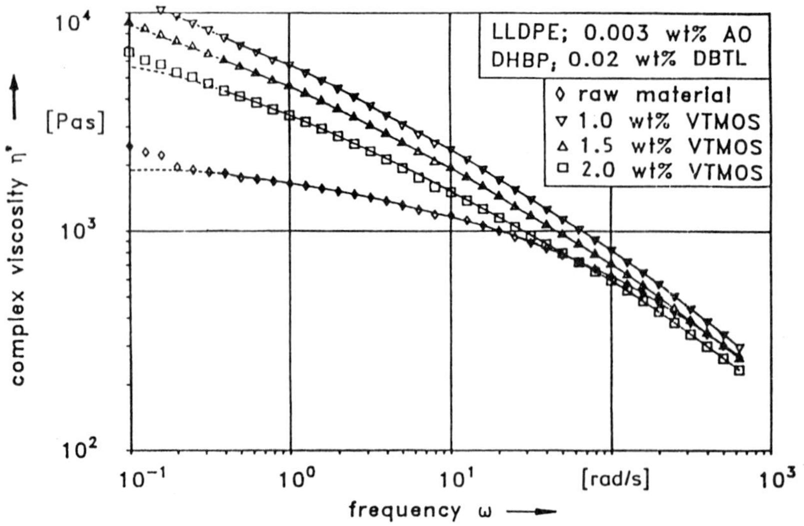

Fig. 6. VTMOS graft polymerization: Viscosity increase as function of VTMOS content

When comparing the potential of the presented on-line measuring technologies for material and quality characterization, it is obvious that totally different measuring methods can complement each other ideally in reactive polymer compounding. While the on-line rheometer particularly enables the additions of radical donor and the melt processability to be quantified in a favourable way, FTIR spectroscopy provides a good solution to the problem of quantitative detection of the organosilane concentration in polymers and of grafting reactions.

RADIOMETRIC DENSITY MEASUREMENT IN THE FIELD OF POLYMER COMPOUNDING

In the production of polymer blends and polymers including fillers or fibre reinforcements as well as in the preparation of pigment master-batches the material density or rather the effective composition of such mixtures is a special point of interest. An in-line monitoring of density values and mass fractions of material compounds allows the design of control strategies to guarantee a long-term consistent composition of such products, independent of external disturbing effects. For these tasks radiometric density measuring systems are an especially suitable means.

Measuring Principles, Instrumentation

Radiometric density measurement is based on the effect, that by the irradiation of materials a portion of the emission is absorbed or deflected into different directions by means of compton scattering. The radiation weakening is determined by the thickness and density of the irratiated material layer. If the layer thickness is kept constant, measurement of the radiation intensity using a scintillation counter allows a direct quantification of material density resp. mixture composition.

Fig. 7 shows a cross-section diagram of our newly developed density measuring instrument. For a detailled description see ref.[4]. Between the barrel of a twin screw extruder and the pelletizing die a heatable barrel section with a preferably rectangular flow channel is installed. A vertically movable carriage carries the shielded radiation source and the scintillation counter. In the irradiated area, both mean pressure p and a representative melt temperature T are measured correspondingly. Ir-192 and Cs-137 are used as radiation sources. An advantage of these emittors is, that the mass absorption coefficient μ as function of the atomic number Z of the elements included in the irradiated melt. The most important evaluation equations are shown in Fig. 8. From continuously measured pressure and temperature values the density of the polymer matrix is calculated on a coupled computer system using the Spencer-Gilmore equation. Together with the value of radiation intensity measured for the mixture I_{st_G}, the timedependent extrapolated radiation intensity for the unfilled channel $I_{st,\Delta t}$ and the previously found values ρ_F, ρ_P, μ_F and μ_P the instantaneous values of the volumetric (k_F, k_P) and weight fractions of the mixtures (a,b) can be continuously determined.

Experimental Results

Experimental studies were performed on a polypropylene material Hostalen for a constant screw speed and extruded through a die with a

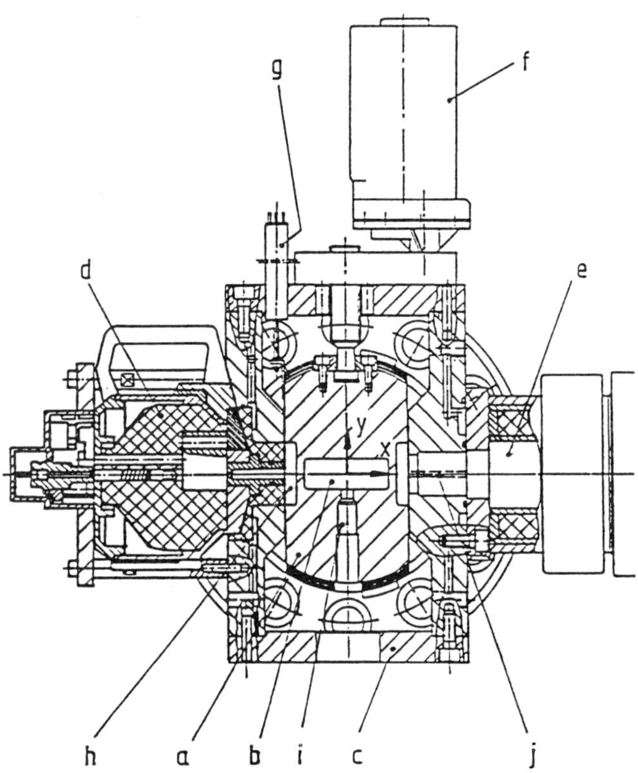

a ... cylinder
b ... melt channel
c ... carriage
d ... radiation source
e ... scint. counter
f ... stepper motor
g ... stroke sensor
h ... recess
i ... p,T-transducer
j ... collimator

Fig. 7. Schematic Diagram of a radiometric density measuring sensor

Intensity of Radiation
$$I = I_0 \cdot \exp(-\mu \cdot \rho \cdot L) \quad (1)$$

Intensity measured with the empty flow channel
$$I_{St,0} = I_0 \cdot \exp(-\mu_{St} \cdot \rho_{St} \cdot L_{St}) \quad (2)$$

Updating
$$I_{St,\Delta t} = I_{St,0} \cdot 2^{-(\Delta t / t_{1/2})} \quad (3)$$

Flow channel filled with melt
$$I_{St,P} = I_0 \cdot \exp[-(\mu_{St} \cdot \rho_{St} \cdot L_{St} + \mu_P \cdot \rho_P \cdot L_P)] \quad (4)$$

Homogeneous Polymer
$$\ln\left(\frac{I_{St,0}}{I_{St,P}}\right) = \mu_P \cdot \rho_P \cdot L_P \ \text{bzw.} \quad (5a)$$
$$\ln\left(\frac{I_{St,\Delta t}}{I_{St,P}}\right) = \mu_P \cdot \rho_P \cdot L_P \quad (5b)$$

Filled and reinforced Polymers

Measured radiation intensity
$$I_{St,G} = I_{St,\Delta t} \cdot \exp[-(\mu_P \cdot \rho_P \cdot L_P + \mu_F \cdot \rho_F \cdot L_F)] \quad (6)$$

$$L_F + L_P = L_K \ ; \quad L_P = k_P \cdot L_K \ ; \quad L_F = k_F \cdot L_K \quad (7)$$

$$I_{St,G} = I_{St,\Delta t} \cdot \exp[-L_K \cdot (\mu_P \cdot \rho_P \cdot k_P + \mu \cdot \rho_F \cdot k_F)] \quad (8)$$

k_P, k_F = volumetric mixture percentages

$$\underbrace{a \cdot \frac{\rho_G}{\rho_P}}_{k_P} + \underbrace{b \cdot \frac{\rho_G}{\rho_F}}_{k_F} = 1 \quad (9)$$

Density of the mixture
$$\rho_G = \frac{\rho_P \cdot \rho_F}{(a \cdot \rho_F + b \cdot \rho_P)} \quad (10)$$

Measured variables	$I_{St,0}$, $I_{St,P}$, $I_{St,G}$, Δt , $\bar{p}$, $\bar{T}$
Input data	W , b^* , p^* , $t_{1/2}$, L_K , μ_P , μ_F

Spencer/Gilmore-equation
$$\rho(\bar{p},\bar{T}) = \frac{W(\bar{p} + p^*)}{[W \cdot b^* \cdot (\bar{p} + p^*) + R \cdot \bar{T}]} \quad (11)$$

Evaluation Equations

$$k_F = \frac{1}{[\rho_F \cdot \mu_F - \rho_P \cdot \mu_P]} \cdot \left[\frac{1}{L_K} \cdot \ln\left(\frac{I_{St,\Delta t}}{I_{St,G}}\right) - \rho_P \cdot \mu_P\right] \quad (12)$$

$$k_P = 1 - k_F \quad (13)$$

$$a_{ist} = \frac{k_P \cdot \rho_P}{[k_P(\rho_F - \rho_P) - \rho_F]} \quad (14)$$

$$b_{ist} = 1 - a_{ist} \quad (15)$$

Fig. 8. Evaluation equations of radiometric density measurement variable die resistance. The radiation intensity $I_{st,p}$ was measured together with the pressure and melt temperature values at the irradiated area. In Fig. 9 the measured radiation intensities $I_{st,p}$ are presented as function of pressure and compared with the calculated data of the actual pressure and temperature dependent density values. These measurement results are in good agreement with the theoretical results calculated using the thermal equation of state of the Spencer-Gilmore type..

Should variations in the filler dosage be detected during compounding or filler and pigment content be measured in-line, the volumetric and thus effective weight concentrations must be determined in an intermediate step from the measured values and the material parameters. In Fig. 10 the results are shown for a compounding process whereby different percentages of talcum were added to PP. One can see, that the amount of filler measured using the radiometric method agrees very well with values resulting from pyrolytic experiments.

STRATEGIES FOR PROCESS CONTROL

In the following it will be shown how the sensor systems described above can be used as principle components integrated in process control concepts.

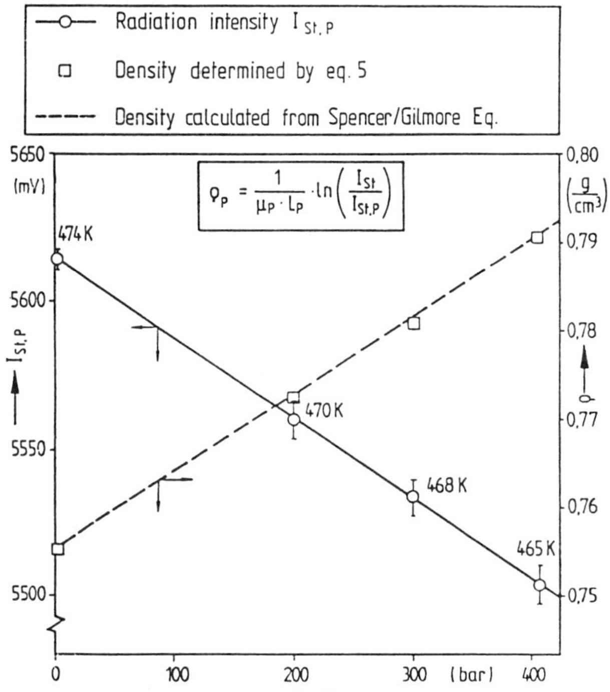

Fig. 9. Measured density values function of pressure

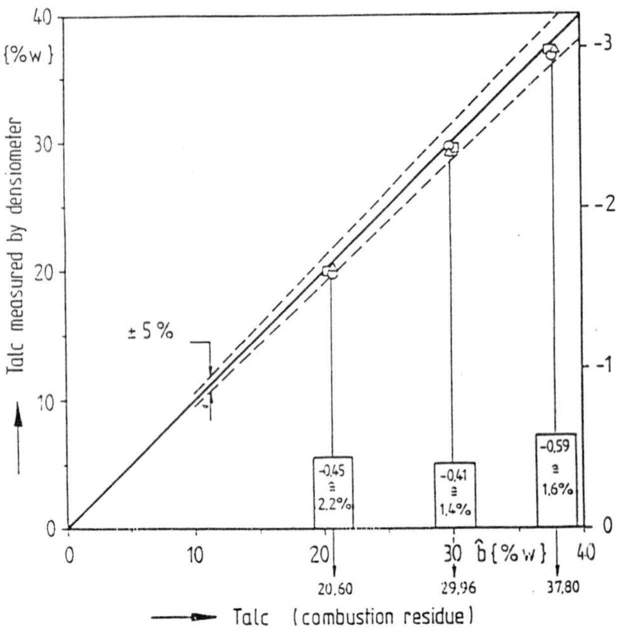

Fig. 10. Talc content of filled PP
detected by a densiometer

Fig. 11 shows a schematic diagram of a control concept with which polymer masterbatches with a constant composition can be produced in long-term production intervals. The actual concentration levels measured in-line using the density measurement sensor can be compared with setpoint values by means of an integrated computer which includes an implimented digital PID controller. Deviations of actual and setpoint values are adjusted in the dosing system, so that the setpoint value can be realized within tight boundaries. In this case, very inexpensive volumetric dosing systems may be used. At the same time, a continuous documentation of the resulting composition of the mixture is possible.

Fig 12 shows a similar process control concept for the controlled peroxide degradation of polypropylene. Hereby, the melt index or the molecular weight are given as setpoint values. A comparison with actual measuring values is performed by using an on-line rheometer system and an integrated data processing unit. The dosage of matrix material and peroxide is controlled in such a way, as that the resulting total output rate remains constant. The following diagram (Fig. 12) shows controlled processing conditions as an example, whereby the peroxide content in the mastermix was changed. It can be seen, that the time needed until the "disturbance" is eliminated is relatively high. This is due to the long machine- and sensor dependent response times, which lead, in connection with the usual time constants of the process, to large controlability ratios. Thus, external disturbances show to a high percentage a strong effect on the process in these relatively simple control concepts.

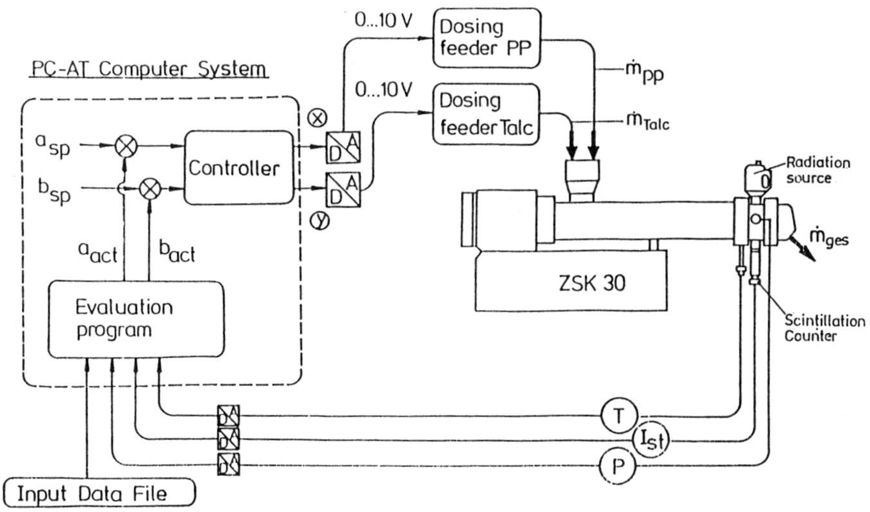

Fig. 11. Closed-loop process control concept for filling and compounding of PP

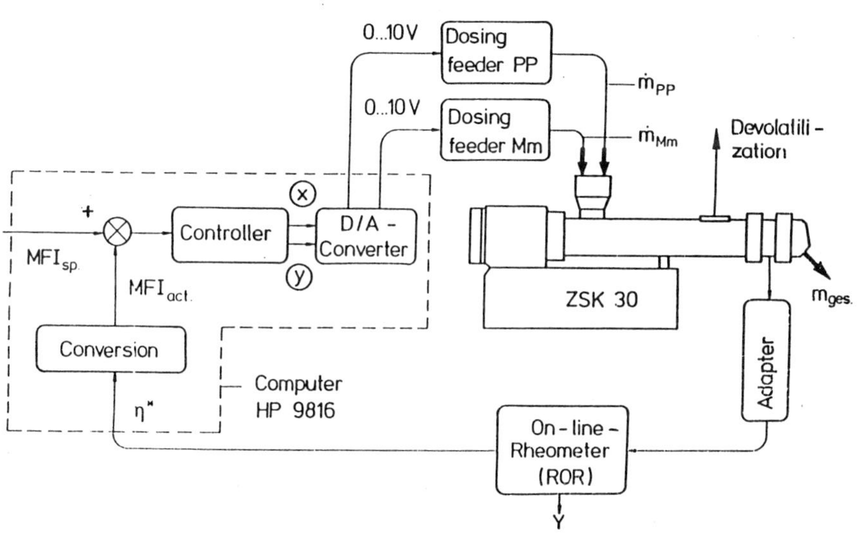

Fig. 12. Closed-loop control concept for peroxide initiated degradation of PP

Such weaknesses can be eliminated by the use of model-supported measuring concepts. Hereby a simplified process model is simulated in real time, whereby additional parameters (temperature, pressure) measured in-line are included. The model is then updated via parameters which can be correlated with the product quality, but measured with some delay. An example for such an important analogical value is the local energy dissipation in a twin screw extruder. The assessment of the specific energy conversion in the individual functional zones of a twin screw device can be realized by the determination of the local torsion angles of the screws by introducing a "delay-time measurement" using an inductive position transducer, whereby for the evaluation procedure the geometry and material parameters as well as the screw speed are incorporated. The aim of such process control concepts is to increase the effectiveness of this highly diversified information, which modern on-line sensor systems make available by real time monitoring, in terms of the resulting product quality.

Outlook

Besides the further improvement of the sensor systems presented, colorimetry and image analysis will have to be included in the activities in the future. The providing of any information which serves the definite quantification of the polymer quality is the basic requirement for the development of automation and process control concepts. In this development preference will increasingly be given to the model-supported methods to improve the response and control behaviour of the concepts.

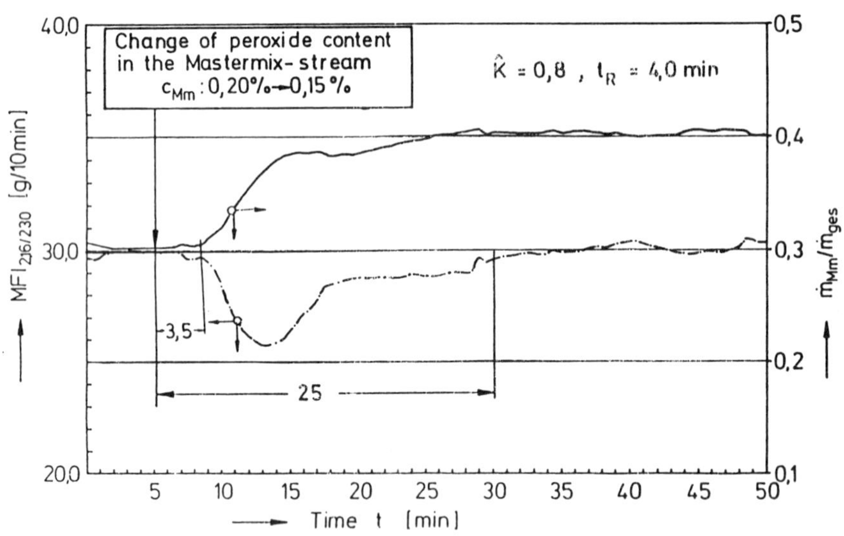

Fig. 13. PP-degradation with a closed-loop control concept (change of peroxide content)

LITERATURE

1. H.-G. Fritz and S. Ultsch, Sensorentwicklung und Automatisierungstendenzen bei der Kunststoffaufbereitung, Kunststoffe 79 (1989) 9, p.785-790.

2. S. Ultsch and H.-G. Fritz, Vernetzung von PE-LLD and PE-VLD mit pfropfpolymerisiertem Vinyltrimethoxylan, Kunststoffe 79 (1989) 10, p.1051-1056.
3. R. Stengler and G. Weis, FTIR-Messungen an Kunststoffschmelzen zur On-line-Qualitätsprüfung, Kunststoffe 79 (1989) 10, p.955-959.
4. H.-G. Fritz and R. Löffel, Radiometrische Dichtemessung in der Kunststoffaufbereitung, Kunststoffe 80 (1990) 2, p.178-182.

THE REQUIREMENTS FOR NONDESTRUCTIVE CHARACTERIZATION IN THE TIRE AND RUBBER INDUSTRY

Joseph W Miller, Jr, Pawan K Handa and Gordon R Schorr

The Goodyear Tire & Rubber Company
Corporate Research Division
Akron, Ohio

INTRODUCTION

The pneumatic tire is a toroidal, high performance polymeric composite, which exhibits characteristics of a flexible-membrane pressure container providing load carrying capability, cushioning, and road handling abilities. It transmits driving and braking torque, cornering forces, and performs all functions essential to locomotion. These features can be attributed to its deformability and damping characteristics. The pneumatic tire is a complex system of interacting composites, each with specific properties for maximum effectiveness.[1] Figure 1 is a schematic of the Goodyear Eagle GT+4 All Season high performance radial passenger tire. Each individual composite in the tire is different from its neighbor in terms of content, properties, shape, and function. Individual composites contain combinations of various synthetic and natural rubbers, reinforcing fillers, steel wire, fabric, sulfur, curatives, and antidegradant materials.[2] Some of these are listed in Table 1. Also, a full spectrum manufacturer, such as The Goodyear Tire & Rubber Co, produces passenger, truck, aircraft,

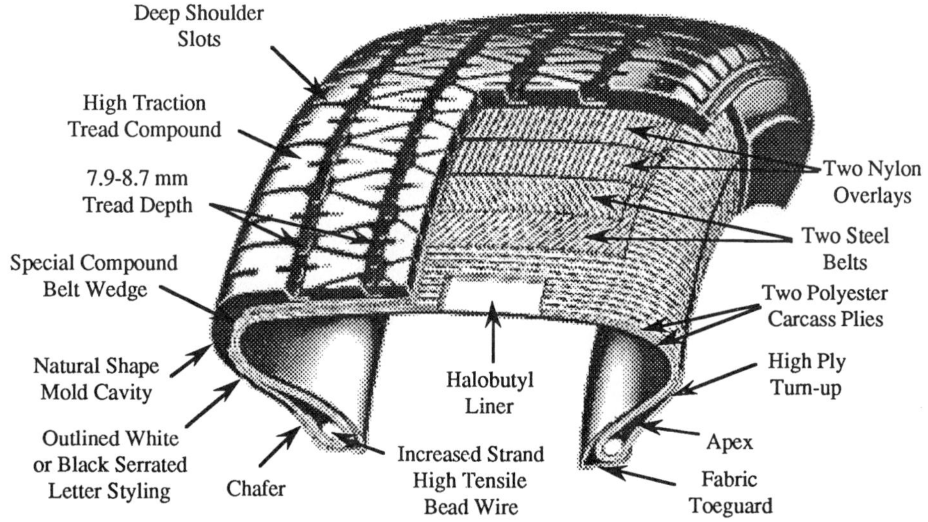

Figure 1 Cross Section of Goodyear's All Season High Performance Passenger Tire, the Eagle GT+4 [1]

Nondestructive Characterization of Materials IV
Edited by C.O. Ruud *et al.*, Plenum Press, New York, 1991

TABLE 1 SOME MATERIALS USED TO MAKE TIRES

• Natural Rubber	• Steel Wire Cord
• Synthetic Rubber	• Zinc Oxide
• Processing Aids	• Stearic Acid
• Organic Accelerators	• Silica
• Carbon Black	• Extender Oil
• Fabric Cord	• Sulfur
	• Stabilizers

earthmover, farm, and race tires. As would be expected, the requirements and specifications for the tire for an F-15 are very different than those for the family sedan. This diversity adds another level of complexity to the process of producing a tire.

This paper addresses the polymer processing aspects of tire manufacturing, and attempts to define some of the NDE requirements for each step in the process.

QUALITY

There are two forces driving quality improvements in the tire industry. The first is ever-increasing performance requirements for all types of tires; eg, better fuel mileage through reduced rolling resistance, better traction, increased wear, higher speed ratings, better durability, etc. These requirements have forced the proliferation of specialty materials for niche applications such as styrene/isoprene/butadiene terpolymer[3] for high performance passenger vehicles. The manufacture of specialty polymers requires a great deal more fundamental process modeling work[4,5] and has implications for better nondestructive characterization techniques for more complex polymeric structures.

The second quality driver is quality itself. Every tire manufactured must be of the highest quality and uniformity, requiring that each individual composite meet the same exacting standards. At the same time, a Total Quality Commitment (TQC) demands that quality must continually improve. The Japanese call this concept Kaizen, or improvement after improvement.[6] This improvement in our processes requires ever improving nondestructive characterization techniques, both at the microscopic and macroscopic level. Control of a process parameter first requires its measurement.[7]

TIRE MANUFACTURING

The complex tire manufacturing process alluded to in the Introduction can be simplified, for the purpose of discussion, into the three step process as described in Figure 2. The manufacture of the toroidal, high performance polymeric composite starts with polymer production. In polymer production, adjustable parameters such as reactor temperature, pressure, mixing, and catalyst composition are utilized to affect polymer structures such as molecular weight distribution, branching, etc. The finished polymer is then mixed and formed, along with many other additives (Table 1), into individual composites that make up the tire. Each of these forming operations involves a different set of adjustable parameters; eg, mixing RPM or power, temperatures, etc, that are used to affect composite dimensions, carbon black dispersion, morphology, etc. The finished, individual composites are assembled into the uncured or green tire and then cured in a tire mold to produce a finished product that has a certain set of final properties; eg, sulfur crosslink density, viscoelasticity, cut growth, etc.

This process has some similarities to other polymeric processes used in the plastics industry, but it is more complex since the final product is composed of many different composites, each requiring mixing and forming of non-Newtonian, viscoelastic polymers and numerous additives. Also, since these composites are all thermosets, a separate cure process is necessary. This process is described in some detail by Bhakuni, et al.[1] Nondestructive characterization requirements are a function of the methodology used to control each of these three types of processes. The state-of-the-art methodology utilized can be called Intelligent Polymer Processing, a phrase used by the National Institute of Standards and Technology.[8]

INTELLIGENT POLYMER PROCESSING

Figure 3 shows a traditional polymer process. Here process control devices receive inputs from various process sensors that measure process variables such as temperature and pressure. Also input are a combination of off-line and on-line measurements that relate to the polymer product properties. These measurements may include molecular weight, microstructure, etc, depending on the unit operation. These product property measurements present two problems: 1) most of these measurements are off-line tests that are very time consuming, introducing inordinate dead time into the process control and eliminating closed loop control; and 2) there is a temptation to measure and use secondary or cured product properties since these are part of the final specification the customer demands. For example, consider the polymer processing involved in polymer production.

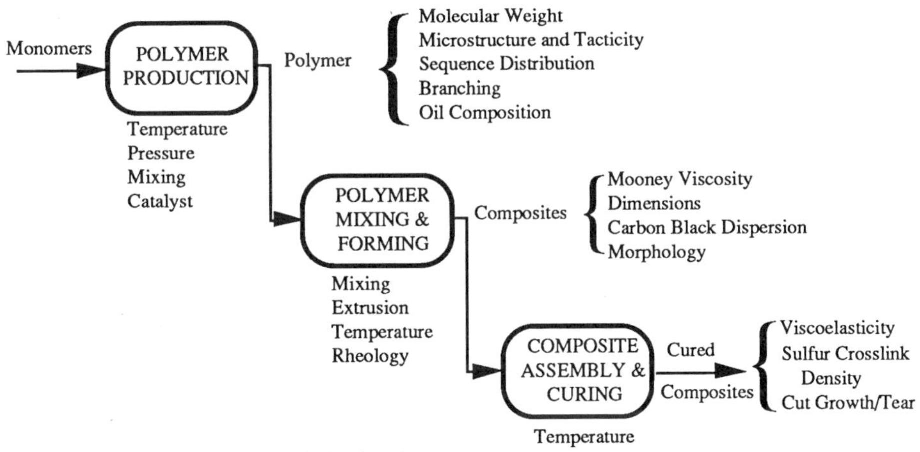

Figure 2 Tire Manufacturing

In producing a polymer, molecular weight is a primary structure that is a function of chemistry and reactor design. The Mooney viscosity is a function of the molecular weight (and other structures), and is important in the downstream polymer mixing and forming operations. The cured viscoelasticity is a critical measurement that is influenced by the molecular weight and molecular weight distribution, and is a critical measurement relating to the physical properties of the tire. Since the molecular weight is usually measured off-line and takes a long time to obtain, Mooney viscosity could be used as an indirect measurement of molecular weight. This methodology has been successfully applied, but if one wants to develop computer-based process controls utilizing process models for dynamic closed loop control, primary structure variables must be measured. Structure/property relations can be used separately from the process control scheme to target primary structures obtained from desired properties. Figure 4 schematically illustrates this simplistic example. By measuring molecular weight distribution on-line with a short response time, along with appropriate process sensors, a process model can be used to control molecular distribution. This assumes an appropriate structure/property relation was used to specify the molecular weight and molecular weight distribution targets to give correct downstream properties.

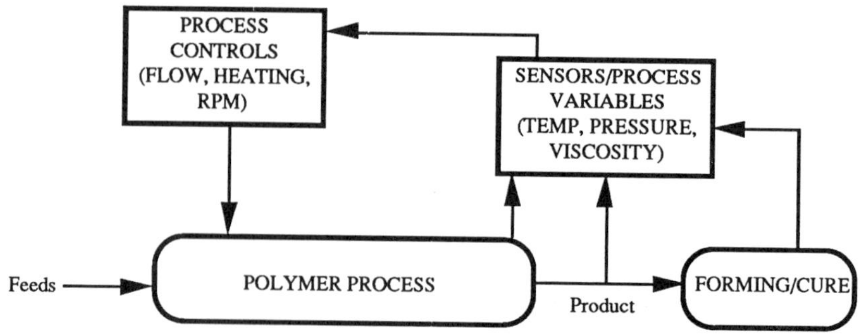

Figure 3 Traditional Polymer Process

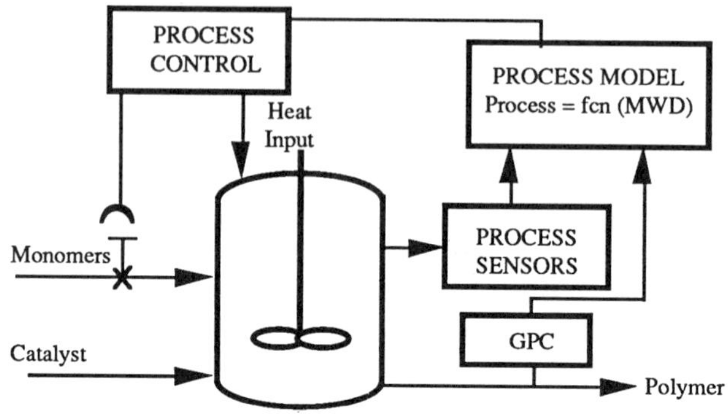

Figure 4 Measurement and Control of Primary Structure Using Process Modeling

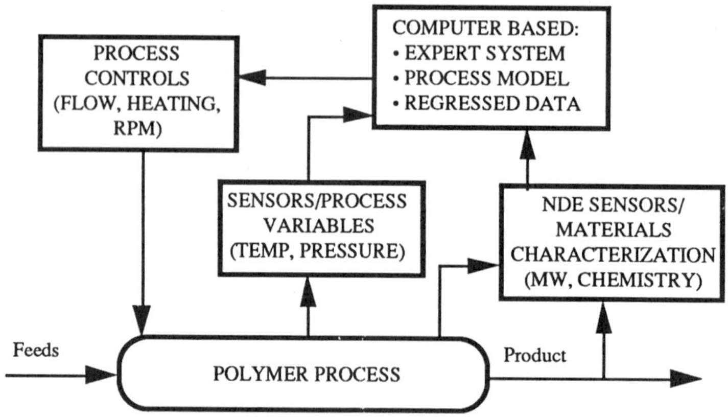

Figure 5 Intelligent Polymer Process

TABLE 2 NONDESTRUCTIVE CHARACTERIZATION PROCESS SENSORS GENERAL PERFORMANCE CONSIDERATIONS

• Accuracy	• Response Time
• Precision	• Stability
• Sensitivity	• Reliability
• Dynamic Range	• Cost

This concept, called Intelligent Polymer Processing, is shown in Figure 5, and is adaptable to any polymer process. A new class of advanced Nondestructive Evaluation (NDE) sensors is utilized to characterize the evolution of microstructure of the polymer in real time. These data and signals from conventional process variable sensors are transmitted to a computerized decision-making system that consists of expert systems that make choices regarding which control parameters to manipulate and process models that quantify actions taken. Where complete models are unavailable, regressed, empirical data can be used or the expert system can use previously developed heuristics to manipulate the control system. Intelligent Polymer Processing is a necessary and sufficient concept for specifying NDE sensor requirements because the process model, the NDE sensors, and control strategy are all interrelated. This measurement technology and model development are best utilized in a complementary relationship. The Intelligent Polymer Processing concept was used in evaluating the requirements for nondestructive characterization sensors in the tire and rubber industry.

NONDESTRUCTIVE CHARACTERIZATION PROCESS SENSORS

Typical performance criteria for sensors are shown in Table 2. Specifics are dictated by the individual processes which we will discuss, but some general comments are in order. Accuracy is the ability of a sensor to measure the "true" value of the desired variable. Precision is a measure of the sensor's reproducibility - how well it reports the same value under identical conditions. Accuracy and precision are the primary requirements for any analytical sensor. However, on-line demands sometimes force compromise; eg, response time may be very important for a particular reactor system with a short residence time which may make a high level of accuracy difficult. Sensitivity or the signal to noise ratio is important for all process control applications because drift in the measured variable could occur without adequate response from the control system to compensate.

Dynamic range or the "working span" of the sensing device is particularly important in unit operations that are used to make a variety of products; sensors must be specified to accommodate a variety of products in cases like these. Stability and reliability are important in all applications; sensors must be robust, requiring as little recalibration and maintenance as possible.[9]

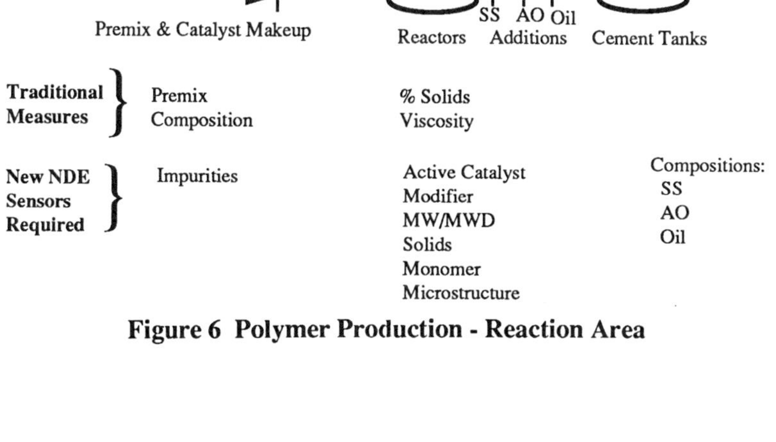

Figure 6 Polymer Production - Reaction Area

NDE IN POLYMER PRODUCTION

As with most polymers, elastomers of interest to the tire industry possess distributions of composition, weight, and structure. These distributions, important to the critical performance characteristics of the compounded polymer, are determined by the nature of the specific reaction mechanism and equipment configuration, and by the control of process variations.

Reaction Area

Figure 6 shows a polymerization process illustrating the NDE sensor requirements for converting from a traditional to an intelligent polymer process.

The liquid (hydrocarbon) side of the process (that is, the monomer and solvent makeup and recovery areas) is typically the best instrumented and controlled operation in the plant. On-line gas chromatographs (GC) having cycle times of the order of a few minutes, are used to monitor concentrations of the various monomers in the reactor feeds. When used in closed loop control of monomer and solvent flow rates, they provide a very effective means for minimizing variations in reactor composition.

Trace impurities present in the monomer, solvent, or catalyst feeds are not so easily measured or controlled. Depending upon the nature of the specific catalyst system, these naturally occurring species can poison the reaction at levels less than 1 ppm. Uncertainties in measurement and control of temperature, residence time (reactor level), catalyst, and monomer rates will alter both the reaction kinetics and the polymer microstructure. Indications of reactor performance may be obtained by monitoring the heat generated by the exothermic polymerization reaction (catalyst activity or impurity level), and by laboratory testing of the cement from the reactors for solids content (conversion) and viscosity (molecular weight).

Downstream of the reactor, the addition of catalyst short-stopping (SS) agents, antioxidants (AO), and extending oils are based upon the polymer content (or monomer conversion) expected at the end of the reactor chain. Residence times of 1 to 4 hours for solution polymers and 10 to 12 hours for emulsion polymers, coupled with the time required for laboratory analyses, create a sizeable lag time in identifying and responding to process perturbations. To minimize variations in product properties and performance, many manufacturers have relied upon extensive blending to achieve "uniformity." While the targeted "bulk" or average properties are obtained by blending (up to 24 hours of production material in emulsion processes), the actual distributions of molecular weight and particle size, and thus final properties, are greatly dependent upon the exact makeup of the blend.

Problems associated with conventionally controlled reaction processes are typified in emulsion SBR polymerizations. First reactor conditions, and primarily those related to catalyst activity, fluctuate with the cyclic variations in impurity levels. This causes continuous changes in conversion, molecular weight, and the primary product specifications - viscosity. To control the conversion, residence times are manually adjusted based upon the off-line analysis of the polymer exiting from the last reactor in the chain. Adjustments are repeated throughout the cycle; always several hours behind and never addressing the cause. While the extremes in the viscosity swings are in fact reduced, cycles of 10 to 20 Mooney points are typical in the industry.

In Intelligent Polymer Processing facilities, molecular conditions of the polymer replace environmental conditions within the reactor as the primary process measurements and controls (Table 3). On-line analytical measurements along the reactor chain including molecular weight and molecular weight distribution, microstructure, catalyst activity and level, and solids concentration of cement (and for emulsion systems, particle size distributions and branching) would provide a first opportunity to monitor growing polymers, and to closely control process and product in a feedforward fashion. Downstream analysis of short stop, antioxidant, and oil concentrations would confirm correct additions and give closed loop feedback control capabilities. Fundamental polymerization models have been

TABLE 3 NDE SENSOR REQUIREMENTS - REACTOR AREA

Parameter	Analysis	Targeted Accuracy
Liquid Feed Streams		
Impurities	Poisons	ppm Level
In-Situ Reactor Conditions		
Activity	Active Catalyst	5% (Relative)
Polymer Cement		
Compositions	Polymer Content	0.5%
	SS and AO	0.05%
	Extending Oil	0.5%
Microstructure	Cis/Trans/Vinyl	2.0%
	Co-monomer Compositions	1.0%
	Sequence Distribution	-
Macrostructure	Molecular Weight Dist	-
	Coupling and Branching	-

demonstrated for emulsion[10] and solution[4,5] processes. These enable the process engineer to relate variations in the process to variations in the product. When dynamic process models are combined with new analytical process measurements, a strategy for predictive control in real time is possible.

Finishing Area

The Finishing Area includes the processes for polymer isolation (steam stripping or coagulation), drying (extrusion or apron driers), weighing, baling, and packaging. For the solution process shown (Figure 7), the solvent and unreacted monomers are removed from the polymer cement by steam stripping. Too mild a condition leaves a high residual hydrocarbon level; too severe a condition forms fine particles which quickly plug the process. The primary stripper control parameters of temperature, steam sparge rate, agitation, and level are fixed; residual hydrocarbons in the exiting crumb slurry are determined in the lab.

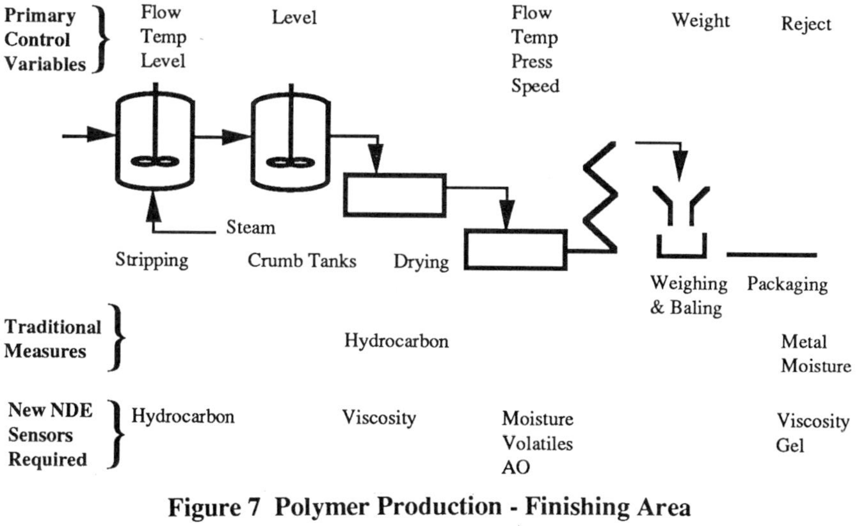

Figure 7 Polymer Production - Finishing Area

The rubber crumb slurry is pumped to the dewatering extruder where free water is squeezed out. The second extruder uses mechanical energy to heat the massed rubber and vaporize trapped water. Hot crumb exits the extruder, onto a cooling conveyor, at 1 to 2% moisture. By the time it is weighed and baled, the moisture level has been reduced to typically 0.5%. Too little work input at the extruder results in wet polymer; too much work overheats the product, consumes a portion of the antioxidant package, and leads to premature aging and polymer degradation. Primary extruder conditions of screw speed, feed rate, and die pressure are fixed; moisture levels are confirmed in the lab.

TABLE 4 NDE SENSOR REQUIREMENTS - FINISHING AREA

Parameter	Analysis	Targeted Accuracy
Stripped Crumb Slurry		
Compositions	Residual Hydrocarbon	0.1%
Dried Rubber		
Compositions	Residual AO	0.1%
	Residual Hydrocarbon	0.1%
	Moisture	0.1%
Properties	Macro- and Microgel	1.0%
	Bulk Viscosity	2.0%

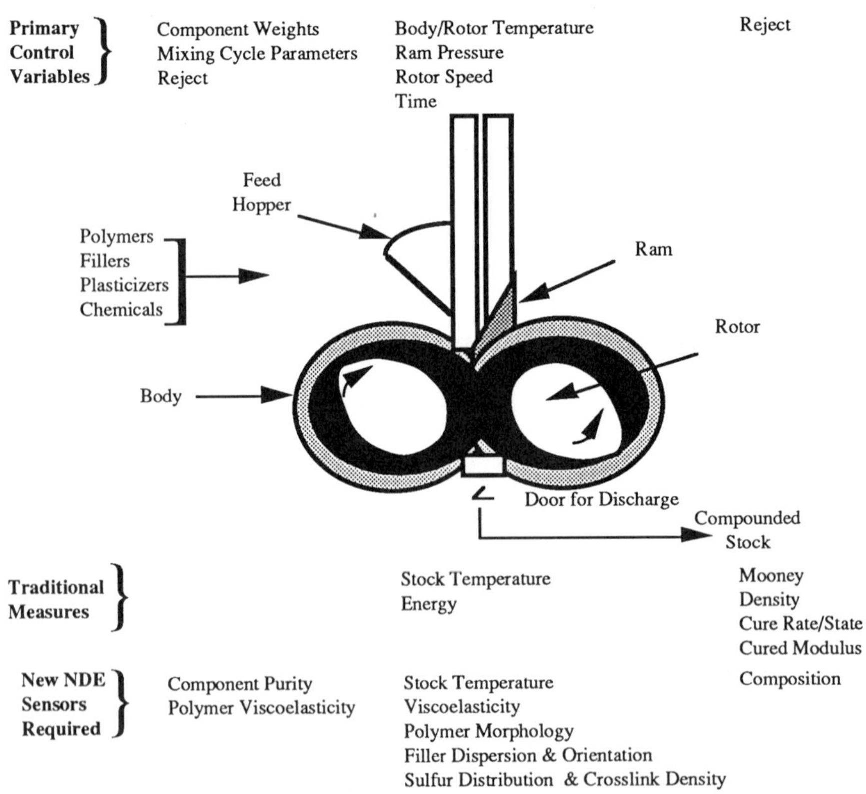

Figure 8 Polymer Forming - Banbury Mixing

In the Intelligent Polymer Processing facility, on-line analytical measurements obtained along the Finishing Line (Table 4) are used in place of infrequent, spot analyses conducted in the QC lab hours after the polymer has been produced. Continuous monitoring of product conditions provides not only important feedback measurements needed for process control, but a first time opportunity for the engineer to fine tune process conditions to each product. Required analyses would include residual hydrocarbon in the stripped crumb, and moisture, volatiles, residual antioxidant, polymer viscosity, microgel, and macrogel in the dried rubber.

NDE IN POLYMER MIXING AND FORMING

Mixing

An internal mixer such as a Banbury mixer (Figure 8) is employed to blend polymers, fillers, oils, and chemicals to produce a rubber compound with desired transport and physical properties; eg, viscosity and dispersion.[11] At the end of mixing cycle, the product is dumped on a mill and stretched into sheets in preparation for additional Banbury mixing or downstream processing such as extruding. The compound properties are affected by type, composition and ratios of ingredients, and mixing cycle parameters: sequence time, shear stress, shear strain and rate of shear strain, rotor speed, ram pressure, and temperature.

TABLE 5 NDE SENSOR REQUIREMENTS - BANBURY MIXING

Parameter	Analysis	Targeted Accuracy
Banbury Inlet Streams		
Compositions	Oil in Polymers	0.2%
	Moisture in Fillers	0.1%
Viscoelasticity	Viscous/Elastic Modulus	1.0%
Banbury In-Situ		
Stock Temperature	Temperature	2°F
Viscoelasticity	Viscous/Elastic Modulus	1.0%
Polymer Morphology	Domain Size/Distribution	*
Filler Dispersion	Particle Size Distribution	*
Filler Orientation	Anisotropy	*
Sulfur Distribution	Spatial Distribution	*
Crosslink Density	Compound Scorch Level	0.1%
Banbury Outlet		
Compositions	Polymers, Fillers, Oils & Chemicals	0.5% (Relative)
*Quantitative Method Required		

In the traditional Banbury mixing process, the feed ingredients are weighed and charged into the Banbury in a given sequence. The mixing process variables such as Banbury body and rotor temperature, ram pressure, and rotor speed are fixed. The batch is mixed to a constant set of temperature, time, and specific energy input conditions. These dump criteria can result in significant differences between batches because of variation in feed component purity and viscoelasticity, variation in initial loading temperature, and inaccuracy in measurement of actual stock temperature. The product is tested off-line, and secondary properties; eg, Mooney (viscoelasticity), density (composition), cure (scorchiness, cure rate and state), and cured modulus (finished properties), are used to either accept or scrap the batch.

Intelligent Polymer Processing applied to the mixing operation would require NDE sensors for on-line measurements of feed component purity and viscoelasticity and product composition, and for in-situ measurements of primary product parameters such as compound temperature, viscoelasticity, polymer morphology, filler dispersion and orientation, sulfur distribution, and sulfur crosslink density. Table 5 lists the desired accuracies for these sensors. The component purity analysis would be utilized to adjust weights of different ingredients charged to the Banbury such that uniformity of composition is maintained. For example, determination of polymer oil content would be utilized to adjust the amount of other plasticizers added to the batch, providing a feedforward control. The polymer viscoelasticity measurement would provide feedforward information to control mixing process variables, or in an extreme case to reject the polymer bale.

The most important in-situ measurements are compound temperature, viscoelasticity, filler dispersion, and sulfur distribution. The compound temperature measurement using the current techniques is often inaccurate due to the large heat-sink capacity of the mixer's metal body. Too high a temperature causes scorchiness and poor subsequent processability. The compound viscoelasticity determines throughput rate and die swell (hence final product dimensions) in downstream processing. The filler dispersion determines the tire performance properties such as tread wear and tear. Since sulfur and accelerators are the last ingredients added to the mix, their uniform distribution ensures uniformity of the overall mix, and hence the final product. These measurements coupled with a mixing process model/expert system would provide feedforward information to control the listed mixing process variables. A dump criterion based on attainment of desired range for all these parameters would result in improved product properties as well as uniformity and waste minimization. On-line product composition analysis would help identify problems with the feed charging systems, and would also be utilized to reject out-of-spec batches.

Extrusion

Extruders are employed to form tire components; eg, apex and sidewall. Either the sheets are masticated on a mill to reduce plasticity and then fed to an extruder (hot feed system) or fed directly to a cold feed extruder. Multiple extruders with a single head are used to form a complex component; eg, tread. A schematic of the extruder system is shown in Figure 9. The extrudate is cooled and then either cut to the required size or wound-up on a roll to be used at the tire building machine. The weight and dimensional measurements of the final product are influenced by compound properties and extruder variables such as viscoelasticity, screw speed, barrel and screw temperature control, number of barrel pins and their configuration (for a pin barrel cold feed extruder), head and die design, and extrudate take-away system.

TABLE 6 NDE SENSOR REQUIREMENTS - RUBBER EXTRUSION

Parameter	Analysis	Targeted Accuracy
Extruder Inlet		
Viscoelasticity	Viscous/Elastic Modulus	1.0%
Polymer Morphology	Domain Size/Distribution	*
Filler Orientation	Anisotropy	*
Crosslink Density	Compound Scorch Level	0.1%
Extruder In-Situ		
Stock Temperature	Temperature	2°F
Viscoelasticity	Viscous/Elastic Modulus	1.0%
Extruder Head Profiles	Pressure Gradients	0.5%
	Velocity Gradients	0.5%
Extruder Outlet		
Dimensions	Contour	0.1%
	Split Location	0.1%
Residual Stress	Relaxation	0.01%
Polymer Morphology	Domain Size/Distribution	*
Filler Orientation	Anisotropy	*
Crosslink Density	Compound Scorch Level	0.1%
*Quantitative Method Required		

In a traditional extrusion process, extruder head pressure and extrudate temperature, specific weight, and contour are monitored on-line, and used to control screw speed and take-away belt speeds in a typical feedback fashion. The contour measurement for non-stationary extrudate is not reliably accurate, and also does not distinguish various components of the extrudate (split locations). The extrudate is cooled for a given length of time and then skived (cut) to final length, whether it is fully relaxed or not.

NDE sensors are required for on-line measurements of feed viscoelasticity, structure and scorch stability, product contour, split locations, residual stress and structure, and for in-situ measurements of stock temperature and viscoelasticity, and velocity and pressure distribution at die head to facilitate extruder feedforward control and to accomplish desired extrudate dimensional stability. Some of the requirements for these sensors are listed in Table 6. The output rate and die swell are very dependent upon compound viscoelasticity. The compound sulfur crosslink density (degree of scorchiness) controls maximum temperature rise that can be tolerated in the extruder. On-line measurement of these parameters would be utilized in a feedforward control mode to adjust extruder screw speed, barrel and screw temperatures, and number of barrel pins and their configuration (through use of hydraulic systems that can adjust each pin position individually during processing), or in an extreme case to reject the feed. The polymer morphology and filler orientation of extrudate affect final tire properties, and are strongly influenced by the shearing mechanism in the extruder, and the extruder head and die design. On-line measurement of these parameters would result in better process design and control.

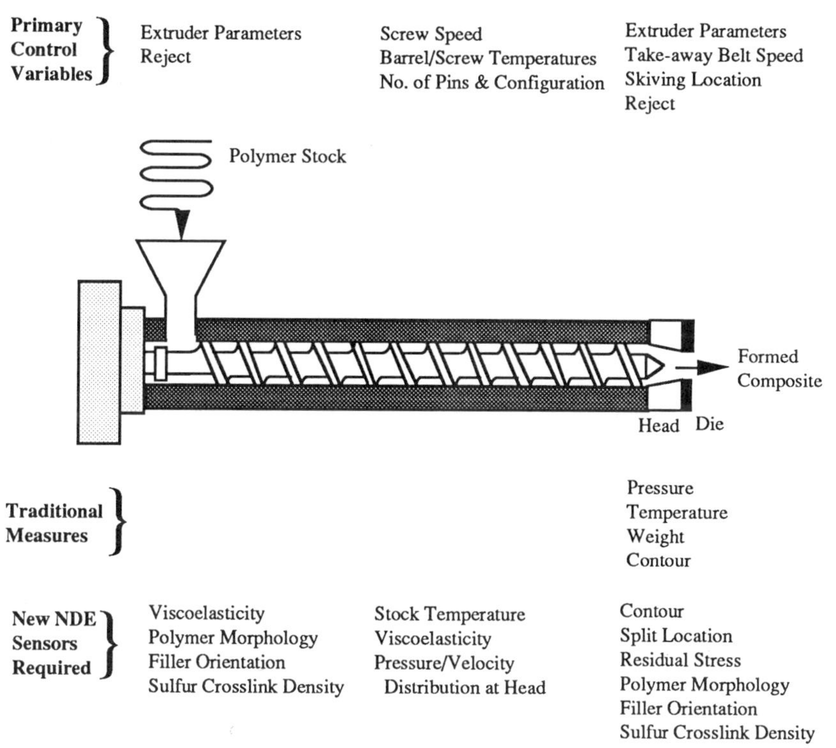

Figure 9 Polymer Forming - Extrusion

The most important in-situ measurement is stock temperature. Excessive temperature build-up causes scorch. This results in lumps, loss of building tack, and variation in cured tire properties. The problems associated with currently available temperature probes are: 1) inaccurate reading due to the strong influence of barrel metal temperature, and 2) inability to measure temperature at a precise radial location. An NDE sensor for measuring actual stock temperature at a given radial and axial location for 3-dimensional temperature mapping in the extruder is very desirable and would provide feedforward information for control of extruder variables. Uniform pressure and velocity at the die are necessary to avoid dead spots (leading to scorched compound) and for extrudate uniformity. The measurement of these profiles would result in better extruder head and die design, and would also minimize die trials. The exact contour and split locations are critical for final tire force and performance uniformity. These measurements would be utilized either to adjust relative screw speeds of multiple extruders and other extruder parameters providing closed loop feedback control, or for product rejection. The residual stress measurement would be used to adjust skiver location so as to ensure that the component is fully relaxed and its dimensions would not change significantly after skiving.

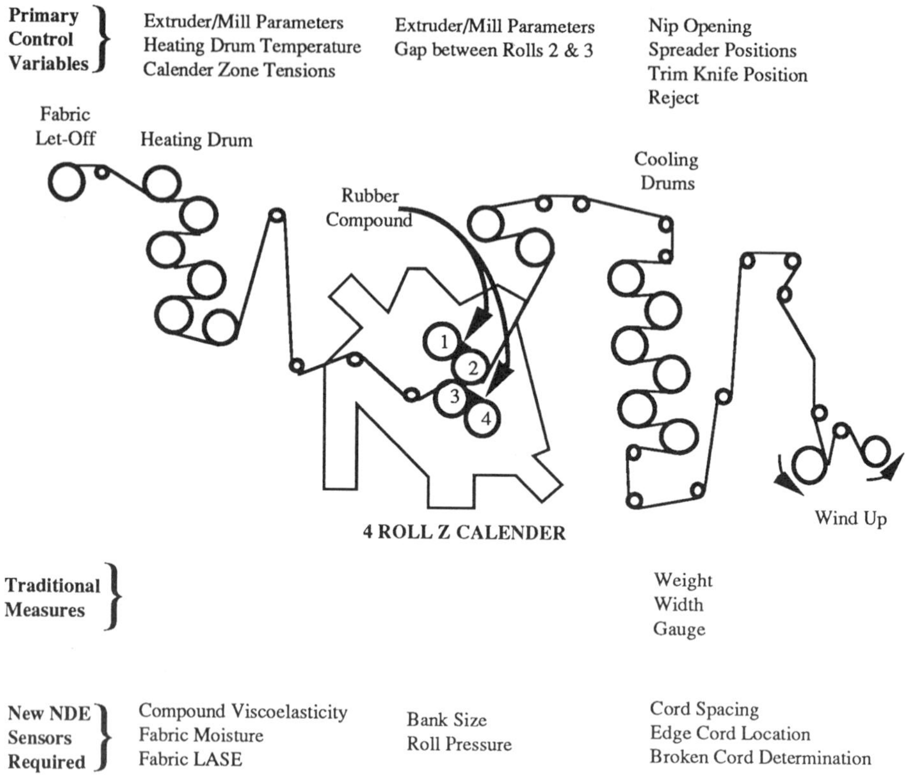

Figure 10 Polymer Forming - Calendering

Calendering

Calenders are employed to produce sheeted gum as well as fabric reinforced tire components; eg, innerliner, fabric plies, and wire belts. A schematic of the calender system is shown in Figure 10. The fabric is supplied to the calender nip using a sophisticated let-off system, and continuous strips of rubber compound are supplied using either a mill or an extruder. The important dimensional measurements for the product are: gum and total weight, width, top and bottom gauge, cord spacing, and distance from edge of gum to first cord. Both calender and rubber compound variables such as nip opening, roll pressure, tension across the calender, compound viscoelasticity and temperature, and bank height affect the product dimensions.

In the traditional calendering process, roll speed ratios and temperature, fabric heating drum temperature, and tension of different zones across the calender are held constant. At the calender outlet, nuclear transmission, x-ray transmission and x-ray backscatter sensors are employed to monitor the listed product parameters, and to feedback the information to control calender roll gaps and spreader and trim knife positions.[12] The cord space measurement and the edge cord detection capability of these sensors is limited to steel wire and does not extend to polyester, nylon, and rayon fabrics. The other parameters in the system such as compound viscoelasticity, bank height, and rubber-fabric interface pressure (roll pressure) are adjusted manually.

In addition to the above listed sensors, the intelligent calendering operation would require on-line sensors to measure feed viscoelasticity, fabric moisture and LASE (load at specified elongation, a stress-strain relationship) at let-off, calender bank size and roll pressure, cord spacing, edge cord position, and broken cord location for the calendered fabric. Table 7 lists desired accuracies for these sensors. Compound viscoelasticity affects its swell characteristics, and hence the product gauge. The tire durability, ride, and handling are affected by calendered fabric moisture content (especially for nylon and rayon fabrics) and LASE. These properties can be altered by adjusting process variables such as hot drum temperature and system tension. On-line analysis of the fabric moisture and LASE at calender let-off would be utilized for process feedforward control. Bank size affects compound scorchiness (leading to different processability and cure rate) and calender roll deflection (leading to different product gauge and profile). On-line sensors for the bank size would provide feedback information to control speed/time and temperature of the rubber supplying extruder/mill. Correct roll pressure fills rivet space between the cords while maintaining the correct cord spacing. Its measurement and control (by adjusting the gap between calender rolls 2 and 3) are critical to obtain uniform treatment. Broken cord and variation in cord spacing in calendered fabric result in weak spots in the tire. The broken cord determination would provide a basis for sample rejection. On-line cord space sensor would provide feedback measure to control spreader positions. The edge of the calendered fabric is normally spliced to another fabric. For splice integrity, edge cord detection to provide feedback control for precise gum edge trimming is critical.

COMPOSITE ASSEMBLY AND CURING

Different tire components; eg, innerliner, toe guard, ply, bead, apex, chafer, sidewall, belt, overlay, and tread, are applied on a building machine to form a green (uncured) tire. The design and complexity of the building machine depends upon the construction of the tire. From the standpoint of tire uniformity, it is critical to place all components on the tire machine at the precise location and without any stretching or snaking.

Curing

In the traditional curing process (Figure 11), the green tire is first shaped and then cured by application of fixed specified temperatures and pressures over a period of time in an automatic curing press such as the Bag-O-Matic press. A constant temperature is maintained by steam in the shell around the tire curing mold, and heat and pressure are

TABLE 7 NDE SENSOR REQUIREMENTS - CALENDERING

Parameter	Analysis	Targeted Accuracy
Calender Inlet		
Viscoelasticity	Viscous/Elastic Modulus	1.0%
Fabric Properties	Moisture	0.1%
	Stress-Strain	0.1 lb
Calender In-Situ		
Bank Size	Compound Volume at Nip	2.0%
Roll Pressure	Pressure	1.0%
Calender Outlet		
Fabric Dimensions	Cord Spacing	0.0005"
	Edge Cord Position	0.0005"
	Broken Cords	Identify All

TABLE 8

NDE SENSOR REQUIREMENTS - TIRE CURING

Parameter	Analysis	Targeted Accuracy
Curing In-Situ		
Temperature	Temperature Profile	2.0°F
Cure State	Sulfur Crosslink Density Profile	1.0%

applied internally in cycles through an attached heat-resistant rubber bladder. The temperature-pressure-time cure cycle is established from rate of cure of different tire components, and their location in relation to bladder and mold (heat transfer considerations).

The cured tire should have the final form and physical properties, including final cured state, required for service. Because of batch-to-batch differences in compound composition, compound time-temperature history (degree of scorchiness), and component gauge, the same component can experience a different rate of cure in different tires. Thus, a prespecified cure cycle leads to cure state non-uniformity in the tire. An intelligent curing process would require NDE sensors for in-situ 3-dimensional temperature and sulfur crosslink density measurements. Table 8 lists requirements for these sensors. These measurements coupled with curing process model would provide feedforward information to control applied temperature, pressure, and time, thereby allowing attainment of optimal state-of-cure for each component in every tire.

SUMMARY

The production of a pneumatic tire is a complex process involving polymer production, mixing, and forming, followed by composite assembly and cure. The application of Intelligent Polymer Processing methodology requires the development of a large number of NDE sensors, many of which would measure polymer and compound in-situ microstructural-type properties on-line and in real time. Implementation of these sensors could dramatically improve uniformity and final tire properties.

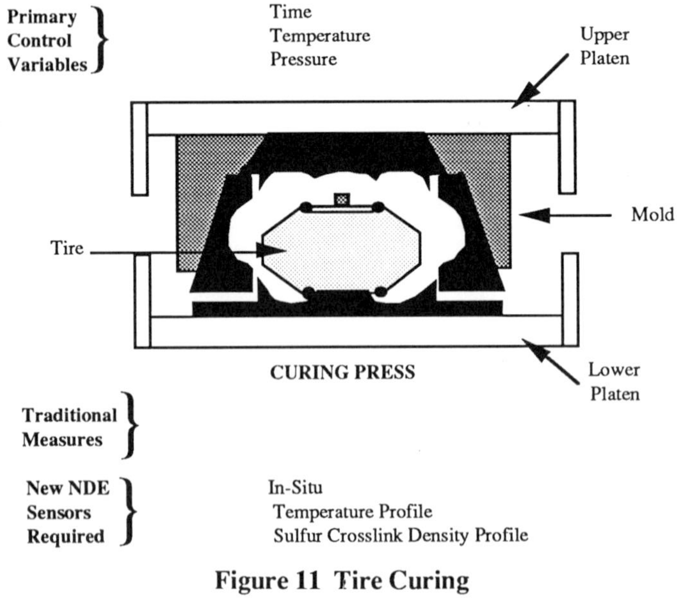

Figure 11 Tire Curing

REFERENCES

1. Bhakuni, R S, et al, "Tires", Encyclopedia of Polymer Science and Engineering, Vol 16, 2nd ed, John Wiley, 1989.
2. Bhakuni, R S, et al, "Pneumatic Tire Compounding", The Vanderbilt Rubber Handbook, Robert Ohm (ed), R T Vanderbilt, Inc, Norwalk, 1990.
3. Halasa, A F, et al, "SIBR for High Performance Tires", Rubber and Plastic News, in press.
4. Chang, C C, et al, "Fundamental Modeling in Anionic Polymerization Systems", Journal of Applied Polymer Science, Vol 39, No. 11, 2395-2417, 1990.
5. Chang, C C, et al, "Modeling Studies of the Controlled Anionic Co-Polymerization of Butadiene and Styrene", ACS Rubber Division, 137th Meeting, Las Vegas, May 30, 1990.
6. Imai, M, "Kaizen", Random House Business Division, New York, 1986.
7. Miller, J W, Jr and A J Bur, "Report on the Working Group on Polymer Processing", Intelligent Processing of Materials, NISTIR 89-4024, US Department of Commerce, January 1989.
8. Yolken, H T and L Mordfin, "Intelligent Processing of Materials - Report of an Industrial Workshop Conducted by the National Institute of Standards and Technology", NISTIR 89-4024, US Department of Commerce, January 1989.
9. Yeh, JTY, "Online Composition Analyzers", Chemical Engineering, January 20, 1986.
10. MacGregor, J F, et al, "Dynamic Modeling of Emulsion Polymerization Reactors: A Review", AIChE National Meeting, Philadelphia, August 12, 1984.
11. Johnson, P S, "Mixing Equipment and the Mixing Process", Basic Compounding and Processing of Rubber, Harry Long (ed), Rubber Division, ACS, Inc., 1985.
12. Embleton, W T, "The Measurex 2002 Wire Calender System Sensor Package", RU7533/4M/10-85/W Coast.

ULTRASONIC CHARACTERIZATION OF POLYMERS UNDER SIMULATED PROCESSING CONDITIONS

L. Piché

National Research Council Canada
Industrial Materials Research Institute
75 De Mortagne, Boucherville, Québec J4B 6Y4 CANADA

ABSTRACT

In contrast to usual materials, polymers are characterized by the non-equilibrium nature of their structure. Polymers are viscoelastic in that their mechanical properties depend not only on pressure (p), volume (V) and temperature (T), but also time (t). Viscoelasticity determines processability, while thermodynamic history during processing influences end-use properties. It is therefore important to perform mechanical measurements in conditions representative of those for manufacturing. This cannot be achieved with usual methods; ultrasonics, however, provides a unique means to realize such experiments. Using this approach, we describe an apparatus which measures sound velocity (v) and attenuation (a) and also incorporates simultaneous monitoring of the specific volume (V) to characterize the thermodynamic state of the material. The instrument allows close control of the thermodynamic history in the temperature range from T = -150 °C to T = 300°C and pressures up to p = 2.0 kbars. Results for the velocity, attenuation and specific volume are translated in terms of a dynamic complex modulus for the description of viscoelasticity. The experiment constitutes an essential step toward on-line measurements for process control in the plastics industry.

INTRODUCTION

The success story of plastics is attributable to the introduction of more advanced processing technologies but more so, to a better understanding of the basic properties of these materials, among which viscoelasticity[1,2] stands out as most important. Viscoelasticity results from the nonequilibrium nature of the material as evidenced by strongly time dependent properties. Consequently end use behavior is conditioned by thermal history during processing where the polymer is subjected to complex temperature, pressure cycles. In view of optimizing productivity it would be a great advantage to assess the viscoelastic state of the material while it is being processed. For this, ultrasonics should prove to be a unique and powerful tool, provided the characteristics for the propagation of sound can be correlated[3,4,5] to viscoelasticity. As a first and essential step in this direction, a laboratory instrument was developed that provides accurate measurements of velocity and attenuation under conditions that are representative of processing. The instrument also measures the specific volume as a means of characterizing the thermodynamic state. First a short overview is presented of processing in relationship to materials properties. Then, the technique is sketched out. Results are shown for typical industrial polymers and some peculiarities of material behavior are illustrated which demonstrate the usefulness of the approach.

POLYMERS AND PROCESSING

Polymers are constituted of large molecular chains comprising a great number of covalently linked repeating units The attraction between chains is provided by secondary bonds, usually van der Waals forces. At high temperature where thermal energy exceeds intermolecular forces, the material appears a viscous liquid with random configuration. As temperature is lowered, the material becomes rubbery and rigidifies to the point where, upon reaching a certain characteristic temperature, the glass transition temperature, T_g, the structure becomes frozen-in and the polymer is a glass. Now, because of the high viscosity, the liquid may be undercooled to various degrees so that T_g is dependent on the cooling rate. The nonequilibrium, or visco-elastic nature of polymers has a strong bearing on materials properties and is manifested in a multitude of characteristic phenomena[1,2,6] at all stages of its life. Here we consider only amorphous polymers where the structure is random. In cases of highly regular chains, strong intermolecular forces promote formation of ordered (crystalline) regions. However, the behavior of semicrystalline materials is still dominated by the amorphous component.

Most fabrication techniques[7] use an extruder, where the raw polymer is conveyed by a rotating screw through a series of zones in which it is heated, compacted and softened. In extrusion processing, the material is forced through a die having a particular shape, after which the extrudate of uniform cross section (tubing, sheeting, insulating coating for wires ...) is cooled to solidify. Injection molding is more common and involves preparing a charge of molten polymer under high pressure in a chamber that directly precedes a hydraulic ram. At the proper time the ram injects the molten charge into a mold that is kept below T_g of the polymer. The mold is then opened to eject the plastic in finished form.

Rapid, economical processing requires more than the easy flow of the molten polymer: it is also necessary to heat the polymer rapidly before flow and to cool it rapidly after flow. Due to the nonequilibrium nature of the material, processing of plastics requires a delicate balance of temperature, pressure and time conditions. For example high molecular weight polymers necessitate higher temperatures in order to improve flow characteristics, however, this costs more and also leads to degradation. On the other hand, when the temperature is too low, processability retrogresses from the liquid into the rubbery flow region producing ripples or discontinuities known as melt fracture. Also, quick release of pressure near T_g results in dimensionality problems. In general, lack of control on processing parameters are manifested in plastic parts in the form of density fluctuations, structural inhomogeneity, crazing, dimensional instability, residual stress, mechanical birefringence, etc.

The viscoelastic nonequilibrium nature of the material is maintained in the final end product. Mainly, polymers relax when submitted to constant strain, and creep under constant stress. The time dependence for mechanical properties is very strongly influenced by temperature, especially near T_g, and determines long time behavior. A dramatic illustration of stress relaxation leading to catastrophic failure was given recently by the Challenger accident. Examples for creep include polymeric piping transporting hot pressurized liquids, automobile engine parts ... compact discs. Finally, not only do polymer properties depend on temperature, pressure and time, but they are also very nonlinear, which by no means simplifies their description.

TECHNIQUE

While the ultimate objective is the design of a technique that performs on-line, it is of the upmost importance to evaluate the approach in the more simple situation where the material is not forced to flow. Although this may not be very appealing, in view of a straight forward application, it is nonetheless a necessary step. For example, it has been argued that because of high frequencies, ultrasonics could not assess long time behavior and also that the ultrasonic movements were not correlated to the essential bulk properties. With this in mind, we constructed a laboratory instrument that uses ultrasonics to measure linear mechanical properties of polymers under well controlled temperature, T, pressure, p, and time, t, conditions representative of those the material encounters during its life, but without macroscopic flow. Also, the instrument provides simultaneous measurement of specific volume, V, as a means to characterize the thermodynamic state of the material.

Here we briefly sketch out our experiment, a complete description was given elsewhere[5,8]. The sample in the form of a thin disc (≈ 3 to 4 mm thick) is confined between two aligned buffer rods at the opposite end of which are attached transducers. The assembly is constructed such that the buffers can be submitted to axial clamping with allowance for axial displacement to compensate thermal expansion of the polymer. The clamping force can be adjusted from 0 to 2 kbars. The axial length of the assembly is monitored (± 1 μm), which allows following changes of specific volume. Provision to scan the temperature in the range from -150 to 300°C is made through heating elements and a liquid nitrogen cooling circuit attached to the buffers. The temperature control system allows one to establish heating or cooling rates from 50°C/min to 1°C/hr, or to stabilize the temperature to ± 0.1°C. The setup is under computer control and the collection of data concerning length, temperature and acoustic signal is accomplished every 10 s and stored on magnetic discs. Further processing allows correcting for the influence of the buffer lines. Measurements can be made with both longitudinal and shear waves. Here, however, we study longitudinal waves which propagate in solids and liquids alike.

Usually viscoelastic behavior is described in terms of a complex modulus. For longitudinal waves, $L^* = L' + iL''$, where the storage modulus L' refers to the elastic component and the loss modulus L" describes the viscous losses. These are related to specific volume, V, velocity, v, attenuation, a, and frequency, f, through:

$$L' = (1/V)v^2 \quad \text{and} \quad L'' = (1/V)v^3\, a/\pi f \tag{1}$$

Depending on whatever is more pertinent, results may be described either in terms of L' and L" or v and a.

THERMAL BEHAVIOR AND INFLUENCE OF MORPHOLOGY

Here, polystyrene serves to illustrate thermal behavior of amorphous polymers. Also, we examine the effect of chain morphology which is known to influence processability and end use properties. For this we consider two samples, PS1 and PS2, having different molecular weights: M(PS1) = 130,000 g/mol and M(PS2) = 9,000 g/mol. The ratio M(PS1) / M(PS2) ≈ 14 measures the relative chain lengths in the materials and it is accepted[1] that the longer chains in PS1 may entangle while M(PS2) stands below the critical value for these effects. The materials were prepared by melting the pellets directly in the apparatus, and then rapidly cooling, dT/dt = -40°C/min, to 20°C at a constant pressure p = 225 bars. Measurements were carried out at the same pressure, using a constant heating rate dT/dt = +2.0°C/min and ultrasonic frequency f = 1.75 MHz.

The specific volume, V, in Fig. (1-A) was initially similar for both glasses. On increasing temperature, the specific volumes suddenly rise much faster. For PS1 this occurs near 105°C and, in the region 105 to 140°C, the results are very sensitive to the heating rate indicating that structural rearrangements lag changes in temperature;it is only near 140°C that the material comes to thermal equilibrium. Near T_{ll} = 175°C, the entanglement network breaks up and the polymer becomes free-flowing-liquidlike. The behavior near 105°C is typical of the glass transition phenomenon which is identified as a kinetic process of volume change. Therefore dilatometry constitutes the basic method for the determination of the glass transition temperature, T_g. Because of the rate aspect, T_g is ill defined. However, for operational purposes, T_g is given by the marked rise of specific volume: T_g (PS1) = 105°C. For PS2, T_g (PS2) = 96°C and already at 120°C, the exansivity equals that of PS1 above T_{ll}. On one hand the larger specific volume, V(PS2), corresponds to excess free volume associated with shorter chains, while on the other hand the absence of a T_{ll} feature in PS2 is in line with the absence of entanglement.

The storage modulus, L' in Fig. (1-B), decreases rapidly with temperature due to anomalously strong anharmonicity in the disordered structure. At T_g, L' suddenly drops while the loss modulus, L" in Fig. (1-C), begins to increase, reflecting the irreversibility of structural rearrangements at the glass transition. Here, the materials properties evolve slowly and the ultrasound probes this evolution, irrespective of frequency. At higher temperature, L' describes a sigmoïd while L" goes through a strong maximum. This is typical of a relaxation process where the time scale is that of the ultrasound, $\tau^{-1} \approx \omega = 2\pi f$. The description for L* proceeds from the model[1,2] for the linear viscoelastic solid:

$$L'(\omega) = L_U - (L_U - L_R) \int H(\tau)\, (\omega\tau)^2 / [1 + (\omega\tau)^2]\, d\tau \qquad (2)$$

$$L''(\omega) = (L_U - L_R) \int H(\tau)\, (\omega\tau) / [1 + (\omega\tau)^2]\, d\tau \qquad (3)$$

where $H(\tau)$ is the distribution of relaxation times, τ. At low temperature, $\omega\tau >> 1$, the material appears an elastic solid with modulus L_U, while at high temperature, $\omega\tau << 1$, it behaves as a liquid with viscosity $\eta = L''/\omega$ and modulus L_R.

Obviously, L* is closely linked to specific volume V. However, Fig. (1) demonstrates that L* also discriminates the influence of microscopic configurational details. In the liquid state where $\omega\tau < 1$, the high molecular material, PS1, is more viscous than PS2 and therefore more difficult to process, this is also true for the rubbery state, $\omega\tau > 1$, where PS1 is a tougher material than PS2 mainly because of the entropy spring associated with the entanglement network. Processing, therefore, implies a tradeoff between mechanical and thermal energy in order to achieve optimum flow rate. Below T_g, the specific volume is the same for both materials, while the modulus is larger for the short chain material, L'(PS2) > L'(PS1). This indicates that vitrification was too fast for the structure to readjust in order to accommodate the extra van der Waals volume associated with the greater number of chain ends in PS2. Therefore PS2 is not in equilibrium with respect to minimum internal energy requirements which may lead to structural instability. This was later demonstrated by further cooling the sample to 15°C where it shattered into a powder. This is another example where the ultrasonic modulus reflects materials particularities associated with processing.

SIMULATED PROCESSING CONDITIONS

In using ultrasonics to monitor viscoelasticity during processing, one would only measure velocity, v, and attenuation, a. In Fig. (2) we show results for V, v, and a obtained at f = 2.5 MHz and pressures up to 800 bars. The sample was the same high molecular weight material, PS1, described above, also the same thermal path was followed. Actual processing involves taking the material through a tortuous route along the various curves of figure 2.

p-V-T path

The behavior for v and a in Fig. (2) is closely related to that of L' and L" in Fig. (1). Also, the curves at different pressures show the same overall features. Mainly, increasing the pressure causes a reduction in volume, thereby increasing intermolecular forces which results in higher modulus (v increases) and lower mobility (τ increases). Correspondingly, T_g increases with pressure, dT_g/dp = 46°C/kbar and this dictates processing conditions. In the region $\omega\tau < 1$ where the material is liquidlike, pressure induced structural changes, as manifested through changes in V, result in strong modifications of materials properties (v, a, τ) that are nonetheless reversible. On cooling the material at constant pressure, say along curve 4 in Fig. (2), the "structure-property relationship" is frozen in at T_{g4}. However, in practice, corresponding to the extrudate coming out of the die, or the molten charge being injected in the mold, or the opening of the mold, the pressure is released at a temperature above T_{g4}, after which the material is cooled to solidify. As was seen above in this region of temperature, $\omega\tau >> 1$, the relaxation times are long so that cooling is not along a pseudo equilibrium isobar as in Fig. (2). Then, the finished product may be far from equilibrium and therefore unstable, unless the process is controlled very closely. Obviously, at this stage, ultrasonic monitoring of viscoelasticity would be most useful. Another application concerns the growing field of computer simulation of processing operations. Here, the ultrasonic technique provides a very accurate determination of the equation of state that is required as input for the calculations. Also, the method measures viscous losses associated with compression and that are usually neglected for lack of available data.

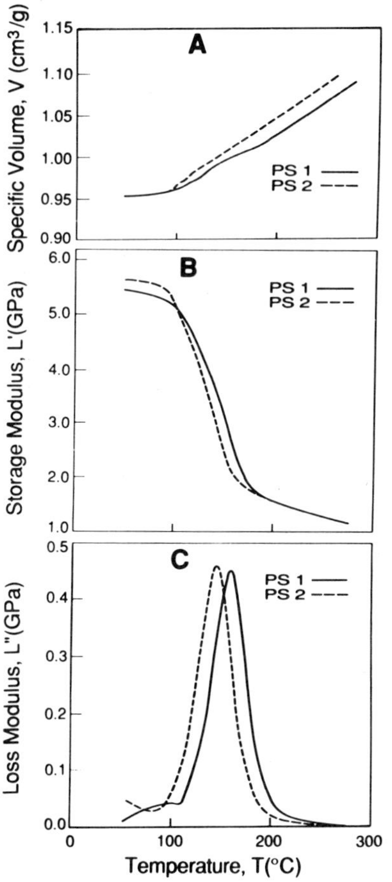

Fig. 1. Specific volume, storage and loss moduli at frequency f = 1.75 MHz and pressure p = 225 bars, for polystyrene materials PS1 and PS2 with molecular weights M_N(PS1) = 130,000 g/mol. and M_N(PS2) = 9,000 g/mol. respectively.

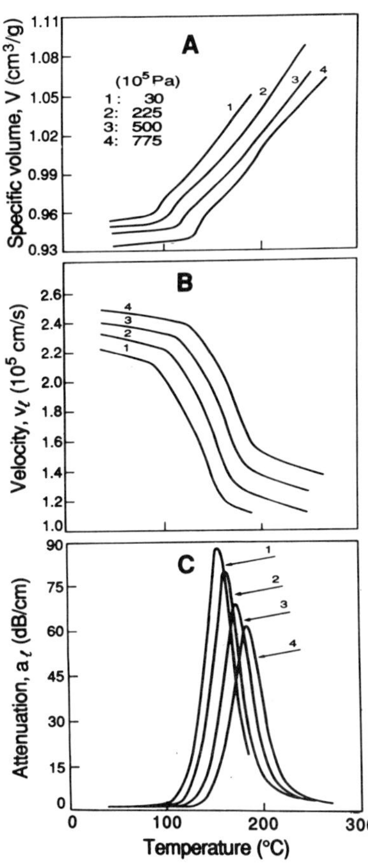

Fig. 2. Influence of pressure on specific volume, longitudinal velocity and attenuation for polystyrene, at frequency f = 2.5 MHz.

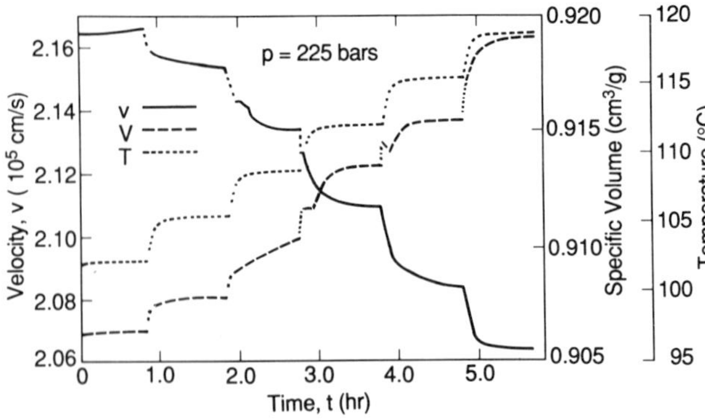

Fig. 3. T-jump experiment showing the evolution with time of temperature, T, specific volume, V, and velocity, v, of longitudinal waves at f = 2.5 MHz, following temperature steps $\Delta T = 3°C$, starting at $T_0 = 102°C$ at constant pressure p = 225 bars.

T-jump stimulated creep

Long time instability is demonstrated in the T-jump creep experiment at constant pressure, p = 225 bars, in Fig. (3). Here, temperature was incremented by $\Delta T = 3.0°C$ every hour and the time response function was recorded for T, V and v. The time for the system to reach thermal equilibrium was 3.5 min. On the first jump from 102°C to 105°C, V and v exhibit a spontaneous short time response followed by long times toward equilibrium. On the following jump, the short time contribution was maintained while equilibration time had greatly diminished although still exceeding 1 hour. This evolution was sustained up to the last step from 114°C to 117°C where the response time was near 3.5 min. The comportment for V is typical[1,6,9] of T-jump creep experiments, where short time behavior is attributed to elasticity and long time behavior identifies viscoelasticity. As seen in Fig. (3), although $\omega\tau >> 1$ in eqs. 2 and 3, the ultrasound measures materials properties as they slowly evolve with time near T_g, and therefore constitutes a probe for equilibrium, non-equilibrium behavior. This result clearly demonstrates that the glass transition phenomenon is independent of the measuring frequency, although T_g depends on the rate of heating for the experiments in Figs. (1) and (2).

P-jump stimulated creep

The experiment closely simulates conditions for ejecting plastic parts from their mold during injection molding. In this case, the sample was cooled along the 750 bars isobar in Fig. (2) down to T = 106°C, well below Tg = 130°C at this pressure. Then, the pressure was released step by step. The time response function for p, V and v below 300 bars is shown in Fig. (4). After each jump, V and v undergo a short time response followed by long times toward equilibrium. As was the case for the T-jump measurement in Fig. (3), this demonstrates the non-equilibrium properties of the material. One may foresee that ultrasonics could be used to monitor materials properties directly in the mold and dictate the proper time for releasing pressure in order to optimize the process. Although p-jump measurements are very important, they are seldom performed because of experimental difficulties with usual techniques[10]; therefore again ultrasonics proves to be of great interest for polymer studies.

Going back to Fig. (4) one notes the very anomalous short time behavior for V which first decreases when releasing pressure. This may be interpreted by that decreasing pressure also makes the material weaker as indicated by the drop in velocity. Then, the structure readjusts and becomes tighter in order that intermolecular forces equilibrate the external pressure. This behavior illustrates the character of nonlinearity associated with polymer properties; also it constitutes strong evidence for a thermodynamic phase transition that involves configurational entropy. In this respect, the long times exhibited in Figs (3) and (4) lead to $\omega\tau >> 1$. According to eqs. (2) and (3), this would mean that $L' = L_U$ is constant. The only possible conclusion is that (L_U - L_R) and τ are time dependent, indicative of an autocatalytic process, similar to a cooperative second order phase transition. This point is worthy of further investigation, for a better understanding of the underlying mechanisms could lead to a more systematic approach for optimizing processing parameters.

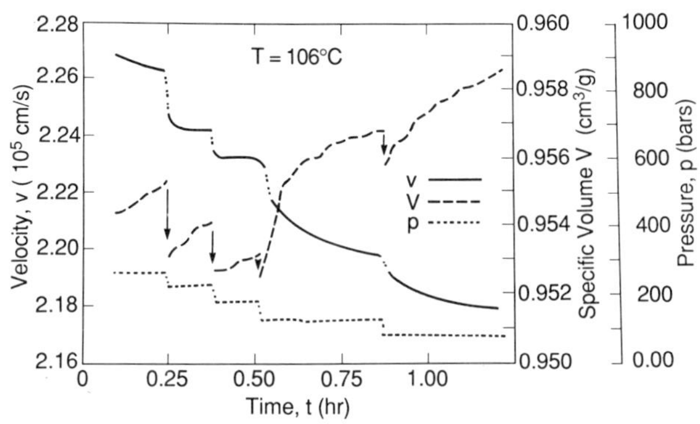

Fig. 4. p-jump experiment showing the evolution in time of pressure, p, specific volume, V, and velocity, v, of longitudinal waves at f = 2.5 MHz, following pressure jumps. Here the sample was cooled to T = 106°C under constant pressure p = 750 bars, and the data is shown starting at 200 bars.

CONCLUSION

We described a novel apparatus for the characterization of polymers, that combines measurement of specific volume to assess structural and thermodynamic changes and ultrasonics to appraise mechanical behavior. The approach is unique in that experiments may be performed under pressure, temperature and time conditions representative of processing operations. Results obtained on a typical amorphous polymer, namely polystyrene, were presented. It was shown that the method probes characteristics of thermomechanical behavior that are associated with the thermodynamic state and details of the molecular structure and which determine processability and end use properties. In particular, the technique measures the time dependence of mechanical properties over the wide spectrum specific of polymer viscoelasticity. Actually we have studied a great number of industrial polymers (amorphous, semicrystalline, elastomers, blends and composites) with similar results. As it stands, the instrument is a powerful tool for the characterization of polymers with applications for modeling properties, and for quality control. On the other hand, the technique may be easily implemented on line to monitor the evolution of materials properties during manufacturing operations in view of process control. In this context we are presently in the course of adapting the technique to an extrusion line in order to study the influence of flow on viscoelastic behavior. Finally, our study suggests the existence of a true thermodynamic transition associated with the glass transition phenomena. Further work along these lines is also ongoing, which could lead to define a more efficient protocole for processing.

REFERENCES

1. J.D. Ferry, "Viscoelastic Properties of Polymers", John Wiley, New York (1980).
2. I.M. Ward, "Mechanical Properties of Solid Polymers", John Wiley, New York (1983).
3. B. Hartman, "Ultrasonic Measurements" in "Methods of Experimental Physics", ed. R.A. Fava, Academic Press (1980), Vol. 16-c, Chap. 12.1, pp. 59-90.
4. R.A. Pethrick, "Acoustical Properties" in "Comprehensive Polymer Science", eds. C. Booth and C. Price, Pergamon, Oxford (1988) Vol. 2, "Polymer Properties", Chap. 17, pp. 571-599.
5. L. Piché, submitted to Polymer (1990).
6. G.B. McKenna, "Glass Formation and Glassy Behavior" in "Comprehensive Polymer Science" eds. C. Booth and C. Price, Pergamon, Oxford (1989) Vol. 2, "Polymer Properties", Chap. 10, pp. 311-361.
7. I.I. Rubin, "Injection Molding, Theory and Practice", Wiley-Interscience Series, New York (1972).
8. L. Piché, F. Massines, A. Hamel, C. Néron, "Method and Apparatus for the Ultrasonic Characterization of Polymers under Simulated Processing Conditions", U.S. Pat. No. 4,754,645, July 5, 1988.
9. D.J. Plazek, M.J. Rosner and D.L. Plakek, J. Polym. Sci. B. Polymer. Phys., 26, 473 (1988).
10. J. J. Tribone, J.M. O'Reilly and J.J. Greener, J. Polym. Sci. B. Polym. Phys., 27, 837 (1989).

ACOUSTIC EMISSION FROM SHORT GLASS FIBRE REINFORCED INJECTION MOULDED PARTS WITH WELD LINE

Bodo Brühl

Institute for Polymer Testing and Polymer Science
University of Stuttgart
Germany

INTRODUCTION

When a thermoplastic polymer melt flow is partioned during the injection moulding process, a weld line is produced where the separated melt streams rejoin. This weld line may result in an area of weakness in injection moulded parts, specially in the case of fibre reinforced materials. The weld line is characterized by a change in the fibre orientation: the fibres are to a large extent oriented perpendicularly to the mould filling direction which coincides with the direction of load. An increase in strength intended to result from the addition of fibres may therefore not be fully achieved.

SPECIMEN AND MATERIALS

The tensile tests were performed with shouldered flat specimen (width 10 mm/height 3,2 mm, shape similar to DIN 43455-S3). The moulding compounds used were polybutylene therephtalate (PBT-GF30) and polycarbonate (PC-GF30) both reinforced with short glass fibres (30% by wt.).

RESULTS AND DISCUSSION

Acoustic emission (AE) and mechanical properties

In tensile tests the cumulative number of AE events monitored up to a certain test load level could be correlated with the mechanical properties as demontrated in figs. 1 and 2. For the PBT-GF30 specimen it is possible to determine a regression line (fig. 1). The PC-GF30 specimen presents a similar tendency but the deviation is significantly stronger (fig. 2).

One goal of this work was to obtain more information about the behaviour of these materials to provide a better understanding and explanation of the background. Due to its high acoustic activity the PC-GF30 makes the measurement more difficult since high event rates must be accommodated. This material has been given priority and a way will be shown how to surmount these difficulties.

Nondestructive Characterization of Materials IV
Edited by C.O. Ruud *et al.*, Plenum Press, New York, 1991

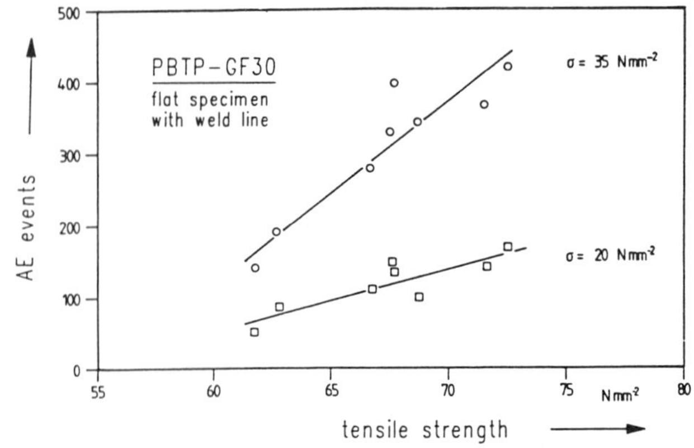

Fig. 1. Cumulative number of AE events at various load levels as a function of the tensile strength.

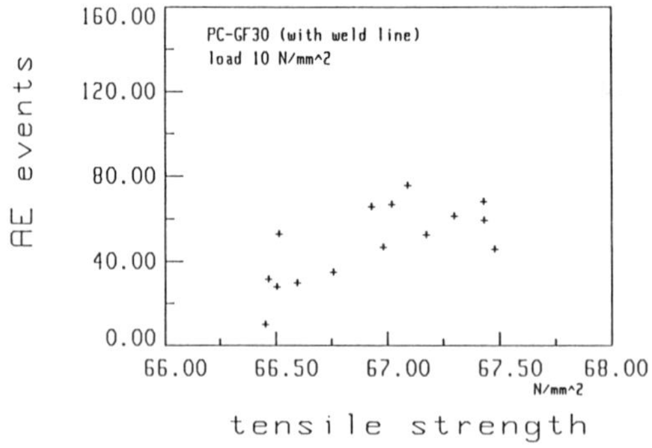

Fig. 2. Cumulative number of AE events as a function of the tensile strength.

Transient recorder based measurement system

The results described in the first part were obtained with a conventional AE system, the SPARTAN device from Physical Acoustics Corp. (PAC). The tests indicated some limitations in the capabilities. Due to the storage transfer of the AE data to the hard disk this system can handle only 120 events/sec in steady state mode. Higher event rates can only be monitored for short periods of time corresponding to the filling of the internal random access memory (RAM). Therefore this system can not monitor AE during the whole test of PC-GF30 specimen with actual event rates (cf. fig. 3). As shown later this system is also not able to produce satisfactory information in location tests.

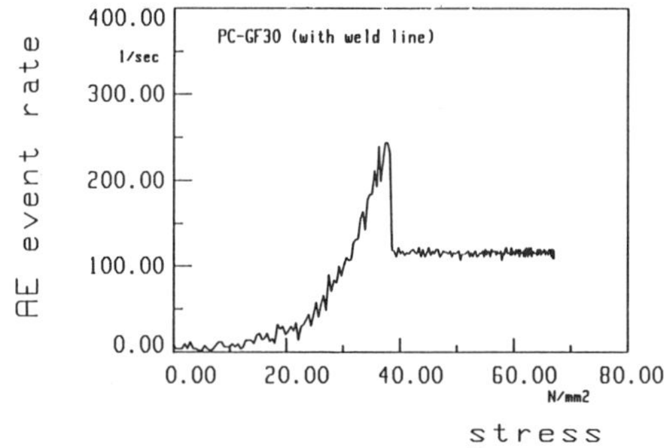

Fig. 3. AE event rate of PC-GF30 Specimen.

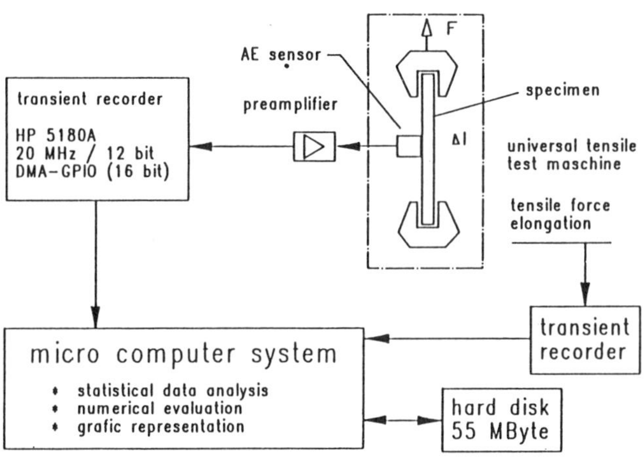

Fig. 4. Schematic diagram of the transient recorder based measurement (TRB) system.

For this reason a transient recorder based measurement (TRB) system was built (details in fig. 4). The microcomputer is a HP310 for data acquisition and a HP340 for all evalution (both Hewlett Packard). Software programs are available for the measurement procedure and all kinds of data evaluation such as time domain analysis (automatic identification of AE events and determination of their signal parameters) and even frequency domain analysis are possible.

For the numerical analysis a resonant transducer (PAC R15 / 150 kHz) was placed directly on the weld line. For location analysis two resonant transducers (PAC u30S /300 kHz) were used at a distance of 44 mm centered to the weld line. The AE signals were preamplified (PANAMETRICS 5660B/ 0,02 - 2 MHz/40 dB) and then stored into the transient recorder. The sampling time was 200 nsec with a record length of 4096 points. A microcomputer reads the data from the transient recorder (GPIO-Interface) and writes them to a hard disk. This setup is a compromise between a maximum in time resolution and a maximum in measurement rate. Therefore the capability of this system (numerical analysis: 25 measurements/sec, location: 14 measurements/sec) allows a statistical interpretation of AE tests.

AE event rate

The TRB system is active for measuring only during a small period of the test duration. But this measurement provides very precise information which is not affected by high event rates. The event rate obtained in a TRB measurement can be obtained by dividing by the actual measurement time.

In fig. 5 a maximum of more than 10 AE events during a TRB measurement of 2 msec is found. This corresponds to a rate exceeding 5.000 AE events/sec. At the end of the test the distributions shift to lower values. This could be due to a smaller AE acitivity or to coalescensce of several AE events. The results of the energy measurements in Fig. 6 confirm that in fact the acoustic activity decreases also since the sum of AE energy obtained during the TRB measurement interval decreases.

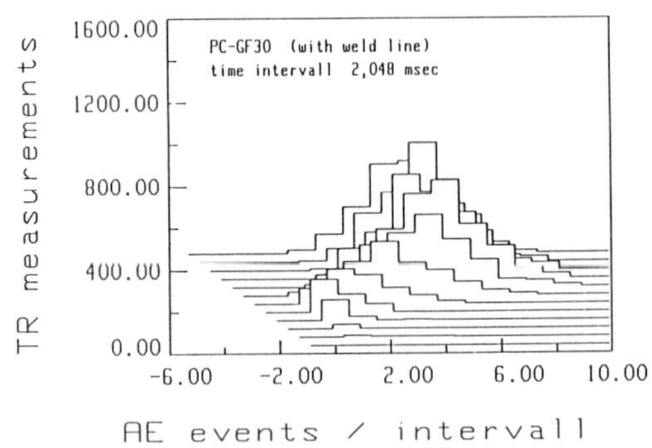

Fig. 5. Distributions of number of AE events found in a TRB measurement as a function of test duration (interval 10 sec).

Location

Due to the limitations in the data processing capabilities of the conventional AE system the results of the location tests on PC-GF30 specimen were not satisfying. Even when a large number of AE events could be recorded, only very few valid locations were found.

Fig. 8 shows the distributions of location during the test. On the left and right side of the diagram the number of non locatable AE events are arranged. To make the number of located events visible a different scale is used in the middle section of the diagram. Because less than 1% of all AE events could be located this result is not a secure basis to discuss the origin of AE in this specimen.

With the TRB measurement system more than a third of the AE events could be located (figs. 9, 10). The distributions in the first half of the test indicate that a major amount of AE comes from regions outside the weld line. The weld line itself emits only few AE events (fig. 10).

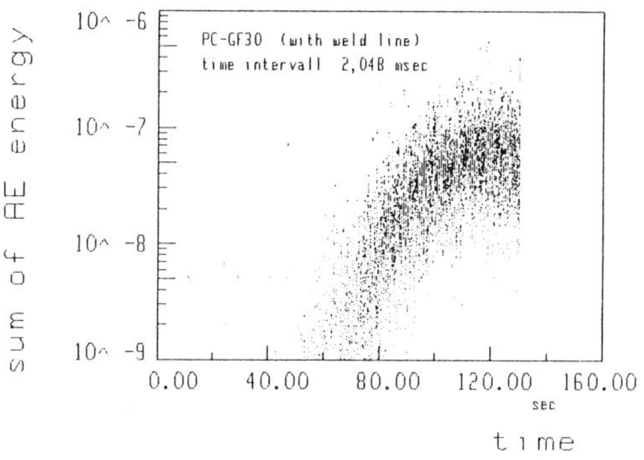

Fig. 6. Sum of AE energy in a TRB measurement as a function of test duration.

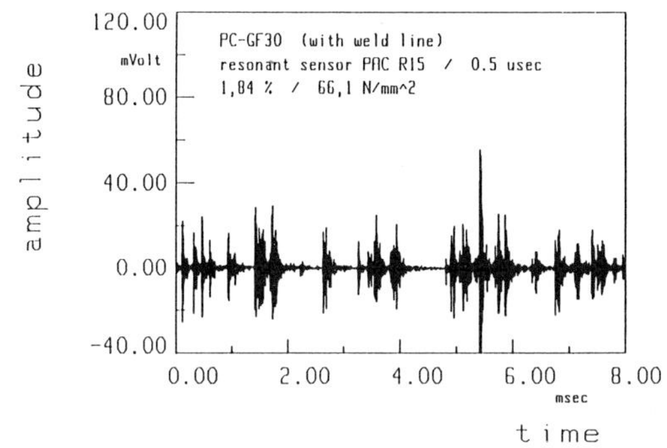

Fig.7. Transient recorder measurement (different setup to achieve longer time interval).

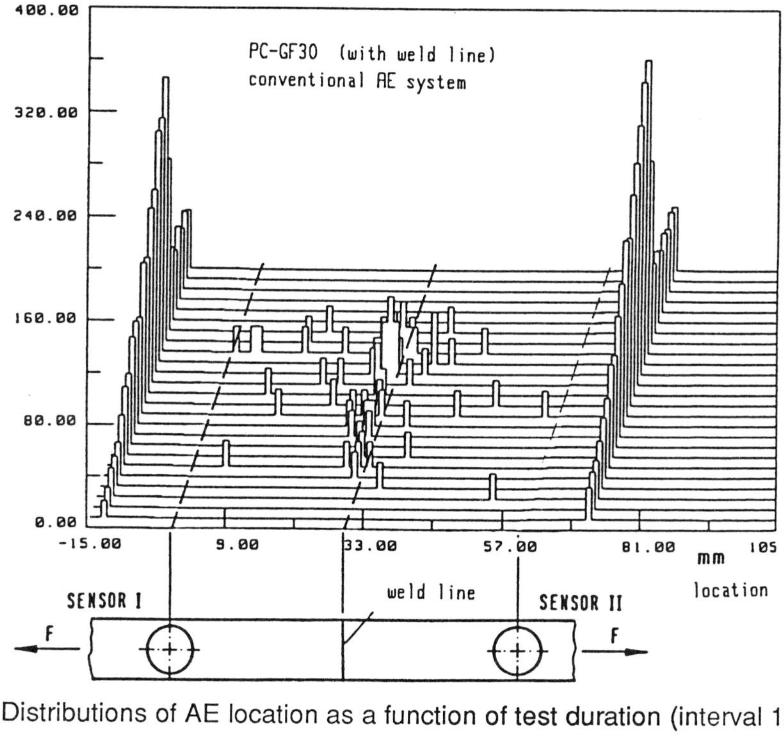

Fig. 8. Distributions of AE location as a function of test duration (interval 10 sec).

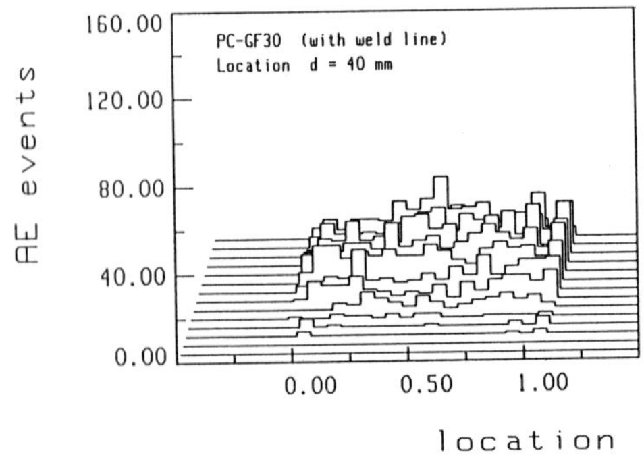

Fig. 9. Distributions of AE location obtained with TRB system as a function of test duration (interval 10 sec).

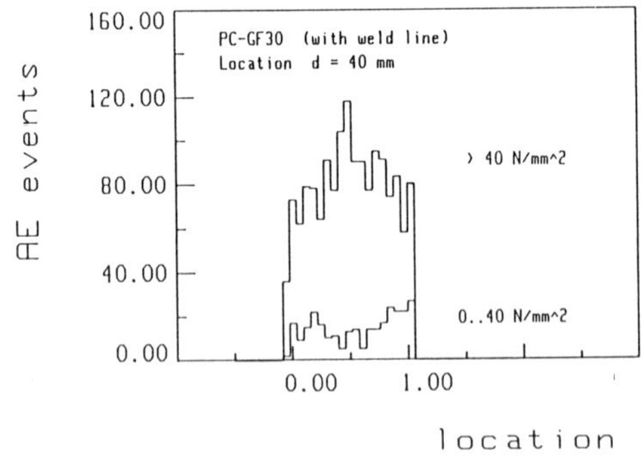

Fig. 10. Collection of distributions of location obtained with TRB system for the first and second half of test duration (cf. fig.9).

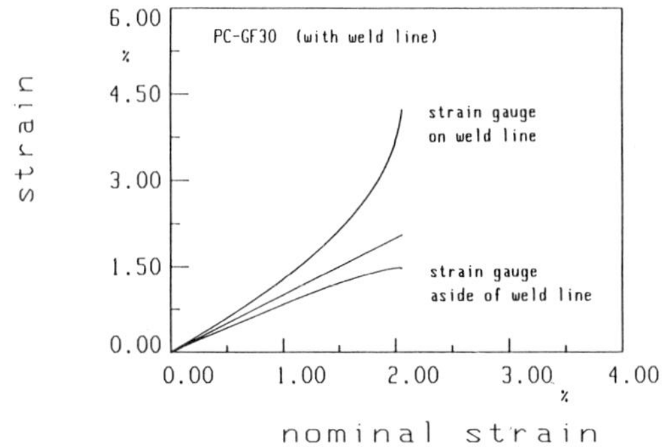

Fig. 11. Measurement of strain on specimen with weld line.

Fig.11 gives an explanation why the weld line is an active AE source only at higher load. The strain of the weld line region shows significant deformation only after the second half of the test is reached and then the weld line comes to failure. All specimen of the tensile tests broke in the weld line region.

Further investigations are required to find out why acoustic emission behaviour of the specimen and mechanical properties of the weld line are correlated, even if the weld line is not the main source of the observed AE information.

SUMMARY

Measurements were performed on injection moulded parts from glass fibre reinforced thermoplastics. The result was that acoustic activity is correlated with weld line strength at low mechanical loads. In further tests a transient recorder based measurement system was used to determine material behavior of these specimen. Therefrom the high acoustic activity at increased material stresses could be recorded and location of acoustic emission (AE) events could be performed. The failure of the weld line is shown by a slight decrease in the acoustic activity. The investigations to locate the origin of AE confirm: At small loads only a fraction of the detected AE signals is emitted from the weld line itself. At increased load and beginning of failure the origin of AE signals is predominantly located in the weld line.

ACKNOWLEDGEMENTS

The author is grateful to the Deutsche Forschungsgesellschaft (DFG) for financial support of the project and to the companies BASF AG and BAYER AG for the injection mould and the moulding compounds.

ACOUSTIC EMISSION SENSORS FOR NDC OF CARBON/CARBON

B. R. Tittmann

Penn State University
University Park, PA 16802

INTRODUCTION

This report demonstrates how acoustic sensors can monitor changes in microstructure during the pyrolysis of Carbon/Carbon. In particular, we have demonstrated on flat plates up to one foot square in size that Acoustic Emissions (AE) are significantly different between normal microcracking associated with the chemo/physical changes during pyrolysis and the catastrophic delaminations that occur if the processing temperature/time treatment is incorrect. About 30 experimental runs have now been carried out, which show that AE provide precursor information in advance of delaminations so that the potential exists for making corrective changes during the processing treatment.

The manufacturing process of carbon-carbon is a long, high-risk procedure which is based on empirical methods. Variations in the constituent materials, the part geometry, or time/temperature profiles during first carbonization may produce damage that is difficult to diagnose and expensive to correct. Recent work[1-9] has shown that during first carbonization, matrix cracking is a consequence of "normal" carbonization. In fact, microcracks are desirable, in that they release stress and create an open network, that allows the release of gases formed by the chemical changes. On the other hand, delaminations are intolerable because they reduce through-thickness stiffness and buckling strength. They are typically caused by excessively fast heating of relatively large and/or thick sheet material and represent "anomalous" carbonization. The goal of the manufacturer is to achieve carbonization in as short a time as possible but without excessive damage from delaminations. The key is the monitoring of internal microstructure during the course of carbonization in order to optimize the process. Recent work on ultrasonic sensors shows that these are especially suitable for monitoring microstructure because the acoustic interaction with the material is noninvasive and because the sensor data provide direct information on the evolution of microstructure.

TECHNICAL DISCUSSION

Experimental Approach

Figure 1 shows the experimental apparatus used in the measurements. It consists of a furnace and an array of sensors, including active and passive acoustic sensors, thermocouple gauges and a gas flow meter. A desktop computer collects the data and provides temperature control. Another computer operates in order to carry out material state monitoring, prognosis, planning and explanation. Only the acoustic sensors are addressed here, whereas the design and operation of the other sensors are described in Reference 3 and the planning, explanation and control station are described in Reference 9.

Nondestructive Characterization of Materials IV
Edited by C.O. Ruud *et al.*, Plenum Press, New York, 1991

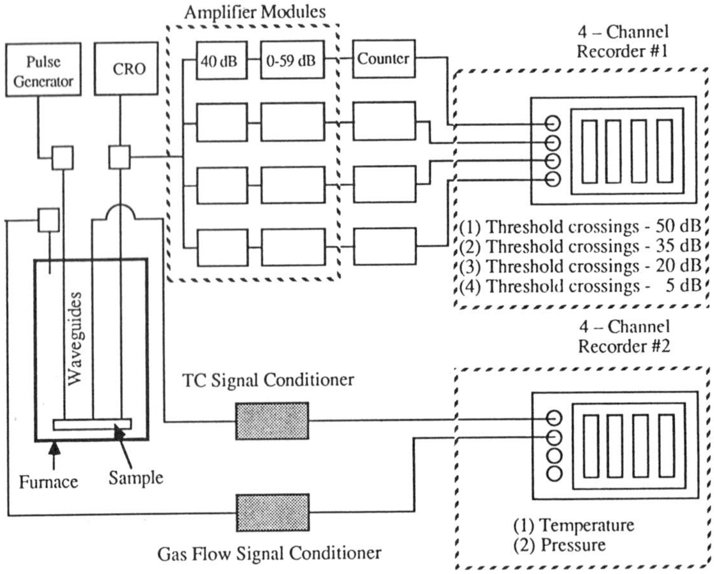

Figure 1. Schematic of apparatus.

Since the carbonization process involves the heating of the component to 900°C, the acoustic sensors are attached to the component via waveguides (1/8 inch thick) fashioned from high temperature alloys (stainless steel or Inconel 718). The waveguides are cylindrical rods which are threaded at opposite ends of the sample. This connection is bonded adhesively using a drop of K-641 resin which is allowed to polymerize at approximately 160°C prior to each run. The transducers are broad band and designed to make use of many vibration modes spread over a broad frequency band. The transducers are clamped to a small platform which is brazed to the top of each waveguide outside the furnace. ransducers and waveguides are coupled acoustically using a high viscosity silicone oil.

In order to tailor the experimental approach to the pyrolysis of carbon-carbon it is useful to consider the AE activity anticipated. Thus the separation of matrix material from the fibers is thought to lead to small stress drops, compared to the joining of many of these

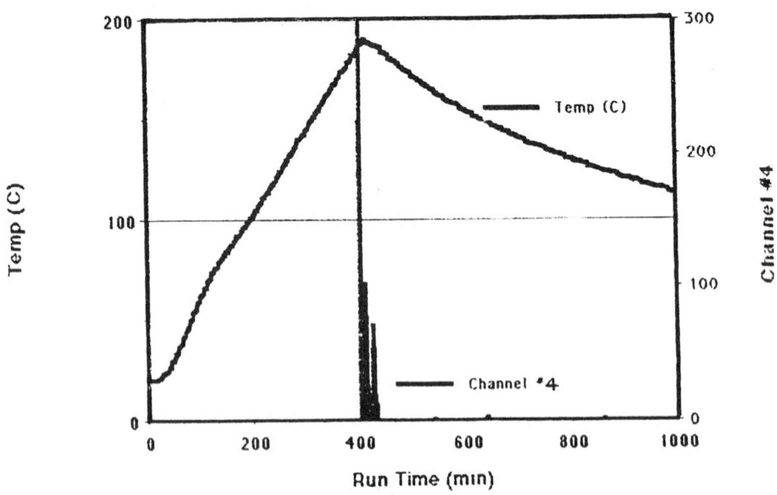

Figure 2. Delamination Run.

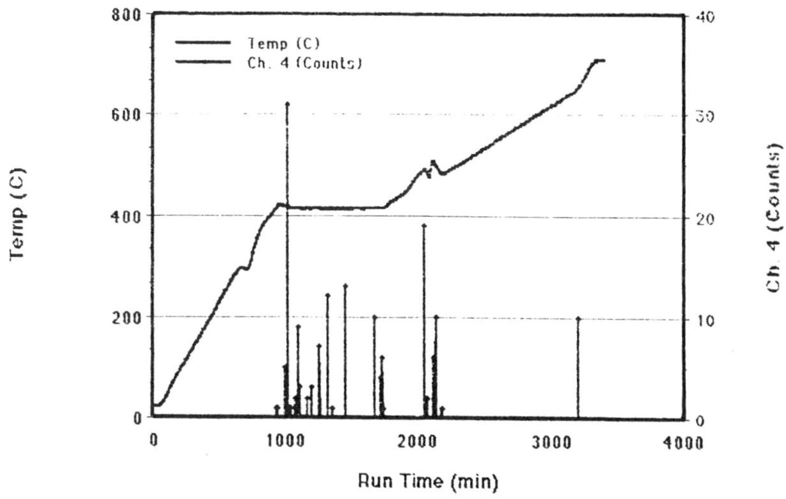

Figure 3. Channel #4-Count/2Min Interval.

separations to form a matrix crack that extends through an entire ply. An even greater stress drop is thought to come from a delamination that separates two plies at their interfact over large areas of the part. Fiber fracture leading to the highest stress drop and therefore the highest amplitude AE event is thought to occur seldom and is most likely associated with delaminations.

Microstructural Information from AE Data

Our method for parameterizing AE activity on a statistical basis involves counting threshold crossings at each of several amplitudes which are evenly spaced on a log scale. The threshold amplitudes selected were 31.6 microvolts, 178 microvolts, 1 millivolt and 5.62 millivolts, all referenced to the transducer output. The most sensitive setting was approximately 7dB above background electronic noise. During a "normal" run without delaminations there are two periods of intense activity, the first while heating between 400°C and 600°C and the second while cooling. These effects are treated in Reference 4. AE is caused by a sudden release of localized stress. Sources of stress during carbonization are the transformation shrinkage of the matrix, differential thermal expansion and then entrapment of evolved gases. The release of stress is typically associated with crack initiation and crack growth/extension.

In carbon-carbon pyrolysis microcracking is normal and desirable, because the goal is to form an open, interconnected network of cracks to enable the escape of evolved gases. Desirable microcracks include fiber-matrix cracks and matrix-matrix cracks. Fiber-matrix cracks arise as a result of the unbonding between the reinforcing fibers and the matrix. Fiber-matrix cracks developing between neighboring sites link up and produce matrix-matrix cracks which eventually extend through the matrix of a given ply typically in a direction perpendicular to the fibers. The AE waveforms associated with fiber-matrix and matix-matrix cracks typically involve short duration times and relatively low amplitudes. In contrast, delaminations typically cover larger areas and therefore the characteristic duration times are comparatively long and the waveform amplitudes are high. Figure 2 shows an example of a run leading to delaminations. These signals were observed in the highest threshold during a run, for which a higher than normal heat-up rate was employed, purposely.

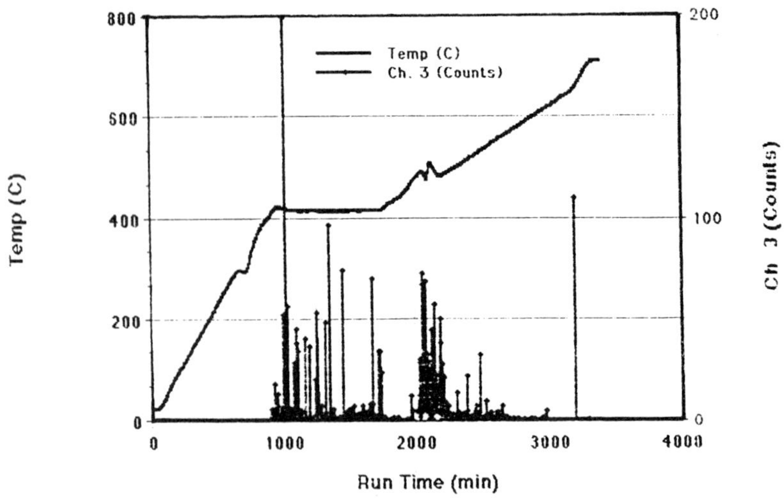

Figure 4. Channel #3-Counts/2Min Interval.

Experimental Results

The goal of this run was to proceed on a moderately high heat-up rate and use the AE sensor to avoid delaminations.

The panel had a nominal thickness of 0.33 inch and lateral dimension of 7.5" x 7. 6". The panel was carefully dried prior to the run. This drying treatment proved to be effective because a rate of 25/Hr could be sustained without any AE events to about 425 C, which was reached in just under 1000 minutes. Figs 3 through 5 shows the temperature/ times profiles with accoustic events in the highest (4) , second highest (3) and lowest (2) AE threshold channels, respectively. Up to about 425 C no AE events were recorded in any channel suggesting no chemically induced shrinking had occurred.

At about 425 C the two highest AE channels began to show pronounced AE activity with one cluster of 30 events in channel 4. Based on previous runs this activity was not sufficient to represent a delamination but signified a probable precursor suggesting a rapid build-up of local stress. Consequently, the temperature increase was put on hold until the AE activity decreased significantly in the two highest channels. This took about 750 minutes. At about 1500 minutes into the run and at about 415 C, the lowest channel (1) indicated the start-up of activity, normally associated with microcracking. This was expected, since the matrix material was probably undergoing a phase transformation (Ref. 4) leading to the chemical induced shrinkage normally occuring at this stage of the process. A heat-up rate of 10 C/H was now set up for the furnace, which was maintained - except for a small operator induced perturbation - until the completion of the heat treatment at about 650 C. After this, the heat was removed and the furnace was allowed to cool at a natural rate over a period of about 48 hours.

The total heat treatment took about 3200 minutes or about 53 hours with little or no acoustic activity above 600 C. This total run time is much shorter than the total run time of a previous run with a similar panel, that run took about 6000 minutes or 100 hrs and still showed AE activity and therefore still some evidence of shrinkage induced cracking at a temperature as high as 50 C. Thus, the results suggest that, by using an AE sensor, it is possible to reduce the run time and probably achieve a better product.

CONCLUSIONS

The use of acoustic emissions (AE) sensors has been shown to be successful for the detection of delaminations during the first carbonization of carbon-carbon panels. The

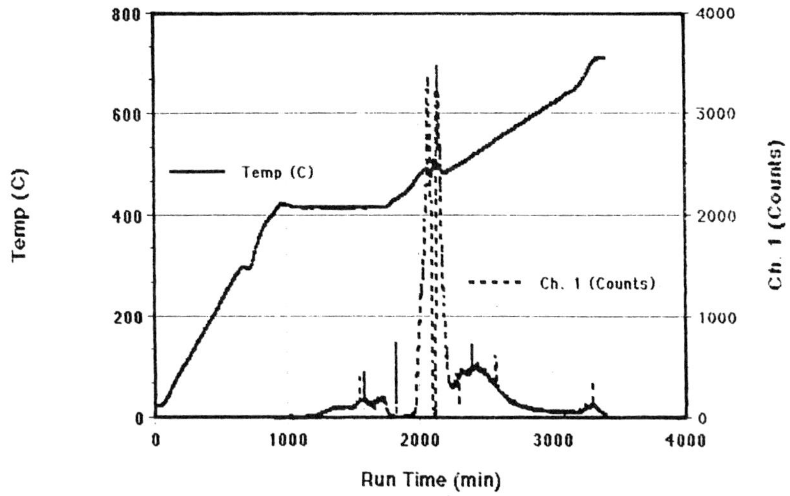

Figure 5. Channel #1-Counts/2Min Interval.

onset of delaminations are characterized by high amplitude, long duration events which are distinct from the low amplitude, short duration events associated with the micro-cracking which constitutes a normal evolution of the matrix microstructure during first carbonization. There is now mounting evidence on the observation of AE precursors to delaminations so that thepotential for warning and prevention of delaminations may be realized. This effect has indeed been tried and successfully realized in one of the recent carbonization runs. These results open the door to the acquisition of a statistically meaningful database for application to process control in the factory whether in process simulation or production. The AE technology appears ready to the next stage of addressing complex part geometries and various constituent materials. Enhancements for diagnostics such as triangulation and 2-D imaging of the stress zone giving rise to the precursor also appear feasible.

ACKNOWLEDGEMENT

The author acknowledges the support of the Rockwell International Science Center through the Office of Naval Research under Contract N00014-87-C-0724.

REFERENCES

1. "Process Science for 2D Carbon-Carbon Exit Losses", prepared by Aerojet Strategic Propulsion Co., AFWAL Report No. TR 86-4028, July 1986.
2. K. T. Clemens and S. C. Brown, "Real-Time Acoustic Emission Monitoring/Control Technique for Carbonization of 2-D Phenolic-Carbon Composites", 8th JANNAF RNTS Mfg., Patrick AFB, October 1986.
3. W. J. Pardee, M. R. Mitchell, A. Gupta, F. Montgomery and J. Sheehan, "Effect of Carbonization Kinetics on In-Process Mechanical Properties", Proceedings of the ASM Conference, Indianapolis, IN, October 1989, in press.
4. J. R. Bulau, "AE Monitoring for Control of Carbon-Carbon Pyrolysis", Proceedings of IEEE 1988 Ultrasonics Symposium (B. R. McAvoy, Editor), IEEE Cat. No. 88 CH 2578-3, pp. 1057-1063 (1988).
5. B. R. Tittmann and J. R. Bulau, "Acoustic Emission and Ultrasonics Sensing for Carbon-Carbon Pyrolysis Monitoring", Proceedings of the ASM Conference, Indianapolis, In, October 1989, in press.
6. B. R. Tittmann, "Acoustical Studies of Damage Mechanism in Carbon-Carbon during First Carbonization", Proceedings of the 1989 IEEE Ultrasonics Symposium, in press.

7. B. R, Tittmann, "High Temperature Young's Modulus of Carbon-Carbon Composite Material" Proceedings of The ASME Conference, San Francisco, Ca 1989, in press.
8. B. R. Tittmann and J. R. Bulau, "NDE of Delaminations during Processing of Carbon-Carbon Composites", Proceedings of the International Symposium on Acoustical Imaging, Santa Barbara, Ca, 1989, in press.
9. W. J. Pardee, R. C. Addison, D. J. Miley, B. S. Seely, M. A. Shaff and B. R. Tittmann of Rockwell International; A. Gupta of Ioptix; and F. Montgomery and J. Sheehan of General Atomics, "Intelligent Control of Carbon-Carbon Pyrolysis", Proceedings of the 1989 TMS Fall Meeting, Indianapolis, IN, October 1989, in press.

NONDESTRUCTIVE CHARACTERIZATION OF SOLIDS LEVELS IN HIGH VISCOSITY EMULSIONS

F. Villamagna and M. C. Lee
ICI Explosives Group Technical Center
801 Richelieu blvd, McMasterville, Québec, Canada J3G 1T9

J. -D. Aussel and C. K. Jen
IMRI, National Research Council of Canada
75 de Mortagne blvd, Boucherville, Québec, Canada J4B 6Y4

1 INTRODUCTION

Emulsion explosives are extensively used in the construction and mining business. The physical consistency of emulsions can range from a stiff, putty-like substance to a fluid, pumpable product, depending on the choice of fuels and process [1]. Emulsions which are the most recent generation of commercial explosives, are a safe, cost-effective and high performance alternative to conventional nitroglycerine and slurry explosives. Their high velocity of detonation and exceptional water resistance offer superior rock breaking properties and wider usage on surface or underground.

Emulsions are made of two non-dissolving liquids, a water solution of nitrate salts, usually ammonium, calcium or sodium, and a liquid fuel. This mixture of liquid oxidizer and liquid fuel is not explosive until sensitizers, such as gas microballoons (MB), are added. In addition, metal powder such as aluminum (Al) may be added to emulsions to increase the fragmentation and heave. Despite recent improvements, homogeneous or heterogeneous nucleation of the disperse phase during storage still remains as an unavoidable consequence of their composition. The performance and shelf life of emulsions are highly affected by the total level of solids. Therefore it is of practical interest to develop a technique which can detect the salt content in the field, or the concentration of MB or Al inclusions as an on-line process.

Ultrasonic pulse-echo methods are widely used as a nondestructive technique for the detection of the flaws, inclusions and voids, and the material characterization [2]. In this paper we describe a simple, fast and nondestructive ultrasonic pulse-echo technique which can be used in the field or on-line to characterize the "inclusions" of emulsions. This technique is based on the measurement of the ultrasonic velocity dispersion, attenuation and statistics of the scattering noise induced by the inclusions in emulsions.

2 SAMPLE PREPARATION

Because of the uncertainty associated with irradiating explosives with ultrasonic pulses (see section 3), all the samples used in this study were sodium chloride based dummy emulsions. The presence of solid crystals was mimicked by adding ground sodium chloride, with an average particle size of 425, 210, 150, 106 and 75 μm. Samples with microballoons (MB) and aluminum particles (Al) are prepared with 3M microballoons and atomized aluminum respectively.

Since trapped air is an ultrasonic scatterer, air entrainment during sample preparation has been carefully minimized, which in some cases included degassing samples under vacuum prior to taking the measurements. The oil phase of the dummy emulsions was composed of a 4:1 ratio of paraffin oil and Tween 80, with the liquor to oil weight ratio kept at 94:6. Each sample was packed in a thin plastic

Nondestructive Characterization of Materials IV
Edited by C.O. Ruud *et al.*, Plenum Press, New York, 1991

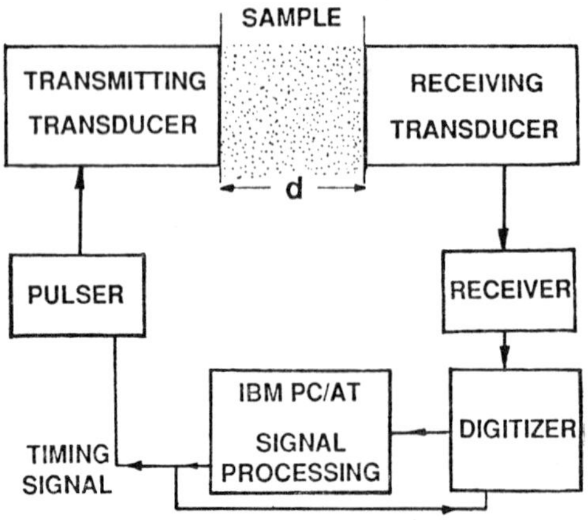

Figure 1 Experimental set-up.

Figure 2 Transmission geometry used for the measurement.

cylinder with a diameter of 25 mm. They were tested within a few days of preparation to avoid any emulsion breakdown, even though all the doped samples were found to be stable for a period of several weeks under ambient storage.

3 EXPERIMENTAL SET-UP

The set-up for ultrasonic dispersion and attenuation measurement is shown in Fig. 1. It consists of a commercial pulser and receiver, a pair of ultrasonic transducers, a digitizer (Tektronix 7854) and an IBM PC/AT. A signal processing software package developed at IMRI [3] is used for the signal acquisition and processing.

Because of the high ultrasonic attenuation in the samples, a transmission geometry was used to keep a sufficiently high signal-to-noise ratio (SNR). An ultrasonic transducer serves as the transmitter and another as the receiver. These two transducers are aligned and separated by a constant distance *d*. Since the packaged samples are soft and deformable, as shown in Fig. 2, they can be squeezed between the two transducers and the ultrasonic path length *d* is kept constant for all the measurements. Ultrasonic transducers centered at frequencies 0.5, 1.0, 2.25 and 5.0 MHz were tested. 2.25 MHz was found to be the most suitable frequency, because it provides the received ultrasonic pulses with a proper SNR and frequency bandwidth.

4 HAZARD ANALYSIS

Ultrasonic pulses irradiating the emulsion explosives can be regarded as hot spot sources, and a detailed hazard analysis is necessary because of the safety concerns. Three possible initiating mechanisms have been identified:

(i) sample heating due to acoustic absorption: heating due to acoustic absorption has been extensively reviewed [5]. It has been shown to increase temperatures only a few degrees per hour even with an ultrasonic power of several watts. Since the typical average power for ultrasonic pulses in our measurements is about a few mW or less, the heating mechanism would not provide an initiation source. The pressure pulse generated from the ultrasonic transmitter used in our measurements is less than 1 cal/cm^2, which is much lower than the minimum shock initiation energy of 50 cal/cm^2 for most high energy emulsion explosives [6].

(ii) compression type event from the pressure pulse and (iii) acoustic cavitation [3]: the compression mechanism of aerated samples is also an unlikely means of initiation. When ultrasonic waves are passed through a fluid, they create regions of compression in front of the pulse, and regions of expansion immediately behind the pulse. During the expansion process a cavity can form in the reduced pressure region by either volatilizing some of the fluid, nucleating the dissolved gasses, or obtaining gas from the crevices on solid particles in the fluid. For emulsion explosives, a small number of air bubbles may also be present in the system due to incorporation of air at the solids blending stage. Submitted to rapid ultrasonic pulsing, the bubbles are exposed to successive compression and expansion cycles. They continually grow and contract, keeping a dynamic balance between the vapor inside the bubble and the liquid outside. In some cases, the ultrasonic waves will sustain a bubble as it simply oscillates in size. In other cases the average size of the bubble will steadily increase, until the bubble can migrate to a surface and break. This is generally observed with high intensity sound waves in low viscosity fluids [7]. For low intensity ultrasonic waves and high viscosity fluids, the cavity is much more likely to oscillate in phase with expansion and compression cycle for long periods of time [7]. Since gas diffusion is a function of surface area, and the pressure differences during compression and expansion are not excessively large, the amount of gas that diffuses into the cavity during expansion is greater than what diffuses out during compression. If the cavity cannot migrate to a surface, it will eventually reach a critical size where it can no longer sustain itself and collapse in less than a μs. The collapse will generate localized temperatures in excess of 2000°C [4], creating hot spots which can then serve as possible initiation mechanisms.

To reduce the initiation possibility to zero, the following choices were made:

(a) transmission geometry instead of reflection geometry. Indeed, the ultrasonic attenuation varies exponentially with the distance, and since the propagation length in transmission is one half of that in reflection, less ultrasonic energy is required to obtain the same SNR.

(b) large diameter transducers to reduce the ultrasonic energy per unit area, and ensure homogeneity. In the experiments since the sample diameter is 25 mm, transducers with 12.7 mm diameter aperture were selected.

(c) minimal repetition rate of the ultrasonic pulses to reduce the average ultrasonic energy present in the sample. Single pulse measurements with enough SNR are feasible with signal acquisition and processing software packages [3]. However, the averaged data obtained with a large number of pulses has higher SNR, thus higher accuracy, than data obtained by single pulse.

(d) the distance *d* between the transducers was kept to minimum, but was large enough to ensure homogeneity of the sample and allow measurement of ultrasonic parameters such as time delay and attenuation.

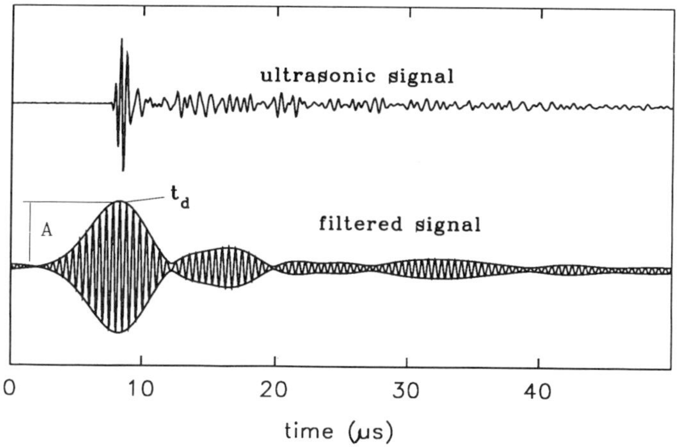

Figure 3 A typical ultrasonic signal unprocessed (upper trace), and narrow-band filtered around 2 MHz (lower trace). The attenuation ***A*** and time delay t_d are measured on the filtered signal. The ***RMS*** and ***SNR*** values are measured on the unprocessed ultrasonic signal.

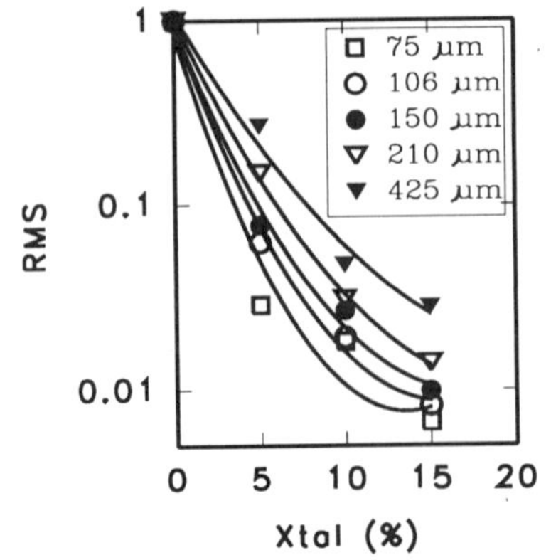

Figure 4 Variations of the ultrasonic RMS value vs the content of crystals. The lines are regressions of the experimental points.

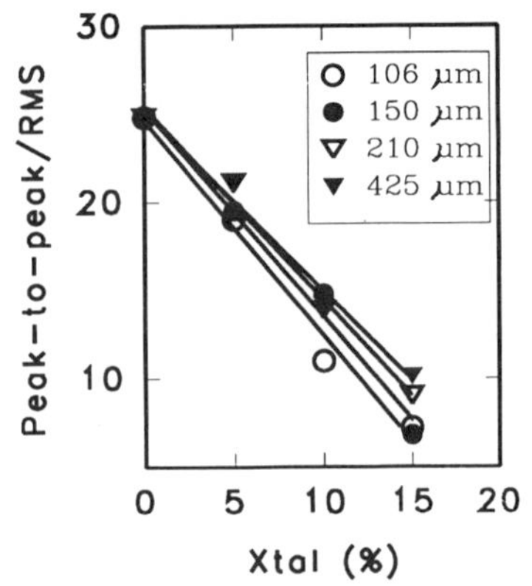

Figure 5 Variations of the ultrasonic SNR (peak-to-peak to RMS value ratio) vs the content of crystals. The lines are regressions of the experimental points.

5 ULTRASONIC FEATURE EXTRACTION

A typical ultrasonic signal measured with our experimental set-up is shown in Fig. 3. It consists of a short-time duration ultrasonic echo followed by a long-time duration ultrasonic noise, which is caused by the interference of the ultrasonic waves scattered by the inclusions. The information on the inclusions size and concentration can be measured both on the ultrasonic echo and the backscatterred noise.

The inclusions cause a frequency dependent attenuation and time delay of the ultrasonic echo. The time delay t_d and the amplitude A of the ultrasonic echo are measured after narrow-band filtering of the signal [8], as illustrated on Fig. 3. The center frequency of the filter is varied by software to measure the attenuation and time delay at various frequencies.

The inclusions also generate a scattering noise, whose statistics depend on the concentration and the size of the inclusions. The measured statistical parameters are: (i) the root-mean-square value ***RMS***, which characterizes the absolute amplitude of the signal, and (ii) the ratio of the peak-to-peak amplitude to the RMS value, which measures signal-to-noise ratio (***SNR***) defined as the ratio of the ultrasonic echo signal to the scattering noise RMS value. Typical results are outlined in the following two sections.

6 CHARACTERIZATION OF THE SALT CONTENT

Figures 4 and 5 show the variations of ***RMS*** value and of ***SNR*** versus the crystallization content, for 5 different crystal sizes ranging from 75 to 425 µm. The ***RMS*** value is normalized to the crystal-free measured RMS value, and it decreases sharply to about 0.01 at 15 % crystal content. This decrease is sharper for smaller particle sizes. The decrease of ***SNR*** is slower and nearly linear, and almost independent of the particle size. The ***SNR*** values corresponding to 75 µm crystal size are not shown in figure 5, because of large error caused by the small ***RMS*** values. Figures 4 and 5 could be used as calibration curves for characterization of the level of crystallization of explosive emulsions, which is an unavoidable consequence of their storage and aging process. The percentage of crystal can be estimated from ***SNR*** (Fig. 5), which is quite independent of the crystal size, and from ***RMS*** with known crystal size. In addition, if the percentage of crystal is known, the crystal size could be obtained from ***RMS***.

7 CHARACTERIZATION OF THE AL AND MB CONTENT

Figures 6 and 7 show the variation of the ultrasonic velocity at 2 MHz and of the ultrasonic attenuation at 1 MHz with the percentage of Al and MB inclusions. The three curves of each figure correspond to Al, MB and Al with equal content of MB; 6% inclusions of Al and MB means 3 % Al mixed with 3 % MB. The velocity dispersion and the attenuation are stronger for MB inclusions than for Al

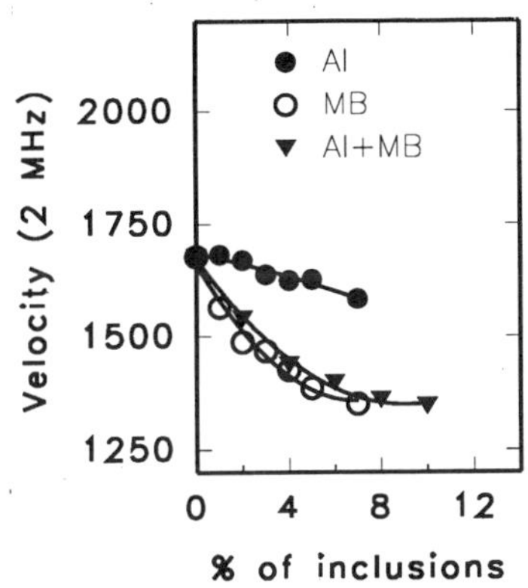

Figure 6 Variations of the ultrasonic velocity at 2 MHz vs the content of Al and MB inclusions. The lines are regressions of the experimental points.

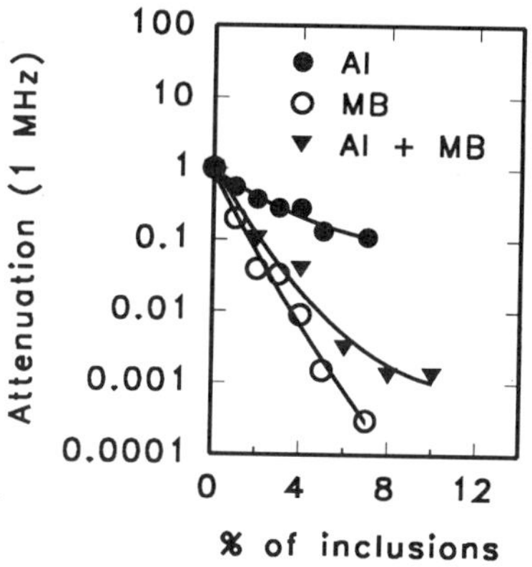

Figure 7 Variations of the ultrasonic attenuation at 1 MHz vs the content of Al and MB inclusions. The lines are regressions of the experimental points

inclusions. These curves can be used as calibration curves for the measurement of the inclusion content, provided that only one inclusion type is present (either Al or MB). However the estimation of the Al and MB content in an emulsion containing both Al and MB is not straightforward and requires additional information, either from the RMS value or the SNR of the signals. A pattern recognition approach is currently under study to perform this simultaneous estimation.

8 CONCLUSIONS

We have presented an ultrasonic method of quantifying solid contents. The method is simple and nondestructive, and relies on measuring the attenuation, time-of-flight, and statistics of the broadband ultrasonic pulse traversing the sample. Means of reducing and possibly eliminating acoustic cavitation include use of a transmission geometry, large diameter ultrasonic transducers, minimal ultrasonic pulse repetition rate and ultrasonic path.

The level of solid caused by the aging process of the emulsion explosives has been successfully measured from the root-mean-square value and the amplitude-to-rms-value ratio of the ultrasonic signals. Such measure could provide a on-field quality testing of the emulsions.

The content of either aluminum or microballoon inclusions alone can be measured from the velocity dispersion or the attenuation of the ultrasonic waves in an emulsion. These measures could be useful as an on-line characterization method during the emulsion fabrication. Further study includes a pattern recognition approach to measure simultaneously the respective contents of aluminum and microballoon inclusions.

REFERENCES

[1] Blasters' Handbook, 6th ed., C-I-L Inc. Explosives, North York, Ontario, Canada, September (1987) pp. 528-529

[2] J. Krautkramer and H. Krautkramer, Ultrasonic Testing of Materials, 2nd ed., Springer Verlag, New York, (1977)

[3] J. -D. Aussel, J. -P. Monchalin, J. F. Bussière, D. Malenfant, P. Roy, J. -F. Cusson and C. Carnois, "*A PC-based data acquisition and signal processing package for ultrasonic nondestructive testing*", Canadian Society of NDT Journal, Vol. 10, (1989) pp. 18-27

[4] K. S. Suslick, "*The chemical effects of ultrasound*", Scientific American, (1989) pp. 80-86

[5] K. S. Suslick, "*Ultrasonic - Its chemical physical and biological effects*", VCH Publishers, New York, (1988)

[6] Private Communication, B. Mohanty, ICI Explosives

[7] C. K. Holland and R. E. Apfel, "*An improved theory for the prediction of microcavitation thresholds*", IEEE Transactions on Ultrasonics, Ferroelectrics and Frequency Control, Vol. 36, (1989) pp. 204-208

[8] J. -D. Aussel and J. -P. Monchalin, "*Study of surface acoustic wave dispersion using laser-ultrasound and application to thickness measurement*", Review of Progress in Quantitative Non Destructive Evaluation, D.O.Thompson and D.E.Chimenti Editors, Plenum Press, New York, Vol.8 (1989) pp. 535-542

IN PROCESS CHARACTERIZATION OF GALLIUM ARSENIDE CRYSTALS BY X-RAY DIGITAL RADIOGRAPHY AND COMPUTED TOMOGRAPHY

J.W. Eberhard, K.W. Mitchell, R.A. Koegl and J.M. Brown

General Electric Corporate Research and Development
P.O. Box 8 Schenectady, NY 12301

INTRODUCTION

Gallium arsenide has promised to revolutionize the semiconductor industry for many years. However, numerous practical difficulties have prevented this potential from becoming reality. As part of a DARPA funded effort to enhance the manufacturability of gallium arsenide crystals, we have designed and implemented an x-ray inspection system for imaging these crystals during growth. The objective of the work is to measure melt height, crystal diameter, meniscus shape, and liquid-solid interface shape of a 3 in. diameter crystal grown using the liquid encapsulated Czochralski process. The shape of these structures is a critical factor in determining whether the growth process results in a single crystal with good electrical properties, or whether twinning or polycrystalline growth is present.

A third generation x-ray computed tomography inspection system using a 640 element linear array detector has been used to achieve both digital radiography and computed tomography images of gallium arsenide crystals during growth. X-ray inspection of crystals during growth has been reported previously[1,2], but we are unaware of any previous attempts to use CT for this purpose. The system can collect data and reconstruct 600 x 600 images with 0.010 in. pixel elements in 5 to 10 seconds. These images have been analyzed to provide critical information about crystal growth conditions over the course of a run, including melt height, crystal diameter, meniscus shape, and liquid-solid interface shape (Figure 1). In addition, the x-ray images provide a standard of comparison for evaluation of other sensor data acquired simultaneously. The usefulness of a CT imaging system for in process inspection is clearly demonstrated.

X-RAY IMAGING SYSTEM

Optimum performance of an x-ray imaging system requires that a custom design be undertaken for a particular application. This is particularly important for an in process inspection system, since the requirements of high speed inspection place greater than normal demands on most key system components. The x-ray system which has been constructed for GaAs imaging is a third generation CT system, composed of several major subsystems and a system executive. The key components are a 420 KVP Phillips x-ray tube, a custom manipulator to position source and detector for imaging appropriate sections of the crystal, a GE patented high pressure 640 element xenon ionization chamber detector, a matching hybrid data acquisition system, a STAR ST-INT array processor with custom hardware backprojection boards, a CAMAC based system controller for providing sense information about the crystal including rotation rate and position, and an image processing and display package, all controlled by system executive software running on a MicroVAX GPX computer.

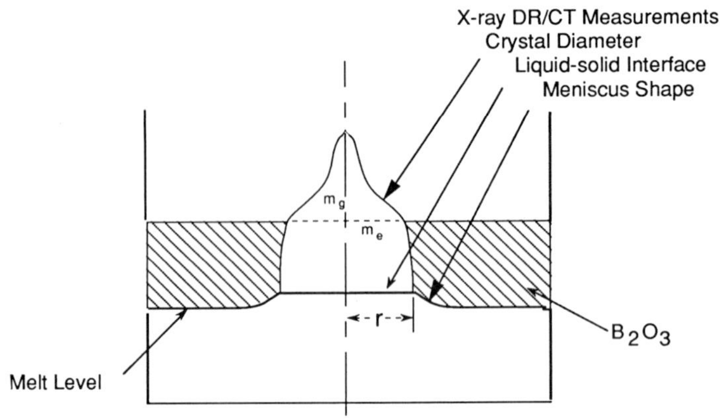

Fig. 1. Crystal Growth Parameters

The system is designed to image crystals of nominal diameter up to 3 in. with spatial resolution of 0.020 in. The system resolution function width is chosen as wide as possible, to allow the greatest possible number of x-rays to enter each detector element. This is particularly important in this application, since Gallium Arsenide is a highly x-ray absorbing material, grown at high temperature in a complex, x-ray absorbing crystal puller. In the final design, the linear array detector contains 640 elements on 0.010 in. centers. The source to detector spacing is 29.25 in., and the number of view angles ranges from 300 to 1200, depending on the rotation rate of the crystal.

To meet the timing requirements imposed by the crystal growth process, a third generation inspection system is required. In such a system, the detector array is wide enough to span the object of interest, and all the data for each view angle are acquired simultaneously. Data from subsequent view angles can thus be acquired in real time as the crystal rotates in the field of view. Both Digital Radiography (DR) and Computed Tomography (CT) images are available. In DR mode, the part is translated vertically through the beam as the detector array is sequentially sampled to produce the image one line at a time. In CT mode, the object is rotated at fixed height and the detector is sampled at a large number of closely spaced angles over a 360 degree range. Because of the special requirements of imaging a rotating crystal during growth, these basic imaging modes are modified slightly. In Spiral DR imaging, the data are acquired line by line in a fast vertical scan past the growing crystal. The rotation of the crystal is ignored, so each line of the image contains data from a different orientation of the crystal relative to the detector. A 4 in. vertical scan past the crystal requires approximately 6.7 seconds.

CT Mode is also slightly modified from conventional CT. In this mode, the rotation of the crystal in the puller provides the rotation required for CT imaging. X-ray data acquisition must be synchronized with the puller rotation rate. Time to acquire and reconstruct one CT image depends on crystal rotation rate and typically varies between 5 and 20 seconds. CT images are reconstructed in parallel pipeline fashion, and the image is complete less than 0.1 sec. after data acquisition is complete. In addition to reconstructing the detector data from all the view angles into a CT image, it can be displayed one line above the other - DR fashion - into a Sinogram image. Certain types of analysis can be done more easily on sinogram images than on DR images. In each imaging mode, a "differential" image capability is also provided. In "differential" mode, an image of the puller is recorded before the crystal growth run is begun. This image is subtracted from the data acquired during the growth run before the image is reconstructed and displayed in order to enhance the features of interest for GaAs characterization. The differential imaging capability removes most of the clutter due to the complex heater pack geometry from the x-ray images. In addition, it provides CT data which goes to zero at the edges of the measured field which is an important constraint for standard CT image reconstruction algorithms.

A cross-sectional view of the interior of the crucible is shown in Figure 2. DR images capture the geometry shown in the figure, including crystal diameter, melt height, and outer meniscus shape. CT images correspond to horizontal slices through the crucible at fixed height. At the height labeled "P18" in the figure, the crucible contains an outer ring of boric oxide, a very narrow ring of liquid GaAs, and a section of the GaAs crystal at the center. Therefore, CT images provide a method for directly determining the liquid-solid interface shape. The liquid-solid density difference has been estimated at 12%, which is difficult to detect due to the large x-ray absorption in the puller pressure vessel and heater pack. To improve the signal to noise ratio in the detector data, a set of heat shields has been modified to reduce the thickness of the crystal puller heater pack. A heat shield configuration with two out of three thin shields was used in the data acquisition reported in this work.

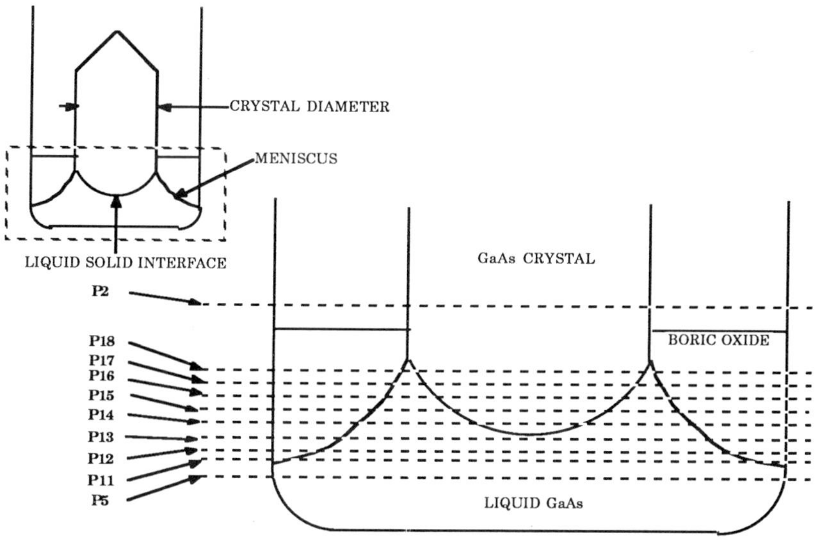

Figure 2. GaAs Meniscus/Interface Geometry

EXPERIMENTAL RESULTS

The images obtained using the x-ray system were used to measure the key crystal growth parameters of interest. Both the melt height and the meniscus shape were obtained from the DR images. Diameter measurements were extracted from the sinogram and CT images. In addition, qualitative information concerning the liquid-solid interface was obtained from the CT images.

Since the density of the liquid gallium arsenic is greater than the density of the boric oxide, a boundary exists between the two regions in the DR image. Figure 3 shows a DR image of the crystal during growth. The boundary between the encapsulant and the melt is clearly visible in this image. This boundary is enhanced and detected using a Sobel filter specifically designed for horizontally oriented edges[3]. Examining the melt height extracted from several DR images taken at different times during the growth process provides a mechanism for tracking the liquid level as a function of time. Figure 4 shows how the melt height varies during the course of the run. The large variation in melt height indicated by the plot in Figure 4 occurred during an early growth run and is not typical. In general, the melt level can usually be controlled to within approximately +/- 0.060 in. of the desired level.

Figure 3. Digital Radiograph Image of GaAs Crystal During Growth

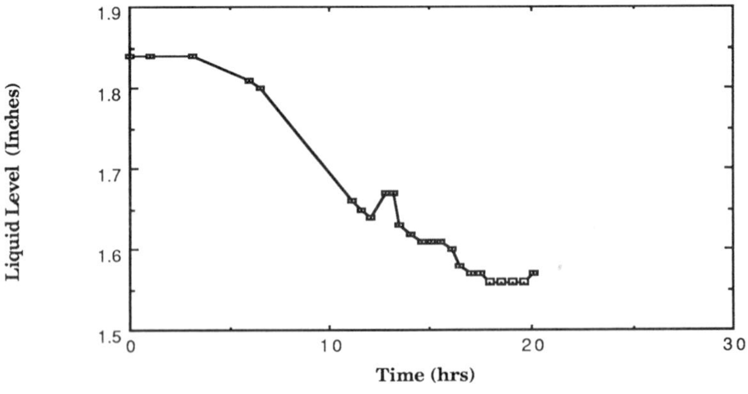

Figure 4. Melt Height Variation as a Function of Time

In addition to melt height level, the outer edge of the meniscus can also be determined from the DR image. In order to extract the shape, the edges of the DR image are once again enhanced using an edge enhancement filter. The distance between parallel sets of edge pixels is then measured. Depending on the height of the edge pixels above the melt level, the distance measured will correspond to either the crystal width or the crystal plus meniscus width. However, this approach is not the optimal method for mapping the outer meniscus shape, because the crystal width varies with respect to the angle at which it is measured. A cross-sectional slice through the crystal has an elliptical shape as opposed to a circular one. The DR image represents a "snapshot" of the crystal at one angle for each height along the object. Consequently the accuracy of the measured width cannot be determined without knowledge of the rotational position of the crystal at the time that the line of the DR image in question was taken.

Crystal diameter measurements can be made accurately from the sinogram images, however. The sinogram image consists of x-ray detector data sets at a fixed height on the crystal taken at different view angles. The sinogram data acquired consisted of 600 lines of data which corresponded to two complete revolutions of the crystal. From each line of the data a crystal width measurement was made. This resulted in crystal width measurements every 1.2° around the crystal. The accuracy of these measurements was then validated with physical measurements. The physical measurements were obtained using a coordinate measuring machine (CMM). Crystal width measurements were taken every 10° around the crystal. A comparison between the CMM measurements and width measurements from the sinogram data for one particular height is shown in Figure 5. Notice that there is a nearly constant offset between the two sets of measurements. This is the result of magnification effects related to slight offsets in crystal position in the x-ray imaging system. However, the important characteristic to note is the relative shapes of the two curves, which are identical. The fact that the 0.050 in. variation between the major and minor axis of the crystal can be accurately measured from the sinogram data demonstrates the benefits of using x-ray imaging techniques in this application.

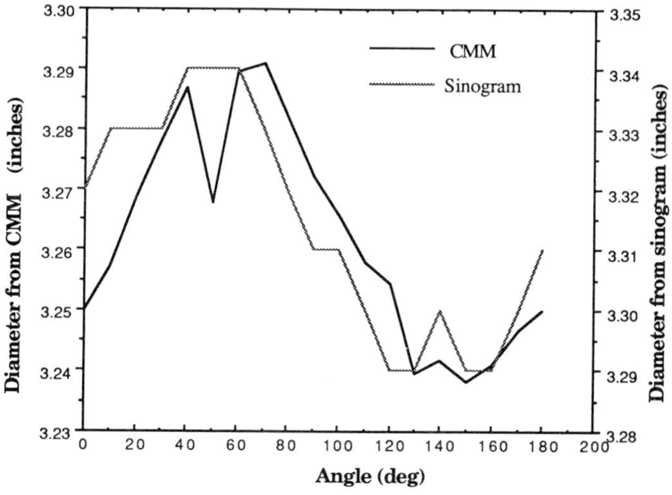

Figure 5. GaAs Crystal Width Measurements using CMM and Sinogram Data

The outer edge of the meniscus can also be measured from the sinogram data. Each sinogram image contains information about the crystal at a particular height. For sinogram images significantly above the melt level (> .100 in), the crystal diameter can be measured directly. In sinogram images taken near the melt, the total width of the crystal plus the liquid is measured. By measuring this width in several sinogram images, each corresponding to a different height along the crystal, it is possible to map out the shape of the outer edge of the meniscus region. Figure 6 presents a plot of the outer edge obtained using this approach.

Neither the DR images nor the sinogram images are appropriate for measuring the width of the meniscus region at a given height. However, high quality CT images provide information which can be used to determine the location of the liquid-solid interface. Figure 7 shows a typical CT image of the crystal and the surrounding encapsulant. After processing the image, qualitative information concerning the size of the liquid region was obtained.

The major problem in quantitatively determining the size was the low signal-to-noise (SNR) ratio of the CT images, caused by the crystal puller pressure vessel and heater pack. To improve the SNR, spatial filtering was used to enhance the contrast of the liquid region of the image with respect to the solid. The processed elliptical crystal image was then mapped into a rectangular array by extracting data along radial lines from the image. The creation of the rectangular array facilitated the use of standard 2D filtering techniques while still retaining almost all of the continuity and resolution of the liquid region present in the original CT image. It is important that the processing techniques not distort or reduce the resolution of the liquid-solid interface. The resulting image was then down sampled to remove the high pixel-to-pixel variation caused by the random noise still present. The image was down sampled such that one line in the resulting image corresponded to the average of the data contained in a 22.5° wedge of the spatially filtered CT image. The down sampling was performed three times with offsets of 10° and 15° for the second and third applications respectively. This resulted in the creation of an array containing 48 rows. Where the first row corresponded to the average of a 22.5° wedge of data centered on 11.25°; the second row was the average of a 22.5° wedge centered on 21.25°; the third row was centered on 26.25° and so on.

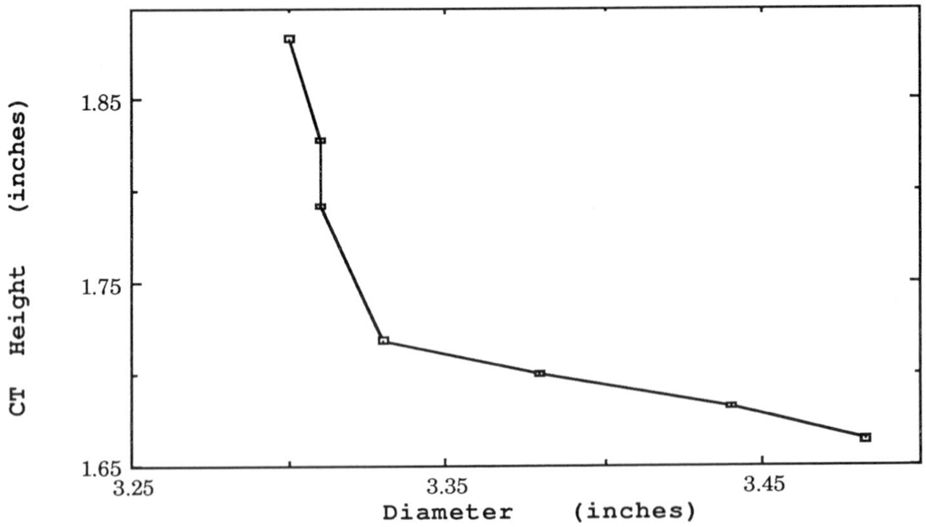

Figure 6. Outer Edge of Meniscus Obtained From Sinogram Data

Although the difference between the density of the liquid and solid GaAs is small, it is possible to see the variation in the size of the liquid region using the processed CT images. This is evident in Figure 8 which shows wireframe representations of two processed CT images. Figure 8b was obtained from a CT image taken 0.140 in. above the melt level. Figure 8a was taken only 0.040 in. above the melt. The liquid regions have been further enhanced in both images using a process called contrast stretching[4]. Clearly the liquid region is much wider in the image corresponding to the CT image taken closer to the melt, as expected (Figure 2).

A second obstacle in quantitatively measuring the width of the liquid is a phenomena known as beam hardening[5]. The result of beam hardening is that even in the absence of a liquid region the CT images would still exhibit higher pixel values near the edge of the crystal than in the interior. A mitigating characteristic of this effect is that it is only a problem when one is trying to accurately characterize liquid regions which are thinner than the region of edge pixels which are normally affected by beam hardening. However, for wide liquid regions such as the one indicated by Figure 8a, it is clear that the high ridge is mostly due to the presence of the liquid. On the other hand in Figure 8b, it is conceivable that a significant portion of the high ridge is due to the effects of beam hardening. Techniques for removing or minimizing the effects of beam hardening are currently being examined. The success of these techniques coupled with improved CT images should allow for quantitative characterization of the liquid-solid interface.

SUMMARY AND CONCLUSIONS

A custom third generation digital x-ray inspection system was used to acquire digital radiography and computed tomography images of gallium arsenide crystals during growth. The images were used to measure melt height, crystal diameter, and outer meniscus shape over a 24 hr. growth run. In addition, the gallium arsenide liquid-solid interface was detected in CT images near the melt level, and the width of the liquid region was seen to increase nearer the melt, as expected. Both beam hardening corrections and thinner heat shields are planned to allow more quantitative determination of the liquid width and the detailed liquid-solid interface shape.

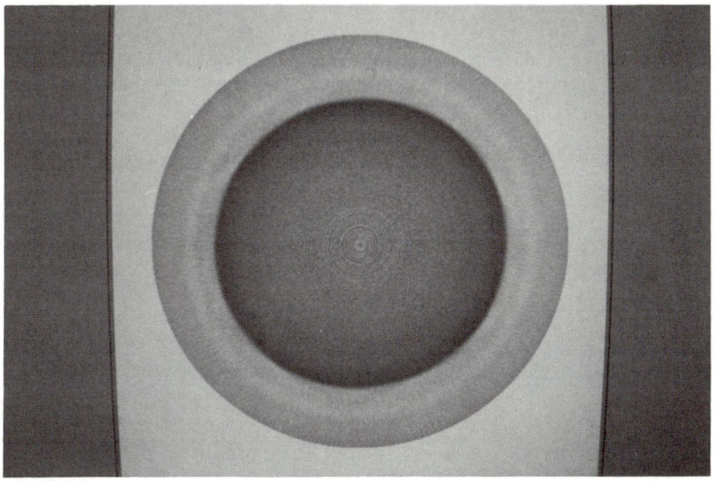

Figure 7. CT Image of GaAs Crystal During Growth

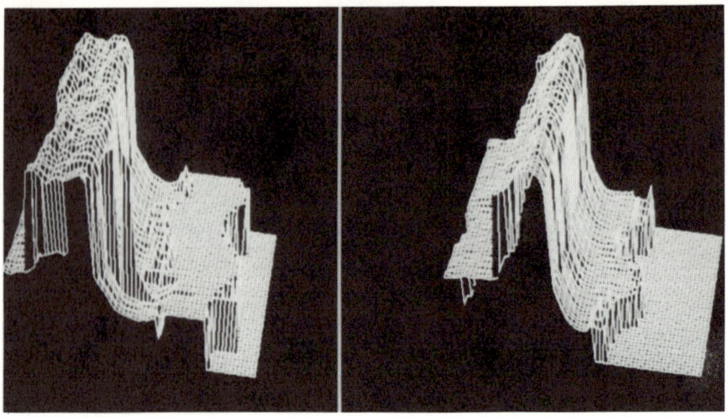

(a) (b)

Figure 8(a).Processed CT Image of Region .040" Above the Melt

Figure 8(b)Processed CT Image of Region .140" Above the Melt

ACKNOWLEDGEMENTS

The authors would like to acknowledge the Defense Advanced Research Projects Agency and the Army Research Office which supported the work under contract DAAL03-86-C-0022.

REFERENCES

1. Barber, P.G., Crouch, R.K., Fripp, Jr., A.L., Debnam, Jr., W.J.,Berry, Jr., R.F., and Simchick, R., "A Procedure to Visualize the Melt-Solid Interface in Bridgman Grown Germanium and Lead Tin Telluride", Journal of Crystal Growth, Vol. 74, pp. 228-230, 1986.
2. Ozawa, S., Miyairi, H., Nakajima, M., and Fukuda, T., "Growth and Characterization of Undoped Dislocation Free GaAs by Arsenic Ambient Controlled LEC with the Aid of X-ray Imaging System", Inst. Phys. Conf. Ser. No. 79, Chapter 1, Adam Hilger, Ltd., pp. 25-30, 1986.
3. Gonzalez. R.C., and Wintz, P., "Digital Image Processing", Chapter 7, pp. 337, Reading, Mass, Addison-Wesley Publishing, 1979.
4. Jain, A.K., "Fundamentals of Digital Image Processing", Chapter 7, pp.235, Englewood Cliffs, NJ, Prentice-Hall, 1989.
5. Kak, A.C., "Computerized Tomography with X-Ray, Emission, and Ultrasound Sources", Proceedings of the IEEE, Vol 67, No. 9, pp. 1254 , Sept. 1979.

COMPUTERIZED TOMOGRAPHY OF HIGH EXPLOSIVES

H.E. Martz, D.J. Schneberk, G.P. Roberson,
S.G. Azevedo, and S.K. Lynch†

Nondestructive Evaluation Section, L-333
Lawrence Livermore National Laboratory
P.O. Box 808, Livermore, CA 94550

INTRODUCTION

Computerized tomography (CT) techniques are being investigated to improve the current manufacturing process flow of high explosives (HE). In order to revise the manufacturing process flow, it must be demonstrated that CT can provide information comparable to the techniques currently used: 1) shadow graphs, to determine gross outer dimensions; 2) radiography, to detect cracks, voids, and foreign inclusions; 3) dye penetrant, to identify surface cracks; 4) a wet/dry bulk-density measurement technique; and 5) destructive core sampling for exhaustive density measurements.

To meet this objective, a series of CT experiments have been designed and were divided into three phases. The phase I experiments are discussed here and in more detail elsewhere.[1] Application of x-ray CT could significantly improve safety, and greatly reduce the process-flow time and costs, since CT could simultaneously determine rough outer dimensions, bulk density, density gradients, and the presence of cracks, voids, and foreign inclusions within HE.

Until now, x-ray CT studies of HE materials have resorted to a study of mock-HE materials, *i.e.*, polymeric materials whose x-ray attenuation is similar to, but not the actual material.[2,3] In this paper, our unique HE operational capability and laboratory-built CT scanners will be used to study actual PBX9502 high explosives and mock HE. In the next section, the PBX9502 chemical components and manufacturing process flow are discussed. The scans performed and discussion of results are given in the CT Experiments, and Results and Discussion sections, respectively. In the conclusion, the CT results presented here are related to the typically used HE investigation techniques mentioned above.

---

† Current address American Smelting and Refining Co., P.O. Box 30200, Amarillo, TX 79120.

†† TATB represents 1,3,5-Triamino-2,4,6-Trinitro-Benzene ($C_6H_6N_6O_6$).[5]

* Kel-F 800 consists of 75% chlorotrifluoroethylene polymer (C_2F_3Cl) and 25% vinylidenefluoride ($C_2H_2F_2$) with a calculated formula of $C_{2.1}H_{0.4}C_{0.6}F_{2.7}$.[5]

EXPLOSIVES (PBX9502) MATERIAL

In the current HE manufacturing process flow, PBX9502 components are prepared by mixing a TATB†† explosive powder with a Kel-F 800* plastic binder to insure a solid pressing which has certain mechanical properties (tensile, compression, etc.). The mixed powder is heated in an oven and poured into a rubber bladder which surrounds an aluminum mandrel. A vacuum is pulled on the rubber/aluminum mandrel assembly, which is placed in a hydrostatic or mechanical press. The mandrel assembly is pressed to 1406 kg/cm^2, which forms a rough pressing.

PBX9502 material properties result from its two constituents and the pressing process. Using Kouris' et al. chemical formula mixture rule,[4] effective-Z is calculated to be 7.1 for TATB and 12 for Kel-F. The PBX9502 pellet bulk proportions are 95% TATB and 5% Kel-F, resulting in an effective-Z of 7.6. Prior to the formulation, the density of TATB is ~1.938 g/cm^3 and ~2.02 g/cm^3 for Kel-F.[5] For the Phase I study, the pressed PBX9502 density value is ~1.9 g/cm^3.

The nondestructive evaluation of HE pressings involves two sets of issues. First, what is the integrity of the pressing? Are there any cracks and voids, or inclusions? Second, what are the material properties of HE? Does the Kel-F mix uniformly with the TATB? Is the component weight fraction correct? What is the HE density? Both sets of issues are addressed in the Results and Discussion section.

To meet facility safety requirements, the total amount of PBX9502 could not exceed 12 g for each experimental apparatus used in this study, and no other HE or detonator could be in the area. Therefore, two different sets of pellets were fabricated. Both sets were cylindrical in shape and differed only in length. One set consisted of four pellets ~1.27 cm diameter and ~1.27 cm long, and will be referred to as the small pellets. The second set consisted of three pellets ~1.27 cm diameter and ~4 cm long, and will be referred to as the medium pellets. The small pellets are of a size such that all four could be placed within the experimental apparatus, since their total weight does not exceed 12 g. This allowed CT experiments to be conducted on all four small pellets simultaneously, while only one medium pellet could be studied at a time.

CT EXPERIMENTS

Two CT scanners were used to evaluate the PBX9502 pellets. The pencil-beam computerized axial tomography (PBCAT) scanner[6] was used to nondestructively evaluate both the small and medium PBX9502 pellets in two-dimensions. The video computerized axial tomography (VCAT) scanner[6,7] was used to inspect one medium PBX9502 pellet in both two- and three-dimensions (2D and 3D).

The PBCAT scanner uses the first-generation ("translate/rotate") scan geometry. In this study, an ~63 mCi ^{109}Cd (energy levels: 22, 25, and 88 keV) radioisotopic source, a high-purity intrinsic-germanium detector, γ/x-ray spectroscopy electronics, and a multichannel analyzer were used to acquire the sinogram data. All PBCAT reconstructed images are presented as values of absolute linear attenuation coefficients (cm^{-1}). The small pellet data were acquired at two different slice planes.[1] Only the slice plane located at the center of the pellets, perpendicular to their longitudinal axes will be discussed. The sinogram data set consisted of 90 projections at 2° intervals over 180°, and 120 ray sums with a 1 mm^2 source and detector aperture and 1/3 mm translational overlap, resulting in a spatial resolution of ~1 mm. The total data acquisition time was on the order of 3 days resulting in, on the average, ~6% and ~9% photon counting statistics using the net counts from the 22 and 25 keV energy peaks respectively. The data were reconstructed into 120 x 120 pixel images (Fig. 1).

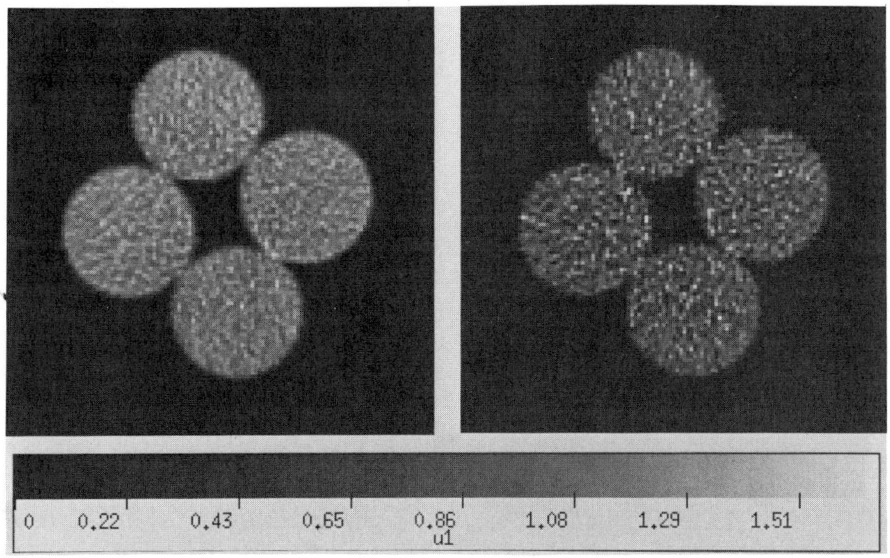

Figure 1. Reconstructed images from the PBCAT 22 (left) and 25 (right) keV sinogram data for the small high-explosive pellets at the center-slice plane. The greyscale is a measure of the attenuation coefficient in cm^{-1}.

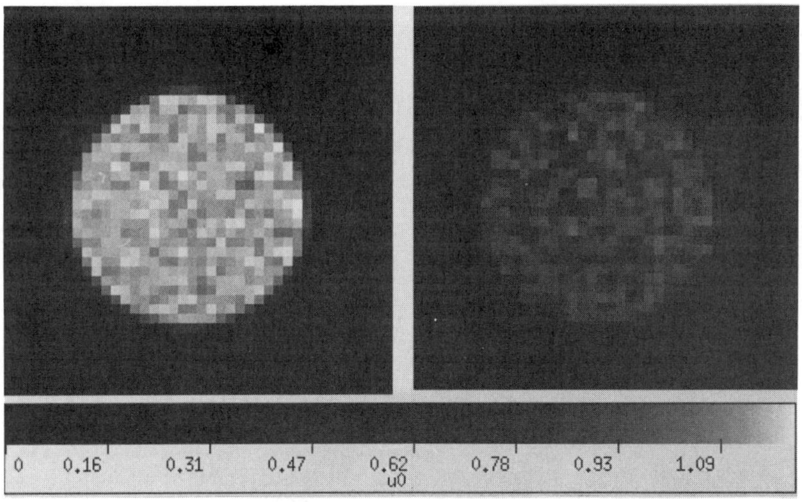

Figure 2. Reconstructed images from the PBCAT 22 (left) and 25 (right) keV sinogram data for the medium-pellet. The greyscale is in cm^{-1}.

One single medium pellet was scanned at two slice planes using PBCAT.[1] Only the slice plane at the pellet center, perpendicular to its longitudinal axis will be discussed. This sinogram data set consisted of 90 projections at 2° intervals over 180°, and 40 ray sums with a 1 mm^2 source and detector aperture and 1/2 mm translational overlap, resulting in a spatial resolution of ~1 mm. The total data acquisition time was 4 days. Photon counting statistics, on the average, were ~1% and ~3% for the 22 and 25 keV ray sum data, respectively. The medium pellet sinogram data were reconstructed into 40 x 40 pixel images (Fig. 2).

VCAT uses a third-generation ("rotate only") scan geometry, a microfocus x-ray tube (~10 μm spot, 25-160 kVp, 100 W) source, and an image-intensifier/CCD-video-camera detector system.[6,7] All VCAT reconstructed images are presented by relative attenuation values. The medium pellet sample was scanned using VCAT to acquire 103 contiguous slice planes from the pellet center to upper end, perpendicular to its longitudinal axis. The data were acquired at 42 kVp and 2.0 mA with 180 projections at 1° intervals and 267 ray sums, resulting in a spatial resolution of ~250 μm. The total data acquisition time was ~1 hr. The data were reconstructed into 103 two-dimensional 267 x 267 pixel images. All 103 images were both surface- and volume-rendered into 3D images (Fig. 3).

RESULTS AND DISCUSSION

The best pressed HE are uniform in both density and elemental composition. The HE pressings used in this study were manufactured to be uniform within each individual pellet, and from pellet-to-pellet. The CT data and multi-energy CT calculations were used to evaluate the PBX9502 pellets. First, the CT linear attenuation coefficient data are discussed in the Nondestructive Inspection section; the resultant atomic number and density data are discussed in the Material Characterization section.

Nondestructive Inspection

A qualitative inspection of and comparison among the CT image data (Figs. 1-3) reveals no major anomalies (*i.e.*, cracks, voids, and/or inclusions) within each PBX9502 pellet or from pellet-to-pellet at the slice plane(s) chosen. On the other hand, a CT study of mock HE pellets using

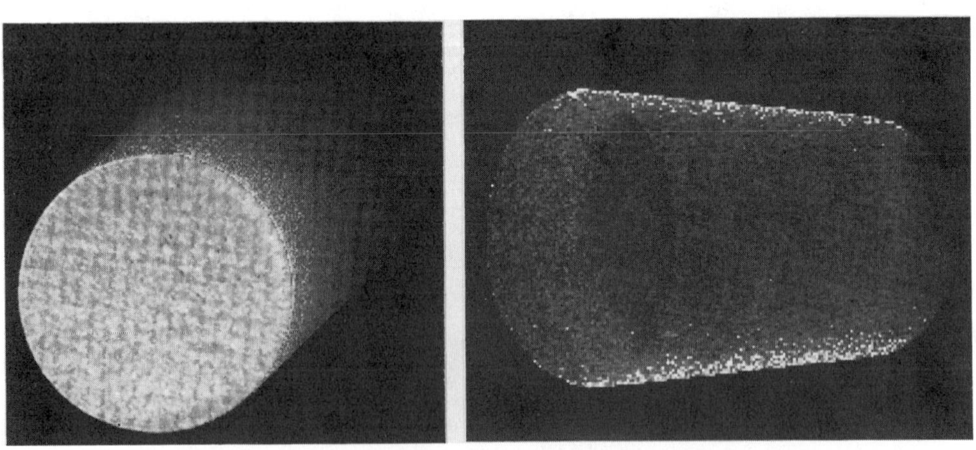

Figure 3. Surface-rendered CT image of the medium pellet (left). The distribution of shading in the image does not represent attenuation coefficient variations in the pellet. The shading is used to provide depth perception (3D) when confined to a 2D display medium. Volume-rendered CT image of the medium pellet (right). The distribution of shading in the image represents attenuation coefficient variations in the pellet.

PBCAT revealed a major anomaly (an inclusion) in one mock pellet (Fig. 4). The corresponding line out shows the amplitude (1.4 cm^{-1}) and size (1 mm^2) of the inclusion present. Thus, the CT results show that an inclusion is easily revealed in the mock HE, but none are present in the PBX9502 pellets for the slice planes evaluated as expected.

The difference from pellet-to-pellet may be quantified by comparing their mean and standard deviation, σ, values for an extracted region from each pellet. The standard deviation is a function of three components: (1) measurement errors (*i.e.*, photon counting statistics and systematic noise); (2) image reconstruction errors; and (3) atomic number, Z, and density, ρ, deviations within the explosives. The average small pellet PBCAT data resulted in mean and σ linear attenuation coefficient values of 1.1±0.1 and 0.9±0.2 cm^{-1} at 22 and 25 keV, respectively. The medium pellet's PBCAT data resulted in mean and σ values of 1.12±0.03 and 0.87±0.7 cm^{-1} at 22 and 25 keV, respectively. These data agree within experimental error with the individual values, and with the predicted values, 1.11±0.03 and 0.86±0.03 cm^{-1}.[1]

For the PBCAT data, a method has been determined to separate the measurement and reconstruction errors from the Z and/or ρ deviations.[1] Using this method a detailed analysis of the CT data in Figs. 1 and 2 (discussed elsewhere[1]) shows that a small proportion of the linear attenuation coefficient pixel-to-pixel deviations is not due to a combination of measurement and reconstruction errors; thus, we attribute the deviations to be due to subtle changes in Z and/or ρ.

Three conclusions are drawn from the CT nondestructive inspection of PBX9502: (1) both small and medium pellets contain a high degree of internal uniformity as measured by the linear attenuation coefficient for the PBCAT data and relative values obtained in the VCAT data; (2) the uniformity is maintained from pellet-to-pellet; and (3) in spite of this overall internal uniformity there appear to be small deviations due to changes in Z and/or ρ (discussed in detail below).

Material Characterization

Overall, the linear attenuation coefficient images are uniform (Figs. 1 and 2), however, this may not be true for the distribution of Z and ρ.

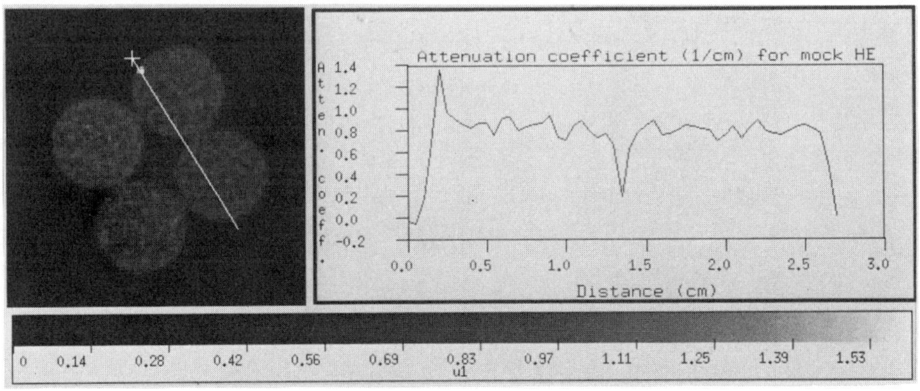

Figure 4. A reconstructed image from the PBCAT 22 keV sinogram data of mock HE pellets (left). The location of an extracted line out is shown by the white line. The line out (averaged over three adjacent pixels) is plotted to the right.

In PBX9502, changes in *Z* are a result of a variation in the distribution of TATB (*Z*=7.1) and Kel-F (*Z*=12) within the pressed pellet. Density variations can be a result of non-uniformities in the pressing process, or from higher amounts of Kel-F, which has a slightly higher density than TATB. Quantitative multiple-energy data generated by PBCAT for the medium pellet and techniques summarized in the Appendix were used to calculate *effective-Z, weight fraction,* and *density* images (Fig. 5).

At the slice plane chosen, the experimentally determined mean effective-*Z* is 7.8±0.5, which is within experimental error of the predicted value, 7.6. The variations in the effective-*Z* image are due solely to changes in the PBX9502 constituents (TATB and Kel-F) not upon density. This is just one component of the linear attenuation coefficient pixel-to-pixel variations (Fig. 2). The weight fraction coefficient calculations involve both tabulated attenuation coefficients and nominal density values for TATB and Kel-F (Appendix). The experimentally determined mean TATB weight fraction is 0.94±0.03. The expected TATB weight fraction is 0.95. The experimental and expected values are well within experimental error. The pixel-to-pixel variation in the weight fraction image tracks with the variation in the effective-*Z* image; *i.e.,* higher weight fraction values for Kel-F correspond with higher effective-*Z* pixels.

The calculated/estimated density image for the medium pellet also is presented in Fig. 5. The experimentally determined slice-plane mean density value is 1.9±0.2 g/cm^3 and is within the wet/dry bulk density measured value of 1.904 g/cm^3. This agreement provides some credibility to the density values on a pixel-to-pixel basis within the density image. There is considerably more variation in the pixel-to-pixel density values than in the linear attenuation coefficient images (Fig. 2). The large (10-20%) variations may be suspect. Currently a more rigorous determination of the pixelwise error variance for density images is being studied.

In comparing all three resulting images (Fig. 5), two small regions in the top left quarter (near the center) of each image stand out. For both regions, the effective-*Z* values are high and the TATB weight fraction values are low implying a large concentration of Kel-F, but the density values are unexpectedly low. The low density values can be explained by either TATB and/or Kel-F with a large void fraction. The effective-*Z* and weight fraction results suggest that the small regions are Kel-F with a large void fraction.

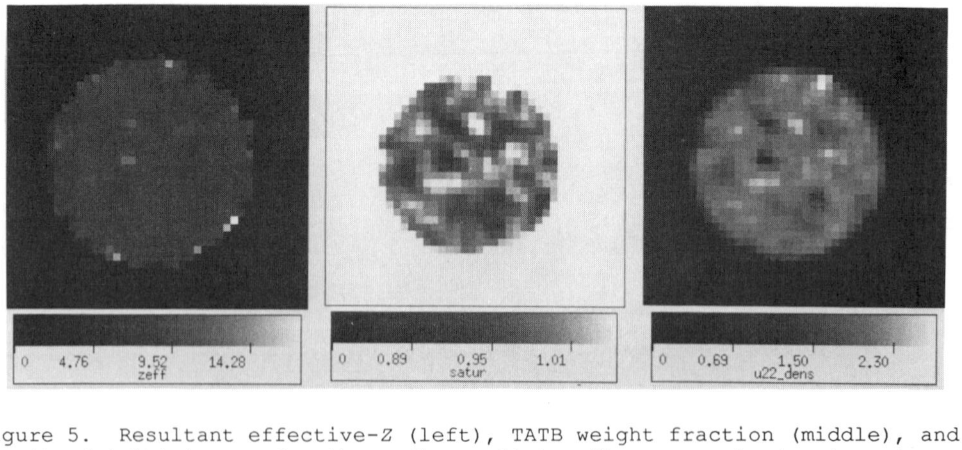

Figure 5. Resultant effective-*Z* (left), TATB weight fraction (middle), and density (right) images for the medium-pellet. The greyscale is given in effective-*Z*, TATB weight fraction, and density (g/cm^3) units, respectively.

CONCLUSION

The CT results have demonstrated the capabilities of CT technology applied to the inspection of HE, for example:

- using VCAT, spatial resolutions down to 250 µm were achieved;
- a foreign inclusion smaller than 2 mm^3 in size was detected;
- material density was determined within error of the reported wet/dry bulk-density measurement technique.

Overall, the linear attenuation coefficient data reveals that the PBX9502 is very uniform. However, there are a few variations in this data that are larger than the calculated error bars (determined from a combination of measurement and reconstruction errors). The experimentally determined effective-*Z*, weight fraction, and density results suggest the following:

- the mean effective-*Z* value is slightly higher than, but within error of, the predicted value;
- the Kel-F weight fraction value is slightly higher than, but within error of, the formulated value; and
- the linear attenuation coefficient pixel-to-pixel variations are a function of a nonuniform distribution of TATB, Kel-F, and density.

In summary, a preliminary comparison between CT and the current manufacturing inspection techniques reveals that CT may be able to replace most if not all of the current techniques used. If the small and medium pellet results are scalable to full size explosives (~45 cm diameter), then CT is a likely candidate for an automated high explosives inspection system. It should be pointed out that CT can be used to determine gradients in not only density but chemical constituents as well. Therefore, it may be necessary to reinvestigate the current bulk density specification with respect to wet/dry measurement techniques.

ACKNOWLEDGMENTS

The authors would like to thank Rich Whipple for providing the mock HE samples, and Linwood Hester and Derrill Rikard for their help in data acquisition and handling of the HE pellets. We also are grateful to Hyman Golopol for several meaningful HE chemistry discussions and to Brian Cabral for help in the 3D image rendering. Work performed under the auspices of the U.S. Department of Energy by the Lawrence Livermore National Laboratory under contract No. W-7045-ENG-48.

REFERENCES

1. H. E. Martz, G. P. Roberson, M. F. Skeate, D. J. Schneberk, S. G. Azevedo, and S. K. Lynch, High Explosives (PBX9502) Characterization Using Computerized Tomography, UCRL-ID-103318, Lawrence Livermore National Laboratory, Livermore (1990).
2. G. A. Sheppard, Development of a High-Resolution Inert High-Explosives Tomography Phantom, submitted to Appl. Radiat. Isot. (1988).
3. E. Östman, Determination of Density Resolution and Smallest Detectable Region in Explosives by X-Ray Tomography Scanning, in: "The American Society for Nondestructive Testing Topical Conference on Industrial Computerized Tomography Topical Proceedings," Seattle (1989).
4. K. Kouris, N. M. Spyrou, and D. F. Jackson, Materials Analysis Using Photon Attenuation Coefficients, in: "Research Techniques in Nondestructive Testing Vol. VI," R. S. Sharpe, ed., Academic Press, New York (1982).
5. B. M. Dobratz and P. C. Crawford, LLNL Explosives Handbook Properties of Chemical Explosives and Explosive Simulants, UCRL-52997, Lawrence Livermore National Laboratory, Livermore (1985).
6. H. E. Martz, S. G. Azevedo, J. M. Brase, K. E. Waltjen and D. J. Schneberk, Computed Tomography Systems and Their Industrial Applications, accepted by Appl. Radiat. Isot. (1990).

7. H. E. Martz, M. F. Skeate, D. J. Schneberk, and S. G. Azevedo, Design, Performance, and Application of a CCD Camera-Based CT System, in: "The American Society for Nondestructive Testing Topical Conference on Industrial Computerized Tomography Topical Proceedings," Seattle (1989).
8. D. G. Schneberk, H. E. Martz, S. G. Azevedo, and M. F. Skeate, Multiple-Energy Techniques in Industrial Computerized Tomography, submitted to Review of Progress in Quantitative Nondestructive Evaluation, D. O. Thompson and D. E. Chimenti, eds., Plenum Press, New York (1990).
9. D. E. Cullen, M. H. Chen, J. H. Hubbell, S. T. Perkins, E. F. Plechaty, J. A. Rathkopf, and J. H. Scofield, Tables and Graphs of Photon-Interaction Cross Sections from 10 eV to 100 GeV Derived from the LLNL Evaluated-Photon-Data Library (EPDL), UCRL-50400, **Vol. 6**, Rev. 4, Lawrence Livermore National Laboratory, Livermore (1989).
10. P. Engler and W. D. Friedman, Review of Dual-Energy Computed Tomography Techniques, Mats. Eval. **48**:623 (1990).

APPENDIX A

This appendix describes the mathematics behind the effective-Z, weight fraction, and density images that can result from dual-energy CT experiments.

The Effective-Z Image

A ratio of two CT reconstructed images at different energies reduces to a ratio of the x-ray photon cross sections, $\sigma(Z,E)$,[8]

$$R(Z,E_1,E_2) = \frac{\mu(Z,E_1)}{\mu(Z,E_2)} = \frac{\sigma(Z,E_1)}{\sigma(Z,E_2)}, \tag{A-1}$$

since $\mu(Z,E) = \sigma(Z,E)\rho\left(\frac{N_A}{A}\right)$, (A-2)

where $\mu(Z,E)$ is the linear attenuation coefficient, N_A is Avogadro's number, and A is the atomic weight. The ratio image is independent of density and only reflects the chemical (Z) constituents in that volume element, as seen by the x-rays. Using x-ray photon tabulated cross sections,[9] a function can be constructed which relates this CT ratio to a Z or effective-Z value. A polynomial interpolation technique is used to convert from the CT ratio value to an effective-Z value.

The Weight Fraction Image

Calculation of a particular component weight fraction image employs techniques similar to those described elsewhere.[10] Denoting the linear attenuation coefficients for the two components TATB and Kel-F as $\mu_T(E)$ and $\mu_K(E)$, the weight fractions as S_T and S_K, respectively, and the values for the reconstructed image as $\mu_R(E)$, we can express the attenuation coefficients at different energies as follows:

$$\mu_R(22) = \mu_T(22)S_T + \mu_K(22)S_K, \tag{A-3}$$
$$\mu_R(25) = \mu_T(25)S_T + \mu_K(25)S_K. \tag{A-4}$$

Also, since the TATB and Kel-F mixture is pressed to ~97% of theoretical density there are very few air voids, thus $S_T + S_K = 1$.

Since we have three equations and two unknowns, any two of these equations can be used to generate estimates of S_T and S_K. In an effort to use all the data at our disposal, we elected to estimate the weight fractions from the difference of the two reconstructed images at different energies using the following equation:

$$\mu_R(22) - \mu_R(25) = S_T[\mu_T(22) - \mu_T(25)] + S_K[\mu_K(22) - \mu_K(25)]. \tag{A-5}$$

From this equation and $S_T + S_K = 1$ we can solve for the respective weight fractions:

$$S_T = \frac{[\mu_R(22) - \mu_R(25)] - [\mu_K(22) - \mu_K(25)]}{[\mu_T(22) - \mu_T(25)] - [\mu_K(22) - \mu_K(25)]}, \tag{A-6}$$

and $S_K = 1 - S_T$. A representative TATB weight fraction image is shown in Fig. 5.

The Density Image

Using the effective-Z image and tabulated mass attenuation coefficients[9] a calculated/estimated mass attenuation coefficient, $[(\mu/\rho)_C(E)]$, may be determined for every pixel in the effective-Z image. Dividing this calculated/estimated image out of the reconstructed linear attenuation coefficient image at energy E results in a density estimate on a pixel-to-pixel basis since

$$\rho = \left\{ \frac{[\mu_R(E)]}{[(\mu/\rho)_C(E)]} \right\}. \tag{A-7}$$

A representative density image (calculated/estimated using the 22 keV data) is shown in Fig. 5.

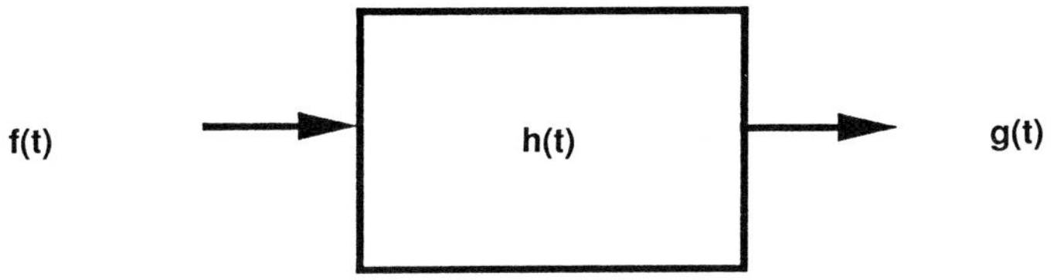

Figure 1. Transfer function identification, f (t), g (t) and h (t) are input signal, output signal and impulse response, respectively.

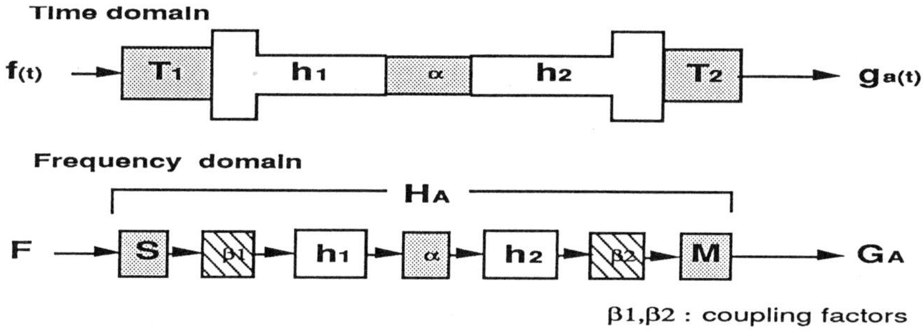

Figure 2. Schematic diagram showing elastic wave transmission the adhesive joint.

EXPERIMENTAL METHOD

The measurement system of ETFuM in schematically shown in Figure 3. Signal F is generated by a pseudorandom noise generator, input to the transmitting transducer, passes through the adhesive joint specimen and is picked up by the receiving transducer and amplifier. It is then fed into a transient recorder along with the input signal. The signals are compared and only information on the adhesive joint is extracted by following equation (5) or (6). The pseudorandom noise is modified to show any aliasing at the FFT analyzer[1]. The entire system is controlled automatically by microcomputer.

Flat type PZT transducers, Models 1045 and 5045 made by Fuji Ceramics Co., Ltd., were employed and the coupling agent between the transducers and the specimen was W-400 (developed by Nippon Steel Corporation). It provides good reproducibility of the mounting condition, for both the amplitude and phase components.

The transducers provide virtually flat characteristics, varying by only ± 5 dB, up to 1.6 MHz and 5 MHz, respectively. The measured data obtained with these transducers were sufficiently reliable for comparison without performing any special sensitivity calibration.

In general transducer sensitivity is affected by the mounting pressure. Figure 4 shows sensitivity of each frequency range as a function of the mounting pressure.

The results presented in Figure 4 clearly indicate that the transmission of the ultrasonic wave improves with higher contact pressure. In addition, the degree of improvement varies from one frequency to the next. The results in the figure show that good reproducibility was obtained in this experiment at a surface pressure of 0.1 Kgf/mm2 provided the frequency was in the vicinity of 5 MHz. In subsequent tests therefore, the surface pressure of the transducer was sent so that it satisfied this condition.

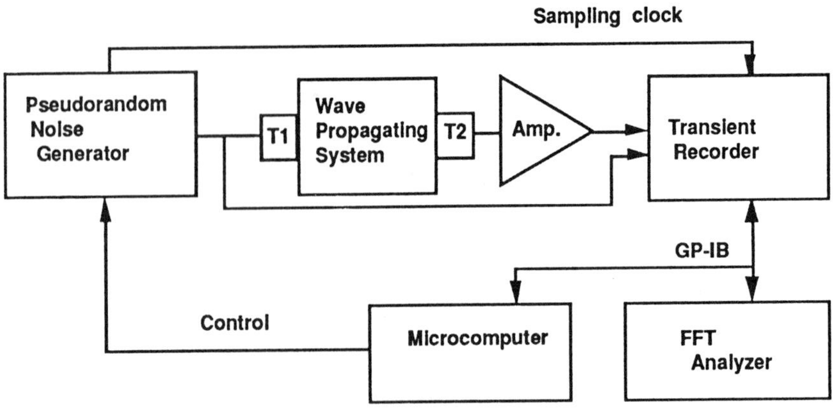

Figure 3. Block diagram of ETFuM measuring system.

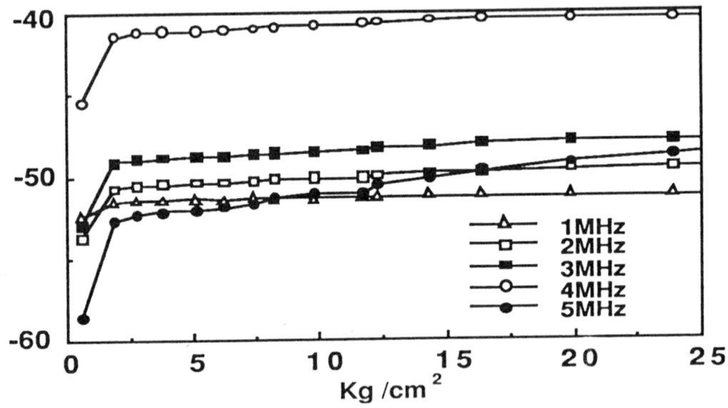

Figure 4. The relation between sensitivity of sensor and its mounting pressure.

Three kinds of epoxy adhesives were used in this work shown in Table 1. One was a clear type, 007, without a compound, another type, 008, contained a ceramic compound, and the third type 009, contained a rubber compound. The third type showed especially high acoustic impedance.

The shape of the test pieces used in the measurements is shown in Figure 5. The specimens were made of SUS304, with each piece being 60 mm in length. Two pieces were bonded together using one of the adhesives. Two kinds of bonded surface being roughly finished by race and smoothly finished by grinder. The two types of surface finish and three types of epoxy adhesive were combined to make six kinds of adhesive joint specimens. We then evaluated their tensile strength using ETFuM. The two test pieces are joined together and transducers, T1 and T2, are mounted at either end.

A pair of jointed specimens were set in an Instron type tensile test device using the jig. The transfer function was measured before the test. And then the tensile test was carried out at a cross-head speed of 1 mm/min. Cross-head speeds of 0.1 mm/min and 0.01 mm/min were also used to compare the effect of the loading rate on the strength of the adhesive joint.

Table 1. Kinds of epoxy adhesive employed.

Ep007	**Epoxy Resin curing in room temperature**
Ep008	**Ep007 with ceramic compound**
Ep009	**Ep008 with rubber particle compound**

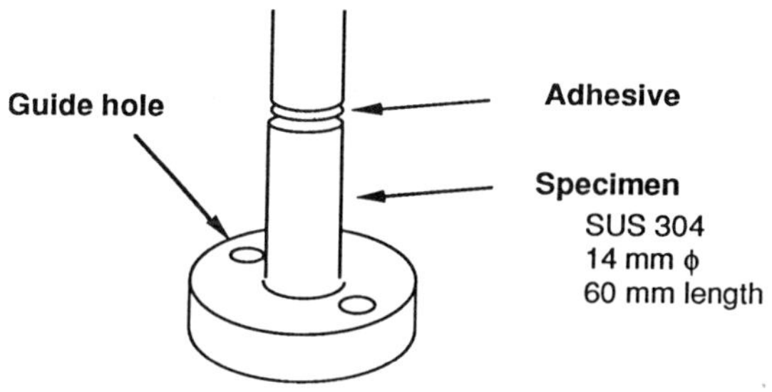

Figure 5. Adhesive joint specimen.

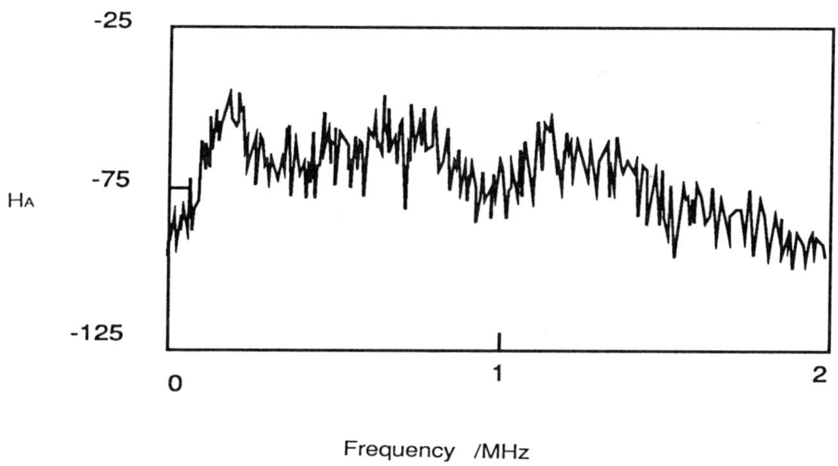

Figure 6. Typical frequency spectrum of output signal G(ω) of the joint.

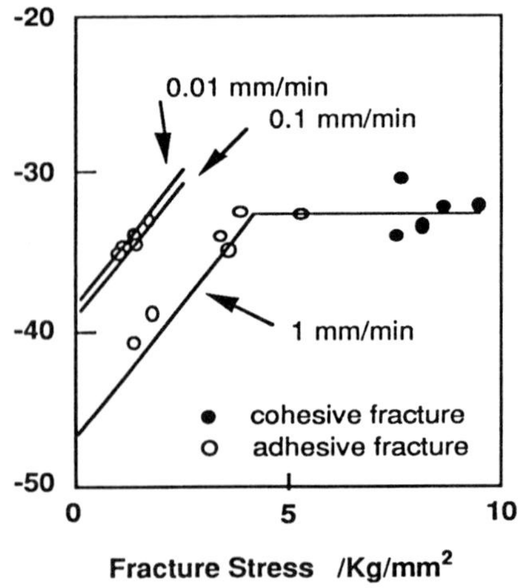

Figure 7. Correlation between the evaluation function Ha and the tensile strengths for the adhesives examined, including the effect of cross-head speed during the tensile test.

EXPERIMENTAL RESULTS AND DISCUSSION

The signal input to T1 is converted to an elastic wave, passes through the adhesive joint and is detected by T2. Figure 6 is a typical spectrum of the received signal. The amplitude of each frequency component was sufficient to permit analysis in a range of 250 KHz to 1.5 MHz. Then the transfer function of the system was obtained using the same frequency range.

Based on the foregoing results, the evaluation function Ha was defined as follows:

$$H_A = \int_{250\ \mathrm{KHz}}^{1.5\ \mathrm{MHz}} (G(\omega))^2\, d\omega \qquad (7)$$

The transfer function, Ha, is shown in figure 7 as a function of the tensile strength of the adhesive joint. Adhesive and cohesive fractures are shown together in the figure. There is a linear relation between Ha and tensile strength up to a tensile strength of 4 kg/mm^2 each of the fracture modes and bonded surface condition were different. After that, Ha is constant because the power of the input signal was only 10 volts due to the limited capacity of the equipment.

Here are also the results obtained with and without a compound. The relation between tensile strength and the transfer function was not affected by the addition of a ceramic or rubber compound for either type of bonded surface finish.

Figure 7 also compares the results for different cross-head speeds. Tensile strength increases with a decreasing load rate, and so the relation between tensile strength and the transfer function shifts, but the same tendency is still seen. In addition it is important to find the most suitable frequency range for evaluate adhesive joint strength so as to avoid getting useless information and thereby enhance reliability for ETFuM.

CONCLUSION

A new nondestructive evaluation method of ETFuM (Elastic-wave Transfer Function Method) was developed and applied to evaluate adhesive joint strength. Two kinds of specimens with smooth and rough surfaces for bonding, and three kinds of epoxy adhesive (clear, with ceramic and rubber compound) were used.

Good correlation was seen between adhesive tensile strength and the Ha, the parameter of ETFuM, within the range of measurement made in this work. It is therefore concluded that the method presented here is useful to a certain extent in making evaluations of adhesive joints intended for use in actual products.

ACKNOWLEDGEMENT

The authors are indebted to Prof. S. Nunomura for his continuous support of this work. The authors are also indebted to Mrs. X.J. Gao for assisting in much of the experimental work.

REFERENCES

1. H. Nakamura and A. Wada, Applied Physics 13, 407 (1978)(in Japanese) or H. Nakamura, Y. Hushimi and A. Wada, Japan J. Appl. Phys. 16, 2301 (1977).

MATHEMATICAL MODELING OF WAVE PROPAGATION IN ANISOTROPIC MEDIA (MATHEMATICAL SOLUTION)

Fares A. Chedid Helou

Department of Mathematical Sciences
Kent State University
Ashtabula Campus
Ashtabula, Ohio 44004

John H. Hemann

Civil Engineering Department
Cleveland State University
Cleveland, Ohio 44115

Abstract

Mathematical modeling of the interaction of ultrasonic waves with anisotropic media (21 constants) is carried out. An anisotropic plate immersed in a liquid medium is acted upon by an ultrasonic incident wave of arbitrary frequency and angle of incidence. The incident acoustic harmonic plane wave originates in the upper half-space fluid. Expressions for reflection and transmission coefficients are derived and presented graphically as a function of the angle of incidence and the product of the frequency and the thickness of the plate . Also, the phase of both coefficients versus frequency are presented. All field variables can be fully specified using this mathematical modeling. This study is very general and can be easily specified to special cases of wave propagation and geometric and material configurations. It can be applied to both Rayleigh and Lamb waves. The numerical computations were done through the use of the VAXIMA/MACSYMA software package.

List of Symbols:

- media (j); j = 1, 2, 3. correspond to upper fluid half-space, plate, and lower fluid half-space respectively.
- $Reflection\ Coefficient\ =\ R\ =\ \frac{REFLECTED\ WAVE\ AMPLITUDE}{INCIDENT\ WAVE\ AMPLITUDE}$
- $Transmission\ Coefficient\ =\ T\ =\ \frac{TRANSMITTED\ WAVE\ AMPLITUDE}{INCIDENT\ WAVE\ AMPLITUDE}$
- σ_{mn}, e_{mn} ; m, n = 1, 2, 3. stress and strain tensor respectively.
- C_{jklm} ; j, k, l, m = 1, 2, 3. Hooke's constants (elastic tensor).
- u_n ; n = 1, 2, 3. particle displacement in the x_n direction.
- A_1, A_2, and A_3 are the incident, reflected, and transmitted wave amplitudes respectively.

Nondestructive Characterization of Materials IV
Edited by C.O. Ruud *et al.*, Plenum Press, New York, 1991

- θ : angle of incidence.
- ϕ : angle between x_1 and in-plane fiber orientation.
- B_j ; j = 1, 2, 3. existing plate wave amplitudes (possibly complex) corresponding to u_j.
- λ_f , ρ_f : Lamé constant and mass density of the fluid.
- γ_f , c_f : plane wave vector and speed in the fluid half-spaces.
- δ: ratio of the x_1 to the x_3 wave numbers.
- c : speed of incident, reflected, transmitted, and refracted waves in the x_1 direction.
- f, ω, h : frequency, circular frequency, and plate thickness respectively.
- $(\,)_{,j}$, $(\,)_{,jl}$, $\ddot{(\,)} = \frac{\partial(\,)}{\partial x_j}$, $\frac{\partial^2(\,)}{\partial x_j \partial x_l}$, $\frac{\partial^2(\,)}{\partial t^2}$; j, l = 1, 2, 3.
- $\gamma \equiv \gamma_f \sin(\theta) = (\frac{\omega}{c_f})\sin(\theta) = \omega\sqrt{\frac{\rho_f}{\lambda_f}}\sin(\theta)$.
- ρ : density of the plate.

1 INTRODUCTION

The high strength-to-weight ratio of composite materials prompted their use resulting in an improved performance of structural components throughout many applications in the aerospace industry and energy related systems. In order to optimize their usefulness, a full understanding of these advanced materials is indispensable. This understanding has been improving through the development and enhancement of many experimental [1, 2] and modeling techniques [3, 4, 5, 6]. References to other studies can be found in those cited above.

Nondestructive evaluation (NDE) techniques have been used to characterize flaws and to estimate materials properties. This ultrasonic approach is based on the generation and detection of ultrasonic waves upon their interaction with the tested part. This experimental approach is sensitive to defects such as flaws, cracks and delaminations, and it proved to be very effective in detecting hydrogen damage in the furnace walls of fossil fired utility boilers [7]. However, experimental data would be of little benefit in the absence of a theory that provides the researcher with a systematic approach to analyze the experimental results and predict the behavior of the material under study.

As expected, theoretical models for advanced materials are much more involved than those corresponding to homogeneous isotropic materials. This is due, in part, to the morphology of composite materials and their directional properties i.e. anisotropy and dispersivity. Because it is almost impossible to fully describe the static and dynamic states of these materials, many approximate theories emerged, all of which approximated the composite with an "equivalent" continuum. Some of these theories include the Effective Stiffness Theories, Effective Modulus Theories, Theory of Interacting Continua and numerous others.

Wave propagation theories in isotropic and anisotropic media have been studied for many years [8, 9, 10, 11, 12, 13, 14, 15]. Some of the recent studies [16, 17, 18] have dealt with plate-like and laminated structures. In reference [16] Nayfeh and Chimenti studied the reflection characteristics of a fluid coupled orthotropic plate subjected to incident acoustic beams at arbitrary angles of incidence and in-plane fiber orientations.

This paper examines the plane wave reflection and transmission coefficients for a plate with general anisotropy (21 constants). These results are very general and can be easily reduced to special cases of wave propagations. Degenerate cases include monoclinic, orthotropic, transversly isotropic, cubic and isotropic materials. Theoretical results for the case of monoclinic-like constants' matrix are presented. This was accomplished by considering an orthotropic plate with general in-plane fiber orientation. The theoretical approach used is well established [3, 12, 16, 17, 18, 19].

2 FORMULATION OF THE PROBLEM

Consider a plate with a general anisotropy (21constants) occupying the region of Eucledian space defined by: $(x_1, x_2, x_3) \in R \times R \times [-h, 0]$ where (x_1, x_2, x_3) is an orthogonal cartesian system of coordinates. Assume that the top and bottom surfaces of the plate are in contact with two fluid half-spaces.

$$\bullet Equilibrium: \quad \sigma_{ij,i} = \rho\, \ddot{u}_j \tag{1}$$

$$\bullet Stress - Strain\ Relation: \quad \sigma_{ij} = C_{ijkl}\, e_{kl} \tag{2}$$

$$\bullet Strain - Displacement\ Relations: \quad e_{kl} = \tfrac{1}{2}\left(u_{k,l} + u_{l,k}\right) \tag{3}$$

Boundary Conditions: (top and bottom surfaces of the plate)

$$\sigma_{33}^{(2)} = \sigma_{33}^{(1)}, \quad u_3^{(2)} = u_3^{(1)}, \quad \sigma_{13}^{(2)} = \sigma_{23}^{(2)} = 0 \quad at \quad x_3 = 0, \tag{4}$$

$$\sigma_{33}^{(2)} = \sigma_{33}^{(3)}, \quad u_3^{(2)} = u_3^{(3)}, \quad \sigma_{13}^{(2)} = \sigma_{23}^{(2)} = 0 \quad at \quad x_3 = -h, \tag{5}$$

The resulting problem is a well posed boundary value problem.

The objectives of this paper are to derive an expression for the reflection and transmission coefficients and to graphically present the above results.

3 MATHEMATICAL DERIVATION

In order to solve this problem the displacement and consequently, the stress fields should be defined in both fluid half-spaces and in the plate. It can be proven that the incident, reflected and refracted plane waves belong to the same plane and that the angles of reflection and transmission are equal to the angle of incidence. These facts result from the satisfaction of boundary conditions at the top and bottom surfaces of the plate. Without any loss of generality we can assume that the incident plane wave vector belongs to the$(x_1 - x_3)$ plane. The $\exp(-i\omega t)$ dependence will be assumed and suppressed.

• *Medium* 1 (upper fluid half-space):

$$u_1^{(1)} = \sin(\theta)\{A_1 \exp\{i\gamma_f[\sin(\theta)x_1 - \cos(\theta)x_3]\} + A_2 \exp\{i\gamma_f[\sin(\theta)x_1 + \cos(\theta)x_3]\}\} \tag{6}$$

$$u_2^{(1)} = 0 \tag{7}$$

$$u_3^{(1)} = \cos(\theta)\{-A_1 \exp\{i\gamma_f[\sin(\theta)x_1 - \cos(\theta)x_3]\} + A_2 \exp\{i\gamma_f[\sin(\theta)x_1 + \cos(\theta)x_3]\}\} \tag{8}$$

$$\sigma_{33}^{(1)} = \lambda_f i\gamma_f\{A_1 \exp\{i\gamma_f[\sin(\theta)x_1 - \cos(\theta)x_3]\} + A_2 \exp\{i\gamma_f[\sin(\theta)x_1 + \cos(\theta)x_3]\}\} \tag{9}$$

• *Medium* 3 (lower fluid half-space):

$$u_1^{(3)} = A_3 \sin(\theta) \exp\{i\gamma_f[\sin(\theta)x_1 - \cos(\theta)x_3]\} \tag{10}$$

$$u_2^{(3)} = 0 \tag{11}$$

$$u_3^{(3)} = -A_3 \cos(\theta) \exp\{i\gamma_f[\sin(\theta)x_1 - \cos(\theta)x_3]\} \tag{12}$$

$$\sigma_{33}^{(3)} = \lambda_f i\gamma_f A_3 \exp\{i\gamma_f[\sin(\theta)x_1 - \cos(\theta)x_3]\} \tag{13}$$

• *Medium* 2 (plate): Let's derive an expression for the displacement field, and therefore, the stress field in the plate. Substituting equations (3) and (2) into (1) results in a system of three *coupled* partial differential equations:

$$C_{ijkl}(u_{k,il} + u_{l,ik}) = 2\rho\ddot{u}_j \tag{14}$$

Expanding equations (14) using the following contracting subscript notation: $1 \rightarrow 11$, $2 \rightarrow 22$, $3 \rightarrow 33$, $4 \rightarrow 23$, $5 \rightarrow 13$, $6 \rightarrow 12$ and the fact that $\frac{\partial()}{\partial x_2} = 0$ results into:

$$\begin{bmatrix} C_{11} & 2C_{15} & C_{55} & C_{16} & (C_{14}+C_{56}) & C_{45} & C_{15} & (C_{13}+C_{55}) & C_{35} \\ C_{16} & (C_{14}+C_{56}) & C_{45} & C_{66} & 2C_{46} & C_{44} & C_{56} & (C_{36}+C_{45}) & C_{34} \\ C_{15} & (C_{55}+C_{13}) & C_{35} & C_{56} & (C_{36}+C_{45}) & C_{34} & C_{55} & 2C_{35} & C_{33} \end{bmatrix} \times \begin{Bmatrix} u_{1,11} \\ u_{1,13} \\ u_{1,33} \\ u_{2,11} \\ u_{2,13} \\ u_{2,33} \\ u_{3,11} \\ u_{3,13} \\ u_{3,33} \end{Bmatrix} = \rho \begin{Bmatrix} \ddot{u}_1 \\ \ddot{u}_2 \\ \ddot{u}_3 \end{Bmatrix} \tag{15}$$

Assume a solution for the displacement field in the plate of the form:

$$u_j^{(2)} = B_j exp[i\gamma(x_1 + \delta x_3) - i\omega t]\ ; \quad j = 1, 2, 3. \tag{16}$$

Substituting equations (16) into equations (15) and dividing throughout by $(i\gamma)^2$ gives the following (in matrix form):

$$\begin{bmatrix} C_{11} + 2C_{15}\delta + C_{55}\delta^2 - \rho c^2 & C_{16} + (C_{14} + C_{56})\delta + C_{45}\delta^2 & C_{15} + (C_{13} + C_{55})\delta + C_{35}\delta^2 \\ & C_{66} + 2C_{46}\delta + C_{44}\delta^2 - \rho c^2 & C_{56} + (C_{36} + C_{45})\delta + C_{34}\delta^2 \\ SYM. & & C_{55} + 2C_{35}\delta + C_{33}\delta^2 - \rho c^2 \end{bmatrix} \times \begin{Bmatrix} B_1 \\ B_2 \\ B_3 \end{Bmatrix} = \begin{Bmatrix} 0 \\ 0 \\ 0 \end{Bmatrix} \tag{17}$$

The above system of equations has a nontrivial solution if and only if its determinant is = 0, which results in a polynomial equation of the sixth order with real coefficients.Therefore, it has six solutions (δ_m; $m = 1, \dots, 6$) in the complex numbers' field. Using the second and third of equations (17), for any (δ_j ; $j = 1, \dots, 6$) we have:

$$\begin{bmatrix} C_{66} + 2C_{46}\delta_j + C_{44}\delta_j^2 - \rho c^2 & C_{56} + (C_{36} + C_{45})\delta_j + C_{34}\delta_j^2 \\ C_{56} + (C_{36} + C_{45})\delta_j + C_{34}\delta_j^2 & C_{55} + 2C_{35}\delta_j + C_{33}\delta_j^2 - \rho c^2 \end{bmatrix} \begin{Bmatrix} B_2^{(j)} \\ B_3^{(j)} \end{Bmatrix} = - \begin{Bmatrix} C_{16} + (C_{14} + C_{56})\delta_j + C_{45}\delta_j^2 \\ C_{15} + (C_{13} + C_{55})\delta_j + C_{35}\delta_j^2 \end{Bmatrix} B_1^{(j)}, \tag{18}$$

Solving the above system of equations for $B_2^{(j)}$ and $B_3^{(j)}$ we get:

$$\begin{matrix} B_2^{(j)} = V_j \times B_1^{(j)}\ ; \\ B_3^{(j)} = W_j \times B_1^{(j)}\ ; \end{matrix} \quad \left\{ j = 1, \dots, 6. \right\} \tag{19}$$

Where:

$$V_j = \frac{N1}{DEN} \tag{20}$$

$$W_j = \frac{N2}{DEN} \tag{21}$$

$$\begin{aligned} N1 = & \quad [\ C_{55} + 2C_{35}\delta_j + C_{33}\delta_j^2 - \rho c^2\ ][\ C_{16} + (C_{14} + C_{56})\delta_j + C_{45}\delta_j^2\ ] \\ & - [\ C_{56} + (C_{36} + C_{45})\delta_j + C_{34}\delta_j^2\ ][\ C_{15} + (C_{13} + C_{55})\delta_j + C_{35}\delta_j^2\ ] \end{aligned} \tag{22}$$

$$\begin{aligned} N2 = & \quad [\ C_{66} + 2C_{46}\delta_j + C_{44}\delta_j^2 - \rho c^2\ ][\ C_{15} + (C_{13} + C_{55})\delta_j + C_{35}\delta_j^2\ ] \\ & - [\ C_{56} + (C_{36} + C_{45})\delta_j + C_{34}\delta_j^2\ ][\ C_{16} + (C_{14} + C_{56})\delta_j + C_{45}\delta_j^2\ ] \end{aligned} \tag{23}$$

$$\begin{aligned} DEN = & \quad [C_{56} + (C_{36} + C_{45})\delta_j + C_{34}\delta_j^2]^2 \\ & - [C_{66} + 2C_{46}\delta_j + C_{44}\delta_j^2 - \rho c^2][C_{55} + 2C_{35}\delta_j + C_{33}\delta_j^2 - \rho c^2] \end{aligned} \tag{24}$$

So:

$$u_k^{(2)} = \sum_{j=1}^{6} B_k^{(j)} \exp[i\gamma(x_1 + \delta_j x_3) - i\omega t]\ ; \quad k = 1,\ 2,\ 3. \tag{25}$$

It is a rather trivial procedure to find the expressions for $\sigma_{i3}^{(2)}$; i=1,2,3. Applying the boundary conditions yields, in general, a system of eight independent equations with eight unknowns, namely: A_2 , A_3 , and $B_1^{(k)}$; k = 1, 2, ... ,6. After rather long algebraic manipulations and reductions, the matrix form of the above system can be written as:

$$\begin{bmatrix} E_1 & E_2 & E_3 & E_4 & E_5 & E_6 & -\frac{\lambda_f}{\sin(\theta)} & 0 \\ E_1H_1 & E_2H_2 & E_3H_3 & E_4H_4 & E_5H_5 & E_6H_6 & 0 & I \\ W_1 & W_2 & W_3 & W_4 & W_5 & W_6 & -\cos(\theta) & 0 \\ W_1H_1 & W_2H_2 & W_3H_3 & W_4H_4 & W_5H_5 & W_6H_6 & 0 & J \\ F_1 & F_2 & F_3 & F_4 & F_5 & F_6 & 0 & 0 \\ F_1H_1 & F_2H_2 & F_3H_3 & F_4H_4 & F_5H_5 & F_6H_6 & 0 & 0 \\ G_1 & G_2 & G_3 & G_4 & G_5 & G_6 & 0 & 0 \\ G_1H_1 & G_2H_2 & G_3H_3 & G_4H_4 & G_5H_5 & G_6H_6 & 0 & 0 \end{bmatrix} \begin{Bmatrix} B_1^{(1)} \\ B_1^{(2)} \\ B_1^{(3)} \\ B_1^{(4)} \\ B_1^{(5)} \\ B_1^{(6)} \\ A_2 \\ A_3 \end{Bmatrix} = \begin{Bmatrix} \frac{\lambda_f}{\sin(\theta)} \\ 0 \\ -\cos(\theta) \\ 0 \\ 0 \\ 0 \\ 0 \\ 0 \end{Bmatrix} A_1 \tag{26}$$

Where:

$$\left.\begin{aligned} E_j &= C_{13} + C_{36}V_j + C_{35}W_j + (C_{35} + C_{34}V_j + C_{33}W_j)\delta_j\ ; \\ F_j &= C_{15} + C_{56}V_j + C_{55}W_j + (C_{55} + C_{45}V_j + C_{35}W_j)\delta_j\ ; \\ G_j &= C_{14} + C_{46}V_j + C_{45}W_j + (C_{45} + C_{44}V_j + C_{34}W_j)\delta_j\ ; \\ H_j &= \exp[-i\gamma\delta_j h]\ ; \end{aligned}\right\} j = 1, ..., 6.$$

$$\begin{aligned} I &= -\left(\frac{\lambda_f}{\sin(\theta)}\right)\exp[i\gamma h\cot(\theta)] = -\left(\frac{\lambda_f}{\sin(\theta)}\right)\exp[2\pi i\left(\sqrt{\frac{\rho_f}{\lambda_f}}\right)(\cos(\theta))(fh)] \\ &= -\left(\frac{\lambda_f}{\sin(\theta)}\right)\exp\left[\frac{2\pi i(\cos(\theta))(fh)}{c_f}\right]; \\ J &= \cos(\theta)\exp[i\gamma h\cot(\theta)] = \cos(\theta)\exp[2\pi i\left(\sqrt{\frac{\rho_f}{\lambda_f}}\right)(\cos(\theta))(fh)] \\ &= \cos(\theta)\exp\left[\frac{2\pi i(\cos(\theta)(fh)}{c_f}\right]; \end{aligned} \tag{27}$$

Using Cramer's rule the expressions for $R = \frac{A_2}{A_1}$ and $T = \frac{A_3}{A_1}$ can be obtained easily.

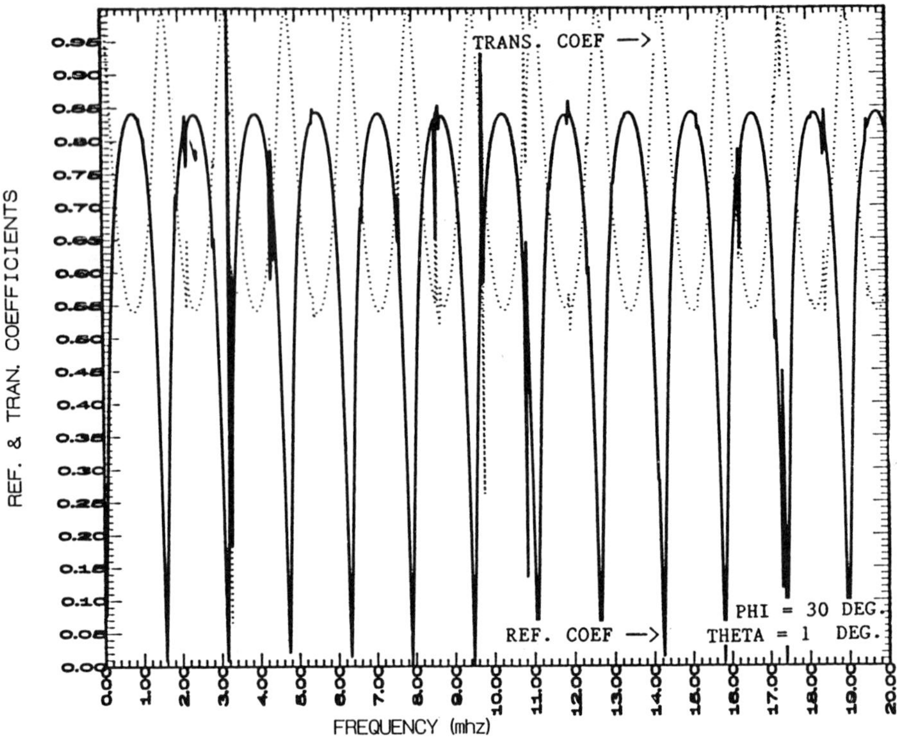

FIGURE 1

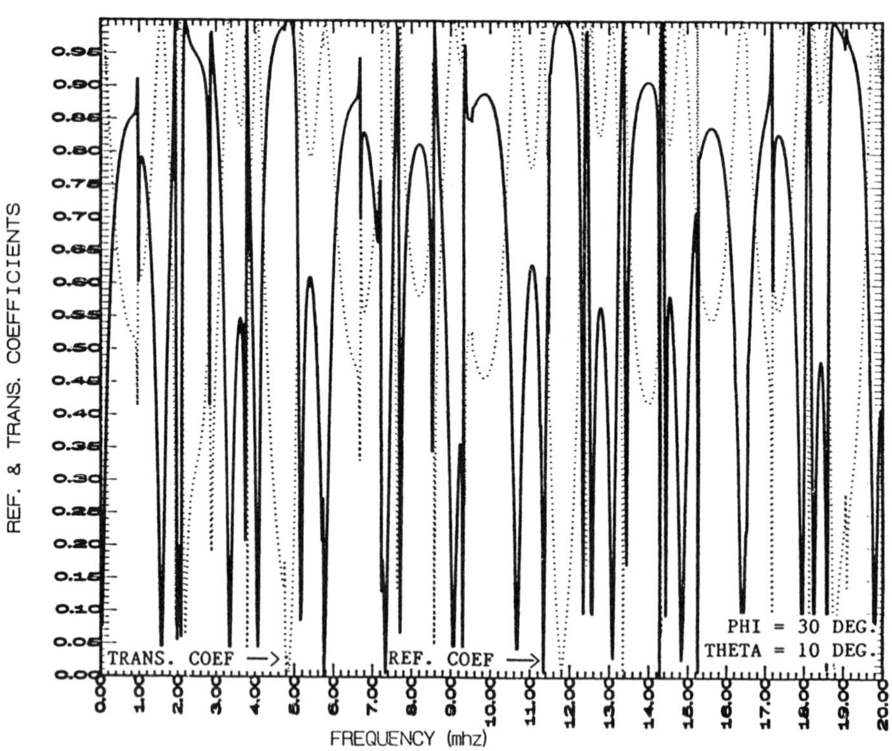

FIGURE 2

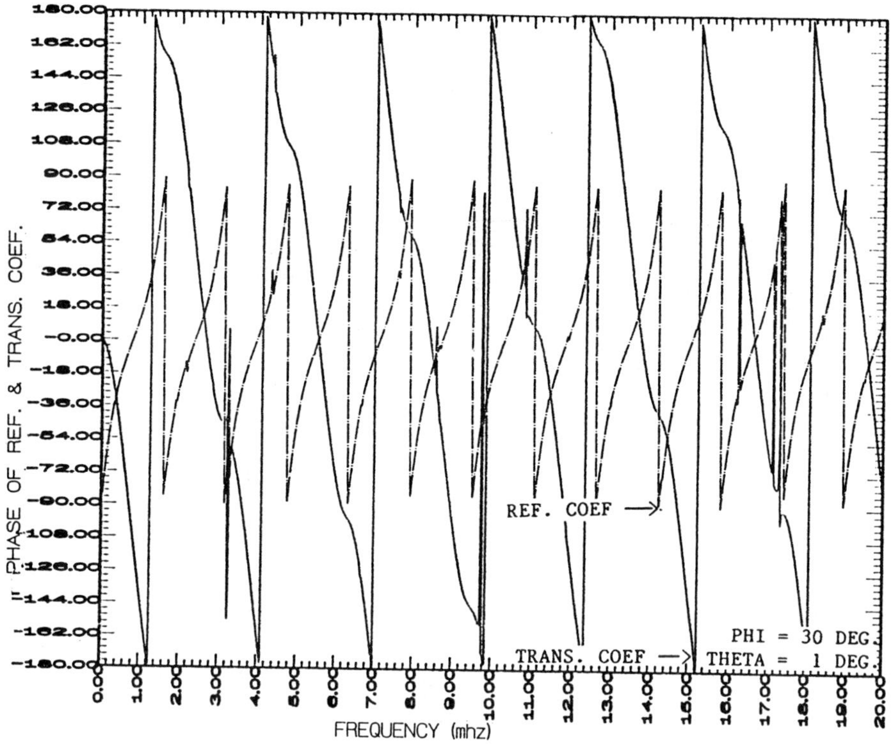

FIGURE 3

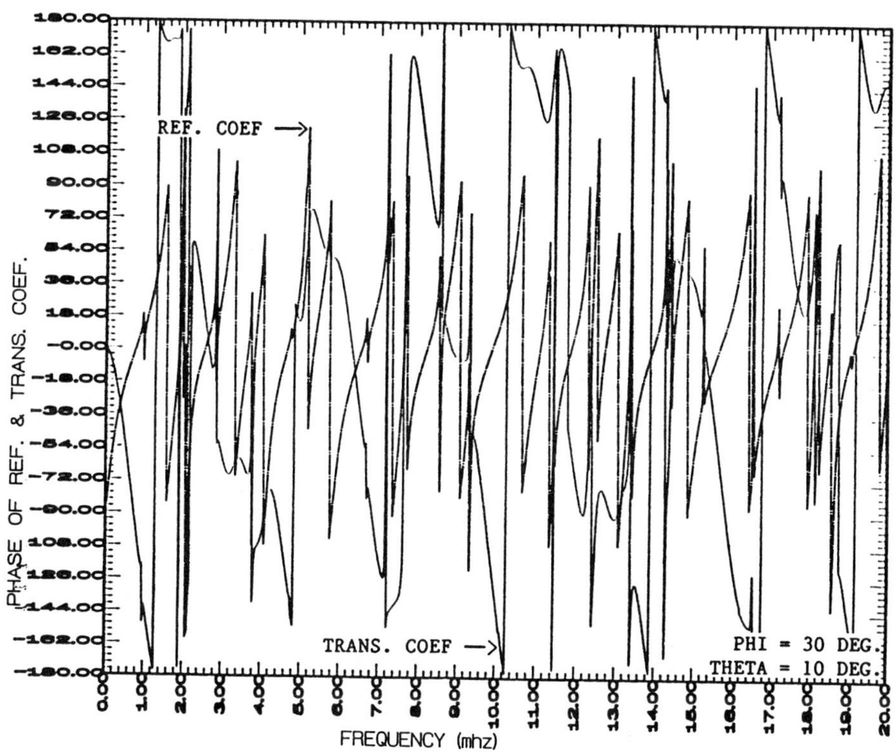

FIGURE 4

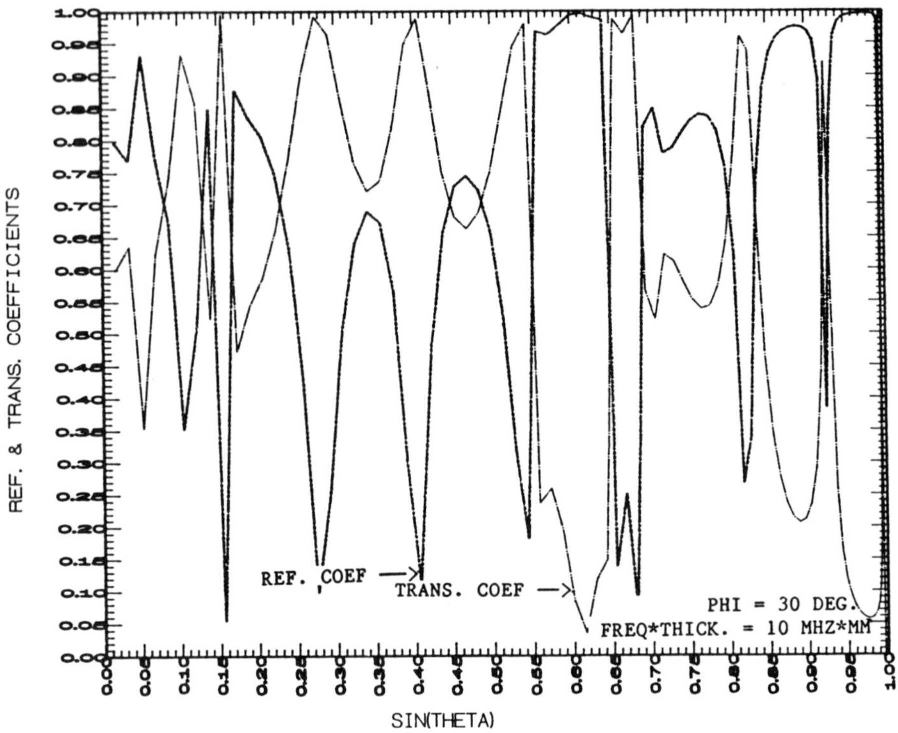

FIGURE 5

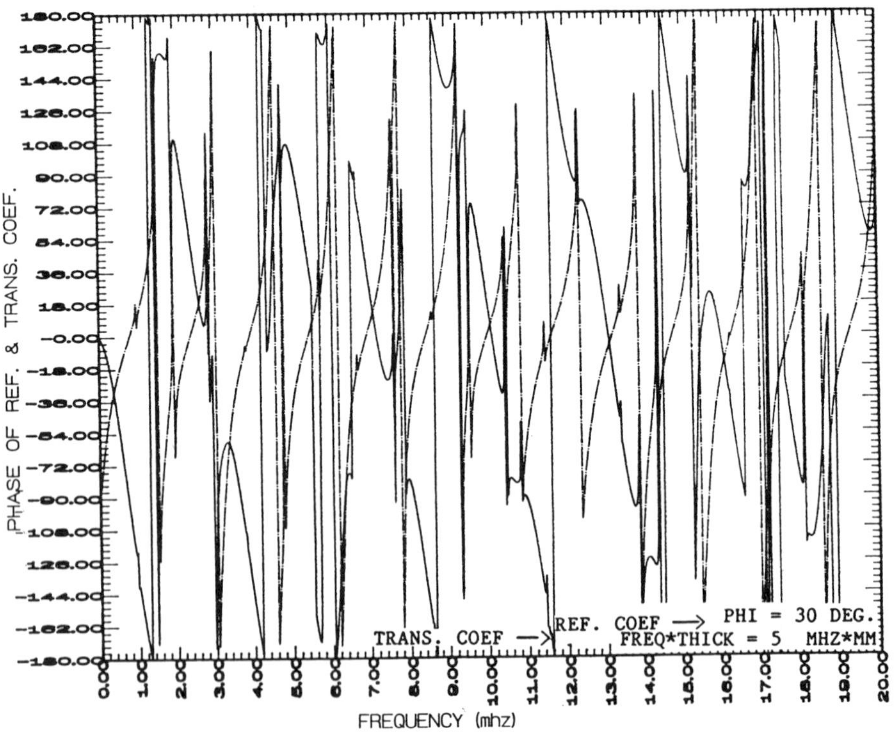

FIGURE 6

4 NUMERICAL ANALYSIS

The reflection and transmission coefficients were evaluated for the case of a fiber composite plate with a general in-plane orientation (ϕ) and angle of incidence (θ). The following plate constants C'_{11}=115.6 GPa, C'_{22}=11.5 GPa, C'_{33}=15.95 GPa, C'_{13}=3.72 GPa, C'_{23}=4.33 GPa, C'_{44}=5.81 GPa, C'_{55}=7.46 GPa, and $\rho = 1.6$ gm/cm^3 along the fiber direction were used. The fiber diameter is assumed to be small as compared to the wavelength of the propagating plane waves. The unprimed material constants are obtained through the usual tensor transformation. Media 1 and 3 are assumed to be water ($c_f = 1480$ m/sec, $\rho = 1$ gm/cm^3).

As θ increases from 1 degree in Fig. 1 to 10 degrees in Fig. 2, the periodic shape of the curves disappears and they become more complicated. This is due to the increased interaction between the normal and tangential components of the upper fluid displacement field as the incident wave interacts with the plate. The branches of the reflection coefficient having relatively small maxima ($|R| < .62$ in Fig. 2) may be attributed to the overlapping of propagating plate modes. Curves similar to those of Figs. 1 and 2 were presented in [16] and the general behavior seems to be identical.

The phase of R and T versus frequency curves of Figs. 3 and 4 correspond to those presented in Figs. 1 and 2 respectively. Once again the periodicity of Fig. 3 disappears in Fig. 4 due to the above mentioned reason.

Generally speaking, Lamb-like modes are excited in the plate when the modulus of the reflection coefficient approaches zero accompanied by a rapid change of phase. Also, $|R|$ and $|T|$ versus (θ) are presented in Figs. 5 and 6. The conservation of energy principle requires that $|R|^2 + |T|^2 = 1$, a relation that is clearly shown in Figs. 1, 2, 5 and 6. Also, it is clearly shown in these figures that the propagation of Lamb-like modes can be studied through the investigation of the near unity values of the transmission coefficient.

5 CONCLUSION and FUTURE DIRECTION

Even though the numerical results presented in this paper correspond to the case of wave propagation in a monoclinic material, the derived expressions for R and T which, to the best of our knowledge, have not been presented in any published work correspond to the general case of material anisotropy. Close examination of R and its phase versus frequency graphs clearly identify Lamb-like modes. Equivalently, the study of T leads to the same results. Future work includes the case of wave propagation with a full 6×6 material constants matrix. Also, it will include the study of some degenerate cases of wave propagation.

ACKNOWLEDGMENT

The expert computer assistance of P. S. Wang, M. Al-Hajery, Y. Doleh, M. O. Rayes and D. Bennett is greatly appreciated.

References

[1] A. Sarrafzadeh, R. J. Churchill, and M. G. Niimura, *Laser Generated Ultrasound*;"Proceedings of a Workshop on Acousto-Ultrasonics: Theory and Application," Blacksburg, Virginia, J. C. Duke,Jr. ed. Plenum Press, New York. pp 201-207 (1988).

[2] P. J. Latimer and H. L. Whaley, *Electromagnetic Transducers for Generation of Ultrasonic Waves*: "Proceedings of a Workshop on Acousto-Ultrasonics: Theory and Application," Blacksburg, Virginia, J. C. Duke,Jr. ed. Plenum Press, New York. pp 209-220 (1988).

[3] D. E. Chimenti, and A. H. Nayfeh, *Leaky Lamb Waves in Fibrous Composite Laminates,* J. of Appl. Phys. Vol. 58, p 4531 (1985).

[4] S. K. Datta, H. M. Ledbetter, and R. D. Kriz, *Calculated elastic Constants of Composites Containing Anisotropic Fibers,* Int. J. Solids Structures Vol. 20, No. 5, pp 429-438, (1984).

[5] A. H. Nayfeh, and G. A. Gurtman, *A Continuum Approach to the Propagation of Shear Waves in Laminated Wave Guides,* J. of Appl. Mech. pp 106-110 March (1974).

[6] A. K. Mal, *Wave Propagation in Layered Composite Laminates - Periodic Surface Loads,* Wave Motion 10 Elsevier Science Publishers B. V.(North- Holland) pp 257-266 (1988).

[7] Ibid 2.

[8] H. Lamb, *On Waves in an Elastic Plate,* Proc. of the Royal soc. of london, Series A, Vol. 93, p 114 (1916-1917).

[9] Lord Raleigh, *On the Free Vibrations of an Infinite Plate of Homogeneous Isotropic Elastic Matter,* Proc. of the London Math Soc. Vol. 20, pp 225-234 .

[10] R. D. Mindlin, *Waves and Vibrations in Isotropic, Elastic Plates,* in : "Structural Mechanics,"J. N. Goodier, and N. J. Hoof, eds. Pergamon Press, New York (1960).

[11] L. E. Pitts, *A Unified Theoretical Description of Ultrasonic Beam Reflection From a Solid Plate in a Liquid,* Ph. D. thesis (Georgetown University, 1975). (unpublished)

[12] H. L. Bertoni, and T. Tamir, *Unified Theory of Raleigh-Angle Phenomena for Acoustic Beams at Liquid-Solid Interfaces,* Appl. Phys. Vol. 2, pp 157-172, Springer-Verlag (1973).

[13] L. M. Brekhovskikh, *Waves in Layered Media,* Academic Press, New York, 1980.

[14] Ibid 4.

[15] S. I. Rokhlin, T. K. Bolland, and L. Adler, *Reflection and Refraction of Elastic Waves on a Plane Interface Between Two Generally Anisotropic Media,* J. of the Acoust. Soc. of America, Vol. 79, pp 906-918, (1986).

[16] A. H. Nayfeh, and D. E. Chimenti, *Ultrasonic Wave Reflection from Liquid-Coupled Orthotropic Plates With Applications to Fibrous Composites,* J. of Appl. Mech., Vol. 55, pp 863-870. (1988).

[17] A. H. Nayfeh, T. W. Taylor, and D. E. Chimenti, *Theoretical Wave Propagation in Multilayered Orthotropic Media,* in : " Wave Propagation In Structural Composites," A. K. Mal and T. C. Ting, eds., pp 17-27, ASME (1988).

[18] D. E. Chimenti, and A. H. Nayfeh, *Ultrasonic Reflection and Guided Wave Propagation in Fibrous Composite Laminates,* in : " Wave Propagation In Structural Composites," A. K. Mal and T. C. Ting, eds., pp 29-37, ASME (1988).

[19] R. Courant, and D. Hilbert, Methods of Mathematical Physics, Volume Two, John Wiley and Sons, (1989).

where v_n is the wave propagation velocity in the nth layer. If $\varepsilon_n/\delta<v_n$ there is no convergence of the iteration equation to the physical solutions. If $\varepsilon_n/\delta>v_n$, then convergence occurs, although more slowly (i.e., the shape of the source wave starts degrading after a smaller number of time steps).[4]

At the interface between layers, the matching between incoming, reflected, and transmitted displacements and stresses must be properly taken into account:

$$\tau^{(inc)} + \tau^{(r)} = \tau^{(t)}$$
$$W^{(inc)} + W^{(r)} = W^{(t)} \qquad (5)$$

As a consequence, if the mesh point i = I corresponds to the interface between two layers of respective impedances Z_1 and Z_2, the iteration equation must be modified. The displacements become

$$W_{I,t+1} = a_I^{(+)} W_{I+1,t} + a_I^{(-)} W_{I-1,t} - W_{I,t-1} \qquad (6)$$

where

$$a_I^{(\pm)} = 1 \mp r_I \qquad (7)$$

and

$$r_I = \frac{Z_1-Z_2}{Z_1+Z_2} \quad \text{(reflection coefficient)} \qquad (8)$$

Eq. (6) can be applied to all the mesh points i by setting

$$a_i^{(+)}=a_i^{(-)}=1$$

for all the points not at an interface. It follows that any kind of layered media can be easily and conveniently treated, always with the same computer code, since Eq. (6) is all that is needed for the iteration procedure (besides the initial conditions). The procedure is the same, independent of the number or characteristics of the layers. All the physical properties of the layers can be input as initial data through the coefficients $a_i^{(+)}$ and $a_i^{(-)}$. Furthermore, the iteration equation (Eq. 6) is so simple that it can be adequately treated even with the relatively small amount of memory associated with each processor of the CM.

II. RESULTS AND DISCUSSION

Preliminary results showing the applicability of the CM approach in the elementary (one-dimensional) case of normal transmission through a single layer have been presented in refs. 4 and 5. They show, for a variety of source wave (pulse) shapes, that even after a large number of time steps (>10,000) the wave does not degrade appreciably if Eq. 4 is satisfied. Other results, including examples of wave propagation in attenuative media, are presented in ref. 7.

Since the focus of the present work is mainly on boundary effects, we will consider, in the following, several examples of wave propagation through three layers. In each case the first layer is water (to simulate immersion in a tank) with a thickness of 15 wavelengths. The second layer is very thin (a fraction of wavelength). The third layer is again water or, alternatively Al, with a thickness of 15 wavelengths.

In Fig. 1 (as in all the following figures, except Fig. 2) we plot the displacements (in arbitrary units) vs. the propagation time (in units of δ). The first line represents the top surface of the multilayered medium, the last one the bottom surface. The other lines represent intermediate levels of depth, not necessarily to scale. That is to say, in Fig. 1, levels 601 and 621 represent the two interfaces (water-Al and Al-water). Although the second layer is much thinner than the other layers, it is represented by several intermediate lines, to show the interference patterns.

Before the first layer and after the last one, we assume, in our simulation procedure, that there is vacuum. As a consequence, we have a doubling of the wave, due to the simultaneous presence of the incoming and reflected waves. This effect can be eliminated by considering additional layers to represent the transducers and/or some absorbing media.

In Fig. 1 the second layer is Al and the third is water. The Al layer has a thickness of $\lambda/2$. Theory predicts[3] that a continuous wave (c.w.) crossing a half wavelength layer does not undergo reflection, due to the destructive interference between the waves reflected by the two (top and bottom) surfaces of the layer. For finite pulses, however, the destructive interference is not complete and, therefore, only partial cancellation of the reflected wave occurs. In fact, we see in Fig. 1a that there is little cancellation of the reflected wave for a short pulse (a 5 cycle multiple sine wave, shown on the left side of the top line). For a longer pulse (15 cycles in Fig. 1b and 25 cycles in Fig. 1c), the reflected wave is almost completely cancelled, except, of course, for the two extremes which are shown as decreasing amplitude signals in the center of the top line and are well separated in Figs. 1b and 1c (almost complete interference for much of the pulse), but joined together in Fig. 1a (little interference). Correspondingly, we observe more transmitted wave for a longer pulse (which is a better approximation of the c.w.) than for a short one.

One can also observe, especially in Figs. 1b and 1c, well-defined interference patterns, both constructive and destructive, between incoming and reflected waves within the thin Al layer. Since the stresses are proportional to the derivatives of the displacements, very large stresses within the layer are generated, as shown in Fig. 2. It may also be worth mentioning that the interference effect, which we have discussed so far, happens only when the thickness of the intermediate layer is strictly one half of the wavelength. In fact, Fig. 3 shows a large reflected wave from the layer, when its thickness is just slightly different from $\lambda/2$ (0.45λ).

When the thickness of the intermediate layer is $\lambda/4$ and its impedance is given by

$$Z_2 = \sqrt{Z_1 Z_3} \tag{9}$$

where Z_1 and Z_3 are the impedances of the first and third layer, respectively, a similar interference effect occurs,[3] with total cancellation of the wave reflected by the intermediate layer. However, this is again true only for continuous waves, with partial cancellation in the case of pulses. This is shown in Fig. 4, in which the intermediate layer is assumed to be a fictitious material, whose impedance satisfies Eq.(9), while the third layer is Al. Figure 4a shows complete cancellation for a long multiple sine wave (10 cycles), except, of course, at the two extremes. The interference is only partial (Fig. 4b) for a gaussian modulated multiple sine wave and is almost nonexistent for a single sine wave (Fig. 4c). It should be noted that the incoming and reflected waves are almost indistinguishable within the intermediate layer because of its thinness. Plots like the ones in Fig. 1 or Fig. 4 may be useful as a guide to tell us under which conditions and for which shapes a pulse wave is a good approximation to a c.w.

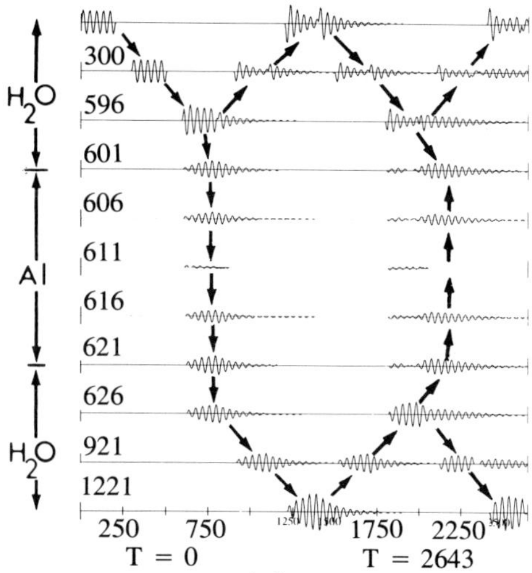

(a)

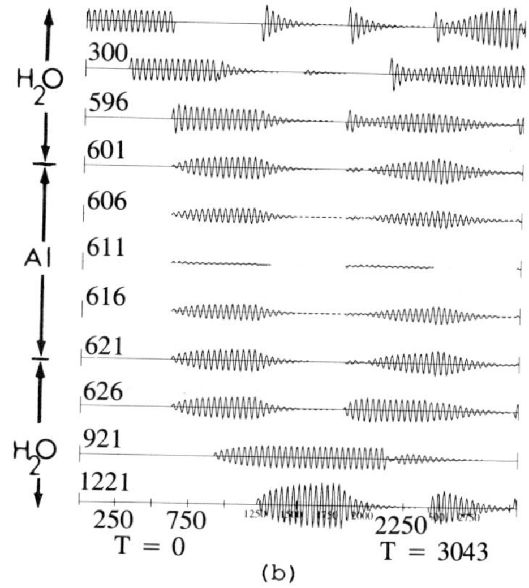

(b)

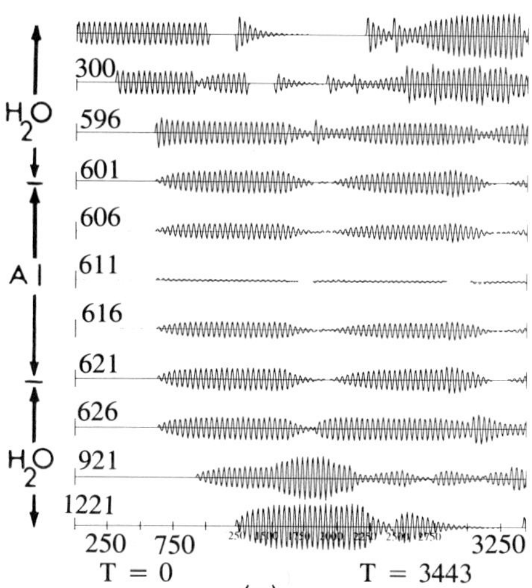

(c)

Fig. 1. Wave propagation through a half wavelength layer. a) 5 cycle source pulse, b) 15 cycle pulse, c) 25 cycle pulse. (Displacements). In fig. 1a arrows have been added as a visual aid to follow the incident, reflected and refracted waves. In each figure the top line corresponds to processor #1; all the other processor numbers are displayed on the corresponding line.

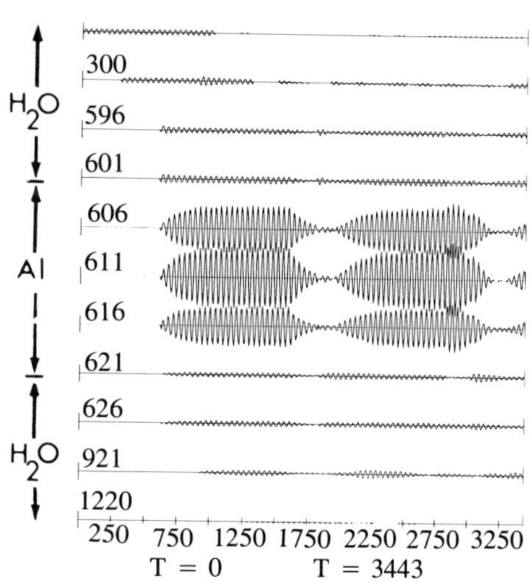

Fig. 2. Wave propagation through a half wavelength layer. (Stresses).

Fig. 3. Wave propagation through an intermediate layer of thickness 0.45λ. (Displacements).

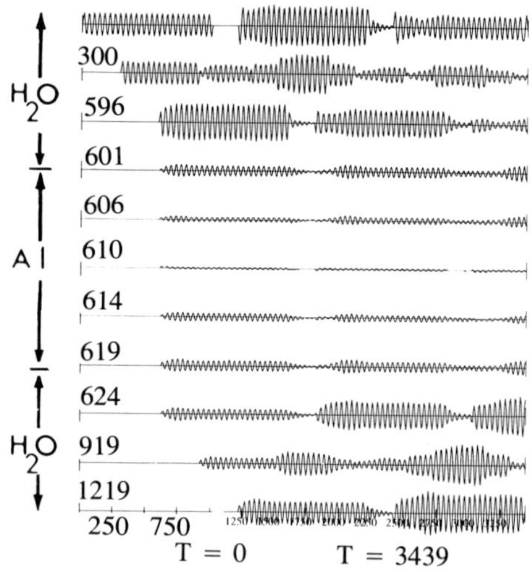

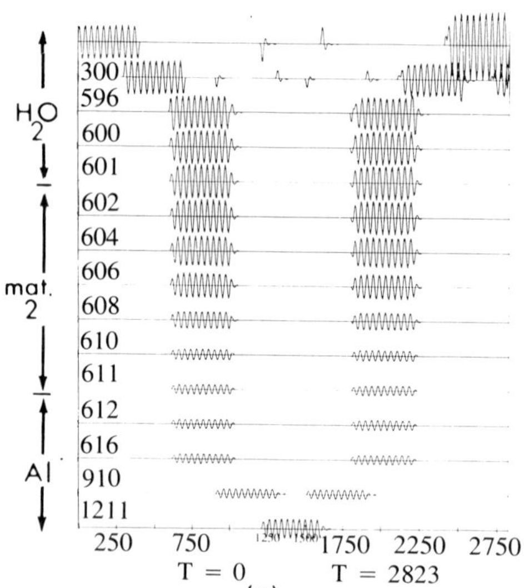

(a)

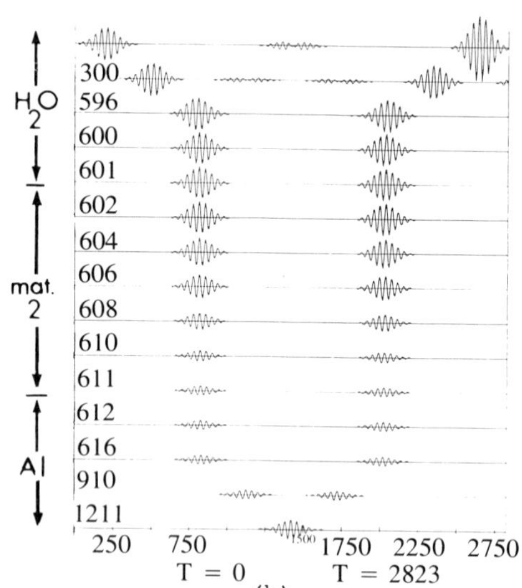

(b)

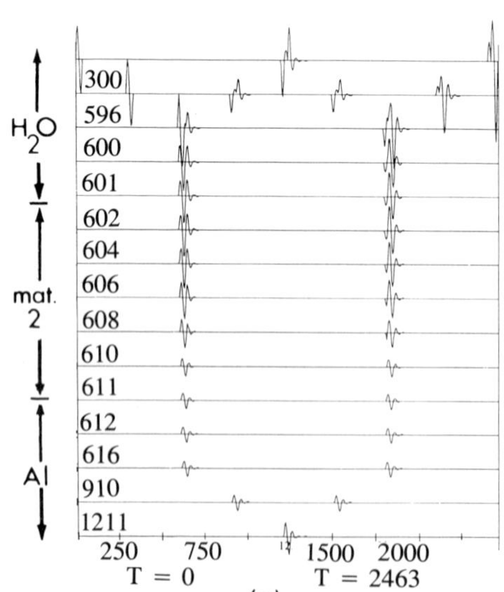

(c)

Fig. 4. Wave propagation through a quarter wavelength layer with impedance $Z_2 = \sqrt{Z_1 Z_3}$. a) multiple sine wave, b) gaussian modulated multiple sine wave, c) single sine wave.

In Fig. 5 we show, for an intermediate layer of Al not satisfying Eq. (9), the effect of decreasing the layer thickness, which is λ/4 in Fig.5a, λ/20 in Fig. 5b, λ/40 in Fig. 5c and λ/200 in Fig. 5d. While in the first two cases the amount of reflected wave is almost the same, in the third case it starts decreasing appreciably and it becomes very small in the last case. This shows that, in order to treat continuously varying (Epstein) layers as linearly varying discrete ones,[7] one should divide Epstein layers in discrete layers not thicker than ≈λ/200. Also, we have performed the simulation shown in Fig. 5c with 1 CM processor representing the intermediate layer. Then we have repeated the calculation with 5 CM processors for the same layer, without finding any noticeable difference in the results. This is very important for the reliability of our CM approach in the case of wave propagation in complex media, because it shows that a thin layer can be correctly represented by a single processor.

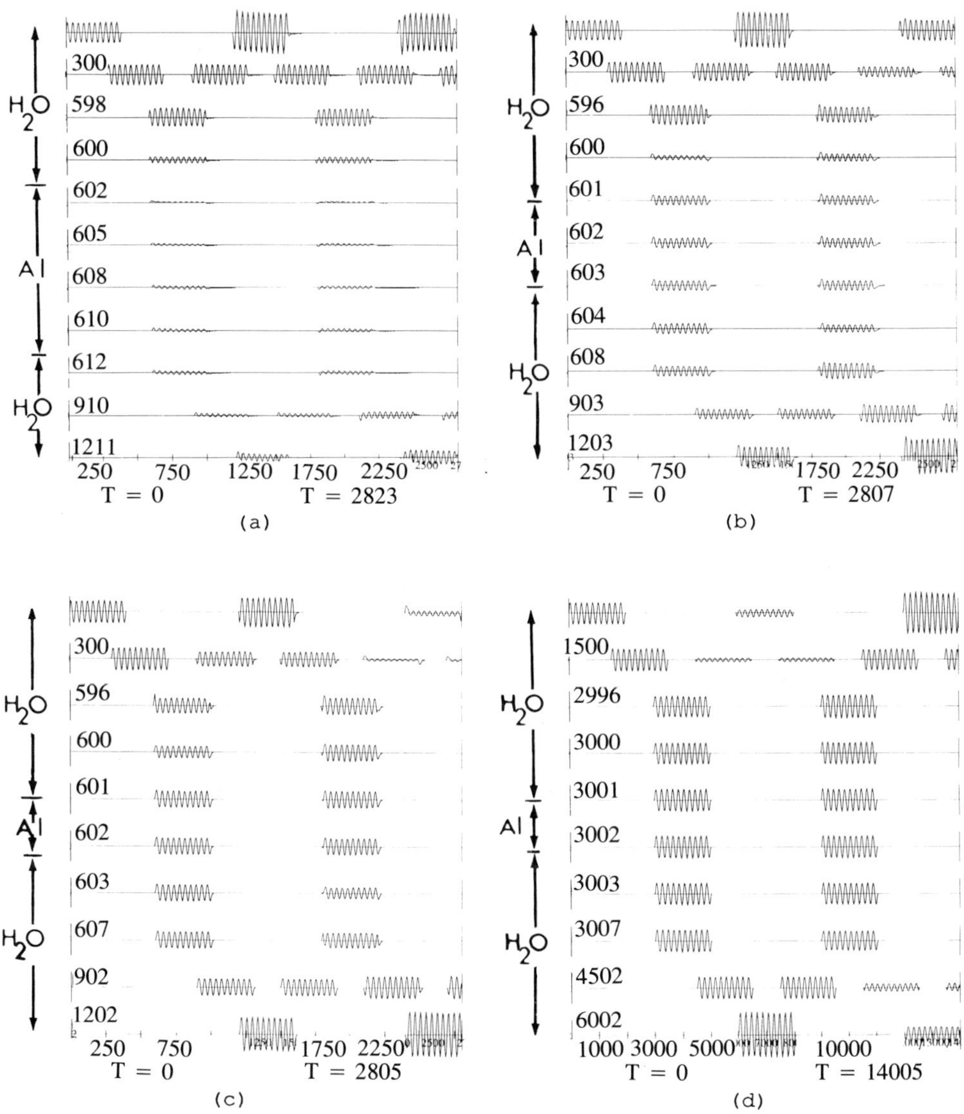

Fig. 5. Propagation through intermediate layers of varying thickness. a) λ/4, b) λ/20, c) λ/40, d) λ/200.

III. CONCLUSIONS

The examples of simulation of the wave propagation mechanism shown in Sec. II demonstrate the practicability of our CM approach in the case of the multilayered media and also its reliability (agreement with analytical predictions). The main features of our approach are its efficiency (very little computer time needed) and its convenience (same computer program for all kinds of different one-dimensional problems).

Although, even in the one-dimensional case several interesting results can be obtained, its most important applications are expected for its extension to 2 and 3 dimensions, which is in progress. It is remarkable that the extension to 2 or 3 dimensions will only require a larger number of processors, but not more computer time, since all of the processors are working simultaneously and independently. The large number of processors required is not expected to be a problem, since a full CM has 65,536 physical processors and as many as 4,000,000 processors, if virtual processors are included. This means 2,000 processors per side in a two-dimensional grid and 158 per side in 3 dimensions (which is sufficient for simple cases). And, of course, it is not unrealistic to expect that the present number of CM processors will increase in the near future.

ACKNOWLEDGEMENTS

We acknowledge the Naval Research Laboratory Connection Machine Facility for support and use of its Connection Machine.

REFERENCES

1. W. Daniel Hillis, "The Connection Machine," The MIT Press, Cambridge, Mass. (1985).
2. "Introduction to Data Level Parallelism" (Thinking Machines Technical Report 86.14), Thinking Machines Corporation, Cambridge, Mass. (1986).
3. L. M. Brekhovskikh, "Waves in Layered Media," Academic Press, New York (1980).
4. P. P. Delsanto, T. Whitcombe, H. H. Chaskelis and R. B. Mignogna, Use of the Connection Machine to Study Ultrasonic Wave Propagation in Materials, in: "Review of Progress in QNDE," Vol. 9A, p. 141, D. O. Thompson and D. E. Chimenti, eds., Plenum Press, New York (1990).
5. P. P. Delsanto, T. Whitcombe, H. H. Chaskelis and R. B. Mignogna, Proc. Conf. "Elastic Wave Propagation in Materials," edited by S. K. Datta et al., North Holland, Amsterdam (1990).
6. P. P. Delsanto, T. Whitcombe, H. H. Chaskelis and R. B. Mignogna, to be submitted to Wave Motion.
7. P. P. Delsanto, T. Whitcombe, N. K. Batra, H. H. Chaskelis and R. B. Mignogna, to appear in the "Proc. 17th Annual Review of Progress in QNDE," Vol. 10, D. O. Thompson and D. E. Chimenti, eds.

COERCIVITY MEASUREMENT FROM ANALYSIS OF THE TANGENTIAL MAGNETIC FIELD

G. Fillion, M. Lord and J. F. Bussière

National Research Council Canada
Industrial Materials Institute
75 De Mortagne, Boucherville, Québec J4B 6Y4 Canada

INTRODUCTION

Coercivity, H_c, is one of the most useful magnetic parameters which can be used to nondestructively characterize the microstructure or mechanical properties of ferromagnetic materials (1-2). However, for in-the-field applications it is generally not possible or practical to have materials of ideal geometry, i.e. slender cylinders around which one can wind coils to measure magnetic induction and deduce H_c. Other techniques must therefore be devised. The most useful are based on locally magnetizing the material studied, while monitoring either eddy current response or Barkhausen noise (BN) (3, 4). These techniques, however, are limited to thin layers because of skin depth effects. Recently Dobmann and Pitsch[5] have shown empirically that from analysis of the harmonic content of the surface tangential field, H(t), it is possible to obtain a field, H_{co}, which is strongly correlated to H_c. This technique, which is based solely on analysis of the tangential field, has the advantage of probing much greater depths than is practical using eddy currents or BN.

In the present paper we have modeled the behavior of the surface tangential field H(t) for the case of a rectangular sample magnetized with a C-shaped electromagnet. We show that from harmonic analysis of H(t) we can reproduce the results of Dobmann and Pitsch. In addition, we show that H_c can also be obtained from a local minimum in $\partial H/\partial t$. The deviation between actual H_c's and H_{co} or H_{cd} is analyzed for different shapes of hysteresis loops.

THEORETICAL ANALYSIS

The experimental configuration to be analyzed in this paper is shown schematically in figure 1. A C-shaped electromagnet is used to magnetize a rectangular shaped sample. A Hall probe is used to monitor the time dependence of the magnetic field H(t) at the surface of the sample while a current I(t) is applied to the electromagnet.

The magnetic circuit will be analyzed based on Ampere's law,

$$NI = \oint H.dl \tag{1}$$

where N is the number of turns in the coil exciting the electromagnet, I is the current in the coil, H is the magnetic field and dl is a line element along the direction of the magnetic circuit. For purposes of simplifying the simulation, we have made the following approximations:

Nondestructive Characterization of Materials IV
Edited by C.O. Ruud *et al.*, Plenum Press, New York, 1991

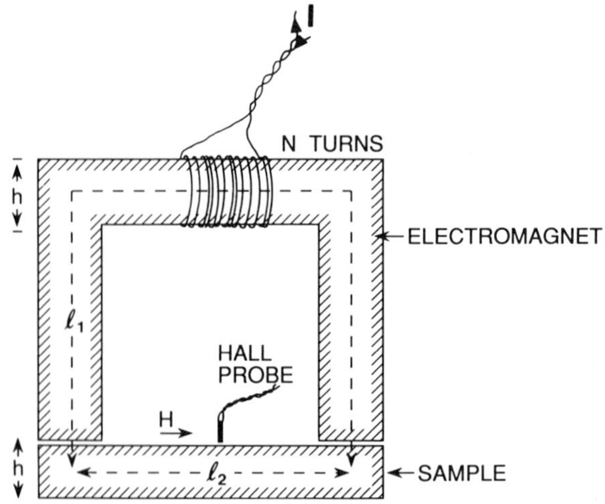

Fig. 1. Schematic of magnetic circuit including an electromagnet and sample.

1) All the magnetic flux is contained within the electromagnet and sample, i.e. there is no flux leakage to the surrounding air, and

2) The flux distribution is uniform within the cross section of both sample and electromagnet.

In general, flux loss will occur to the surrounding air, particularly near sharp angles. However, the above approximations should provide an adequate picture of the behavior of the magnetic circuit. Using equation (1) we therefore obtain for the case shown in figure 1,

$$NI = H_1 l_2 + H_2 l_2 \tag{2}$$

where l_1 is the length of the magnetic path in the electromagnet and l_2 in the sample. Similarly H_1 and H_2 denote respectively the field in the electromagnet and sample. Because of flux conservation, we may also write

$$\Phi = B_1 A_1 = B_2 A_2 \tag{3}$$

where Φ is flux, B is magnetic induction, A is area and the indices 1 and 2 refer respectively to electromagnet and sample.

In order to calculate the behavior of the magnetic circuit, it is also necessary to model the hysteretic behavior of B vs. H. We have chosen to use a model proposed by Ponomarev[6], which gives simple analytic expressions for the magnetization M:

$$M^{\pm}(H) = \frac{2M_s}{\pi} \tan^{-1}\left[\frac{\pi}{2}\left(\frac{H \pm H_c}{H_s}\right)\right] \pm \frac{\delta M}{2}, \tag{4}$$

where

$$\delta M = \frac{2M_s}{\pi}\left\{\tan^{-1}\left[\frac{\pi}{2}\left(\frac{H_{max} + H_c}{H_s}\right)\right] - \tan^{-1}\left[\frac{\pi}{2}\left(\frac{H_{max} - H_c}{H_s}\right)\right]\right\}, \tag{5}$$

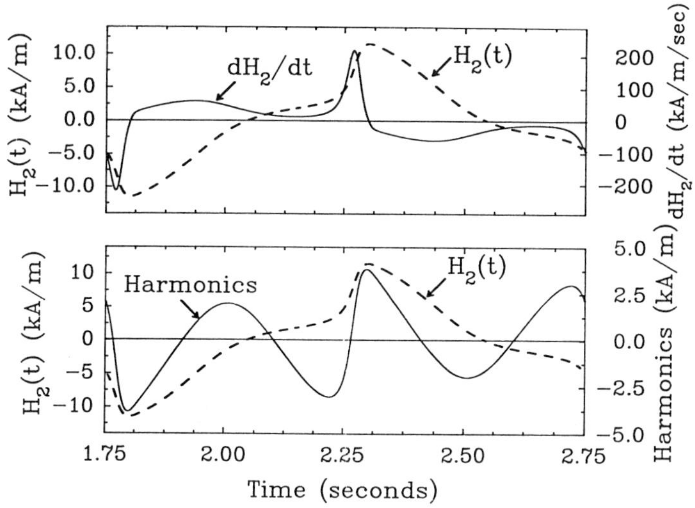

Fig. 2. Calculated time dependence of the sample tangential field, $H_2(t)$, its time derivative dH_2/dt and the harmonics of $H_2(t)$ for the case of sinusoidal voltage during the electromagnet. The Ponomarev model was used for the sample with H_c = 2000 A/m, H_s = 1500 A/m and H_{max} = 11.5 k A/m.

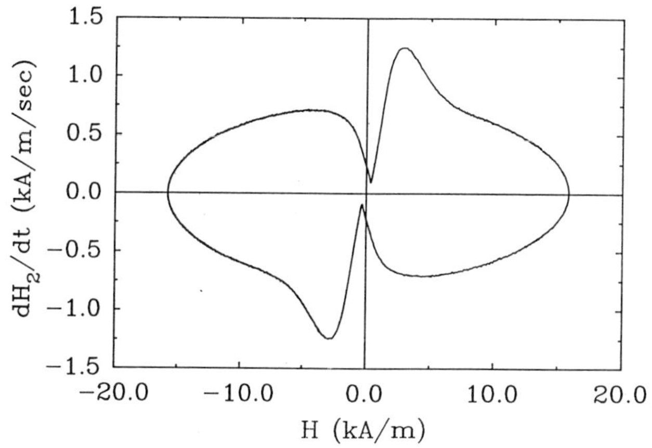

Fig. 3. Calculated value of dH_2/dt plotted versus H_2 for the same case as figure 2.

M_s is the saturation magnetization, H_c is the coercivity, H_s is a parameter describing the squareness of the hysteresis loop, and H_{max} is the magnetic field amplitude for a given cycle. The upper(lower) sign is used when the magnetic field is increasing(decreasing).

The experiment simulated consists in magnetizing the electromagnet / test piece by applying either a sinusoidal voltage or current waveform to the exciting coil of the electromagnet. The simplest case to simulate is to apply a current waveform to the exciting coil of the electromagnet. In this case the tangential magnetic field of the test piece, $H_2(t)$ can be obtained by solving equations (2) and (3) for each instantaneous current. If a voltage waveform is applied to the electromagnet, obtaining $H_1(t)$ and $H_2(t)$ is only slightly more difficult. In this case the current flowing in the circuit is not known and must be solved using

$$V(t) = N\frac{d\Phi}{dt} + RI \tag{6}$$

where R is the DC resistance of the coil and the first term on the right hand side of equation (6) is the induced voltage in the exciting coil due to the change in magnetic flux enclosed by the coil.

Determination of coercivity from $H_2(t)$

When a sinusoidal current or voltage waveform is applied to the electromagnet, the $H_1(t)$ and $H_2(t)$ waveforms are not perfectly sinusoidal. Deviations from sinusoidal waveforms are particularly noticeable near the coercivities H_{c1} and Hc_2 of electromagnet and test sample respectively. From experimental observation we have found that dH_2/dt reaches a local minimum at a value Hcd, which is close to H_{c2}. Calculations and experimental measurements of H_{cd} will therefore be presented below and correlated with H_c for various types of hysteresis loops.

In addition, it has been reported by Dobmann and Pitsch[5] that the field of first zero crossing of the harmonics of H(t), H_{co}, is also closely correlated with H_c. To obtain the time signal of the harmonics of H(t), its Fourier transform was calculated, the fundamental component was filtered out and the reverse FFT was performed on the filtered data. The value H_{co}, as defined by Dobmann and Pitsch[5], is the magnetic field at the time when the upper harmonics signal crosses the zero axis after remanence. Under certain cases the zero axis crossing of the upper harmonics signal occurs just before remanence. In this case we have modified the definition of H_{co} to be the magnetic field at the zero crossing of the upper harmonics signal which is nearest to remanence. Using this definition both positive and negative H_{co} can occur.

Simulation Results

For simulation purposes the magnetic circuit shown in Figure 1 was chosen with magnetic circuit path length l_1 = 0.32 m, l_2 = 0.16 m. The cross-sectional area of both test piece and electromagnet was taken to be A = 0.0016 m^2. The exciting coil of the electromagnet had N = 100 turns.

The material properties chosen for the electromagnet and test piece were modeled using the Ponomarev model described above. The electromagnet parameters were $M_{s1} = 2.39 \times 10^6$ A/m, H_{c1} = 100 A/m and H_{s1} = 150 A/m. these values were chosen, somewhat arbitrarily, so that the electromagnet would have a very small coercivity and high saturation magnetization. The high saturation magnetization is necessary to ensure that the electromagnet does not saturate magnetically. Since the results are not very sensitive to the characteristics of the electromagnet, similar results would have been obtained with an electromagent with a smaller M_s but larger area. The test pieces had $M_{s2} = 1.194 \times 10^6$ A/m, H_{s2} = 1500 A/m while the coercivities, Hc_2, ranged from 500 A/m to 4000 A/m. For the Ponomarev model the maximum differential magnetic susceptibility is M_s/H_s, which for these materials is constant at approximately 800.

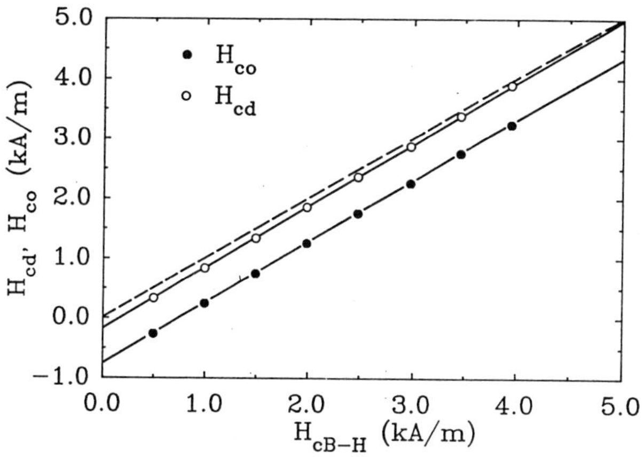

Fig. 4. Calculated values of H_{cd} and H_{co} plotted versus $H_{cB\text{-}H}$. The maximum field H_{max} was 16.4 k A/m for all cases.

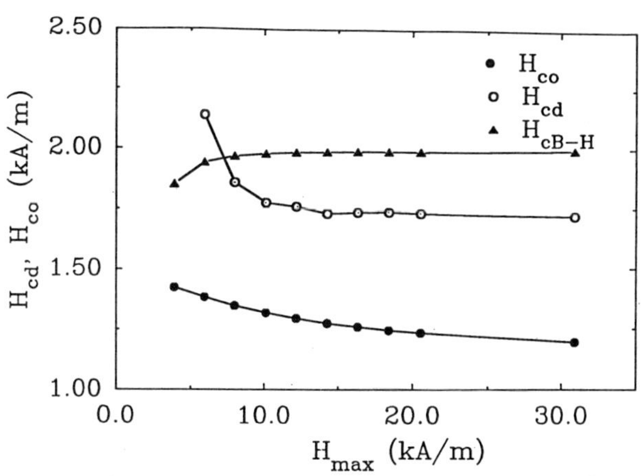

Fig. 5. Effect of magnetic field amplitude H_{max} on H_{co}, H_{cd} and $H_{c\,B\text{-}H}$.

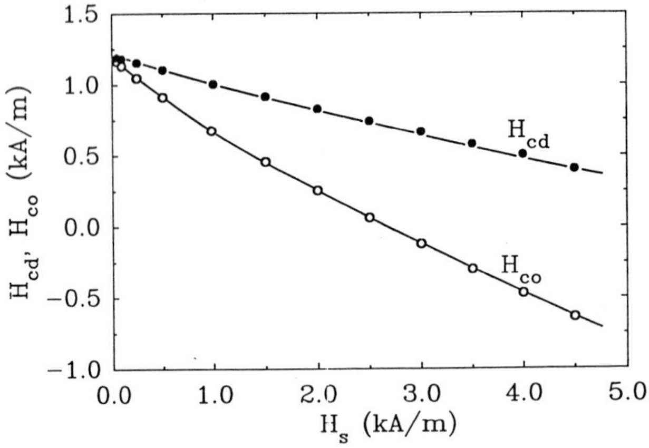

Fig. 6. Effect of squareness of hysteresis loop, as determined by the parameter H_s, on H_{cd} and H_{co}.

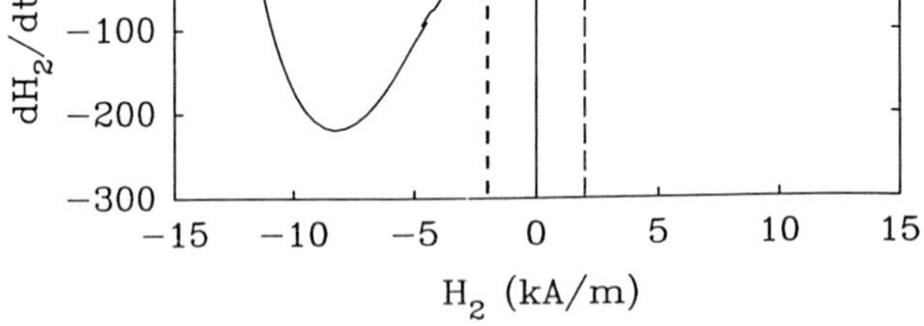

Fig. 7. Experimental measurement of dH_2/dt vs H_c for a 1025 steel sample.

Typical calculated results of the time dependence of $H_2(t)$, dH_2/dt and the harmonics of $H_2(t)$ are presented in figure 2 for the case where $H_c = 2000$ A/m and $H_{max} = 11.5$ k A/m and sinusoidal voltage excitation. Note that $H_2(t)$ deviates considerably from a sinusoidal waveform; this is generally not the case for sinusoidal current excitation. The harmonics goes through zero at a field H_{co} close to $H_{cB\text{-}H}$ after the field crosses the zero value. Similarly, the value of dH_2/dt goes through a local minimum at Hcd which is close to $H_{cB\text{-}H}$. This is more clearly visible in figure 3, where dH_2/dt is plotted versus H_2 for a complete cycle. Values of H_{co} and H_{cd} were calculated for different values of $H_{cB\text{-}H}$ for both sinusoidal voltage and sinusoidal current excitation. Results for voltage excitation are presented in figure 4. It is seen that both H_{co} and H_{cd} are linearly correlated with $H_{c\ B\text{-}H}$, with the value of H_{cd} closer in absolute value to $H_{c\ B\text{-}H}$. than H_{co}. Nearly identical results were obtained for sinusoidal current excitation.

The effect of the amplitude of the cycle H_{max} was also investigated, and is presented in figure 5 where $H_{c\ B\text{-}H}$, H_{co} and H_{cd} are plotted versus H_{max} for the case where $H_c = 2000$ A/m. As expected, the value of $H_{c\ B\text{-}H}$, i.e. the field at which the induction drops to zero after field reversal, decreases as the H_{max} decreases. Somewhat surprisingly however, both H_{co} and H_{cd} increase as the amplitude is decreased, the effect being more pronounced in the case of H_{cd} at low H_{max}. At large H_{max} however, the value of H_{cd} is less dependent on H_{max} in better agreement with $H_{cB\text{-}H}$ than H_{co}.

The difference in absolute value between $H_{c\ B\text{-}H}$ or H_{co} and $H_{c\ B\text{-}H}$ also depends on the squareness of the hysteresis loop. This is illustrated in figure 6 where H_{cd} and H_{co} are plotted versus H_s for the case where $H_c = 1200$ A/m and sinusoidal current. It is seen that for very square hysteresis loops ($H_s \approx 0$) both H_{co} and H_{cd} agree with $H_{c\ B\text{-}H}$. However, as H_s is increased or as the hysteresis loop becomes more "tilted" both H_{cd} and H_{co} decrease, the value of H_{co} decreasing more rapidly with Hs. It should be noted here that typical values of H_s are usually of the order of H_c for typical structural steels.

EXPERIMENTAL MEASUREMENTS

The experimental arrangement was essentially as shown in figure 1. The magnetic field is measured using a Hall probe connected to a gaussmeter. In addition, an analogue differentiator is used to generate the dH/dt time signal of the magnetic field. The gaussmeter output and the derivative of the gaussmeter output are digitized using a Data Precision D6000A transient signal recorder/analyzer. The numberical derivative and the FFT can be performed either using the D6000A or stored on disk and analyzed later on a microcomputer. The electromagnet used was a C-shaped electro-magnet made from a laminated transformer core. Its dimensions are 75 mm in length, 50 mm in height and a cross section of 12.5 x 31 mm. The 450 turn coil of the electro-magnet is powered by a voltage controlled power supply. We were unable to use a current controlled power supply for this electromagnet. The magnetic flux B was monitored simultaneously using a pickup coil wound directly on the samples in order to determine $H_{c\ B\text{-}H}$. Most of the samples were pearlitic steel bars cut from railroad rail heads with typical dimensions of 72 x 7.5 x 32.5 mm^3. In addition, samples of 1025, 1045 and 1090 steel were also used.

The excitation frequency used was 25 m Hz or lower in order to minimize any eddy current effects in the sample. The value of H_{max} was roughly the same for all samples and equal to 16 k A/m. A typical dH/dt vs H result is shown in figure 7, obtained for 1025 steel. The behavior is similar to that calculated in figure 3, however with a sharper minimum near H_c.

A comparison between $H_{c\ B\text{-}H}$ and H_{cd} is summarized for all samples in figure 8. We note that the behavior is essentially as obtained theoretically, i.e. the values of Hcd are lower than $H_{c\ B\text{-}H}$ with an excellent linear correlation between the two (see figure 4).

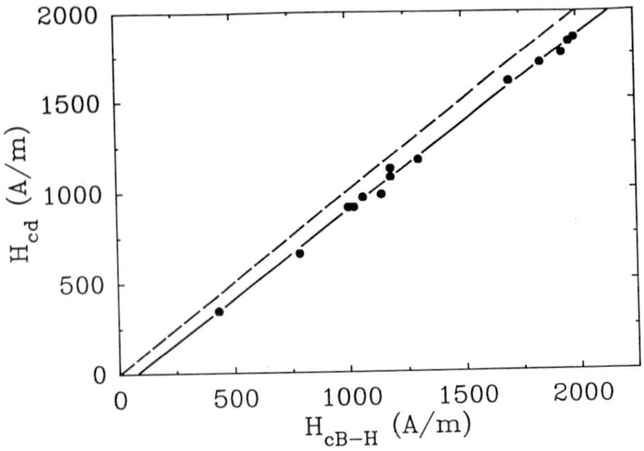

Fig. 8. Experimental comparison between H_{cd} and $H_{c\,B\text{-}H}$ for a number of steels. The dashed line indicates $H_{cd} = H_{c\,B\text{-}H}$.

CONCLUSION

Using Ampere's law and a simple model for magnetic hysteresis, the behavior of the surface tangential field, H(t), was modeled for the case of a material magnetized by an electromagnet driven by a sinusoidal current or voltage. Hysteresis in the sample leads to deviations from sinusoidal behavior which are accentuated near the coercivity, H_c, of the material. These have been analyzed both by Fourier analysis of H(t) and by analyzing dH/dt, leading to characteristic fields H_{co} and H_{cd} respectively. Both H_{co} and H_{cd} have been shown to be strongly linearly correlated with $H_{c\,B\text{-}H}$, the value obtained from B-H hysteresis loops. For the case of an infinitely steep hysteresis loop, all values are equal, i.e. $H_{co} = H_{cd} = H_{c\,B\text{-}H}$. However, as the overall slope of the B-H hysteresis loop is decreased, both H_{cd} and H_{co} decrease to values which are lower than $H_{c\,B\text{-}H}$, the effect being more pronounced for Hco. These results are in good agreement with experimental measurements of H_{co} by Dobmann and Pitsch[5] and measurements presented in this paper for Hcd of a number of steels spanning a broad range of coercivity values.

Analysis of the tangential field appears to be a simple method to deduce parameters which are closely related to the coercivity of the material and can therefore be used to characterize the material nondestructively. It offers the advantage of probing deeper into the tested material than other techniques based on eddy current response or Barkhausen noise, and of being more representative of bulk behavior.

REFERENCES

1. J.F. Bussière, Materials Evaluation 44, No. 5, 560-567 (1986).
2. D.C. Jiles, NDT International 21, 311 (1988).
3. M. Mayo, S. Segalini, M. Putignani, QNDE Vol. 6B, Plenum, p. 1691 (1987).
4. G. Dobmann, W.A. Theiner, R. Becker, in Nondestructive Characterization of Materials III, edited by P. Höller et al., Springer-Verlag 1989, p. 516.
5. G. Dobmann, H. Pitsch, in Nondestructive Characterization of Materials edited by P. Höller et al, Springer Verlag 1989, p. 636.
6. Yu, F. Ponomarev, Defektoscopia, 21, pp. 417-422 (1985).

DETERMINATION OF ELASTIC-PLASTIC BOUNDARY BY SPECKLE PATTERN CORRELATION

F.P. Chiang*, C.J. Tay** and Y.Z. Dai***

*Lab. for Experimental Mechanics Research
SUNY at Stony Brook
***Dept. of Mech. Eng.
SUNY at Stony Brook
NY 11794-2300
**Dept. of Mech. & Indus.
Nat. Univ. of Singapore
Singapore

ABSTRACT

An optical-numerical correlation technique has been developed for detecting the elastic-plastic boundary on the surface of a material which has undergone plastic deformation. Laser speckle pattern produced from a surface is measured before and after plastic deformation which produces a change in their cross correlation. The speckle images are recorded and processed using a computer based vision system. Experimental results obtained for aluminum alloys have demonstrated the high sensitivity of this method. A comparison with theoretical and finite element results is presented.

INTRODUCTION

It has been known for sometime that the surface roughness of material increases with increase in surface strain [1-3]. This is mainly due to the process of dislocation movement. This phenomenon suggests methods to relate plastic strain quantitatively with variation in surface roughness. Surface roughness can be measured by a stylus-type profilometer. But this method tends to damage the material surface under examination. There are optical techniques to quantify surface plastic strain by measuring surface texture parameters. Some[4,5] employed coherent light source and obtained information about surface texture from the light beam reflected from the test surface. Others [6,7] used a white light source and a computer vision system to determine the plastic strain.

In this paper a nondestructive, non-contacting computer based method is described for the determination of plastic deformation. In particular the technique is applied to the determination of the elastic-plastic boundary around a central hole in a plate of finite width. The method employs a converging laser beam which detects the change in surface roughness introduced by plastic deformation. The diffracted images are recorded and digitized sing a computer based vision system and a correlation technique is employed to analyze the digitized images.

THE SPECKLE PATTERN CORRELATION TECHNIQUE

The basic principles of the speckle pattern correlation techniques is based on the phenomenon that the surface of a piece of metallic specimen roughens as it is subjected to

Nondestructive Characterization of Materials IV
Edited by C.O. Ruud *et al.*, Plenum Press, New York, 1991

sufficient amount of plastic deformation. Consequently the resulting speckle pattern changes accordingly. If g(x) is a function which describes the surface of a given area before deformation and f(x) describes the surface after the deformation, the zero shift cross correlation $C_o(g,f)$, which is a measure of the resemblance between the two functions, is defined as:

$$C_o(g,f) = \int_{-\infty}^{\infty} g(x) \times f(x)dx \tag{1}$$

Normalization of C_o yields the cross correlation coefficient:

$$C_c(g,f) = \frac{\int_{-\infty}^{\infty} g(x) \times f(x)dx}{\sqrt{\int_{-\infty}^{\infty} g^2(x)dx \times \int_{-\infty}^{\infty} f^2(x)dx}} \tag{2}$$

For digitized images, g(x) and f(x) take discrete form and the above expression becomes[8]:

$$C_c(g,f) = \frac{\Sigma_{i=1}^{N} g(i) \times f(i)}{\sqrt{\Sigma_{i=1}^{N} g^2(i) \times \Sigma_{i=1}^{N} f^2(i)}} \tag{3}$$

For two dimensional images the expression for cross correlation coefficient becomes:

$$C_c(g,f) = \frac{\Sigma_{i=1}^{N} \Sigma_{j=1}^{N} g(i,j) \times f(i,j)}{\sqrt{\Sigma_{i=1}^{N} \Sigma_{j=1}^{N} g^2(i,j) \times \Sigma_{i=1}^{M} \Sigma_{j=1}^{N} f^2(i,j)}} \tag{4}$$

where g(i, j), f(i, j) are the discrete intensity levels at points i, j of image g and f, respectively and M and N are the dimensions of the arrays g(i, j) and f(i, j).

EXPERIMENTAL PROCEDURES AND DATA ACQUISITION

Plate specimens of 3.2mm thick with a central hole were cut from aluminum alloy 6061-T6 in the rolling direction. The elastic modulus is $72GP_a$; the poisson's ratio is 0.32 and the yield strength is $275MP_a$. The geometry of the specimens are shown in Figure 1 where h = 76mm, a = 6.4mm and R = 12.7mm. One side of the specimen surface was polished until an initial roughness value of approximately $R_a = 0.05\mu m$ was obtained, where R_a is the arithmetic average roughness.

The specimen was loaded axially in a universal testing machine at $\sigma = 0.765\sigma_o$, where σ is the remote normal stress along the X axis, and σ_o is the yield strength of the material. The loaded specimen was then mounted on a X-Y translation stage which is capable of translation in each direction in 0.006mm increments.

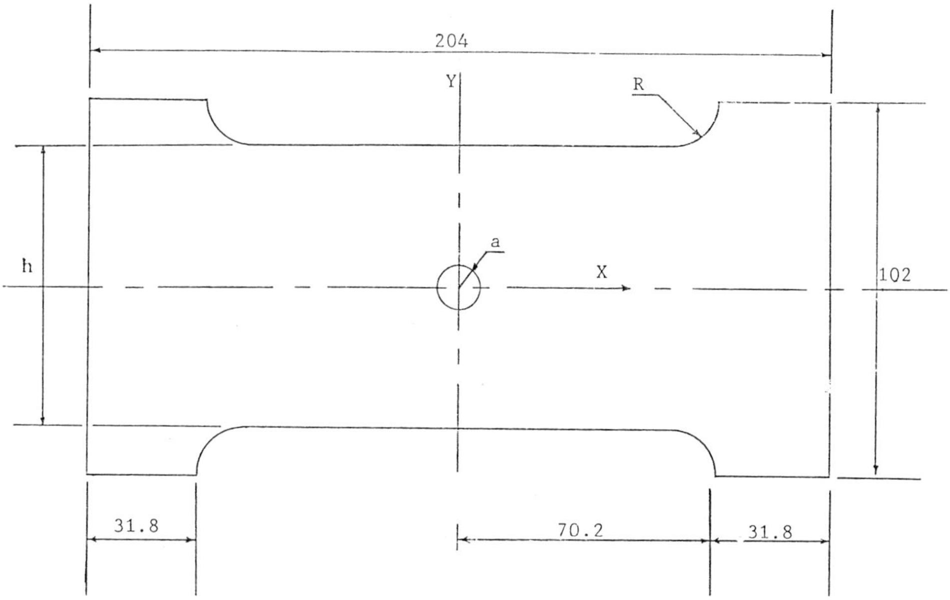

Figure 1 Dimensions and geometry of the specimen

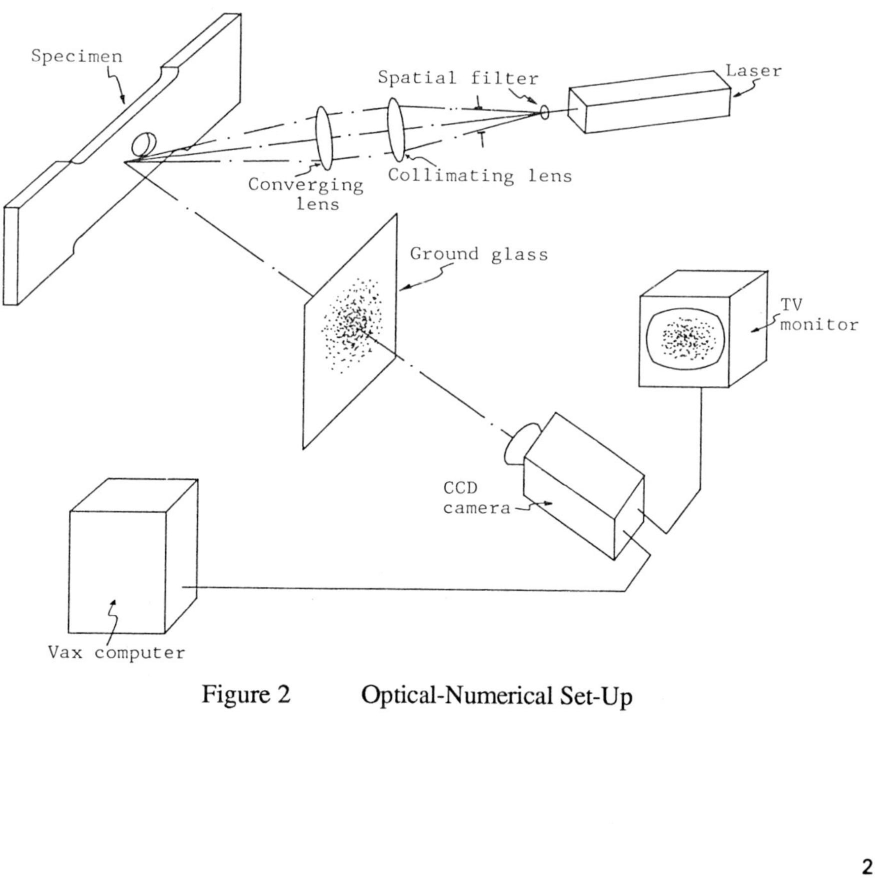

Figure 2 Optical-Numerical Set-Up

The point of interest on the test object was illuminated with a 20mw He-Ne laser. To increase the resolution of the technique, the diameter of the laster beam was reduced through a system of optical arrangement consisting of a spatial filter, an aperture, a collimating lens and a converging lens as shown in Figure 2. The actual size of the laser beam directed on the test piece was estimated to be 0.05mm. A piece of ground glass was placed at a distance of 0.8m from the test object such that the diffraction pattern was observed in the Fraunhofer diffraction zone[9]. The diffracted speckle pattern was recorded by a digital camera placed at 0.36m from the ground glass. The camera (resolution of 256 x 256 pixels, each pixel has 256 gray levels) was connected to a TV monitor and supported by a computer which together with the developed software provided full image processing and analysis capabilities.

RESULTS and DISCUSSION

Figure 3 shows the diffraction images of the test piece along the vertical axis (section C-C). Figure 3a is the diffraction image in the elastic zone. As the light source moves towards the central hole, the strain level increases and as illustrated in Figure 3b & 3c there is a continuous drop in the sharpness of the images and the image boundary becomes less distinguishable. The observation suggests that the correlation level of each subsequent image will decrease.

Figure 4 shows a plot of the cross correlation coefficient along section C-C. As can be seen, the initial images correlate well with that of the elastic zone at B and the correlation coefficient remains fairly constant over a distance of approximately 5.4mm. However, beyond this point (as indicated by point P) the correlation coefficient begins to decrease and reaches a minimum value at the edge of the hole where the maximum plastic strain occurs. The decrease of the correlation coefficient (point P) would therefore provide an indication of the incipient yielding and hence the location of the elastic-plastic boundary.

The threshold value of cross correlation at which the elastic-plastic boundary is to be determined should be chosen according to the cross correlation-plastic strain calibration curve and the definition of elastic-plastic boundary for each specific problem. Suppose the elastic-plastic boundary is defined at where the effective residual strain is 1%; while from the cross correlation-plastic strain relation we find 1% effective strain corresponds to 0.7 cross correlation value, then 0.7 should be used as the threshold to determine the boundary.

Figure 5 shows the elastic-plastic boundary identified using the method described. The results shown are those obtained in region A of the specimen as indicated in Figure 5. The results obtained using the finite element method is also included. A numerical calculation which assumed a state of plane stress and adopted the von Mises yield criterion was carried out using a finite element package (ABAQUS version 4-7-21). The solid line represents a computed effective strain of 0.01%. As can be seen both the experimental and computed results agree reasonable well. An attempt was also made to compare the above results with those obtained using the Kirsch's solution for the stress distribution around a small circular hole in an infinite flat plate subject to uniform tension[10] and the von Mises yield criterion. As can be seen the theoretical results (broken line) underestimated the size of plastic zone. The discrepancy is maybe due to the simplification and assumptions made in the theoretical solution.

CONCLUDING REMARKS

In this paper, it has been demonstrated that speckle pattern cross correlation technique can be applied to the detection of the incipience of plasticity. In a material which had undergone elastic-plastic deformation, the method described was able to detect plastic strain of the order of 0.01%. The technique was supported by a computer vision system via which the recorded images were analyzed and processed. With further automating the technique, it has the potential of being developed into a practical tool.

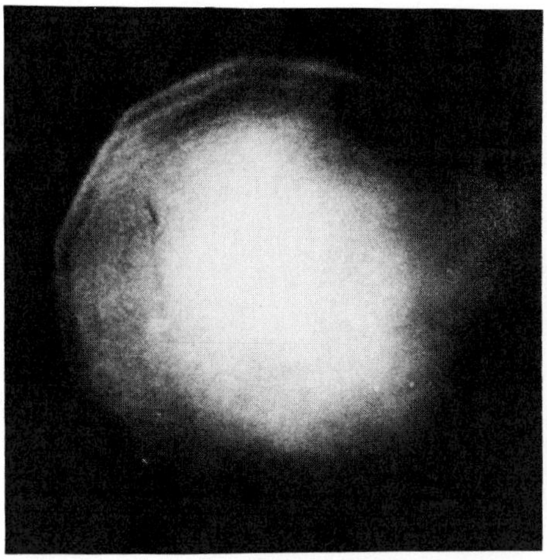

Figure 3a Diffraction Image of Point B

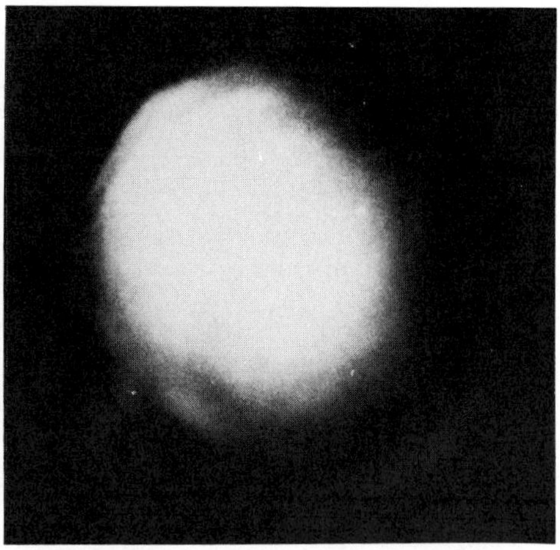

Figure 3b Diffraction Image of Point P

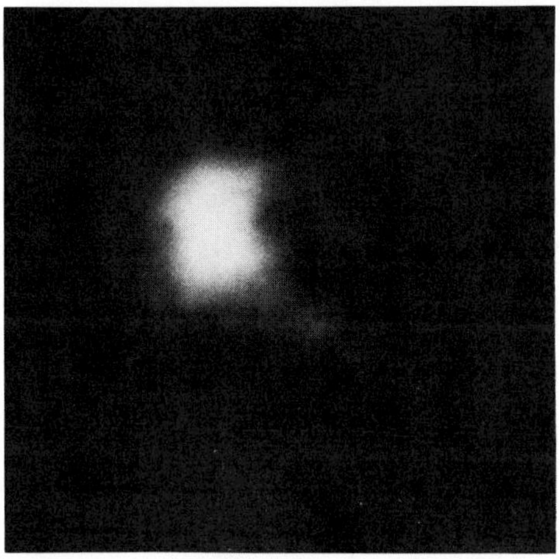

Figure 3c Diffraction Image of Point Q

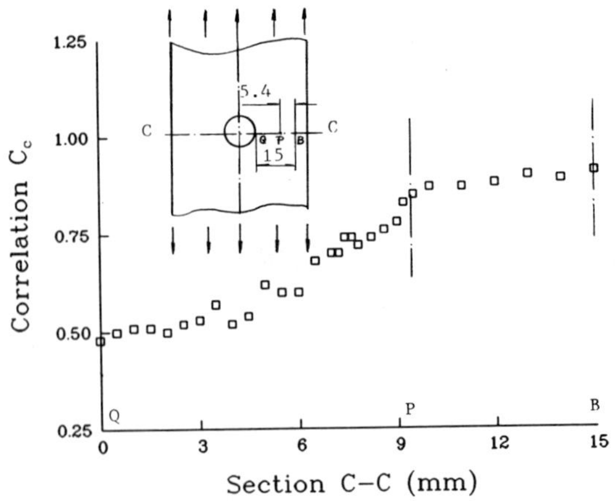

Figure 4 Correlation Coefficient Across the Plastic Zone (along section C-C)

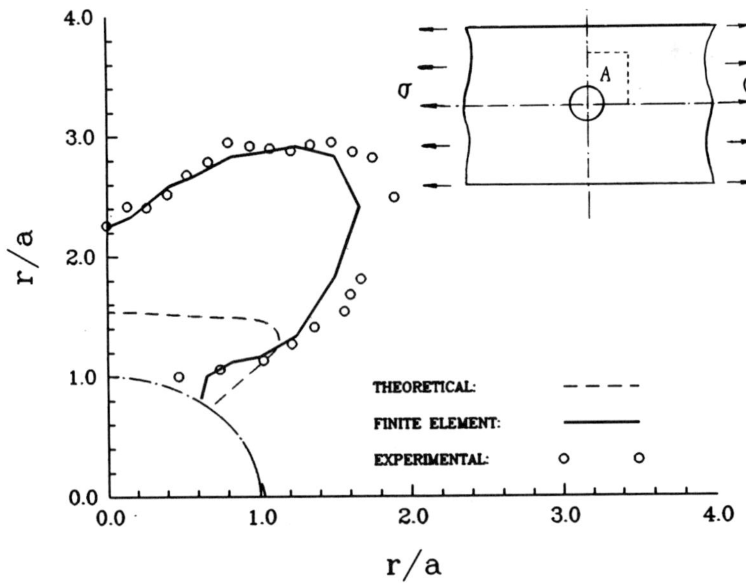

Figure 5 Elastic-Plastic Boundary at $\sigma = 0.765\sigma_0$

ACKNOWLEDGEMENT

The authors wish to thank the Army Research Office, Engineering, Service Division for providing the financial support through contract No. DAAL0388K0033.

REFERENCES

1. K. Yamaguchi, P.B. Mellor, "Thickness and grain size dependence of limit strains in sheet metal stretching," Int. J. Mech. Sci. 18, 85-90 (1976).

2. P.F. Thomson, B.V. Shafer, "The roughening of free surface during plastic working," Int. J. of Mach. Tool Des. Res. 22(4), 261-264 (1982).

3. J.H. Dautzenberg and J.A.G. Kals, "Stress state and surface roughness," in: Advances Technology of Plasticity, Japan (1984).

4. M. Miyagawa and A. Azushima, "Measurement of plastic deformation by means of a laser beam," in: Proceedings of the 9th World Conference on Nondestructive Testing, Session 4a-4, pp.1-7, Melbourne, Australia (November 1979).

5. D. Sigler and W.L. Haworth, "Strain measurement by optical correlation," J. of Nondestructive Evaluation 2(2), 125-132 (1981).

6. C. Lee, Y.J. Chao, W.H. Peters and M.A. Sutton, "Surface texture and plastic deformation," in: Proceedings of the 13th Southeastern Conference on Theoretical and Applied Mechanics, pp. 275-283, Columbia, South Carolina (1986).

7. C. Lee, Y.J. Chao, M.A. Sutton, et al., "Determination of plastic strain at notches by image-processing methods," Experimental Mechanics 29(2), 214-220 (1989).

8. A. Rosenfeld and C.A. Kak, Digital Picture Processing, Academic Press, pp. 19 (1982).

9. M. Francon, Laser Speckle and Related Applications in Optics, Academic Press, pp. 21-24 (1979).

10. D.J.P. Hartog, Advanced Strength of Materials, McGraw Hill Book Co. (1952).

THE EFFECTS OF IN-PLANE STRESS ON WAVES IN THIN FILMS

George C. Johnson and Shih-Emn Chen

Department of Mechanical Engineering
University of California
Berkeley, CA 94720

ABSTRACT

The application of acoustoelasticity for the nondestructive evaluation of residual stress has, for the most part, been limited to waves which may be classified as being of short wavelength. For many thin films, whose thicknesses may be on the order of 1 μm or less, the frequency required to obtain a "short wavelength" is well into the gigahertz regime. Thus, for acoustoelasticity to be applicable to such thin films, an improved understanding of the effect of stress on the wave response in the intermediate to long wavelength regimes must be available. This paper considers the effect of stress on the response of waves propagating in plate-like structures. The frequency equation is derived in the context of three-dimensional elastodynamics in which a small disturbance is superposed on a finite deformation of the nonlinearly elastic plate. Particular attention is directed toward the lowest order antisymmetric (flexural) mode which exhibits membrane response in the long wavelength limit. Results showing the details of the transition from the long wavelength limit to the short wavelength limit (bulk wave response) are presented for this mode.

INTRODUCTION

Residual stresses play an important role in determining the performance of many thin film devices. The presence of large residual stresses can cause a number of detrimental effects, including delamination, fracture, buckling, and electromigration[1]. Unfortunately, these devices are typically very small by conventional stress measurement standards. Thus, many of the "traditional" techniques used for the nondestructive evaluation of stress at the "macro-" level are not always applicable at this "micro-" level. Among the nondestructive techniques that have been successfully used for stress evaluation of thin film structures are diffraction methods[2,3], wafer or beam curvature[1,4], micromechanical devices (beams and membranes)[1,5–7], and ultrasonic techniques[8].

The ultrasonic technique for residual stress evaluation, acoustoelasticity, is based on the observation that the speed at which a wave travels through a material depends upon the state of stress present in the material[9]. In order to adequately describe the observed behavior, the material's constitutive response must include a slight nonlinearity, typically incorporated by allowing the material's stored energy function to be cubic in the strain[9]. The theory for acoustoelasticity is well developed for bulk waves and surface waves in semi-infinite media (the short wavelength limit). This is not the case for waves in thin films or layered structures.

This paper presents an analysis of the acoustoelastic effect in thin films, focusing attention on the simplest case of a free film for presentation of numerical results. The analytical and numerical approach taken here is strongly motivated by the recent experimental work of Meeks and coworkers[8] in developing the instrumentation for making the necessary measurements. Their work has led to a new "Scanning Phase-Measuring Acoustic Microscope" (referred to as the SPAM system) in which the stress-induced variation in wave speed can be determined with precision sufficient for stress analysis and with a spatial resolution of between 5 μm and 10 μm.[8]

The next section of this paper presents an overview of existing acoustoelastic techniques for stress evaluation, indicating the difficulty in applying any of these "traditional" techniques to thin films. Following this, the basic equations of acoustoelasticity in the context of a small deformation superposed on a finite deformation of a nonlinear elastic material are presented in Lagrangian form. This referential description is preferred over the spatial formulation because of the relative ease with which the traction boundary conditions can enforced. Finally, results for a plate with traction free surfaces are presented in terms of dispersion curves and in terms of the changes in velocity which are caused by loading the plate in the plane. It is shown that the traditional acoustoelastic effect, given as a small linear variation in velocity with applied stress, is wholly inadequate. The dependence of the acoustoelastic effect on frequency (or equivalently, wave number) is dramatic, with the long wavelength response undergoing a fundamental change in character.

OVERVIEW OF ACOUSTOELASTICITY FOR BULK WAVES

The concept of using acoustoelasticity to evaluate stresses in films is quite recent[8]. The majority of work in acoustoelasticity has dealt with bulk materials through which longitudinal, shear, or surface waves travel[9]. The variation in the propagation speeds of these various bulk waves has then been used to evaluate the stress state. In all cases, the stress state in question has been taken to be planar.

Within the area of acoustoelasticity, there are several different approaches involving different wave modes. These include: (1) longitudinal waves propagating normal to the plane of the stress[10,11], (2) shear (SV) waves propagating normal to the plane of the stress[9,12], (3) shear (SH) waves propagating in the plane of the stress[13,14], and (4) surface (Rayleigh) waves[15,16]. Because of the geometries involved, these waves are non-dispersive, with the velocities depending upon the state of stress in a manner determined by the material parameters and the wave mode.

The analytical treatment of acoustoelasticity involves the determination of the effect of stress on the speed at which waves propagate through a material. The wave is treated as an infinitesimal disturbance superposed on a nonlinearly elastic material subjected to a finite, static deformation. The equations of motion for the material in its final state, including the wave, are linearized about the statically deformed state, and the characteristic equations for the wave speeds are obtained by assuming a particular wave mode.

The acoustoelastic relation for each of the four approaches listed above gives a linear relation between the in-plane stress and a change in the phase velocity. We consider a homogeneous static deformation and choose coordinates such that the principal stresses in the plane lie in the X_1 and X_2 directions, with X_3 being measured normal to the plane of the stress. The velocity of a wave propagating in the X_i direction with its particle motion in the X_j direction is denoted V_{ij}. Velocities of longitudinal, shear, and Rayleigh waves in the undeformed state are denoted V_l, V_s, and V_r, respectively.

For the longitudinal wave technique, the change in wave speed in an isotropic material is directly proportional to the sum of the principal stresses,

$$\frac{V_{33} - V_l}{V_l} = A_l\,(\sigma_{11} + \sigma_{22})\,, \tag{1}$$

where A_l is the acoustoelastic constant, and σ_{11} and σ_{22} are the normal stresses in the plane. It has been shown that measurements of $\Delta V_l/V_l$ made over an entire structure subject to residual stresses is sufficient to completely evaluate the stress field[11].

The SV wave technique is a differential technique in which the velocities of two shear waves polarized in the principal stress directions are compared. For an initially isotropic material, the difference in the wave speeds is proportional to the difference in principal stresses,

$$\frac{V_{31} - V_{32}}{V_s} = A_s\,(\sigma_{11} - \sigma_{22})\,, \tag{2}$$

This relation does not require precise knowledge of the velocity of the wave in the unstressed state in order to allow the estimation of the difference in principal stresses.

These first two acoustoelastic techniques would not appear to be applicable for evaluating stresses in thin films because, in these cases, the waves must travel in the direction normal to the film (the smallest dimension of the film). In order for a bulk longitudinal or shear wave to have a wavelength of 1 μm, the frequency of excitation must be in the 3 GHz to 6 GHz range, depending upon the material. This is well above the upper limit of acoustic microscopes currently available.

The SH wave technique is also a differential technique in which the velocities of two shear waves are again compared. In this case, each wave propagates in one of the principal directions of stress and is polarized in the other principal direction. The velocities of these waves are related as

$$\rho_o \,(V_{12}^2 - V_{21}^2) = \sigma_{11} - \sigma_{22}\,, \tag{3}$$

where ρ_o is the mass density of the material in the undeformed state, This relation is particularly interesting in that there is no acoustoelastic constant involved. Consequently, measurements of the difference in SH wave speeds can be related directly to the stress difference. Unfortunately, the difficulty in exciting SH waves has thus far prevented the technique from providing any degree of spatial resolution.

The last of the "traditional" acoustoelastic techniques, the surface wave technique, is most similar to that employed in the SPAM system for stress microscopy[8]. It can be used in either an absolute mode or a differential mode, with measurements of surface wave speed being made for propagation along the principal directions. The change in velocities for waves traveling in the X_1 and X_2 directions are related to the stresses as

$$\frac{V_{r1} - V_r}{V_r} = A_{r\parallel}\sigma_{11} + A_{r\perp}\sigma_{22}\,, \qquad \frac{V_{r2} - V_r}{V_r} = A_{r\perp}\sigma_{11} + A_{r\parallel}\sigma_{22}\,. \tag{4}$$

The use of "parallel" and "perpendicular" in the designation of the acoustoelastic constants indicates the orientation of the associated stress component with respect to the direction of wave propagation. The sum of these velocity shifts is then proportional to the sum of the principal stresses, while the difference is proportional to the difference in the stresses.

ACOUSTOELASTIC RELATIONS IN LAGRANGIAN FORM

In analyzing the acoustoelastic response of a thin film subject to in-plane stresses, the same basic approach as indicated above is used. We consider three configurations of the body: the undeformed reference configuration; the deformed configuration, in which the body is subject to a state of finite, homogeneous deformation; and the current configuration, in which the infinitesimal wave is superposed on the finite deformation. Linearized equations of motion for the wave are derived by expanding the various kinematic relations, balance laws, and constitutive relations about the finitely deformed state.

Let $\mathbf{X}$, $\mathbf{x}$, and $\mathbf{x}^*$ identify the locations of a material point in the reference, deformed, and current configurations, respectively. These position vectors may be related to one another through displacement vectors $\mathbf{U}$ and $\mathbf{u}$ as

$$\mathbf{x} = \mathbf{X} + \mathbf{U}\,, \qquad \mathbf{x}^* = \mathbf{x} + \mathbf{u}\,, \tag{5}$$

where $\mathbf{u}$ is taken to be infinitesimal. The deformation gradients associated with the deformed and current states are

$$\mathbf{F} = \frac{\partial \mathbf{x}}{\partial \mathbf{X}}\,, \qquad \mathbf{F}^* = \frac{\partial \mathbf{x}^*}{\partial \mathbf{X}}\,, \tag{6}$$

and the Lagrangian strains are

$$\mathbf{E} = \tfrac{1}{2}(\mathbf{F}^T\mathbf{F} - \mathbf{I})\,, \qquad \mathbf{E}^* = \tfrac{1}{2}(\mathbf{F}^{*T}\mathbf{F}^* - \mathbf{I})\,. \tag{7}$$

For the remainder of this paper, we find it convenient to adopt the index notation, where, in terms of components, the position vectors are given as X_A, x_i, and x_i^*, the deformation gradients as F_{iA} and F_{iA}^*, and the strains as E_{AB} and E_{AB}^*. Upper case Latin indices refer to the undeformed reference configuration, while lower case indices refer to the deformed or current configurations. With this notation, the strains can be related to the displacements in the usual way as

$$E_{AB} = \tfrac{1}{2}(U_{A,B} + U_{B,A} + U_{C,A}U_{C,B})\,, \tag{8}$$

where comma denotes partial differentiation and the summation convention on repeated indices is employed. A similar expression clearly holds for E_{AB}^* with U_A replaced by $U_A + u_A$.

Because of the need to satisfy traction boundary conditions on the lateral surfaces of the film (traction free boundaries for the plate analysis), it is convenient to express the equations of motion in terms of the second (symmetric) Piola-Kirchhoff stresses S_{AB} and S_{AB}^*. For the deformed configuration, the relation between these components of stress and the components of the Cauchy stress σ_{ij} is

$$\sigma_{ij} = \frac{1}{\det\mathbf{F}}\, F_{iA}F_{jB}S_{AB}\,. \tag{9}$$

In the absence of body forces, the equations of motion may be written as

$$[S_{BC}(\delta_{AC} + U_{A,C})]_{,B} = \rho_o \ddot{U}_A \, . \tag{10}$$

The traction boundary conditions on a surface whose normal in the reference configuration is N_B are

$$[S_{BC}(\delta_{AC} + U_{A,C})]\, N_B = Q_A \, , \tag{11}$$

where Q_A is the applied traction. Again, the extension of these equations to the current state is clear.

Note that expressions corresponding to Eqs. (10) and (11) may be written in terms of Cauchy stress as $\sigma_{ij,i} = \ddot{u}_j$ and $\sigma_{ij} n_j = q_i$, respectively, where n_j is the normal to the deformed surface on which traction q_j acts. The advantage of using the referential (Lagrangian) equations instead of those associated with the current configuration is that the normal N_A is known and does not change, while the current normal n_i depends upon the deformation.

In order to complete the description of the problem, the constitutive relation for the material must be specified. As in other acoustoelastic analyses, the nonlinearity which leads to the change in wave speeds is incorporated in a strain energy function Ψ, which takes the form

$$\rho_o \Psi = \frac{1}{2} C_{ABCD} E_{AB} E_{CD} + \frac{1}{6} C_{ABCDEF} E_{AB} E_{CD} E_{EF} \, , \tag{12}$$

where C_{ABCD} and C_{ABCDEF} are referred to as the second-order elastic constants (SOEC) and third-order elastic constants (TOEC), respectively. The relation between stress and strain is then

$$S_{AB} = \rho_o \frac{\partial \Psi}{\partial E_{AB}} = C_{ABCD} E_{CD} + \frac{1}{2} C_{ABCDEF} E_{CD} E_{EF} \, . \tag{13}$$

The complete set of equations in the current state are linearized in the infinitesimal displacement **u** in order to obtain the behavior of the wave. Assuming that the deformed state is static ($\ddot{\mathbf{U}} = 0$) and homogeneous, the field equations for u_A are

$$(S_{BD}\delta_{AC} + \bar{C}_{EBCD} U_{A,E} + \bar{C}_{ABFD} U_{C,F} + \bar{C}_{EBFD} U_{A,E} U_{C,F}) u_{C,BD} = \rho_o \ddot{u}_A \, , \tag{14}$$

where

$$\bar{C}_{ABCD} = \rho_o \frac{\partial^2 \Psi}{\partial E_{AB} \partial E_{CD}} = C_{ABCD} + \tfrac{1}{2} C_{ABCDEF} E_{EF} \, . \tag{15}$$

The linearized boundary conditions are

$$(S_{BD}\delta_{AC} + \bar{C}_{EBCD} U_{A,E} + \bar{C}_{ABFD} U_{C,F} + \bar{C}_{EBFD} U_{A,E} U_{C,F}) u_{C,D} N_B = Q_A^* - Q_A \, , \tag{16}$$

where Q_A^* is the traction in the current configuration. We note that these referential equations are given in slightly different form by Pao, et al.[9].

ACOUSTOELASTIC SOLUTIONS FOR WAVES IN FILMS

Given the field equations and boundary conditions, Eqs. (14) and (16), respectively, the approach taken to obtain the frequency equation for the waves in the stressed film is similar to the traditional approach for evaluating guided waves in unstressed media. Consider waves propagating in the X_1 direction in a plate of thickness $2h$ subject to stresses S_{11} and S_{22}. All other stress components are taken to be zero. The displacements u_1 and u_3 are assumed to be harmonic, with $u_2 = 0$, and appropriate boundary conditions are imposed on the surfaces at $X_3 = \pm h$. Thus,

$$u_\alpha(X_1, X_3, t) = \bar{u}_\alpha(X_3)\, e^{j(kX_1 - \omega t)} \, , \quad \alpha = 1,3 \, , \tag{17}$$

where k is the wave number and ω is the frequency.

Using Eq. (17) in Eq. (14) leads to a pair of coupled, second-order, ordinary differential equations for $\bar{u}_1(X_3)$ and $\bar{u}_3(X_3)$, the coefficients of which involve the elastic constants, initial displacements, frequency, and wave number. The general solution of these equations involves four constants which are identified by the boundary conditions associated with the particular problem being considered.

In the case of a free plate, for which the tractions at $X_3 = \pm h$ are $Q_A = Q_A^* = 0$, the boundary conditions associated with the differential equations governing $\bar{u}_\alpha$ are homogeneous, so that nontrivial solutions can exist only if a particular relationship, between frequency and wave number is satisfied. This relationship, the frequency equation, has been derived exactly by Chen[17] for this case. As in the case of an unstressed plate, the solution naturally separates into modes which are symmetric about $X_3 = 0$ and modes which are antisymmetric (the flexural modes of the plate).

The frequency equation obtained by Chen[17] was solved numerically for the frequency spectra of several modes at various stress levels. Material constants (SOEC, TOEC, ρ_o) were taken to be typical of aluminum, with C_{1111} = 110 GPa, C_{1212} = 27 GPa, C_{111111} = -1350 GPa, C_{111122} = -240 GPa, C_{112233} = -55 GPa, and ρ_o = 2700 kg/m^3. Note that because of the assumed isotropy, the two SOEC and three TOEC given above are sufficient to completely characterize the material.

We focus attention on the response of the lowest-order flexural mode. When the film is unstressed, the resulting spectrum is that for a plate, with the phase velocity approaching zero as the wave number approaches zero and approaching the Rayleigh wave speed V_r as the wave number approaches infinity. When a biaxial stress $S_{11} = S_{22} = S$ is applied, the long wavelength behavior undergoes a fundamental change in character from the plate response to the membrane response in which the phase velocity approaches the speed of the membrane wave, $\sqrt{S/\rho_o}$, as the wave number approaches zero. In addition, the usual acoustoelastic response for Rayleigh waves is recovered in the short wavelength limit. This transition can be observed by plotting the phase velocity V as a function of nondimensional wave number, kh, for the two cases as in Fig. 1. We observe that in addition to the major change in velocity which occurs for small kh, the sign of the velocity change actually changes at around kh = 0.68. Below this value, the applied tension causes the velocity to increase, while above it, the velocity decreases.

Figure 2 shows the change in velocity versus kh for various levels of stress. The absolute velocity change is given rather than the usual relative change because of the dispersion present in the unstressed plate. The point at which the sign change takes place appears to be independent of the magnitude of the stress. As would be expected, the increase in stress causes an increase in the magnitude of the velocity change at all kh. However, the magnitude of the change is much greater at small values of kh than at large values.

As indicated earlier in this paper, all of the acoustoelastic relations for bulk waves in initially isotropic materials involve linear relations between velocity change and applied stress. (Note that for normally achievable stress levels, Eq. (3) for the SH wave response can be written as a linear relation between $V_{12} - V_{21}$ and stress difference.) This is not the case for thin films. The membrane velocity is proportional to the square root

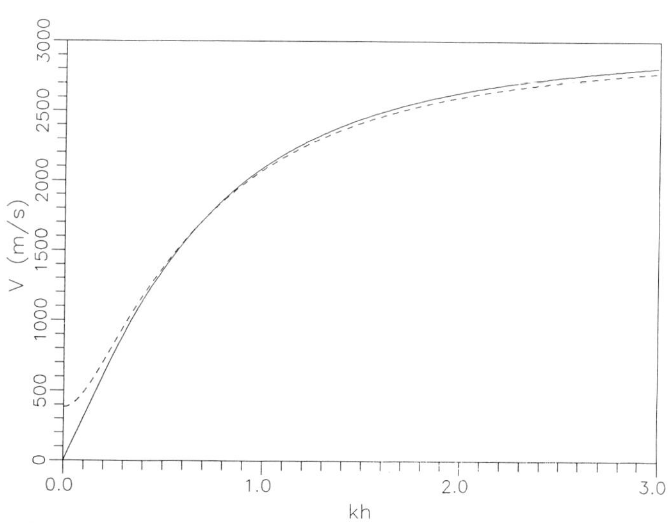

Figure 1. Dispersion curves for the lowest-order flexural mode in an aluminum plate which is stress-free (solid) and which is subject to 400 MPa biaxial tension (dashed).

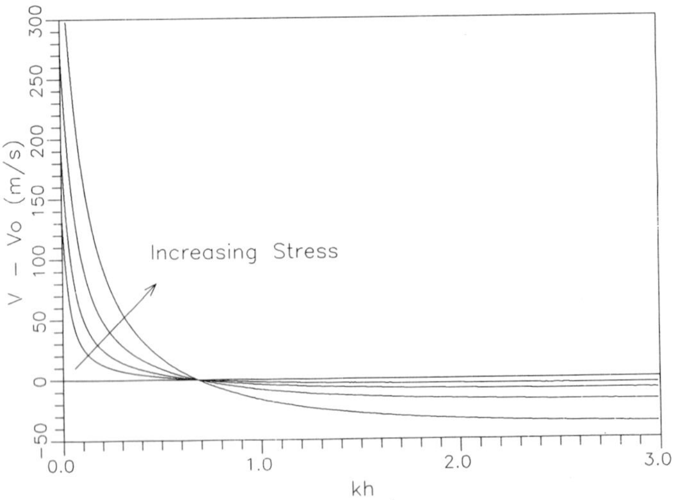

Figure 2. Flexural wave velocity change versus dimensionless wave number for an aluminum plate subject to biaxial tension. Curves are given for stress levels of 50 MPa to 400 MPa.

of the stress, so that the velocity change, from the initially zero velocity in the plate, is similarly proportional to the square root of the stress. As wave number increases, the velocity change becomes increasingly more linear, finally approaching the Rayleigh wave behavior of Eq. (4). This evolution is shown in Fig. 3 as velocity change versus applied biaxial stress for various kh.

Eq. (4) indicates that the acoustoelastic effect for surface waves involves two acoustoelastic constants associated with the stresses in the directions parallel and perpendicular to the direction of wave propagation. Figure 4 shows the corresponding effect for the flexural mode of the plate. For small kh, the component of stress in the propagation direction completely dominates the acoustoelastic response. As kh increases, this "parallel" effect decreases in magnitude, while the "perpendicular" effect initially increases and then decreases. The velocity changes caused by both stress components eventually become negative, though at different wave numbers. These two individual effects exhibit a rather complicated dependence upon wave number, with the "parallel" effect reaching a minimum at $kh \approx 1.5$ and a local maximum at $kh \approx 4$ before attaining its short wavelength (Rayleigh) limit. The "perpendicular" effect exhibits similar local maxima and minima before reaching its limiting value.

Because the membrane wave speed is independent of elastic constants, it is of interest to examine the extent to which the overall acoustoelastic response can be recovered without using the TOEC. The velocity change caused by a biaxial stress in a material with all of the TOEC set to zero is compared with the change caused by the same stress in the actual material in Fig. 5. As expected, the long wavelength limit is correctly captured without the TOEC, but the predicted response becomes increasingly poor as kh increases. Without the TOEC, the analysis fails to predict the change in sign which occurs in the velocity change. This observation highlights the cause of this sign change: There are two basic effects at work - in the long wavelength regime, the effect of the stress term in Eq. (14) dominates, while in the short wavelength regime, the change in stiffness $\overline{C}_{ABCD}$ due to the TOEC is a significant factor.

The acoustoelastic effect in thin films and layers is considerably more complicated than in bulk materials. Experimentally, the presence of dispersion in the response of the unstressed body could lead to substantial difficulty in separating the velocity changes caused by stress from those caused by changes in frequency. Fortunately, this difficulty appears to be avoidable by operating at fixed frequency, as is done in the SPAM system[8].

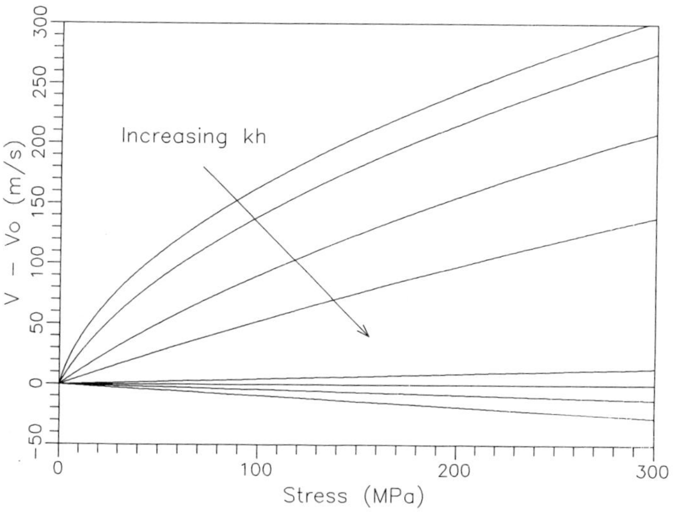

Figure 3. Flexural wave velocity change versus stress for an aluminum plate subject to biaxial tension. Curves are given for dimensionless wave numbers from $kh = 0.01$ to $kh = 3$.

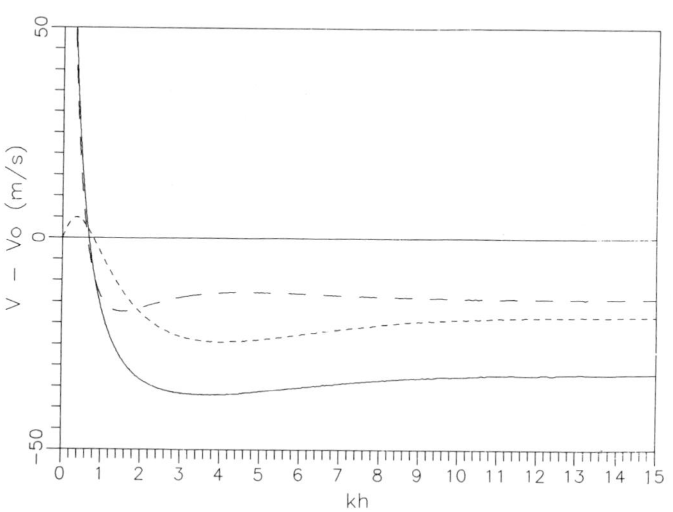

Figure 4. Flexural wave velocity change versus dimensionless wave number for an aluminum plate subject to biaxial tension (solid), uniaxial tension in the direction of wave propagation (dashed), and uniaxial tension perpendicular to the direction of wave propagation (dotted). Stress level in all cases is 400 MPa.

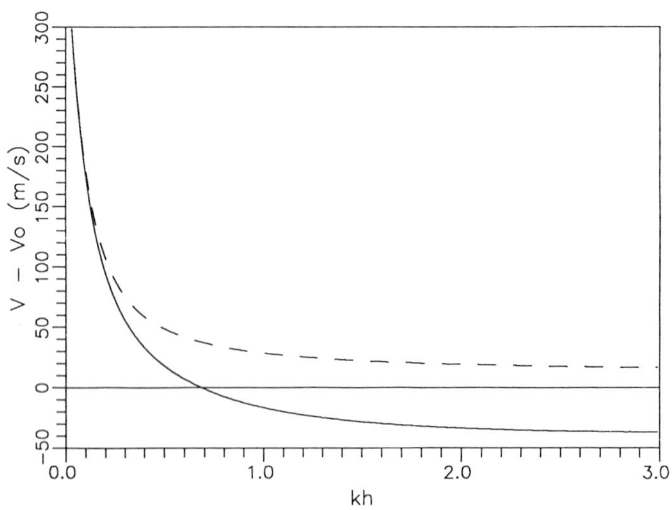

Figure 5. Flexural wave velocity change versus dimensionless wave number in an aluminum plate subject to 400 MPa biaxial tension including the effect of the third-order elastic constants (solid) and neglecting their effect (dashed).

ACKNOWLEDGEMENT

The authors wish to acknowledge the support of this research by the IBM Almaden Research Center through a predoctoral fellowship for Shih-Emn Chen and sponsored research contracts.

REFERENCES

1. Nix, W.D., *Met. Trans.*, 20A, 2217-2245 (1989).
2. Segmueller, A. and M. Murakami, "X-ray Diffraction Analysis of Strains and Stresses in Thin Films," in *Treatise in Materials Science and Technology*, 19A, ed. by H. Herman (Academic Press, New York, 1988) pp. 143-200.
3. Rickerby, D.S, B.A. Bellamy, and A.M. Jones, *Surface Engineering*, 3, 138-146 (1987).
4. Flinn, P.A., *Mat. Res. Soc. Symp., Proc. Vol.* 130, 41-51 (1989).
5. Mehregany, M., R.T. Howe, and S.D. Senturia, *J. Appl. Phys.*, 62, 3579-3584 (1987).
6. Guckel, H., T. Randazzo, and D.W. Burns, *J. Appl. Phys.*, 57, 1671-1675 (1985).
7. Howe, R.T., *Thin Solid Films*, 181 235-243 (1989).
8. Meeks, S.W., D. Peter, D. Horne, K. Young and V. Novotny, *Appl. Phys. Lett.*, 55, 1835-1837 (1989).
9. Pao, Y.H., W. Sachse and H. Fukuoka, in *Physical Acoustics: Principles and Methods*, 17, ed. by W.P. Mason and R.N. Thurston, (Academic Press, New York, 1984) pp. 61-143.
10. Kino, G.S., J.B. Hunter, G.C. Johnson, A.R. Selfridge, D.M. Barnett, G. Herrmann, and C.R. Steele, *J. Appl. Phys.*, 50, 2607-2612 (1979).
11. Dike, J.J. and G.C. Johnson, *J. Appl. Mech.*, 57, 12-17 (1990).
12. Hsu, N.N., *Experimental Mechanics*, 14, 169-176 (1974).
13. Thompson, R.B., S.S. Lee, and J.F. Smith, *J. Acoust. Soc. Am.*, 80, 921-931 (1986).
14. Man, C.S. and W.Y. Lu, *J. Elasticity*, 17, 159-182 (1987).
15. Husson, D., *J. Appl. Phys.*, 57, 1562-1567 (1985).
16. Liang, K., S.D. Bennett, B.T. Khuri-Yakub, and G.S. Kino, *Appl. Phys. Lett.*, 41, 1124-1126 (1982).
17. Chen, S.E., "The Influence of Stress on the Dynamic Response of Elastic Layers," Ph.D. Dissertation, University of California, Berkeley (1989).

THE IMPORTANCE OF X-RAY DIFFRACTION IN THE NON-DESTRUCTIVE CHARACTERISATION OF HIGH-STRENGTH ALUMINIUM ALLOYS

A.W. Bowen

Materials and Structures Dept
Royal Aerospace Establishment
Farnborough, Hants GU14 6TD, UK

INTRODUCTION

X-ray diffraction is a long-established technique for the non-destructive characterisation of crystalline and non-crystalline materials [1-3] and, as such, might appear to lack the glamour and immediacy of newly-developed instrumentation. However, this would belie the major advances that have been made in the technique with the introduction of rapid response detectors, intense tunable synchrotron sources, and, most importantly for the majority of users, the incorporation of computers into both data collection and data processing. These developments have resulted in what can only be described as a minor revolution since modern computer-controlled X-ray diffraction equipment can now provide, non-destructively and quickly, most of the basic structural information that is essential for materials characterisation. This paper will present the types of information that can be extracted from the technique, using as examples some of the high-strength aluminium alloys currently being used, or being considered for use, in the aerospace industry. Whilst the paper will limit itself to X-ray diffraction, it must be recognised that no one technique can provide all the necessary information in the non-destructive characterisation of materials and that full use must be made of any other technique that can complement X-ray diffraction.

ADVANTAGES OF X-RAY DIFFRACTION

The non-contact, non-destructive nature of X-rays is ideally suited as a probe to characterise materials. A possible disadvantage is that only the near-surface is analysed, because of the low penetrating power of most commonly available X-ray sources. However, this can be an advantage as it will result in good depth resolution, particularly if glancing-angle incidence is used. Because of this surface-specific analysis, X-ray and neutron diffraction are excellent complementary analytical tools.

The advantage of using diffracted X-rays is that, knowing the wavelength and diffraction angle, one can make use of Bragg's law [1-3] to extract information on the crystalline condition of the sample (non-crystalline materials can also be analysed [2] by X-rays but this is beyond the scope of the present paper). Thus from the intensity of the X-rays diffracted from a set of crystal planes it is possible to use:

- peak position to calculate the lattice parameter and elastic strain,

- peak intensity to calculate phase proportions, degree of texture or preferred orientation; and to refine the crystal structure using Rietveld methods [4], and

- peak width to calculate plastic strain and particle size.

The interested reader is referred to refs 1-3 for more information on the background to X-ray diffraction.

THE IMPORTANCE OF X-RAY DIFFRACTION FOR THE ANALYSIS OF HIGH-STRENGTH ALUMINIUM ALLOYS

High-strength aluminium alloys fall into three broad categories, viz:

- conventional heat-treatable alloys eg 2000, 7000 and Al-Li alloys
- rapidly-quenched alloys eg Al-Fe alloys
- metal-matrix composites eg Al+SiC(particles),

and their physical metallurgy and mechanical properties are described in detail in recent texts (refs 5-8) and conferences (refs 9-11).

In most applications of these materials, it is imperative that they are in the microstructural condition most appropriate for their efficient production and subsequent safe use. Since these materials can be viewed as consisting of:

- a matrix that is hardened by solutes in solution, dislocations and grain size, and

- a series of "second" phases consisting of constituent particles and precipitate phases,

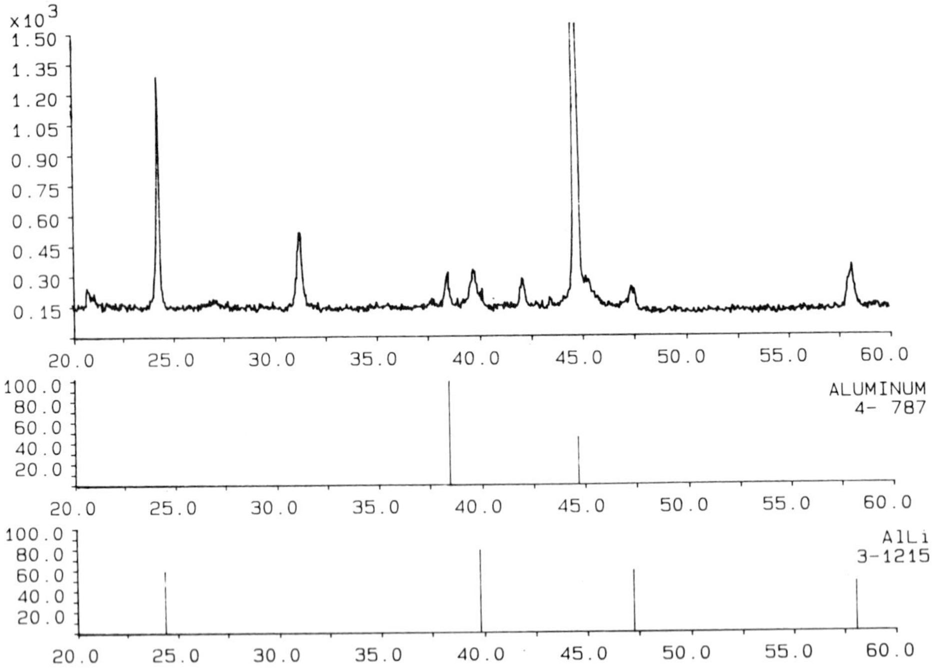

Fig. 1. X-ray diffractogram for an 8090 Al-Li alloy aged for 100h at 190°C, and matched with the reference patterns for aluminium (4-787) and the compound AlLi (3-1215). (Other peaks are from the δ' hardening phase). Vertical axis shows intensity and horizontal axis Bragg angle 2Θ from 20 to 60°.

X-ray diffraction analysis can provide a wide range of microstructural information that may be conveniently categorised as follows:

Constituent	Information that can be provided by X-ray diffraction
"second" phase	identity size
matrix	extent of solid solution elastic and plastic strains texture (preferred orientation) surface-specific reactions

Examples of this microstructural information are given in the following section, which are taken from recent and on-going research by the author and his colleagues.

APPLICATIONS OF X-RAY DIFFRACTION TO THE MICROSTRUCTURAL ANALYSIS OF HIGH-STRENGTH ALUMINIUM ALLOYS

Second phases

Phase identity. When more than one phase is present in an aluminium alloy, a comparison can be made with one's own data base, or with that compiled by JCPDS [12]. With the use of computer search-match routines it is now possible to establish quickly the identity of all the phases in an alloy; provided that volume fractions and particle sizes are not

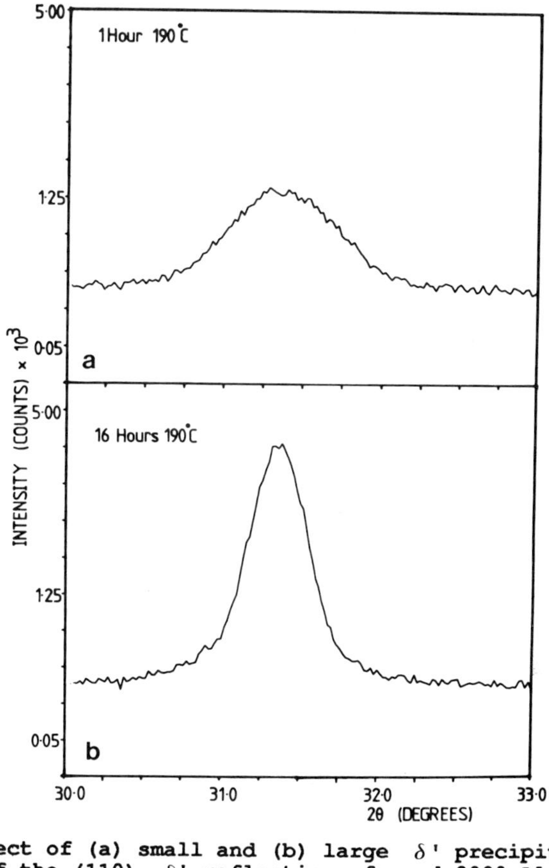

Fig. 2. The effect of (a) small and (b) large δ' precipitates on the width of the (110) δ' reflection of aged 8090 Al-Li alloy sheet. (a) 1h at 190°C, (b) 16h at 190°C.

extremely small. An example is shown in Figure 1, for the 8090 Al-Li alloy, indicating the presence of the matrix (Al ref pattern 4-787) and the compound AlLi (ref pattern 3-1215). (The remainder of the peaks are due to the δ' phase, which is not yet in the JCPDS data base). Note that each aluminium alloy will have a unique signature since although they all have aluminium as the matrix, each "second" phase has a unique crystal structure and/or lattice parameter. The clarity of this signature will, however, depend on such parameters as aged condition and texture.

Particle size. When particles are small (<200nm) their size can be calculated from their contribution to the breadth of a diffraction line [1-3]. The change in line broadening of the (110) δ' reflection in the 8090 Al-Li alloy on ageing at 190°C is indicated in Figure 2. Calculation of the size of the precipitates from such line broadening can be carried out conveniently and quickly using the method of Langford [13] to separate out the particle size and strain contributions (using a Voigt function to deconvolute the line broadening into the component Cauchy (size) and Gauss (strain) components). A comparison of particle sizes calculated by this line broadening method and those observed by electron microscopy is shown in Figure 3, where not only are the trends with ageing time in very good agreement, but the absolute sizes are also very close.

Matrix

Degree of solute supersaturation. Alloying elements in solid solution, depending on their relative atomic sizes, increase or decrease the lattice parameter of aluminium [1-3,14]. In these situations a linear relationship often exists between lattice parameter and the proportion of solute in solution (Vegard's Law, see refs 1-3), which can be used to calculate the degree of supersaturation in rapidly-quenched aluminium alloys. An example is shown in Figure 4 for an Al-Cr-Fe alloy, where the three curves shown indicate that as ageing proceeds the excess solute (in this case predominantly Cr) is rejected, with two consequences: (a) the aluminium lattice expands from 0.4017 (curve 1) to 0.404 (curve 2) and finally 0.405nm (curve 3) (having contracted in the first place because of the inclusion of the smaller Cr atoms in the lattice); and (b) precipitate peaks (in this case of $CrAl_7$ or more accurately Cr_2Al_{13}) appear. Note that whilst the composition of the matrix is changing (and can be readily detected by the matrix peak shifts), that of the Cr_2Al_{13}

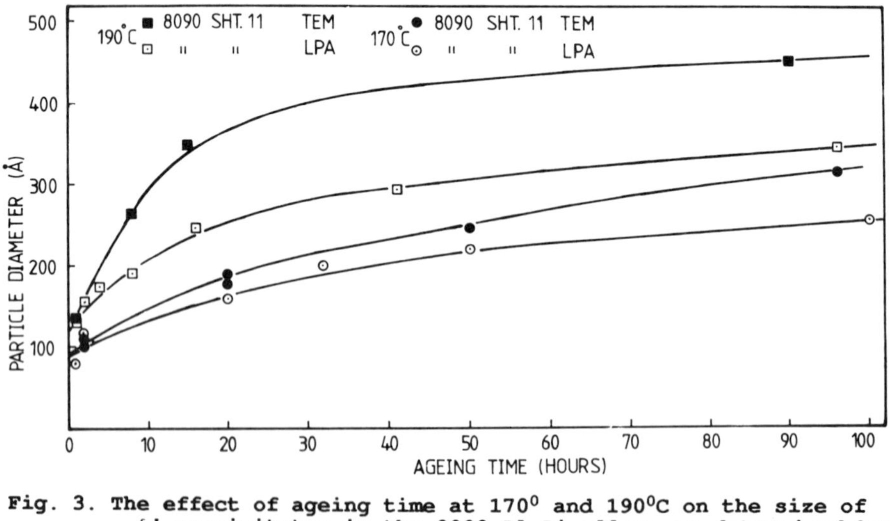

Fig. 3. The effect of ageing time at 170° and 190°C on the size of δ' precipitates in the 8090 Al-Li alloy, as determined by X-ray diffraction line profile analysis (LPA) and by transmission electron microscopy (TEM).

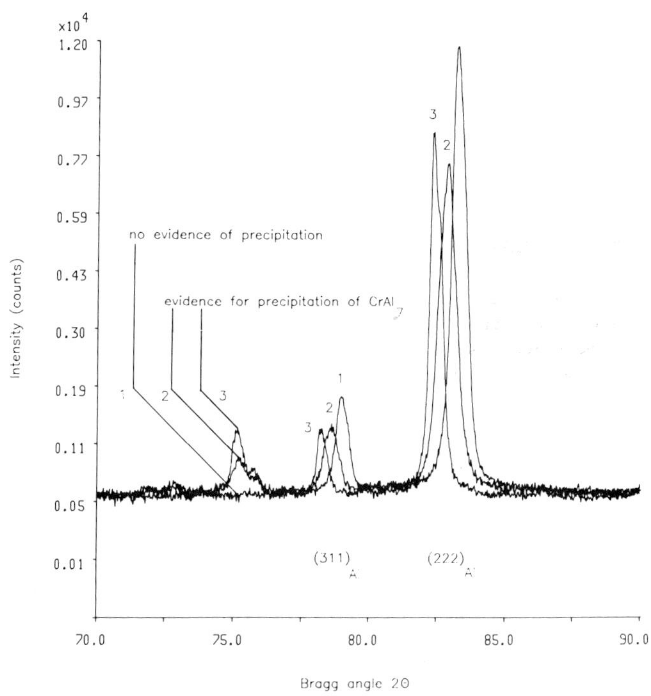

Fig. 4. The effect of ageing on both the matrix and precipitate diffraction peaks of a rapidly-quenched Al-Cr-Fe alloy, in the as-deposited condition (curve 1) and after ageing at 450°C for 0.5h (curve 2) and 5h (curve 3). Retained solute levels:
curve 1 7.7wt%
" 2 4.7 "
" 3 0.0 "
using the empirical equation: 0.405-0.00043nm per wt% retained solute. Note that the diffraction (Bragg) angle decreases as the lattice expands.

precipitate is invariant (ie there are no peak shifts). (Note also that precursors to Cr_2Al_{13} need not have the same composition).

<u>Plastic strain</u>. Rapidly-quenched alloys in the as-prepared condition show high levels of plastic strain because of the very fast cooling rates used. This can be measured readily using line profile analysis. Results from such a study are shown in Figure 5, which shows that the high level of strain in the as-deposited condition is quickly relieved by ageing at 350°C, but that the rate of change falls off quickly with ageing time. (Further details on the microstructure of this rapidly-quenched Al-Cr-Fe alloy can be found in ref 15).

<u>Residual stresses (elastic strains)</u>. Residual stresses can enhance or mitigate applied stresses in components, and are therefore of considerable interest to engineers [16,17]. Since these residual stresses arise because of elastic strains, they can be measured readily by detecting the shifts in the maxima of high angle X-ray diffraction peaks [16,17]. One example from current research concerns the mapping of stresses around fastener holes that have been cold worked to induce a compressive residual stress which is beneficial for fatigue resistance (P Holdway and A W Bowen, unpublished work 1990). Remote from such a 5mm diameter hole drilled centrally in a 7475 aluminium alloy fatigue test piece the values of stress are low, of the order of 15MPa compressive (due to residual strains arising from the original surface preparation of the sheet), whereas within 1mm of the hole the compressive stress is over 200MPa. Current work is concerned with mapping the elastic and plastic stress gradients in more detail.

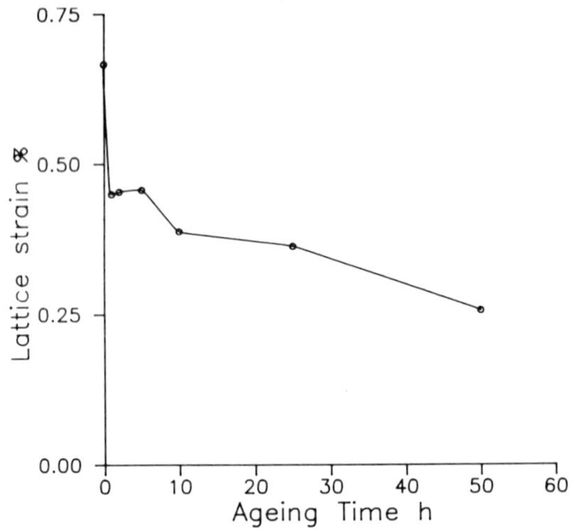

Fig. 5. The effect of ageing at 350^{0}C on the level of plastic strain in a rapidly-quenched Al-Cr-Fe alloy. (220) matrix reflection used for the calculation.

Degree of texture or preferred orientation. Because aluminium alloys have to be mechanically worked to their final form, they will possess a certain degree of texture as a result of the working process (see ref 18 for a recent review). It is very important to ascertain the type and intensity of the texture in order to establish the extent of deformation and any recrystallisation, and to predict the type and level of mechanical anisotropy. Whereas old compositions such as 2024 are not strongly textured, newer compositions such as 8090 are very strongly textured. A comparison of their textures is shown in Figure 6. These are (111) pole figures, which do not allow any quantitative data to be extracted. However, if the pole figures are processed to produce three-dimensional orientation distribution functions (ODF) (see refs 3, 18 and 19 for more details), then an extensive range of quantitative data can be determined. Examples of some of this information are shown in Figures 7 and 8, which report the extent of the texture variability through the thickness of un-recrystallised and recrystallised 8090 Al-Li alloy sheet [20]. These figures indicate that quite different textures (which are expressed in terms of ideal orientations {hkl}<uvw>; see ref 3, 18 and 19 for more details) and intensities exist in the two types of sheet. Since these sheets will have different degrees of mechanical anisotropy [21], knowledge of their textures is essential in order to control this anisotropy in practice [22].

Surface-Specific Reactions. During high temperature treatments used for solution treatment prior to the ageing of aluminium alloys, surface-specific phenomena can occur. Two examples of these phenomena are:

(a) surface recrystallisation. Since surface grains are not as constrained as sub-surface grains, given sufficient driving force they can recrystallise preferentially. In these circumstances, the type of grains at the surface can assume great importance (in eg formability) because of their different properties. An example of such surface recrystallisation for the 8090 Al-Li alloy is shown in Figure 9, as a function of time at 530^{0}C [20], where the surface recrystallisation that has occurred consists predominantly of the cube orientation ({001}<100>).

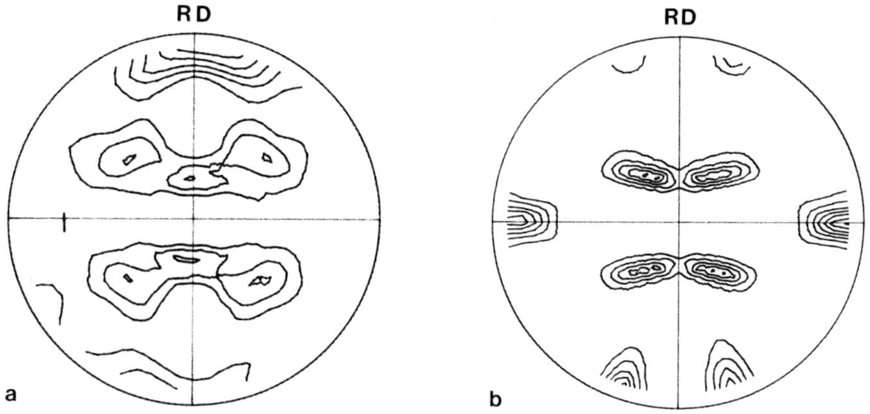

Fig. 6. (111) pole figures for (a) 2024 and (b) 8090 aluminium alloy sheet at the mid-thickness position. RD - rolling direction. Contour levels: (a) 1,1.5,2 etc. (b) 1,3,6 etc (x random).

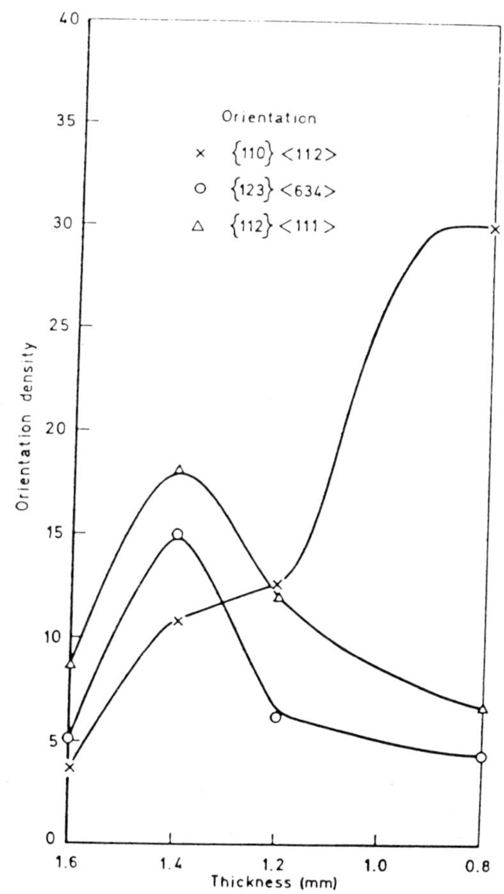

Fig. 7. Variations in the intensities of texture components in 1.6mm un-recrystallised 8090 Al-Li alloy sheet.

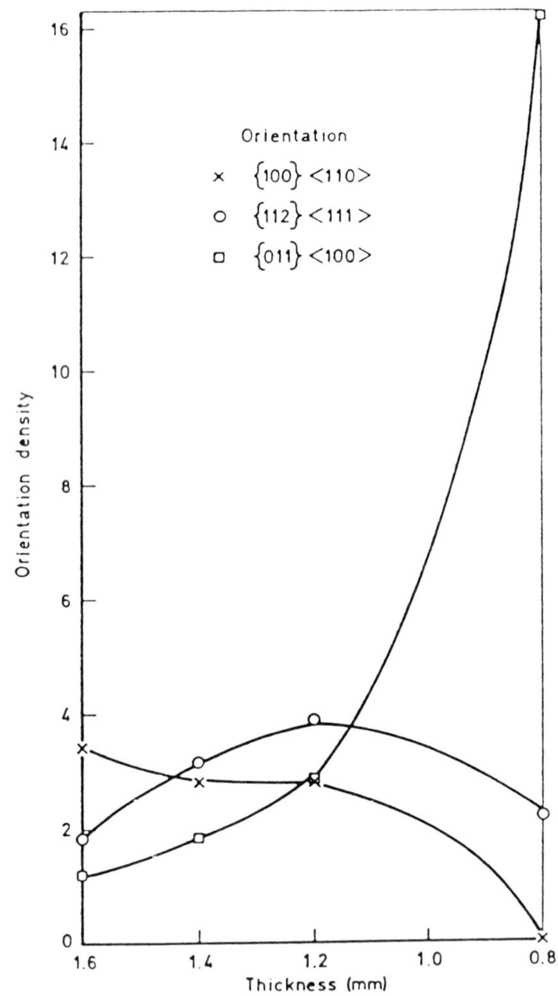

Fig. 8. Variations in the intensities of texture components in 1.6mm recrystallised 8090 Al-Li alloy sheet.

(b) oxidation products. Examples of the reaction products that can form on 8090 Al-Li alloy surfaces are shown in Figure 10, where Li_2CO_3 and γ-$LiAlO_2$ grow with increasing exposure time in air at 530^{0}C. Note that changes due to exposure for as short a time as 15m can be readily detected. (And even after as short an exposure time as 1m (P Holdway and A W Bowen, unpublished work 1990)). Because this results in a loss of lithium from the surface, and hence a loss in surface strength, it is important to monitor the loss of lithium. It has been proposed [23] that X-ray diffraction can be used to measure this loss by measuring the amount of Li_2CO_3 on the surface and relating it to lithium depletion depth using Figure 11 (where the intensities of three Li_2CO_3 peaks are plotted). (Figure 11 was constructed using hardness measurements as an intermediate stage [23]).

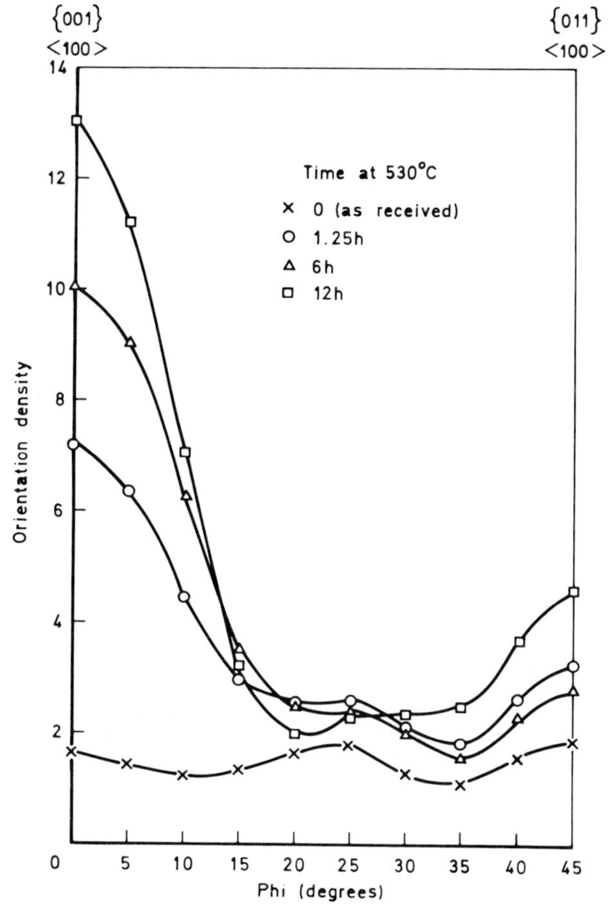

Fig. 9. Effect of exposure at 530^{0}C on surface recrystallisation on 8090 Al-Li alloy sheet. Data are represented as orientation density along the rotated rolling-direction fibre texture of the ODF where Phi is one of the three Euler angles used to represent grain orientations (see refs 18-20 for further details).

CONCLUSIONS

In this brief illustration of where X-ray diffraction can be used to provide information on the constitution and structure of high-strength aluminium alloys, examples have been given of the following:

1. the identity of constituent phases
2. the calculation of precipitate sizes
3. the degree of solute supersaturation
4. the degree of plastic strain
5. the degree of residual stress
6. the type and intensity of texture (preferred orientation)
7. the extent of surface recrystallisation
8. the types of reaction products and their use as a measure of solute loss (in this case lithium) from the surface.

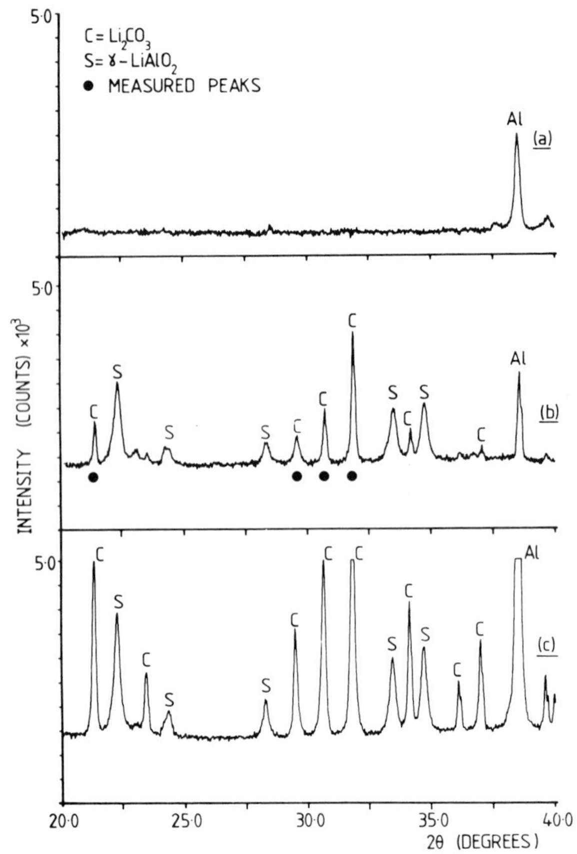

Fig. 10. X-ray diffractograms for the 8090 Al-Li alloy for the (a) as-received condition, and after solution treatment at 530^{0}C for (b) 15m and (c) 1h.

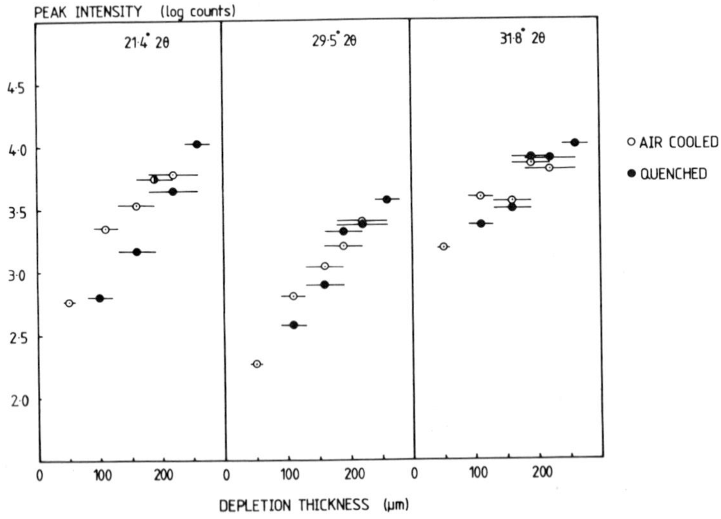

Fig. 11. Calibration plots of peak intensity versus depletion depth for three Li_2CO_3 peaks for the 8090 Al-Li alloy.

ACKNOWLEDGMENTS

The author would like to thank his colleagues Dr P Holdway and C J Gilmore for providing many of the figures and for valuable discussions during the course of the investigations summarised in this paper.

REFERENCES

1. B. D. Cullity, "Elements of X-ray Diffraction", Addison Wesley, Reading, Ma., USA (1979).
2. H. P. Klug and L. E. Alexander, "X-ray Diffraction Procedures for Polycrystalline and Amorphous Materials", Wiley, NY, USA (1974).
3. A. W. Bowen, Ch 5 of "Microstructural Characterisation of High Temperature Materials", Inst of Metals, London, UK (1988) 227.
4. Int. Workshop on the Rietveld Method, Petten, The Netherlands, 13-15 June 1989.
5. J. E. Hatch (Ed), "Aluminum Properties and Physical Metallurgy", ASM, Metals Park, Ohio, USA (1984).
6. A. K. Vasudevan and R. D. Doherty (Eds), "Aluminum Alloys - Contemporary Research and Applications", Academic, Boston, USA (1989).
7. "Advances in the Physical Metallurgy of Aluminum Alloys" Symp., Met. Trans. A, 6A:625 (1975).
8. L. E. Mondolfo, "Aluminum Alloys Structure & Properties", Butterworths, London, UK (1976).
9. E. A. Starke Jr and T. H. Sanders Jr (Eds), "Aluminum Alloys Their Physical and Mechanical Properties", EMAS, Warley, West Midlands, UK (1986).
10. Aluminium Technology '86 Conference, Inst of Metals, London, UK (1986).
11. E. A. Starke Jr and T. H. Sanders Jr (Eds), Proc 5th Int Conf on Al-Li Alloys, AIME, Warrendale, Pa., USA (1988).
12. JCPDS (Joint Committee for Powder Diffraction Standards-International Centre for Diffraction Data, 1601 Park Lane, Swarthmore, Pa. 19081-2389, USA).
13. J. I. Langford, J. App. Cryst., 11:10 (1978).
14. W. B. Pearson, "Handbook of Lattice Spacings and Structure of Metals and Alloys", Pergamon, Oxford, UK 1:(1958), 2:(1967).
15. C. J. Gilmore and A. W. Bowen, Nondest. Test and Eval., 5:379 (1990).
16. SAE, Residual Stress Measurement by X-ray Diffraction: SAE J748a, Soc. of Auto. Eng. Inc., Warrendale, Pa 15096, USA (1971).
17. Ed. W. B. Young, "Residual Stress in Design, Process and Materials Selection", ASM, Metals Park, Ohio, USA (1987).
18. A. W. Bowen, "Texture Development in High-Strength Aluminium Alloys", Mater. Sci. and Tech., (1990) in the press.
19. H. Wenk (Ed), "Preferred Orientation in Deformed Metals and Rocks: An Introduction to Modern Texture Analysis", Academic, Orlando, Fla, USA (1985).
20. A. W. Bowen, in Proc 8th Int. Conf. on "Textures of Materials", J. S. Kallend and G. Gottstein (Eds), AIME, Warrendale, Pa., USA (1988) 971.
21. C. J. Peel and D. S. McDarmaid, Aerospace, 16:18 (1989).
22. A. W. Bowen in Proc Int. Conf. on "Recrystallisation in Metallic Materials", T. Chandra et al (Eds), AIME, Warrendale, Pa., USA (1990) (to be published).
23. P. Holdway and A. W. Bowen, J. Mat. Sci., 24:3841 (1989).

FATIGUE DAMAGE ASSESSMENT BY X-RAY DIFFRACTION AND NONDESTRUCTIVE LIFE ASSESSMENT METHODOLOGY

Robert N. Pangborn and Sam Y. Zamrik

Department of Engineering Science & Mechanics
The Penn State University
University Park, PA 16802

INTRODUCTION

X-Ray diffraction (XRD) has been employed to evaluate cumulative damage incurred under both high cycle fatigue (HCF) and low cycle fatigue (LCF) conditions. Additional objectives of the study were to correlate the x-ray diffraction data to the microstructural characteristics and deformation mechanisms activated by various fatigue regimes and to develop a methodology for nondestructive assessment of remaining life.

Two x-ray diffraction techniques were used in the investigation. X-ray line profile analysis, also referred to as line broadening analysis, is a conventional approach that can be used to disclose the deformations introduced in crystalline materials by various mechanical loadings and processing. The changes in line breadths have been monitored in a number of previous studies for the purpose of monitoring fatigue damage.[1-8] The resulting line broadening can be deconvoluted into contributions caused by microstrains and those due to substructure formation, known as particle or crystallite size broadening. The second technique involves a modification of x-ray double crystal diffractometry for application to polycrystalline materials.[9-11] Originally used to study rocking curves, a measure of the perfection of single crystals, the method has been adapted to disclose the rocking curves and associated distortion for constituent grains in commercial alloys. Known as computer-assisted rocking curve analysis (CARCA), this approach affords a sensitive measure of the nonuniform deformation typically induced by fatigue loading.

EXPERIMENTAL MATERIALS AND FATIGUE TESTING

The x-ray diffraction results derive from fatigue studies of two different nickel base superalloys. The goal is to demonstrate and compare the nondestructive damage evaluations obtained for widely varying deformation responses; i.e., from the small-scale lattice distortions that serve as precursors to late-life crack initiation for HCF to the large-scale plasticity associated with LCF.

The HCF testing was conducted using Inconel 718 as the experimental material. Cylindrical fatigue specimens were prepared with a 3.2 mm gage diameter and 10 mm gage length from the solutionized alloy. A low stress grind was used to prepare the surfaces, followed by lapping with successively finer grits and electropolishing of a layer of about 35µm to remove any residual surface damage. Aging was performed at 718°C for 8 hours in argon followed by air cooling. Load control fatigue tests were conducted in axial tension-tension (R=0.1) at maximum stress amplitudes that ranged from 0.6 to 1.1 times the 0.02% yield strength of 834 MPa.

Nondestructive Characterization of Materials IV
Edited by C.O. Ruud *et al.*, Plenum Press, New York, 1991

Table 1. Materials and Testing Conditions

ALLOY	FATIGUE MODE	REGIME	TEMPERATURE
Waspaloy	Torsion (Biaxial)	LCF (Plastic Strain Control)	HT (649°C)
Inconel 718	Axial (Tension-Tension, R=0.1)	HCF (Load Control)	RT

The LCF testing, by comparison, was conducted in torsion using strain control. Tubular specimens with 12.7 mm OD and approximately 30 mm gage length were prepared from a forged pancake disk. Heat treatment was applied to produce a coarse grain structure (ASTM #3) with small γ' microstructure. After surface preparation similar to that used for the Inconel, fatigue testing was carried out using a fully reversed triangular wave form. During the testing, the temperature was maintained at 1200°F (649°C) by induction heating and a cyclic frequency of 10 cpm was employed to minimize creep effects at the plastic strain amplitudes of 0.52 to 0.92. The testing conditions used in the studies of the superalloys are summarized in Table 1.

RESULTS

Deformation Response to Fatigue

Figure 1 combines the fatigue data obtained for the Inconel 718 in axial loading and the Waspaloy in torsional loading. While the data derive from different alloys, the life curve is typical of that obtained for fatigue under conditions which range from high cycle, stress control tests involving stresses in essentially the elastic range, to tests in the low cycle regime featuring substantial plastic strain. For the former, a log-log plot of the stress amplitude, expressed in terms of the corresponding reversed strain, $\Delta\varepsilon/2$, versus cyclic reversals to failure can be expressed according to Basquin[12] as:

$$\Delta\varepsilon E / 2 = \sigma_a = \sigma_f'(2N_f)^b \quad (1)$$

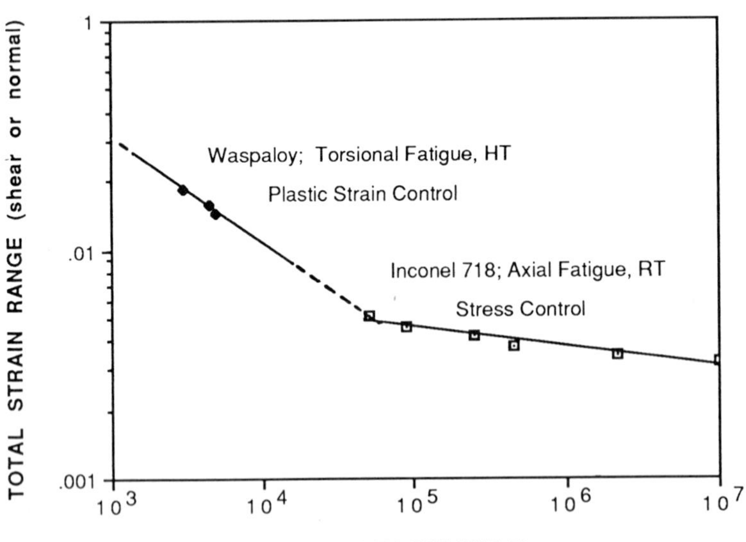

Fig. 1 Fatigue life curves for Waspaloy and Inconel 718.

where E is the elastic modulus and σ_f' and b are the fatigue strength coefficient and exponent, respectively. For the latter, the data follow the Coffin-Manson[14] relation:

$$\Delta\varepsilon_p / 2 = \varepsilon_f'(2N_f)^c \tag{2}$$

where $\Delta\varepsilon_p$ is the plastic strain amplitude and ε_f' and c are the fatigue ductility coefficient and exponent, respectively.

From the scanning electron micrographs in Figures 2 and 3, it can be seen that the deformation response for the two materials is different under the respective fatigue test conditions employed. The Inconel 718 discloses nonuniform slip (Fig. 2a) and evidence that the deformation is primarily limited to those grains oriented for high resolved shear stress. Hysteresis curves for the room temperature tests indicated that cyclic hardening occurs early in the life, with no subsequent softening. Surface "rumpling" (Fig. 2b) develops progressively throughout the life and crack initiation, as expected for HCF, occurred late in the life at life fractions exceeding 0.9.

For the Waspaloy, intense planar slip is activated early in the life. This leads to persistent slip band formation and inhomogeneously distributed intrusions and extrusions (Figure 3a, b) which serve as sites of crack initiation early in the life. The LCF carried out at elevated temperature is characterized by cyclic hardening early in the life and then softening as ultimate failure is approached.

X-Ray Diffraction Analysis

The challenge, in so far as nondestructive monitoring of fatigue damage is concerned,

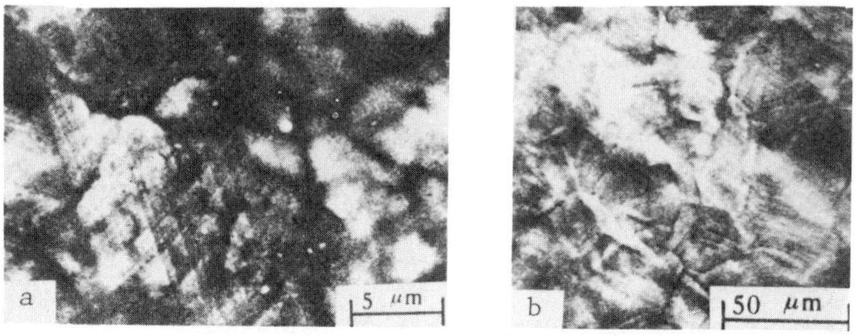

Fig. 2. Nonuniform slip, (a), and surface rumpling, (b), of Inconel 718 fatigued in axial tension-tension.

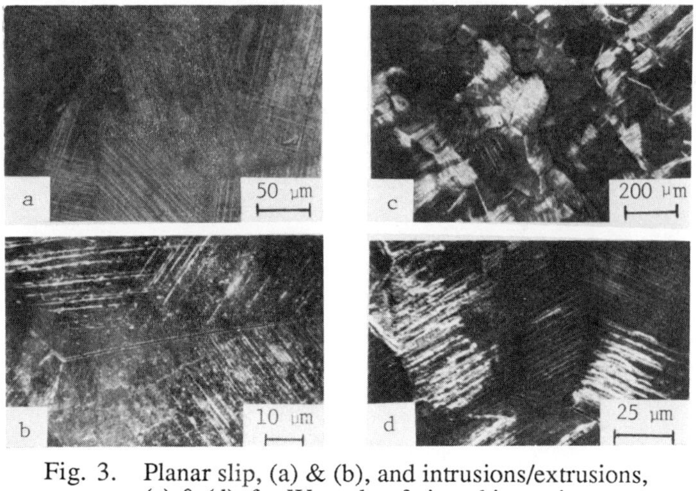

Fig. 3. Planar slip, (a) & (b), and intrusions/extrusions, (c) & (d), for Waspaloy fatigued in torsion.

resides in choosing techniques which will disclose both the subtle, nonuniformly distributed deformation incurred during HCF as well as the equally inhomogeneous, but much more intense deformation introduced by LCF.

Because line broadening analysis gives a measure of the average deformation imposed by mechanical testing, the Waspaloy LCF specimens were evaluated with this technique. Figure 4 shows typical line profiles for virgin and cycled specimens. CuKα radiation was employed and a Rachinger correction was applied to the as-collected profile to strip the Kα_2 component from the doublet. The resultant integral breadths, B, of the profiles (integrated intensity divided by peak height) reveal a 10% increase in breadth, from 0.22 to 0.24 degrees of arc, after cycling for approximately a tenth of the fatigue life. Over the full life of the specimens, Figure 5a shows that the breadths nearly double for tests conducted at elevated temperature. The amount of broadening is comparable for all the plastic strain amplitudes investigated. At lower temperature, the reduced plasticity results in less broadening of the profiles, giving on the order of a 50% increase in breadth. Figure 5b shows the same data plotted as a fraction of fatigue life, which then represent calibration curves for the two test temperatures.

When the same approach was applied to the Inconel 718 HCF specimens, however, the results were quite different. Shown in Figure 6 are the data obtained for several constant load amplitudes and two diffraction profiles, (111) and (222). At this much lower, essentially elastic strain range the broadening is incurred early in the cycling with little subsequent broadening during the remaining 90% of the fatigue life. Although the level of the broadening could be used to indicate the magnitude of the stress amplitude to which a specimen had been subjected, damage assessment is clearly prescribed. Therefore, the double crystal diffraction arrangement was explored as a method by which a more sensitive disclosure of fatigue deformation could be obtained.

In the CARCA technique, the primary x-ray beam is a first reflected from a flat, monochromating crystal to produce a parallel beam incident on the specimen. This reduces the number of reflecting grains such that they are individually resolved using a position sensitive detector, as shown in Figure 7. By rotating the specimen in increments of 3 to 5 minutes of arc and recording the reflected intensities at each position (as illustrated by the multichannel analyzer displays in the figure) the rocking curves for each of the grains can be generated. Although the illustration depicts only five grain reflections, labeled A through E, a typical analysis will give rocking curves for 200 to 2000 grains depending on the number required to generate an adequate statistical sampling of the deformation. Figure 8 shows a reconstruction on the rocking curve profiles for a virgin sample (a) and a cycled specimen (b). The intense, sharp peaks for the virgin sample are consistent with the relatively undistorted grain population. The lower intensity, broad and multimodal profiles for the fatigued specimens can be explained , as depicted in Figure 9, by the introduction of misaligned subgrains and the distortion within them, both a manifestation of the fatigue damage.

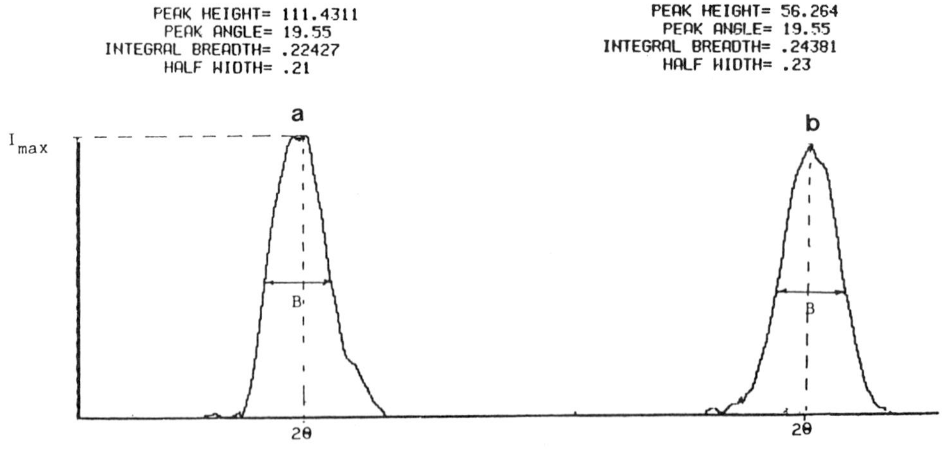

Fig. 4. X-Ray line profiles for virgin specimen, (a), and fatigued Waspaloy, (b).

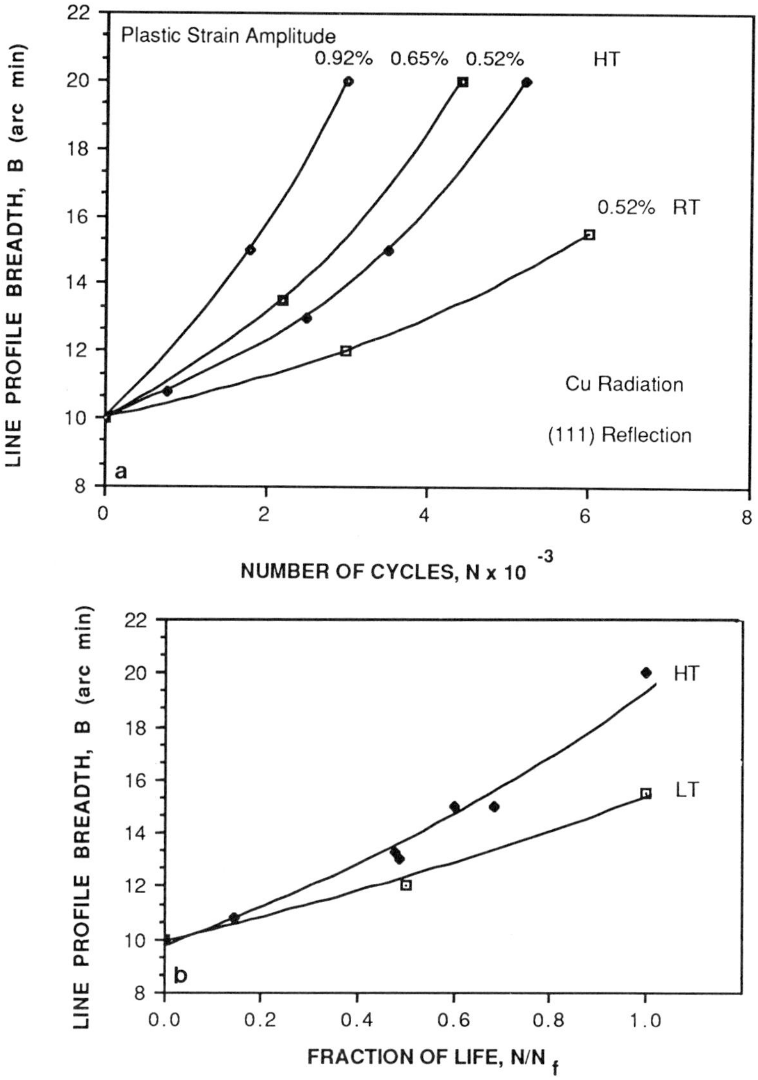

Fig. 5. Line broadening of Waspaloy fatigued in torsion, (a), and vs. fraction of life, (b).

The key to generating a steep dependence of the average rocking curve breadth, β, on fatigue life fraction is the capability to exclude those grains which, at low stress amplitudes, undergo little or no distortion during the cycling. Thus, only those grains exhibiting broadening above a chosen threshold (in this case 1.1 times the average breadth for the virgin specimens, β_0) are included in the analysis. The results, shown in Figure 10 in terms of the normalized average rocking curves breadth, show a steep, two-fold increase over the fatigue life. Included are data from specimens cycled at several constant stress amplitudes as well as data from two spectrum loaded specimens cycled under variable amplitudes. This calibration curve for HCF is comparable to that shown in Figure 5b for LCF.

Engineering Analysis

The x-ray line broadening data can be defined as a damage parameter in terms of the fraction of life, N/N_f, in the form:

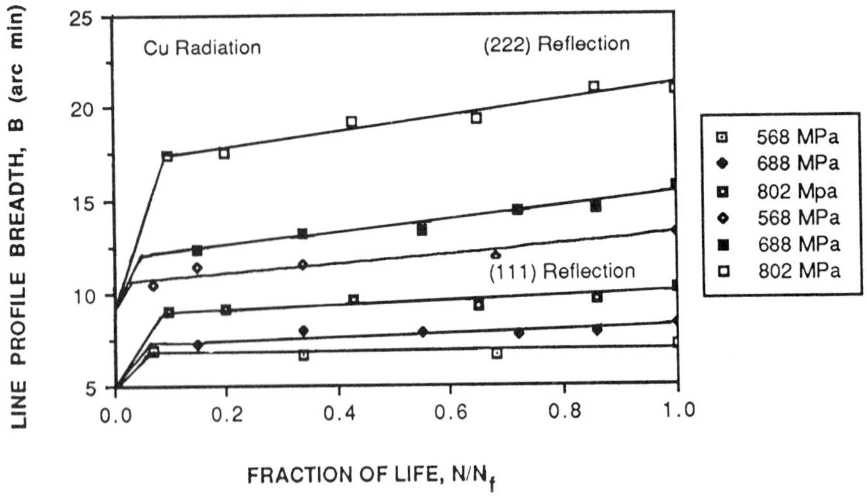

Fig. 6. Line broadening of Inconel 718 fatigued in tension-tension.

$$dB/B = k(N/N_f) \qquad (3)$$

where k is a material constant. Therefore:

$$B = B_o e^{k(N/Nf)} \qquad (4)$$

where B_o is the breadth of the virgin specimen. With progressive cycling, the damage accumulation rate is given by:

$$dB/dN = (k/N_f)B_o e^{k(N/N_f)} \qquad (5)$$

$$= Bk/N_f \qquad (5a)$$

or $$dB/dN/B = k(1/N_f) \qquad (6)$$

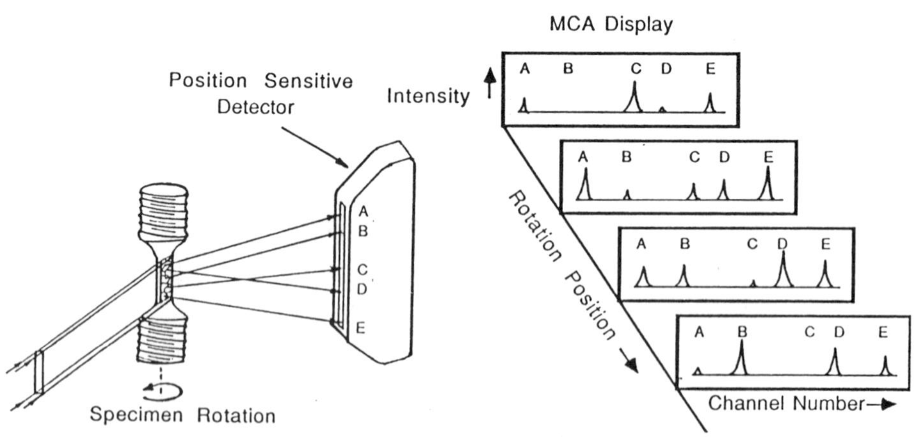

Fig. 7. Construction of rocking curves for reflecting grains from a polycrystalline fatigue specimen.

Recasting this expression in a form similar to the Coffin-Manson relation gives:

$$(dB/dN)/B = 2k\,(2\gamma_f')^{1/c}(\Delta\gamma p)^{-1/c} \qquad (7)$$

$$= k'(\Delta\gamma p)^{\alpha} \qquad (7a)$$

or $$\int dB/B = k'(\Delta\gamma p)^{\alpha}\int dN \qquad (8)$$

where $\alpha = -1/c$ and $k' = 2k(2\gamma_f')^{1/c}$

Finally, integrating gives:

$$B = B_o e^{k'(\Delta\gamma p)^{\alpha}N} \qquad (9)$$

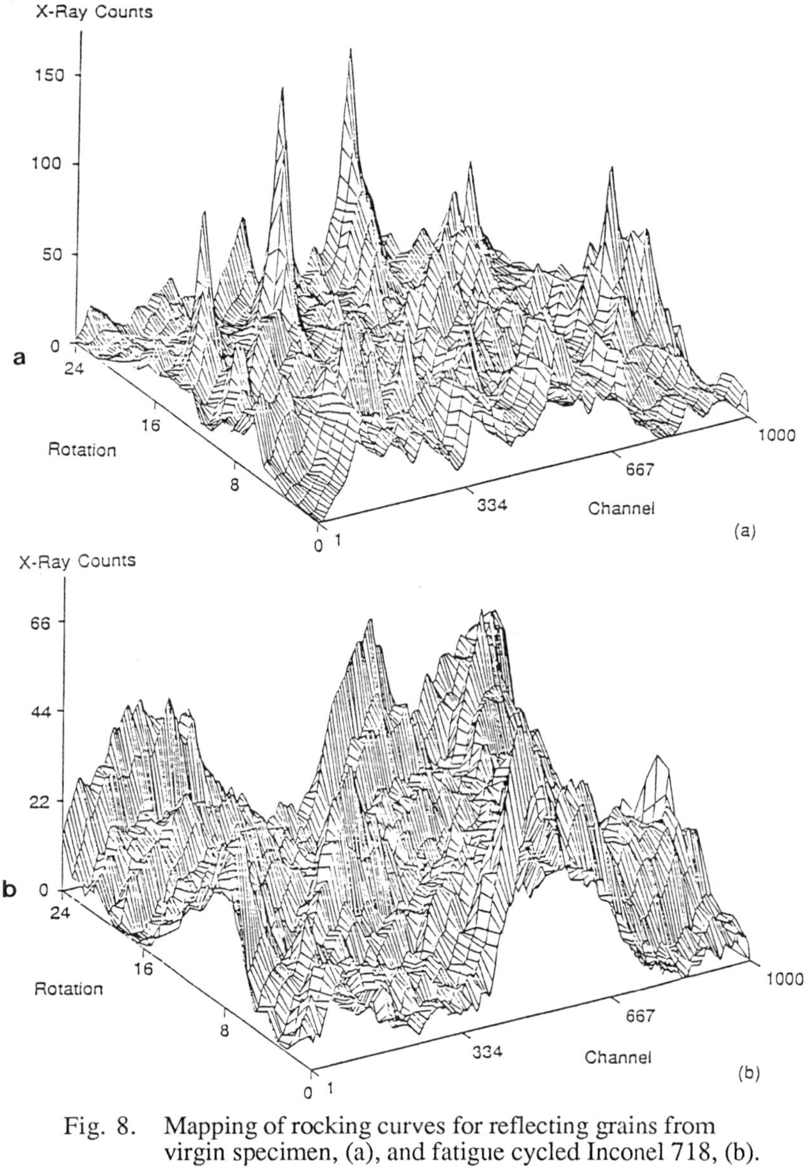

Fig. 8. Mapping of rocking curves for reflecting grains from virgin specimen, (a), and fatigue cycled Inconel 718, (b).

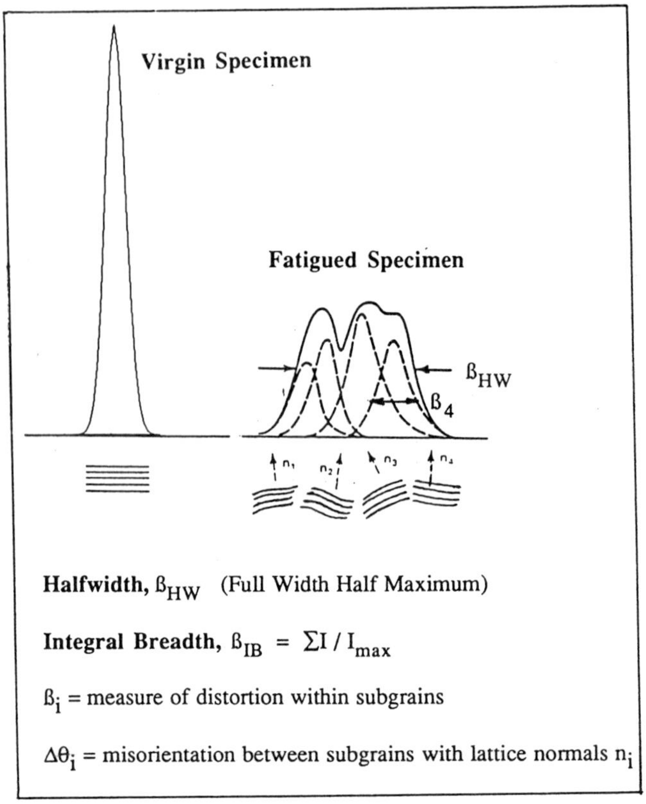

Fig. 9. Effect of fatigue deformation on rocking curve profiles.

The constants k' and α can be obtained from the intercept and slope of a log-log plot of equation 7a. For the Waspaloy, then, α is found to be 1.068 and k' is 0.033. Substitution of these values into equation 9 allows the number of cycles consumed, N, to be calculated, given a measured damage level, B. A similar approach for estimating the expended fatigue life, and hence the remaining life in HCF, is afforded by double crystal diffractometry, using the average rocking curve breadth as the damage parameter.

Finally, by using the data from LCF of Waspaloy, the life curve is computed as:

$$\Delta \gamma p/2 = 26.3(2N_f)^{-0.936}$$

For the HCF of Inconel 718, using the corresponding Basquin relation gives:

$$\sigma_a = 1480(2N_f)^{-0.126}$$

Both are in good agreement with actual life curves for the alloys.

CONCLUSIONS

The results of the study can be summarized as follows:

(1) X-ray line broadening analysis can be used to nondestructively evaluate cumulative damage under LCF conditions, but its utility for HCF resides in its disclosure of the stress amplitude of the fatigue.

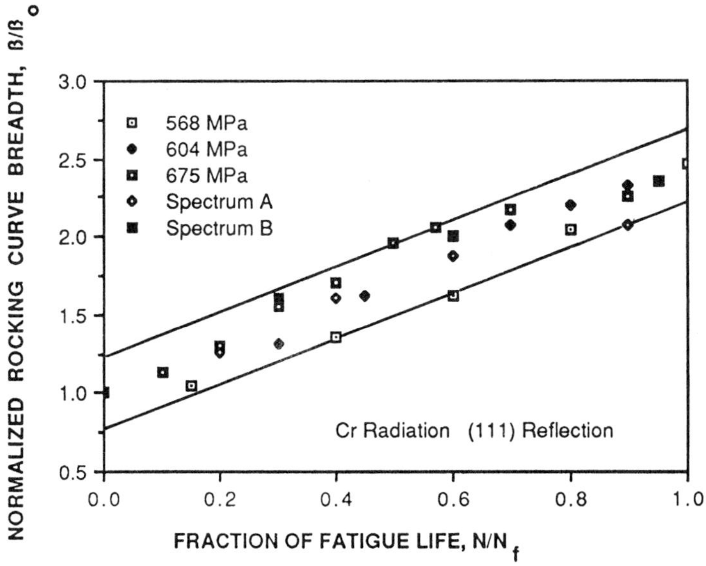

Fig. 10. Dependence of normalized rocking curve breadths on fraction of fatigue life for Inconel 718.

(2) Rocking curve analysis by double crystal diffractometry affords a sensitive measure of the more subtle and localized cumulative damage that serves as the precursor to crack initiation late in the life for HCF.

(3) Both techniques reveal the development of substructure (subgrains) and fatigue induced microstrain. However, the average deformation for HCF is limited; thus, a better manifestation of the expended fatigue life is provided by the double crystal diffraction technique by virtue of its capability to identify the distortions incurred by individual grains.

(4) The line profile and rocking curve broadening are a manifestation of an increase in the dislocation density due to cyclic deformation. Either cyclic hardening (dislocation interaction, pileup and entanglement) or cyclic softening (dislocation rearrangement, networks, cell or subgrain formation) may accompany the broadening.

REFERENCES

1. Taira, S., Gotto, T., and Nakano, Y., 1969, X-Ray Investigation on Low-Cycle Fatigue of Low Carbon Steel, in: "Proc. 12th Japanese Congress on Materials Research," Society of Materials Science, Kyoto, Japan, 8.
2. Gould, R.W., and Pitella, C.F., 1973, An X-Ray Investigation of Fatigue Behavior of Cold Worked Aluminum, Adv. X-Ray Anal., 16:354.
3. Wan, C.M., and Byrne, J.G., 1975, Correlation of X-Ray and Electron Microscope Studies of Fatigue, Int. J. Fracture, 11:251.
4. Koves, G., 1976, Correlation of Residual Stress Level and Fatigue Damage in Face Centered Cubic Metals, in: Microstructural Science, American Elsevier Publishing Co., 4:233.
5. Alexopoulos, P., and Byrne, J.G., 1978, Positron Lifetime & X-Ray Particle Size Measurements During High Cycle Fatigue of AISI 4340 Steel, Met. Trans. 9A:1892.
6. Weiss, V., Oshida, Y., and Wu, A., 1980, Towards Practical Non-Destructive Fatigue Damage Indicators, J. Nondest. Eval., 1:207.
7. Takechi, H., Namba, K., Fujiwara, K., and Kawasaki, K., 1981, Evaluation of Subsurface Fatigue Damge in Strip Mill Rolls by An X-Ray Diffraction Method, Trans. Iron Steel Inst. Jpn., 21:92.

8. Field, J.L., Behnaz, F., and Pangborn, R.N., 1983, Characterization of Microplasticity Developed During Fatigue, in: "Fatigue Mechanisms: Advances in Quantitative Measurement of Physical Damage," ASTM-STP 811, 71.
9. Pangborn, R.N., Weissmann, S., and Kramer, I.R., 1981, Dislocation Distribution and Prediction of Fatigue Damage, Met. Trans., 12A, 109.
10. Pangborn, R.N., Yazici, R., Tsakalakos, T., Weissmann, S., and Kramer, I.R., 1979, Determination of Prefracture Damage in Fatigued and Stress-Corroded Materials by X-Ray Double Crystal Diffractometry, in: "Proc. Symp. Accuracy in Powder Diffr.," National Bureau of Standards, Spec. Pub. 567, 443.
11. Yazici, R., Mayo, W., Takemoto, T., and Weissmann, S., 1983, Defect Structure Analysis of Polycrstalline Materials by Computer-Controlled Double Crystal Diffractometer with Position Sensitive Detector, J. Appl. Cryst., 16:89.
12. Hertzberg, R.W., 1989, "Deformation and Fracture of Engineering Materials," 3rd ed., John Wiley & Sons, NY, 500.
13. Zamrik, S.Y. and Pangborn, R.N., 1989, Fatigue Damage Assessment Using X-Ray Diffraction Methods, Nuclear Engineering Design, 116:407-413.

ACOUSTIC EMISSION STUDY ON FRACTURE TOUGHNESS IN TI-6AL-4V ALLOY WITH VARIOUS ACICULAR α MICROSTRUCTURES

S. Mashino*, T. Kishi**, T. Horiya*** and H.G. Suzuki***

* Engineering Research Laboratory, Japan Tobacco Inc.
1-31 Kurobeoka, Hiratsuka, Kanagawa 254, Japan
** Research Center for Advanced Science and Technology
The University of Tokyo
4-6-1 Komaba, Megro-ku, Tokyo 153, Japan
*** R & D Laboratories-II, Nippon Steel Corporation
5-10-1 Fuchinobe, Sagamihara, Kanagawa 229, Japan

INTRODUCTION

The Ti-6Al-4V alloy is widely used as aerospace structural material. However, the toughness is decreased by increasing strength. It has been reported that the Ti-6Al-4V alloy with coarse acicular α structure has high fracture toughness[1,2], but the fracture mechanism of high toughness in coarse acicular α structure has not been fully understood. In Ti alloys, it has been reported that the nucleation, growth and coalescence of voids or microcracks at the tip of main crack influence the fracture process[3,4].

Acoustic Emission (AE) can detect dynamic deformation and microcracking in materials with high sensitivity. Especially, AE source characterization is one of the most effective methods, which can quantitatively evaluate the size, nucleation velocity and location of microcracks[5,6]. The purpose of this investigation is to make clear the fracture mechanism of Ti-6Al-4V alloy with coarse acicular α structure by AE source characterization.

AE SOURCE CHARACTERIZATION

The relationship between a detected wave, V(t), and a source function, D(t), is expressed in the linear response treatment as follows[7]:

$$V(t) = S(t) * G(t) * D(t) \tag{1}$$

where G(t) is the dynamic Green's function of a medium, S(t) is the response function of a transducer and the symbol, *, indicates the convolution operation. Using G(t) and S(t) both obtained experimentally with simulated signals, D(t) is evaluated as the inverse problem of eq. 1. In the case of mode I tensile cracking, D(t) is given by

$$D(t) = (\lambda + 2 \mu) \Delta A(t)\Delta\phi(t) \tag{2}$$

where λ and μ are Lame's constants[8]. $\Delta A(t)$ and $\Delta\phi(t)$ are the area and the

Nondestructive Characterization of Materials IV
Edited by C.O. Ruud *et al.*, Plenum Press, New York, 1991

opening displacement of a microcrack, respectively. The fracture volume $\Delta A(t)\Delta\phi(t)$ can be estimated from this equation.

EXPERIMENTAL PROCEDURE

Measuring System

Figure 1 shows the block diagram of the AE measuring system. The output from each transducer was amplified, then digitized and stored by the wave memory. The digitized waveform data were then transferred to the computer and AE source locations and source functions were computed.

Materials and Specimens

The material used was the Ti-6Al-4V alloy. The chemical composition and heat treatment condition of the alloy are listed in Table 1 and Table 2, respectively. The optical micrographs of the alloy are shown in Fig. 2. The microstructures consist of fine or coarse acicular α. Acicular α becomes coarser as the cooling rate is lower. The Mechanical properties are given in Table 3. The fracture toughness tests were performed using compact tension specimens of 12.7 mm in thickness.

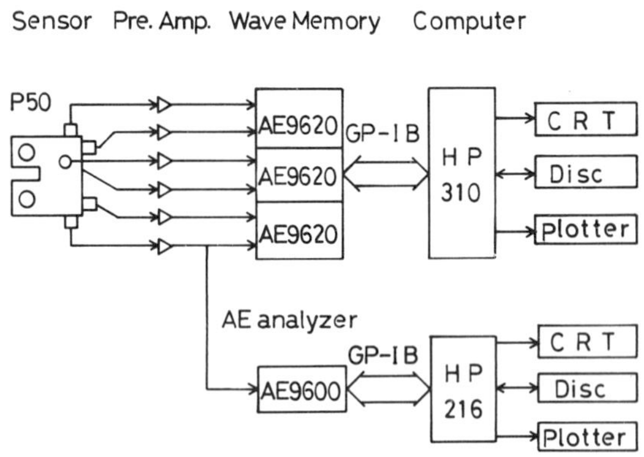

Fig. 1. Block diagram of AE measuring system

Table 1. Chemical composition (mass %)

Al	V	Fe	C	O	N	H
6.59	4.13	0.07	0.01	0.05	0.01	0.0071

Table 2. Heat treatment conditions

Holding condition	Cooling condition
	Water quench (65K/s)
1323K - 0.5h	Air cooling (1.3K/s)
	Furnace cooling (0.016K/s)

Table 3. Mechanical properties

	$\sigma_{0.2}$ (MPa)	σ (MPa)	El (%)	RA (%)
WQ	905	1020	7	19
AC	805	880	10	19
FC	725	780	14	28

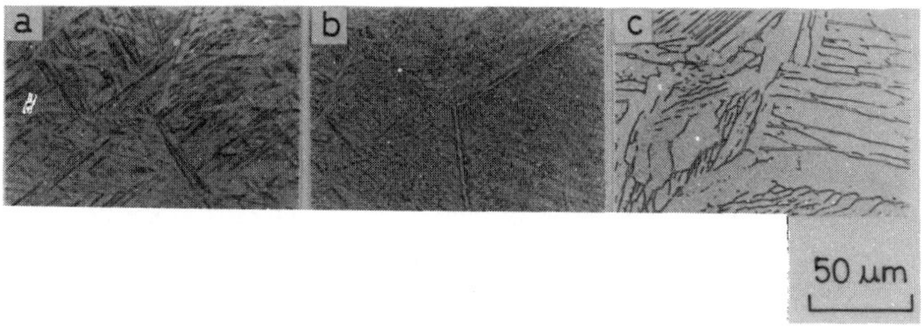

Fig. 2. Optical micrographs of (a) WQ, (b) AC and (c) FC specimens

RESULTS and DISCUSSION

Fracture Toughness and AE Events

Figure 3 shows load-COD (crack opening displacement) curves and AE activities during the tests. Critical crack tip opening displacement (CTOD; δ_c) are 0.22 mm for water quenched (WQ) specimen, 0.39 mm for air cooled (AC) specimen and 0.44 mm for furnace cooled (FC) specimen. These results indicate that the fracture toughness increases as the acicular α structure becomes coarser. In the AE behavior, there is a transition point A where the count rates of large-amplitude AE increase rapidly. The load corresponding to the point A becomes higher as the fracture toughness increase.

Three Dimensional Location

The location of each source event was determined by measuring the differences in the P-wave arrival time between two transducers[9]. Suppose Δt_{i6} is the difference in the P-wave arrival time between the i-th and 6th transducers. Let r_i denote the transducer positions and r the location of the source. The equation for source location can be written as follows:

$$C\Delta t_{i6} = | r - r_i | - | r - r_6 | \qquad 1 \le i \le 5 \qquad (3)$$

where C is the longitudinal velocity. A nonlinear least-square method was used to solve the eq. 3[10]. The location results of large-amplitude AE detected before the maximum load are shown in Figure 4. It is observed that microcracks distribute more widely in the Z direction as the acicular α becomes coarser. In this figure, type I is a smaller microcrack with higher nucleation velocity and type II is a larger microcrack with lower nucleation velocity. These results indicate that microcracks nucleated around the fatigue pre-crack front before the maximum load.

Microcracking Size and Mean Nucleation Velocity

Assuming that a microcrack is a penny-shaped crack subject to a uniform tensile stress, the crack volume can be estimated as follows[11]:

$$\Delta V = \Delta A \; \Delta\phi = 16 \; (1 - \nu^2) \; a^3 \; \sigma \; / \; 3E \qquad (4)$$

where a is the crack radius, E is Young's modulus, ν is Poisson's ratio. The crack radius a was calculated from eq. 2 and eq. 4 by approximation of tensile stress σ to $3\sigma_{ys}$ (σ_{ys} is yield stress)[10], i.e.

$$a = ((1 - 2\nu)D/16(1 - \nu^2)\sigma_{ys})^{1/3} \qquad (5)$$

The relationship between diameter and mean nucleation velocity of microcracks is shown in Fig. 5. There are two types of microcracks I and II. Type I is a smaller one with higher nucleation velocity. Type II is a larger one with lower nucleation velocity. Type I started at the beginning of large-amplitude AE and type II began to be detected at almost the same load level as those corresponding to the transition point A.

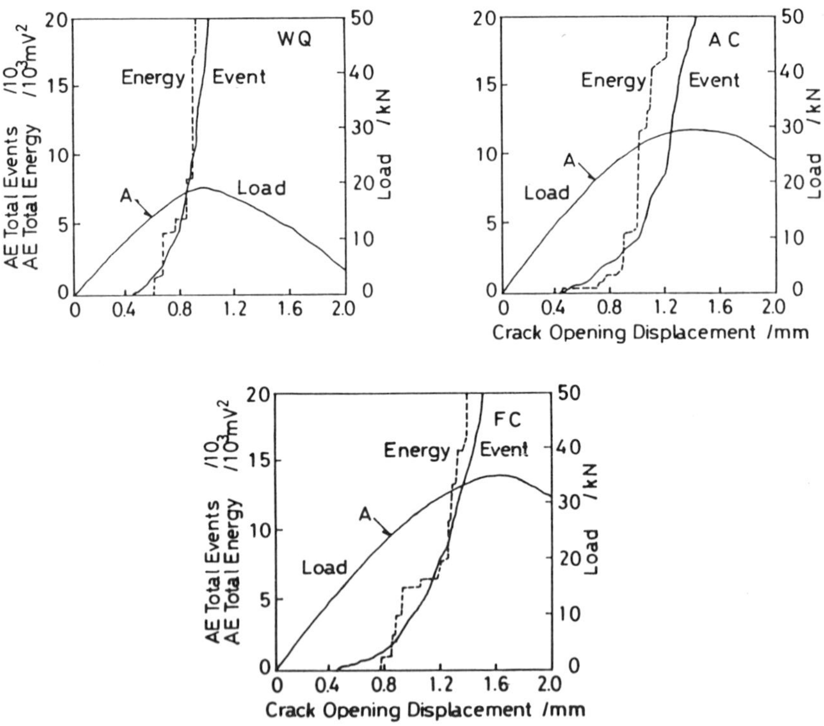

Fig. 3. The load-COD curves with AE events and energy during fracture toughness testing

AE Source Identification

Microscopic fracture surfaces are shown in Fig. 6. In WQ specimen, intergranular cracking with the size of 50 to 200 μm are observed. In AC specimen, several quasi-cleavage microcracks with the size of 50 to 200 μm are observed. Fracture surface in FC specimen is formed with dimple patterns containing elongated dimples, of which the length are about 300 μm and the width are 50 to 80 μm. These microcracking or elongated dimple size are consistent with that of type I microcracks. Therefore, type I microcracks correspond to these microcracks or elongated dimples. In order to examine type II microcracks, some specimens were loaded up to the load near the transition point A and then unloaded. The unloaded specimens were sectioned in the middle of thickness and the crack tip configurations were observed by scanning electron microscope (SEM). The result is shown in Fig. 7. Main crack extension are observed in all specimens. Therefor, it is considered that the stable crack growth occurred at the transition point A, and that the type II micro-cracks correspond to the main crack extension caused by the coalescence of type I microcracks.

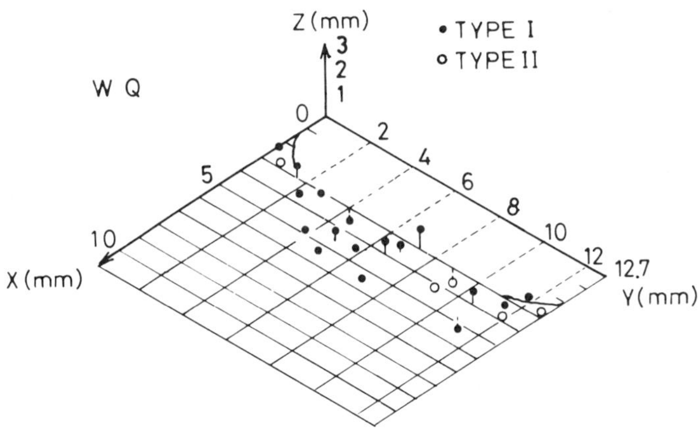

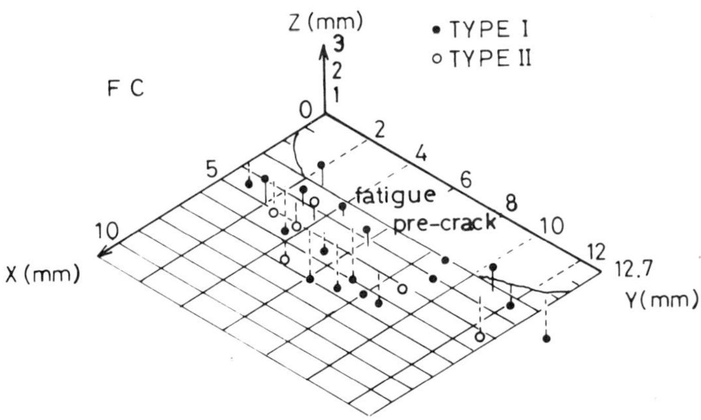

Fig. 4 The location results of large-amplitude AE signals

Microfracture Mechanism

From the above results, the microfracture mechanism of Ti-6Al-4V alloy could be explained as follows:

At lower load level, many microvoids corresponding to the small-amplitude AE nucleated at the phase boundaries. With increasing load, microcracks or elongated voids corresponding to the large-amplitude AE occurred by the growth of these microvoids. Then, at the transition point A (Fig. 3), main crack begin to extend by coalescence of these microcracks or elongated voids.

The nucleation sites of these microcracks are considered to influence the fracture toughness of specimens. In FC specimen, which has the highest fracture toughness, the microcracks distributed widely in the direction normal to fracture surface. In this case, more energy must be required for the coalescence of microcracks. In addition, the stable crack in FC specimen was hindered or distorted by coarse acicular α phase boundaries (Fig. 7). Therefore, high toughness of FC specimen can be attributed to the wide distribution of microcracks and the high crack extension resistance of coarse acicular α phase boundaries.

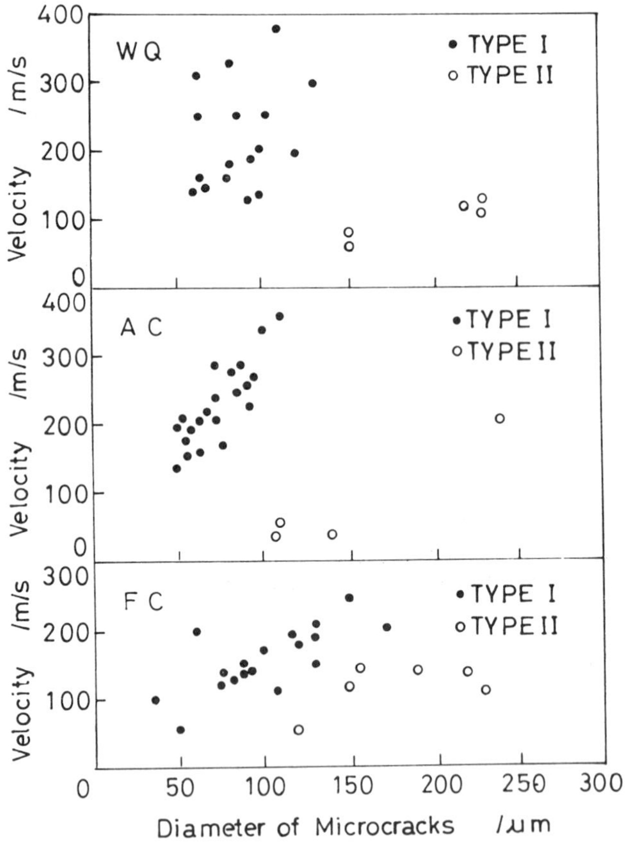

Fig. 5 Relationship between diameter and velocity of microcracks

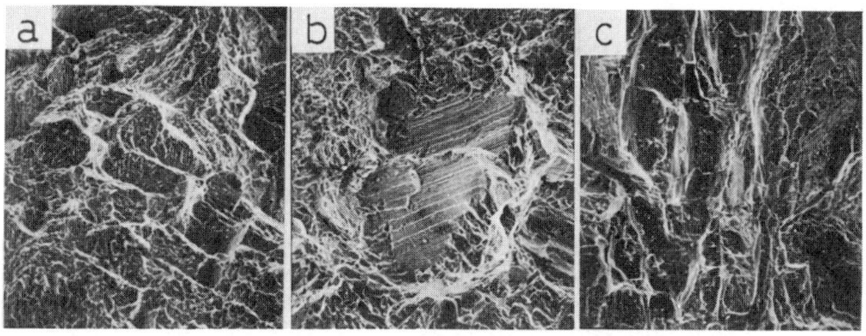

Fig. 6 Scanning electron fractographs of (a) WQ, (b) AC and (c) FC specimens

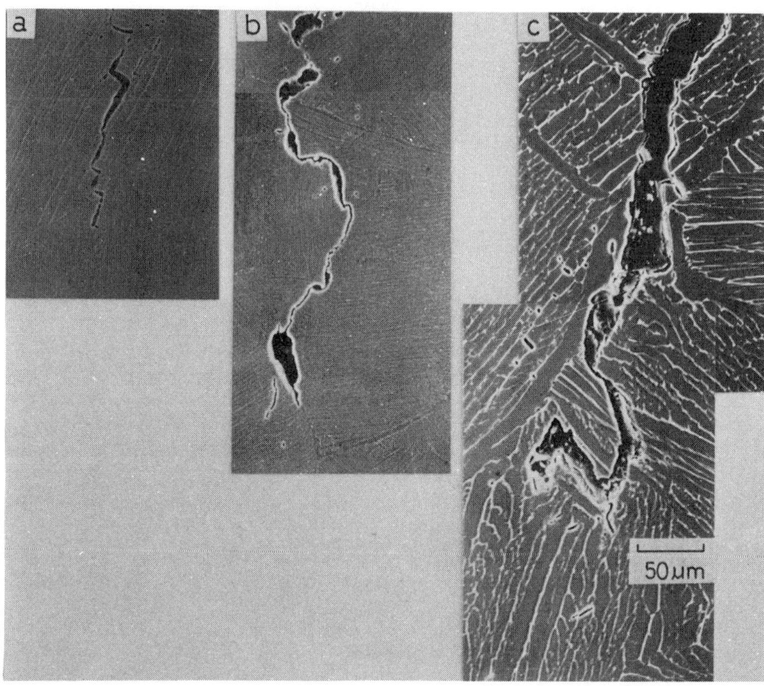

Fig. 7 Scanning electron micrographs at the crack tips of unloaded (a) WQ, (b) AC and (c) FC specimens

CONCLUSION

Microfracture process in the Ti-6Al-4V alloy with various acicular α microstructures has been investigated by AE three-dimensional location, AE source characterization and SEM observation. It is confirmed that AE source characterization is very useful for quantitative evaluation of microcracks. The main conclusions are summarized as follows:

1. Microcracks detected during fracture toughness test were classified into two groups (type I and type II) by AE source characterization. Type I was a smaller microcrack with higher nucleation velocity. Type II was a larger one with lower nucleation velocity.
2. The estimated size of type I microcrack was in good agreement with that of the intergranular cracking in WQ specimen, the quasi-cleavage cracks in AC specimen and the elongated voids in FC specimen.
3. It was considered that type I corresponded to the microcracks around the fatigue pre-crack front and type II corresponded to the main crack extension by coalescence of type I microcracks.
4. The high toughness in coarse acicular α microstructure can be attributed to the wide distribution of microcracks out of fatigue pre-crack plane and the high crack extension resistance of coarse α phase boundaries.

REFERENCES

1. A. Gaysler and G. Luetjering, Proc. 4th Int. Conf. on Titanium, 3, 2001 (1984).
2. H. Margolin, et al, Proc. 5th Int. Conf. on Titanium, 1, 169 (1980).
3. T. Kishi, H. Ohyama and K. Kim, J. Iron Steel Inst. Japan, 61, 123 (1986).
4. T. Horiya, H.G. Suzuki, T. Kishi, Tetsu-to-Hagane, 75, 114 (1989).
5. T. Kishi, K. Ohono and K. Kuribayashi, J. Japanese Soc. Nondestructive Inspection, 30, 911 (1981).
6. T. Kishi, Proc. 7th Int. Acoustic Emission Symposium, 302 (1984).
7. A.N. Ceranoglu and Y.H. Pao, J. Appl. Mech., 48, 125 (1981).
8. R. Burridge and L. Knopoff, Bul. Seismological Soc. Am., 54, 1875 (1964).
9. C. B. Scruby and G.R. Baldwin, J. Acoustic Emission, 3, 125 (1984).
10. M. Enoki, T. Kishi, J. Kihara, J. Japan Inst. Metals, 52, 405 (1988).
11. J. W. Hutchinson, Acta Metall.,35, 1605 (1987).

GRAIN SIZE MEASUREMENT USING BARKHAUSEN NOISE METHOD

M. Ishihara, T. Sakamoto and H. Minami

System Engineering Division
Sumitomo Metal Industries, Ltd.
1-3 Nishinagasu Amagasaki 660 Japan

ABSTRACT

Grain size is one of the most important parameters which determine the mechanical properties, e.g. tensile strength and yield strength of steel sheet. Therefore, it is necessary for quality control in the steel sheet production process to measure the grain size continuously.

Nondestructive methods for measuring grain size, such as the coercive force method, the ultrasonic attenuation method, and the ultrasonic backscattering method have been reported. However, these methods are not suitable for continuous measurement in the production line.

An advanced magnetic Barkhausen noise method has been developed for evaluating the grain size of steel sheet. The steel sheet is magnetized at low frequency (5Hz) by an electromagnet on one side of the sheet, and a ferrite cored coil, set on the opposite side, picks up the magnetic flux. The detected magnetic flux consists of the flux transmitted through the sheet and the high frequency flux generated by the Barkhausen effect. Since these fluxes have different frequency components, each flux is easily discriminated by frequency filters. By combining the Barkhausen noise and the transmitted flux a grain size measuring method that is highly accurate and insensitive to disturbances has been developed. The experimental results show the measuring accuracy of ± 0.5 in ASTM units.

INTRODUCTION

In the production process of steel sheet, it is important for quality control and quality assurance to evaluate mechanical properties continuously. In general, the grain size of steel sheet is one of the main factors for determining mechanical properties. Therefore, measurement of the grain size in the process is very useful for real time evaluation of mechanical properties.

Several nondestructive methods for evaluating the grain size of steel have been reported. For example, ultrasonic wave attenation(1), backscattering of ultrasonic waves(2), and magnetic hysteresis(3) have been used. These methods are suitable for static measurement, but are difficult to apply to dynamic and continuous measurement in the production line. Grain size measurement using Barkhausen noise has also been reported[3,4]. However these methods are not suitable for high speed measurement, since a very low magnetizing frequency (0.01 ~ I Hz) is adopted.

In this paper, an advanced method for continuous Barkhausen noise measurement in the production process is described. The signal processing method for highly accurate measurement and the quantitative analysis of error factors are also described.

PRINCIPLES

The Barkhausen noise method is one of the most effective magnetic methods for investigating characteristics of steel such as grain size, strain and other mechanical properties. Barkhausen noise is generated by discontinuous changes in magnetic flux (Figure 1), which occur with irreversible movements of the magnetic domain walls during a magnetization process. Barkhausen noise intensity depends on external magnetic field strength. The typical patterns of intensity change are shown in Figure 2. In a coarse grained steel, peak intensity of Barkhausen noise is observed at low magnetic field strength. In a fine grained steel, the peak intensity is observed at high magnetic field strength. The RMS voltage of Barkhausen noise, when measured during a magnetization process, is affected by both grain size and magnetization amplitude. Therefore, it is necessary to choose an optimum external field strengh for evaluating the grain size of steel. In order to obtain high accuracy, a new technique in which an electromagnet and a sensing coil are set on both sides of the steel sheet, as shown in Figure 3, was adopted. The sensing coil detects Barkhausen noise and the transmitted flux through the steel sheet. The transmitted flux is also used for measurement. The transmitted flux depends on the magnetic permeability of steel. The coarse grained steel has greater permeability than the fine grained steel, so the transmitted flux also depends on the grain size of steel. Since Barkhausen noise and transmitted flux are both effective for evaluating the grain size of steel, they have been detected at the same time.

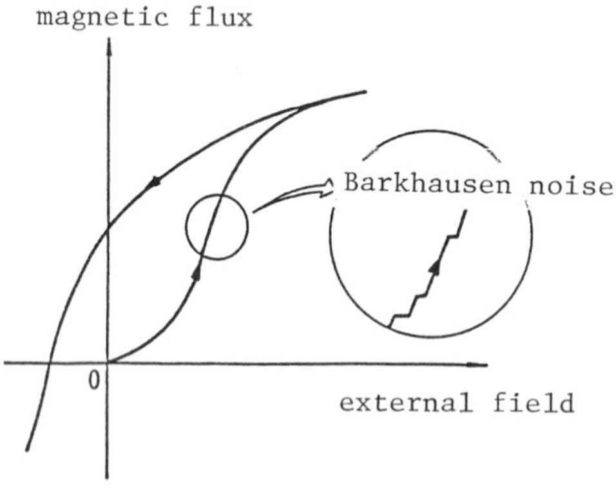

Fig. 1. Schematic illustration of Barkhausen noise.

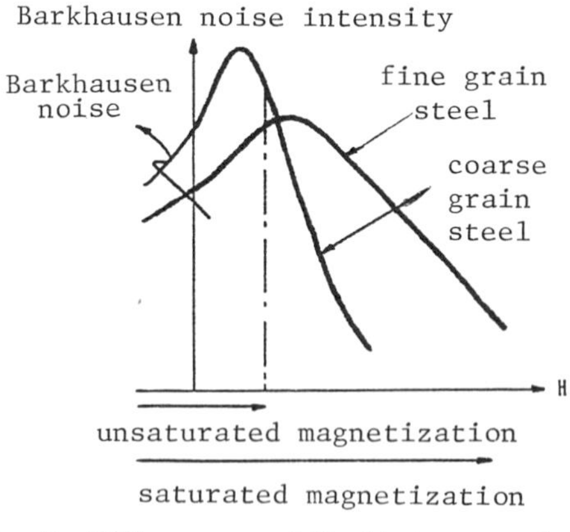

Fig. 2. Differences of Barkhausen noise intensity patterns for different grain size steel.

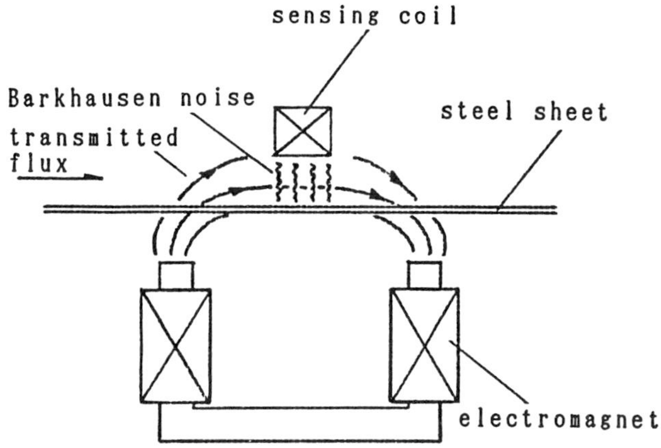

Fig. 3. Schematic illustration of transmitted flux and Barkhausen noise.

EXPERIMENTAL PROCEDURE

Figure 4 shows the Barkhausen noise measuring system. The electromagnet is excited by AC currents and the ferrite cored sensing coil is shielded from external electromagnetic noise by a copper case. The magnetic flux detected by the sensing coil is amplified, and discriminated into low frequency components equal to the excited frequency and high frequency components. RMS voltage of the low frequency component (V_{trans}) corresponds to transmitted flux through the steel sheet, and RMS voltage of the high frequency component (BNV) that has the frequency of some kHz, corresponds to Barkhausen noise. The grain size number of a given steel sheet is obtained by the division of BNV by V_{trans}, (BNV/V_{trans}).

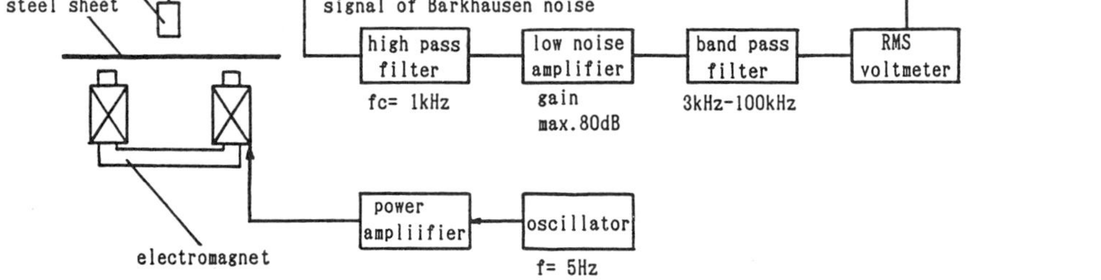

Fig. 4. Block diagram of experimental system.

EXPERIMENTAL RESULTS

In the case of real time measurement in the production line, high frequency magnetization is required for high speed measurement of BNV and V_{trans}. As shown in Figure 5, the increase of magnetizing frequency reduces the correlation between grain size number and BNV, and in saturated magnetization, the tendency is remarkable. So we select unsaturated magnetization with the frequency of 5Hz for online measurement. In this case, BNV decreases linearly with the increase of grain size number.

On the other hand, V_{trans} shows a reversed dependence on grain size number for unsaturated magnetization, i.e. a decrease of grain size number reduces V_{trans}. If (BNV/V_{trans}) is selected as the output, a highly sensitive measurement can be performed.

Some error factors in measuring BNV and V_{trans} are as follows.

- thickness of the steel sheet
- tensile stress
- variation of path line and vibration of steel sheet
- line speed

We will now discuss the influences of error factors and how they are reduced by using (BNV/V_{trans}).

First, the effect of the thickness of steel sheet on BNV and V_{trans} is shown in Figure 6(a) for constant grain size. The thickness dependence of V_{trans} has the same tendency as that of BNV. Therefore the output (BNV/V_{trans}) is insensitive to thickness variations as shown in Figure 6(b) for different grain size numbers N_G.

The next error factor is tensile stress change. In the production line, tensile stress is loaded on steel sheet and is variable. Since magnetic properties of steel depend on stress, BNV is affected by stress as in the case of V_{trans}. Although the output (BNV/V_{trans}) is relatively insensitive to changes of stress as shown in Figure 7, a change of ± 10 MPa in tensile stress induces an error of ± 0.2 in ASTM units. Since this influence is not negligible for the measurement, it may be necessary to take it into account.

Frequency (Hz)	1	2	3
saturated magnetization	□	△	○
unsaturated magnetization	■	▲	●

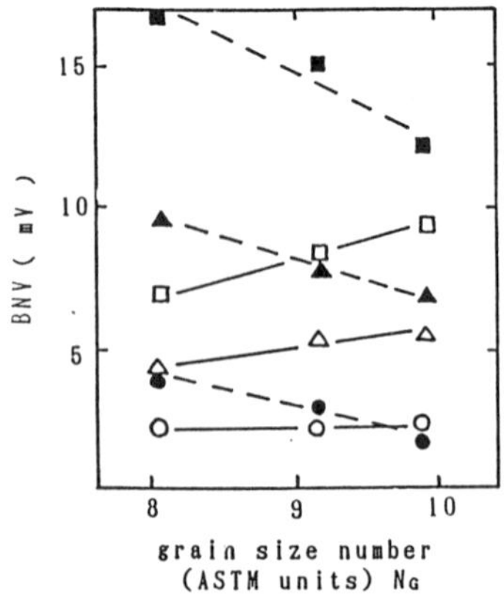

Fig. 5. Dependence on grain size number of BNV for different magnetization conditions.

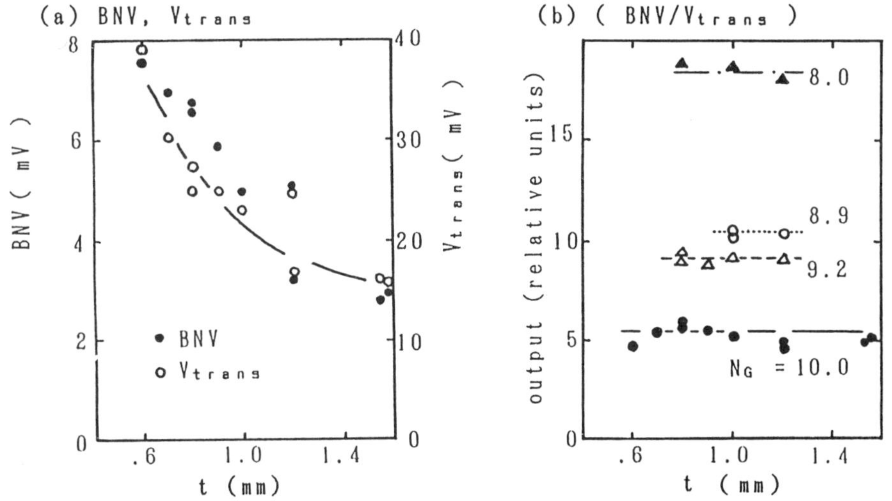

Fig. 6. The relation between thickness of steel sheet and BNV, V_{trans} for the same grain size number.

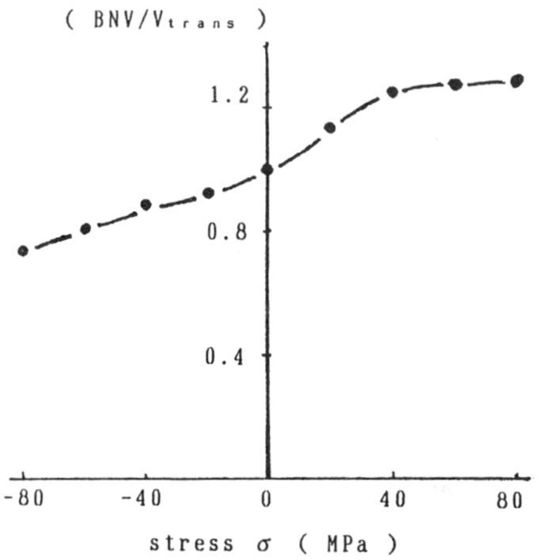

Fig. 7. Stress versus (BNV/V_{trans})

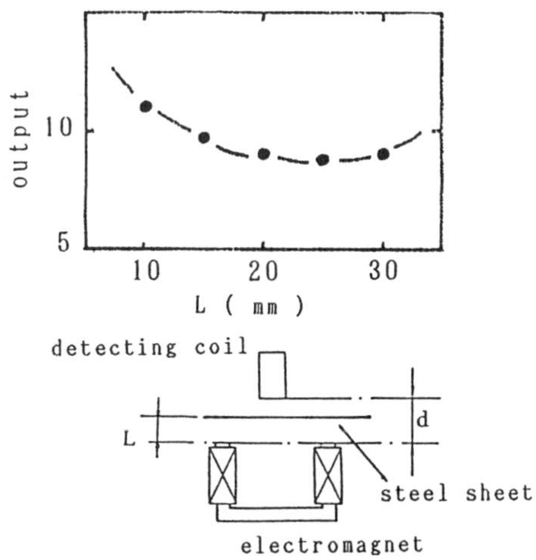

Fig. 8. Influence of variation of path line.

Furthermore, BNV and V_{trans} are affected by the variation of path line and vibration of steel sheet. The output (BNV/V_{trans}) is also affected by path line change as shown in Figure 8. It is important for highly accurate measurement to suppress the variation of path line to less than ± 10 mm. The vibration of steel sheet induces a false signal like Barkhausen noise. This is generated by the inverse magnetostrictive effect(5), and it is difficult to distinguish from Barkhausen noise, so it is necessary to minimize vibrations of the steel sheet.

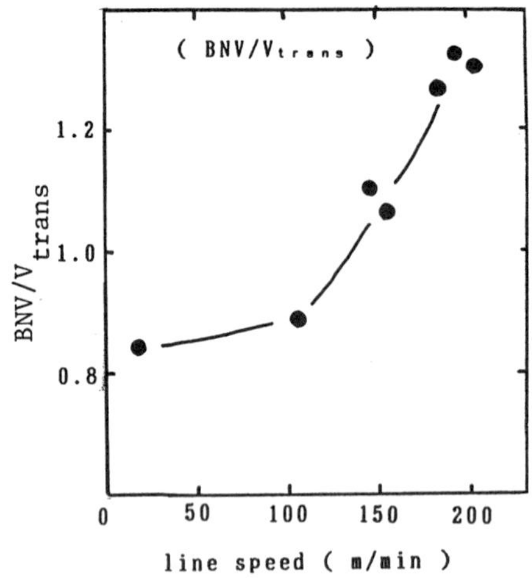

Fig. 9. BNV(V_{trans}) vs. line speed.

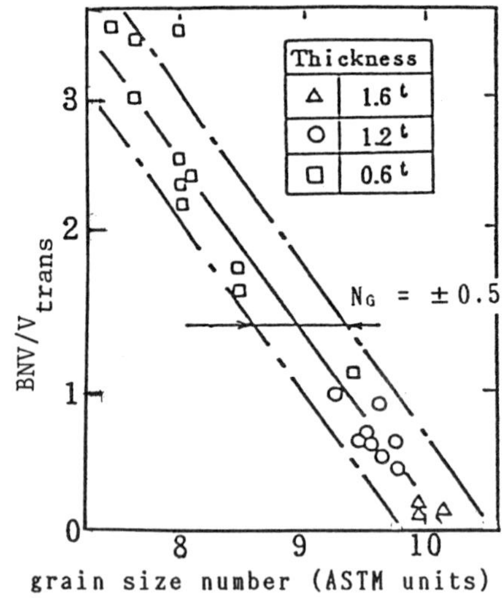

Fig. 10. Measuring accuracy in various kinds of steel coils.

Therefore, pinch rolls have been adopted for suppressing path line change and the vibration of steel sheet. The last error factor is line speed change. The output (BNV/V_{trans}) is varied by line speed change as shown in Figure 9. The output is compensated for line speed by the signal processor as shown in Fig.4.

The production process measurement showed that the relation between the output and grain size number were very good, as shown in Figure 10. The measuring accuracy for various kinds of low carbon steel strips produced in continuous annealing line, is ± 0.5 in ASTM units.

CONCLUSIONS

An advanced measuring method using Barkhausen noise and transmitted flux for evaluating the grain size of steel sheet has been developed. The following results were obtained:

- (BNV/V_{trans}) is useful for evaluating grain size because this output is insensitive to both thickness change and tensile stress change.
- In production line measurement, it is important for accurate measurement to keep the path line stable and suppress the vibration of steel sheet. Furthermore, it is necessary to compensate the output for line speed changes.
- The measuring accuracy of ± 0.5 in ASTM units has been obtained for various kinds of steel strips.

REFERENCES

1) A.Vary, 12th Sympo. on NDE (1979), 33.
2) K.Goebbels and P.Holler, Proc. of 1st Int. Sympo. on Ultrasonic Materials Characterization (1980), 67.
3) R.Ranjan, D.Jiles, and P.Rastogi, IEEE Trans.Mag. 23:18 (1987).
4) S.Segalini and M.Putignami, Mem.Etud.Sci.Rev.metall 84:655 (1987).
5) W.Polanschutz; NDT Int. 19:249 (1986).

APPLICATION OF SCATTERING THEORY TO PLASTIC STRAIN ESTIMATION

Y.Z. Dai and F.P. Chiang

Laboratory for Experimental Mechanics Research
State University of New York at Stony Brook
Stony Brook, NY 11794-2300

ABSTRACT

Light scattering theory and plastic deformation induced surface roughening were studied in an effort to provide some theoretical background to a nondestructive, non-contact plastic strain evaluation method. Experimental investigation shows that plastic strain on the surface of a metallic specimen is linearly proportional to surface root-mean-square roughness and inversely proportional to surface profile correlation length. Utilizing these relations along with Beckmann's light scattering theory, we obtained an expression which describes the scattered light intensity distribution in terms of plastic strain. This expression was applied to plastic strain evaluation and favorably verified by some experiments on aluminum alloy and brass materials.

INTRODUCTION

The phenomenon that plastic deformation changes metallic surface roughness as well as the scattered light intensity distribution has been investigated in the field of experimental mechanics for developing a plastic strain evaluation method.[1-3] However, to the authors' knowledge, all the related work in the published literature are experimental and there has been little theoretical analysis relating plastic strain with scattered light intensity distribution.

Obviously, surface roughness is the key factor relating plastic strain with scattering. Therefore the approach should be on two aspects: the plastic deformation induced surface roughening and the roughness caused light scattering.

In the present work, the relation between plastic strain and surface profile wavelength was experimentally studies. Then these relations were combined with Beckmann's light scattering theory[4] such that the scattered light intensity distribution was expressed in terms of plastic strain. It is hoped that this paper will provide some better understanding on plastic deformation induced light scattering variation and its application to plastic strain evaluation.

PLASTIC STRAIN & SURFACE ROUGHNESS

Study of plastic deformation induced surface roughening was carried out on tensile specimens made of aluminum alloy and brass materials. Specimens were cut from stock sheets along rolling direction and then polished to an initial surface roughness of approximately $\sigma = 0.05\mu m$, where σ is root-mean-square (RMS) roughness. After being loaded on a Tinius Olsen 1000 testing machine to a certain amount of plastic deformation

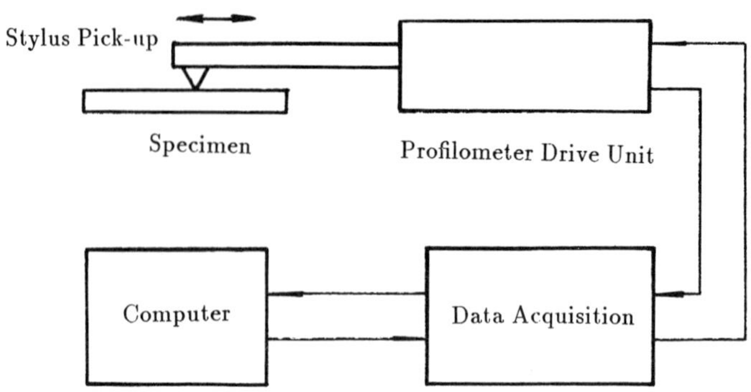

Figure 1. Surface profile measuring system

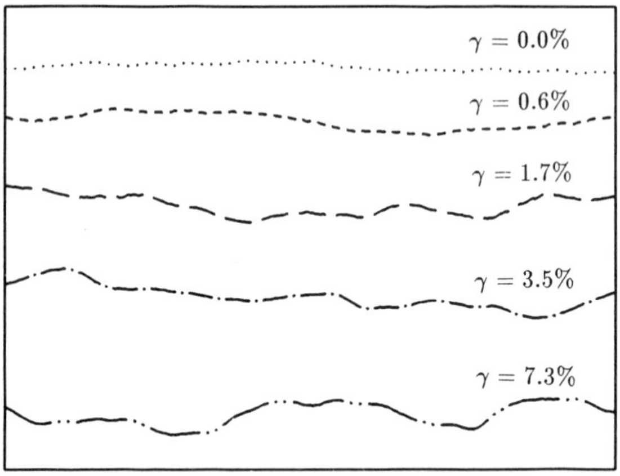

Figure 2. Surface profiles at different plastic strain levels

(measured by strain gage and moiré method) the specimen surface profile was recorded by a system as shown in Fig. 1, where the tip radius of the stylus of the profilometer is 5μm and the surface profile was digitized at a frequency of 1000 points/mm. The surface profiles of aluminum 1100-00 at different amount of plastic deformation are shown in Fig. 2. It is seen that the surface height variation increases while the profile wavelength decreases with plastic strain.

Plastic Strain & RMS Roughness

The zero-mean surface profiles were processed to yield the RMS roughness by the following expression:

$$\sigma = \sqrt{\frac{1}{N}\sum_{i=1}^{N} f^2(i)} \tag{1}$$

where N is the number of data point; f(i) is the discrete profile height at i_{th} point.

Five measurements of surface profile were taken at each plastic strain level. After the maximum and minimum values were eliminated, the rest of the data were averaged and plotted versus plastic strain as shown in Fig. 3 in which the vertical bars represent the standard deviation of the data. The relation between plastic strain and RMS roughness may be expressed by a linear function as:

$$\sigma = \alpha\gamma \tag{2}$$

where α is a constant. This result confirms some early experimental work.[5,6]

Plastic Strain & Profile Correlation Length

The profile correlation length is defined as the lag length T at which the autocorrelation of the profile is 1/e (0.368) of the zero-lag autocorrelation value. The value of T should satisfy the following condition.

$$\frac{\sum_{i=1}^{N-T} f(i) \times f(i+T)}{\sum_{i=1}^{N} f^2(i)} = \frac{1}{e} \tag{3}$$

The obtained profile correlation lengths were processed by the same method as described in the above section and the results are shown in Fig. 4. The relation between plastic strain and surface profile may be approximated by:

$$T = \frac{\beta}{\gamma} \tag{4}$$

where β is a constant.

The constants α, β in the above equations can be determined by using the least square method to fit the experimental data. Tests carried out on brass tensile specimens show that $\alpha \approx 7$um, and $\beta \approx 0.75$μm.

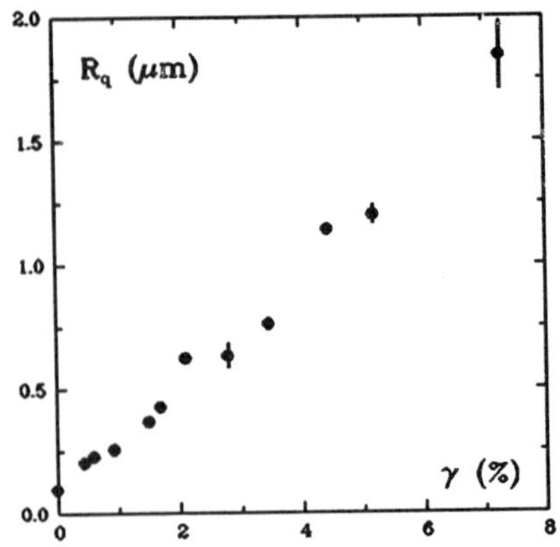

Figure 3. Plastic strain induced surface RMS roughness change

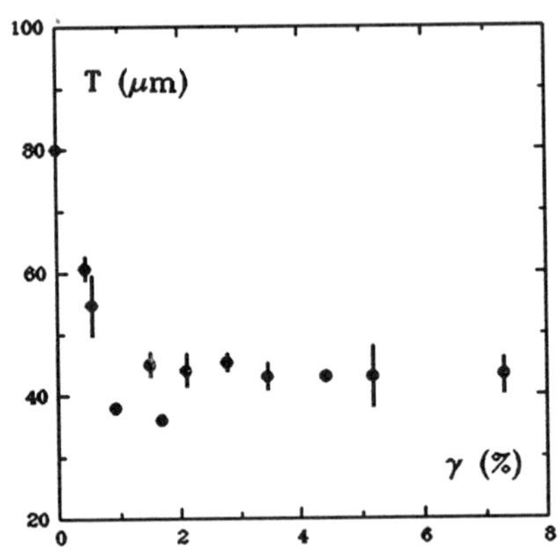

Figure 4. Plastic strain induced surface correlation length variation

In general, Eq. 2 and Eq. 4 should take the form

$$\sigma = \alpha\gamma + c \quad \text{and} \quad T = \frac{\beta}{\gamma + a} + b \tag{5}$$

indicating that the surface roughness σ and profile correlation length T are finite at zero plastic strain due to the limitation of polishing process.

<u>Plastic Strain & Substructure Size</u>

Profile correlation length in general is proportional to profile wavelength, the magnitude of which is observed to be of the order of tens of microns. It might be worthwhile to point out that another plastic deformation induced surface substructure, which is of the order of less than ten microns, is related to plastic strain in a similar way as Eq. 4.

According to Holt's derivation[7], the dislocation substructure size d is related to dislocation density ρ by:

$$d \approx K_c \rho^{-1/2} \tag{6}$$

Applying linear regression to Chiem and Duffy's experimental data[8] on shear strain γ and dislocation density ρ, we get:

$$\gamma = 1.85\sqrt{\rho} - 11.52 \tag{7}$$

Substitute Eq. 7 into Eq. 6 and write the resulting expression in a more general form as:

$$d \approx \frac{pK_c}{\gamma} + q \tag{8}$$

where p, q, K_c are constants. Eq. 8 indicates that shear strain γ is inversely proportional to substructure size d which is an analog to Eq. 4.

BECKMANN'S LIGHT SCATTERING THEORY

Beckmann[4] discussed extensively the scattering of electromagnetic waves from rough surfaces. The solution describing the scattered light intensity distribution, also called power-spectral-density (PSD) for some circumstances, from a square area is:

$$<\rho\rho^*> = e^{-g}\left(\rho_o^2 + \frac{\pi T^2 F}{A}\sum_{m=1}^{\infty}\frac{g^m}{m!m}e^{-v_{xy}^2 T^2/4m}\right) \tag{9}$$

where the left hand side term is the ensemble average of light intensity $\rho\rho^*$ which was originally described as a ratio proportional to the mean scattered power; A is the area of the surface being illuminated; T is the profile correlation length and the other variables are defined as:

$$\sqrt{g} = k\sigma(\cos\theta_1 + \cos\theta_2) \tag{10}$$

$$\rho_o = \frac{\sin(v_x X)}{v_x X}\,\frac{\sin(v_y Y)}{v_y Y} \tag{11}$$

$$F = \frac{1 + \cos\theta_1\cos\theta_2 - \sin\theta_1\sin\theta_2\cos\theta_3}{\cos\theta_1(\cos\theta_1 + \cos\theta_2)} \tag{12}$$

$$v_{xy} = k\sqrt{\sin^2\theta_1 - 2\sin\theta_1\sin\theta_2\cos\theta_3 + \sin^2\theta_2} \tag{13}$$

where $k = 2\pi/\lambda$; λ is the wavelength of the light source and θ_1 is the light incident angle and θ_2, θ_3 are observation angles (see Ref. 4 for detail).

In practice, normal incidence ($\theta_1 = 0$) is often used which makes F = 1 and $v_{xy} = k\sin\theta_2$. Then Eq. 9 can be rewritten as:

$$<\rho\rho^*> = e^{-g}\left(\rho_o^2 + \frac{\pi T^2}{A}\sum_{m=1}^{\infty}\frac{g^m}{m!m}e^{-(kT\sin\theta_2)^2/4m}\right) \tag{14}$$

This expression can be extended to the case of a circular illuminated area with radius r, then the area A is πr^2 and

$$\rho_o = \frac{2J_1(Z)}{Z} \tag{15}$$

where $Z = kr\theta$ and $J_1(Z)$ is the first order Bessel function of the first kind. Utilizing Eq. 2 and Eq. 4, Eq. 14 can be written in terms of plastic strain γ as:

$$<\rho\rho^*> = e^{-g}\left(\rho_o^2 + \left(\frac{\beta}{r\gamma}\right)^2 \sum_{m=1}^{\infty} \frac{g^m}{m!m} e^{-(k\beta \sin\theta_2)^2/4m\gamma^2}\right) \tag{16}$$

where

$$\sqrt{g} = k\alpha\gamma(1 + \cos\theta_2) \tag{17}$$

Therefore, the PSD function has been expressed in terms of shear plastic strain. Once this distribution and the necessary constants are known, then the plastic strain can be evaluated by Eq. 16; or, given a plastic strain value, its corresponding PSD distribution can be predicted (Fig. 5).

APPLICATION TO PLASTIC STRAIN EVALUATION

In this part of the discussion, Eq. 16 shall be used to predict the relation between PSD width and plastic strain followed by an experimental verification.

Analytical Approach

PSD width is defined as the width of the scattered light intensity distribution when its value is a certain fraction of its maximum. This width is approximately proportional to the scattered angle θ. In the following discussion, θ will be used to represent PSD width. Denote the intensity ratio as W, we have

$$W = \frac{<\rho\rho^*>\theta_2 = \theta}{<\rho\rho^*>\theta_2 = 0} = \frac{\rho_o^2 + \left(\frac{\beta}{r\gamma}\right)^2 \Sigma_{m=1}^{\infty} \frac{\left[k\alpha\gamma(1+\cos\theta)\right]^{2m}}{m!m} e^{(-k\beta\sin\theta)^2/4m\gamma^2}}{1 + \left(\frac{\beta}{r\gamma}\right)^2 \Sigma_{m=1}^{\infty} \frac{(2k\alpha\gamma)^{2m}}{m!m}} \tag{18}$$

The term $e^{-k\sigma(1+\cos\theta)}$ has been omitted because the error it introduces is less than 1.0% for $\theta \leq 4°$, which is the maximum scattering angle observed in experiments. Fig. 6 shows the PSD width at both W = 1/2 and W = 1/e versus plastic strain. It can be seen that the theoretically predicted relation is approximately linear.

Experimental Verification

The linear relation between the width of the PSD distribution and plastic strain was verified by a system as described elsewhere.[3] An experimentally obtained PSD distribution for brass material is shown in Fig. 7. As can be seen, it takes quite a similar form compared to the one (Fig. 5) predicted by the light scattering theory. The scattering in

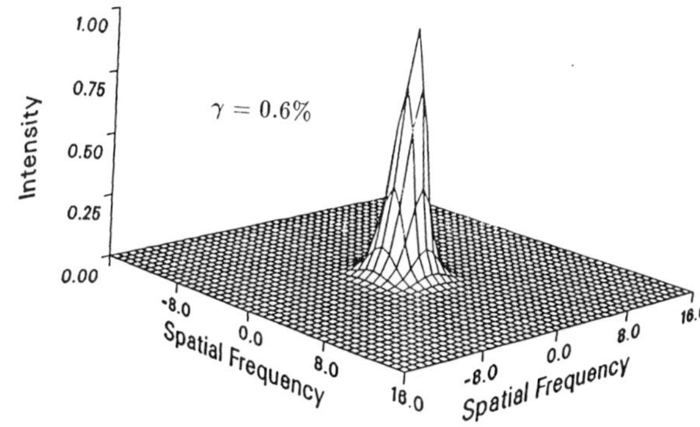

Figure 5. Analytical PSD distribution ($\gamma = 0.6\%$)

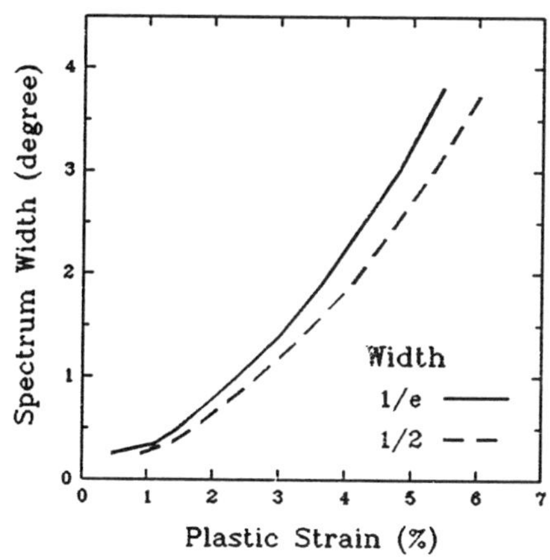

Figure 6. PSD width versus strain (theoretical)

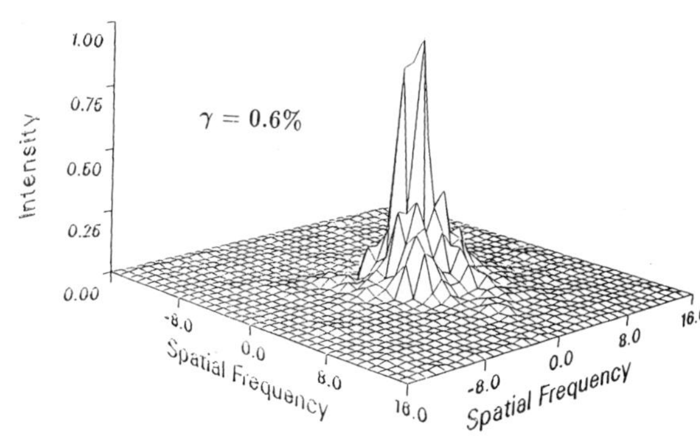

Figure 7. Experimental PSD distribution ($\gamma = 0.6\%$)

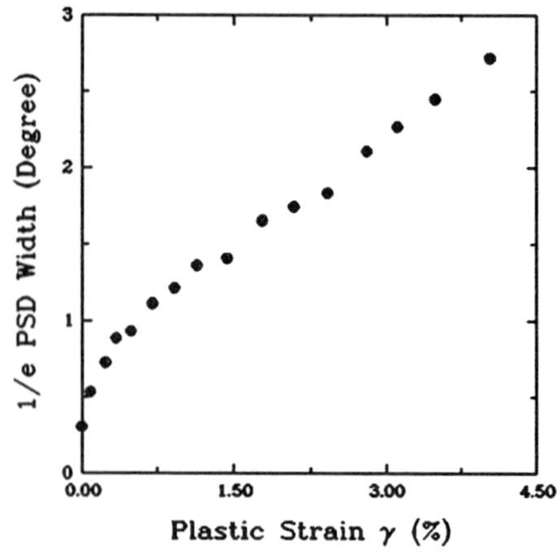

Figure 8 PSD width versus strain (experimental)

experimentally obtained PSD is caused by speckle effects and is believed to vanish by taking an ensemble average on macroscopically similar but microscopically different surfaces undergone the same amount of plastic deformation.

Fig. 8 shows some experimental results on the relation between plastic strain and PSD width (1/e). A linear relation, which is in support of the analytical result as shown in Fig. 6, can be clearly seen.

DISCUSSION & CONCLUSION

It has been shown that metallic material surface roughens as the specimen undergoes plastic deformation in such a way that surface RMS roughness increases linearly while surface profile correlation length decreases inversely proportional to plastic strain for the material tested. The range of validity of this result is 10% plastic deformation even though a much wider range was reported.[5,6] The initial surface roughness plays an important role in the determination of the constants in Eq. 2 and Eq. 4.

No explanation can be offered at this point to the similarity between Eq. 4 and Eq. 8. The knowledge of the internal relationship of these two equations may provide us with a better understanding of the plastic deformation induced metallic surface roughening mechanism.

The expression for describing PDS distribution as a function of plastic strain seemed to be correct. Comparing the analytical solution with the experimental data, we find a good agreement for PSD width method. However, we also see some discrepancy in the early stages of plastic deformation which may be improved by using a more general model like Eq. 5. The accuracy of using this technique to estimate plastic strain can be improved by taking an ensemble average of PSD distributions.

Further experimental development and theoretical investigation on plastic deformation induced surface roughening mechanism may provide us with a practical non-contact, nondestructive method for evaluating plastic strain. The method has a potential to be used under severe environmental conditions such as high temperature and/or high strain rate.

ACKNOWLEDGEMENT

Financial support provided by the Army Research Office, Engineering Service Division through contract No. DAAL0388K0083 is acknowledged.

REFERENCES

1. K. Miyagawa, A. Azushima, Measurement of plastic deformation by means of a laser beam, 9th World Conference on Nondestructive Testing, Australia (1979).

2. A. Azushima, M. Miyagawa, Measurement of plastic zone around a fatigue crack of a stainless steel structure member with a laser beam, Proceedings of the Joint Conference on Experimental Mechanics, Hawaii (1982).

3. F. P. Chiang, Y.Z. Dai, B.Q. Xu, A. Kato, Study of surface roughening under different stress modes by correlation, to appear in Proceedings of SEM Fall Conference, Baltimore (1990).

4. P. Beckmann, A. Spizzichino, "The scattering of electromagnetic waves from rough surfaces," Pergamon Press (1963).

5. M. Fukuda, K. Yamaguchi, N. Takakura, Y. Sakano, Roughening phenomenon on free surface of products in sheet metal forming, J. Jap. Soc. Technol. Plast. 10, 994 (1974).

6. P.F. Thomson, B.V. Shafer, The roughening of free surfaces during plastic working, Int. J. Mach. Tool Des. Res. 22(4), 261 (1982).

7. D.L. Hold, Dislocation cell formation in metals, Journal of Applied Physics 41 (8), July (1970).

8. C.J. Chiem, J.D. Duffy, Strain rate history effects and observations of dislocation substructure in aluminum single crystals following dynamic deformation, Brown University Report MRLE-137, Rhode Island (1981).

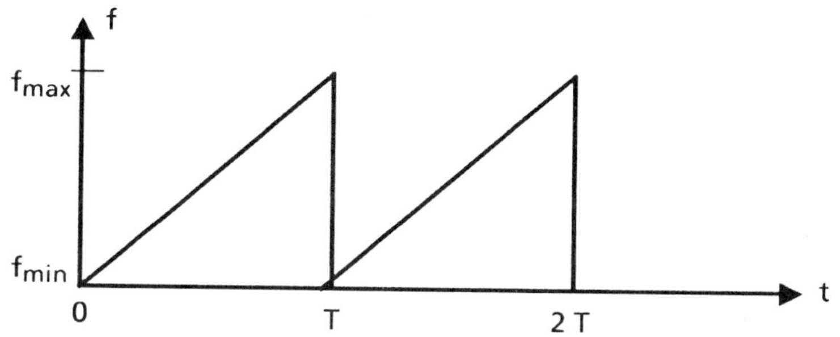

Fig. 3. FM chirp: frequency evolution

If the excitation modulation signal X(t) is an FM chirp or a white noise then $S_{xx}(f)$ = Const. and $C_{xx}(\tau) = \delta(\tau)$ where $\delta(\tau)$ is the Dirac function. Hence $C_{xy}(\tau) = h(\tau)$ and $S_{xy}(f) = H(f)$.

The relationship between X(t) and Y(t) can be written:

$$Y(t) = \alpha X(t - \theta) \tag{2}$$

where θ is the wave propagation delay and a the attenuation. θ and the phase $\Phi_{xy}(f)$ are related by

$$\theta = \frac{1}{2\pi f} \Phi_{YX}(f) \tag{3}$$

So the delay θ can be measured in the time domain by using the maximum of $C_{xy}(\tau)$, and in the frequency domain with $S_{yx}(f)$.

Since FM chirp and pseudo-random noise modulation are theoretically formally equivalent, we restricted the experiment to the first modulation method, that shows higher dynamic range.

Experimental results

All frequency functions reported in this work were recovered by calculations carried out with a Nicolet Sci. Corp. 660 B FFT analyser. The 514 nm Ar+ laser beam was focused on the sample surface with a microscope objective (laser power ~ 700 mW).

Impulse response modification due to a defect. A subsurface defect was simulated by drilling a hole in a cylindrical sample (Fig. 4). We used a pure aluminium sample to relate the physical parameters to a theoretical model.

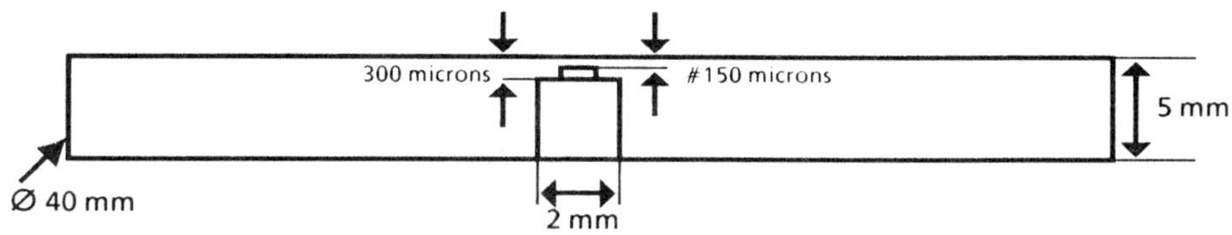

Fig. 4. Pure aluminium sample geometry

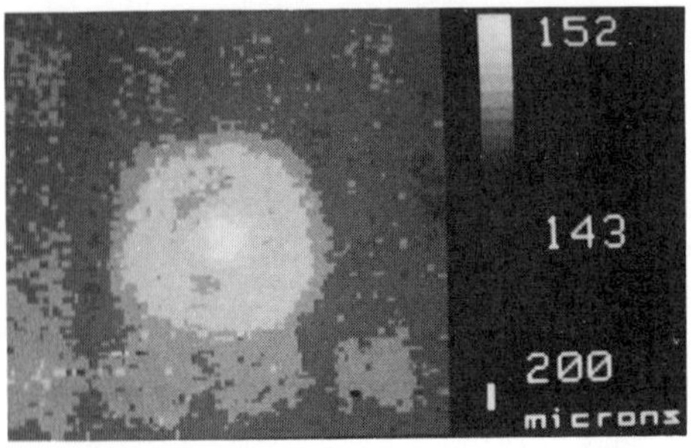

Fig. 5. Phase image of the sample obtained with a CW modulation (7 kHz).

The samples have been imaged to precisely localize the defect (sinusoïdal modulation and X-Y scanning). The phase image (Fig. 5) shows the shape of the defect. The total phase rotation is about 17 degrees at a 7 kHz modulation frequency.

The impulse response of the same sample has been measured at several points of the sample surface. Figure 6 shows two phases $\Phi_{xy}(f)$ of the cross-spectrum $S_{xy}(f)$. One curve was obtained over the defect and the second one a few millimeters away so that the defect does not influence the response. The slope difference is related with the defect depth at low frequencies. At higher frequencies, the sample deformation (due to machining) acts on the phase response.

<u>Influence of the residual static stresses.</u> It has been recently demonstrated that a thermoelastic stress measurement enables a nondestructive evaluation of the distribution of static stress[11]. We show here that the thermoelastic testing gives information on residual static stress.

The sample is a 5 mm-thick pure Cu plate. It has been annealed and one face was polished by chemical attach. The other face was polished by machining (diamond powder and polisher).

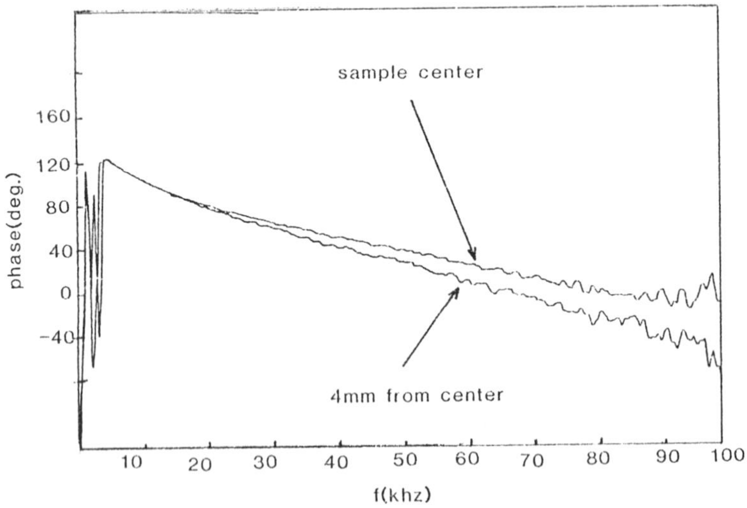

Fig. 6. Cross-spectrum phase of the pure aluminum sample:influence of a drilled hold

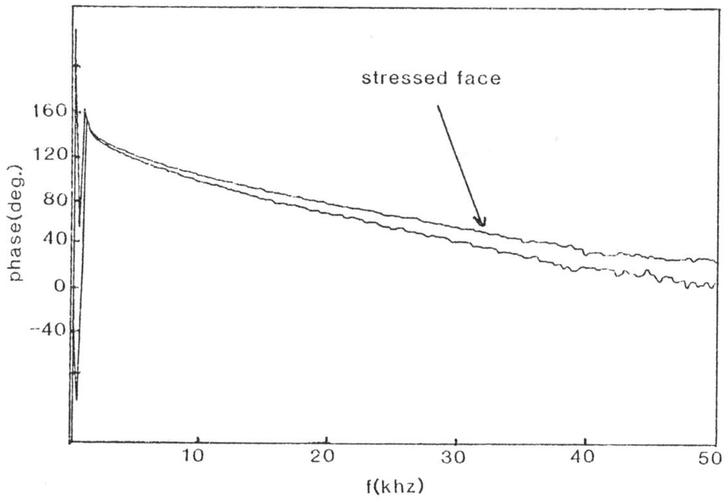

Fig. 7. Cross-spectrum phase of the polished Cu plate

Figure 7 shows the cross-spectrum phases obtained on the two sample faces. The slope variation is obvious above 10 kHz. The corresponding penetration depth is $\mu_T \approx 60\ \mu m$ ($\mu_T \approx = \sqrt{2a_T} / (\omega_M$ where α_T is the thermal diffusivity and ω_M the angular frequency). The depth profiling of this method shows that the resldual stresses are very close to the surface.

CONCLUSION

Thermoelastic testing is a remote technique that presents several advantages: non-destructive, noncontact sensing using optical excitation and detection beams, monitoring of lateral resolution and penetration depth (in the 1 - 100 kHz frequency range, the penetration depth is typically 100 μm - 10 μm).

The two operating modes (imaging with CW modulation and impulse response measurement) give a complete information characterizing the tested sample: defects are localized with imaging and quantitative data are obtained with impulse response. So the two modes are complementary and they must be jointly used to identify forging defects or damaging. A theoretical model should permit to evaluate the real size and the type of many detected faults.

REFERENCES

1. C.O. Ruud, G.H. Penningtom, E.B. Brauss and S.D. Weedman, "Simultaneous residual stress and retained austenite measurement by Xray diffraction", Proc. in "Nondestructive characterization of materials", Springer-Verlag, Saarbrucken, FRG, October 3-6, pp. 406-412 (1988).
2. C.F. Quate, A. Atalar, H.K. Wickramasinghe, "Acoustic microscopy with mechanical scanning. A review", Proc. ofthe IEEE, 67 (8),1092 (1979).
3. A. Rosencwaig, "Thermal wave microscopy with photoacoustics", J. Appl. Phys. 51(4), 2210 (1980).
4. W. Jackson, N.M. Amer, "Piezoelectric photoacoustic detection: theory and experiment", J. Appl. Phys., 51(6), 3343 (1980).

5. Y. Martin, H.K. Wickramasinghe, E.A. Ash, "Thermo and photodisplacement microscopy", Proc. IEEE Ultrasonics Symp., p. 563 (1982).
6. G.C. Wetsel, Jr., "Photothermal generation of thermoelastic waves in composite media", IEEE Trans. on Ultrasonics, Ferroelectrics and Frequency Control, UFFC-33 (5), 450 (1986).
7. A. Rosencwaig, G. Busse, "High-resolution photoacoustic thermal-wave microscopy", Appl. Phys. Lett. 36 (9), 725 (1980).
8. B. Cretin, D. Hauden, "Thermoacoustic microscopy using optical excitation and detection", Proc. SPIE, vol. 809, Scanning imaging technology, p. 64 (1987).
9. D. Royer, E. Dieulesaint, Y. Martin, "Improved version of a polarized beam heterodyne interferometer", Proc. IEEE Ultrasonics Symposium (1985).
10. A. Mandelis, L.M.L. Borm, J. Tiessinga, "Frequency modulated (FM) time delay photoacoustic and photothermal wave spectroscopies. Technique instrumentation and detection", Rev. Sci. Instrum. 57 (4) (1986).
    Part. I pp. 617-621
    Part II pp. 622-629
    Part III pp. 630-635.
11. M. Kasai, T. Sawada, "Non-destructive evaluation of the distribution of stress by means of the photoacoustic microscope"., Springer Series in Optical Sciences, vol. 62, Photoacoustic and Photothermal Phenomena II, Editors: J.C. Murphy, J.W. Machlachlan-Spicer, L. Aamodt, B.S.H. Royce, Springer-Verlag, pp. 33-36 (1990).

FEATURE BASED STUDIES ON GUIDED ULTRASONIC WAVE MODES FOR ANOMALY ESTIMATION IN COMPOSITE STRUCTURES

Krishnan Balasubramaniam*, Yimei Huang and Joseph L. Rose

* Mississippi State University
Dept. of Aerospace Engineering
Engineering Mechanics
Mississippi State, MS 39762

Drexel University
Dept. of Mechanical Engineering and Mechanics
Philadelphia, PA 19104

ABSTRACT

This paper reports some of the recent studies on the utilization of surface waves and plate waves resonances for a quantative estimation of anomalies such as porosity, fiber volume fraction changes, fiber mis-orientation, hydro-thermal degradation in fiber reinforced composite materials. Models describing anomalies within an arbitrary layup of fiber reinforced composite laminates are coupled to generalized wave propagation models for plate waves and surface waves in such structures thus providing useful NDE tools. A feature based approach was also adopted here, for the analysis of the wave mechanics interaction with anomalies.

In the case of Plate wave resonance, the critical frequency values (minimas) were found to provide new features which provide additional sensitivity (when compared to traditional methods) to subtle anomalies such as porosity. Careful experiments were shown to support theoretical predictions. The new features were tested on a wide range of specimens with programmed anomalies.

The surface waves were successfully generated on composite structures using a novel line source wedge transducer (due to Snell's law limitations, the regular angle shoes were not found practical). Features from the azimuthal velocity variations are useful to isolate specific anomaly influence from the often occuring velocity changes due to anisotropy and inhomogenieties.

INTRODUCTION

While dealing with quantitative measurements of subtle imperfections especially in today's complex materials, several of the existing methods have to be modified and several new oblique incidence techniques need to be established. There is an abundance of information available in field of oblique incidence and guided waves, mostly from an acoustical point of view. The strong theoretical background provided during the last two decades will have to serve as stepping stones for translating the available oblique incidence technology into nondestructive applications for modern materials and complex structures.

ANOMALY MODELLING

Here we will consider Greszczuk's theory, given by Martin [1]. He reports that for fiber fractions (FF) in the range 50 to 73 percent, the estimation of material properties agree to within 10% of the values obtained from a more rigorous treatment. Not all kinds of fibers can be treated as isotropic. For example, carbon and graphite fibers exhibit a transversely isotropic class of anisotropy. For such composites, the elastic properties cannot be calculated on the base of Greszczuk's theory. The formulae are provided in detail by Martin [1] and Balasubramaniam [2].

Porosity Content Modeling

If imperfections of a composite can be modeled as a change in properties of one of the constituents of a binary fiber-reinforced composite, or by their different ratios, such tasks can be relatively easy using an effective elastic constant approach. Let us consider for example the effect of small voids. It has been established by a morphological study of porosity defects in graphite-epoxy composites by several investigators that voids tend to be small and spherical at a low porosity volume fraction (less than 2%) and at higher volume fractions interlaminate voids due to trapped air ten to be much larger, flattened and elongated in shape. Therefore, for the case of a low void content (i.e., where the void volume fraction in the matrix is < 0.1), Hashin's [3] expressions for the elastic modulus of isotropic, homogeneous solids containing voids can be applied. Here, the influence of voids is considered as an exclusive property of the matrix alone and hence, it is justifiable to degrade the matrix properties alone. Then for all composites with specific volume fiber fractions (FF), its porosity degree (PC) can be written as

$$PC = P(1 - FF) \tag{1}$$

Thus the density of the degrade matrix (ρ_m) is related to the porosity content and fiber volume fraction as

$$\rho_m = \rho_0[1 - PC/(1 - FF)] \tag{2}$$

The density ρ_c of a composite containing voids from the law of mixtures can be obtain as follows

$$\rho_c = \rho_f FF + \rho_m (1 - FF) \tag{3}$$

Once the material properties of the matrix with voids as a function of porosity degree (PC) are known, the composite material properties can be also calculated. We have thus obtained, for a unidirectional fiber-reinforced composite with isotropic or transversely isotropic fibers, expressions for effective elastic constants as a function of porosity degree (PC) and volume fiber fraction (FF).

The modelling of ply mis-orientation can be achieved by transforming the material properties of the individual ply/ply group to appropriate orientations.

SURFACE WAVES IN ANISOTROPIC MEDIA - Theoretical Development

A computer program has been developed to calculate surface wave propagation in a half space. The traction-free surface may take any arbitrary orientation which need not coincide with any crystal plane of the material. In the program, Euler angles are used to identify the traction free surface. The work is detailed extensively in Ref. 4.

A plane wave form solution, below, is assumed

$$u_k = \alpha_k \exp\{ i P (\mathbf{n} . \mathbf{r})\} \exp\{ i [\mathbf{Q} (\mathbf{e}.\mathbf{r}) - wt]\} \tag{4}$$

where α_k is the displacement vector, **Q** is the magnitude of the propagation vector in free surface, **P** is the decay parameter, $(\omega/Q) = v$ is the phase velocity. To be the solutions of surface wave, the quantity P in each term of the solution must be such that the amplitudes of all the displacement components vanish as **(n.r)** approach to $-\infty$.

The propagation vector is always assumed to be parallel to the free surface even though there may be a real part in P. Similarly, the planes of constant phase are perpendicular to the free surface and to the defined propagation vector, but the amplitude of the displacement term u_i varies in the **-n** direction over a plane of constant phase. The characteristic equation is obtained as follows

$$| C_{jklm} (Q^2 e_j e_m + PQ n_j e_m + QP e_j n_m + P^2 n_j n_m) - \rho\omega^2 \delta_{kl} | = 0 \tag{5}$$

where $\omega = Qv$.

Instead of solving for the phase velocity of the surface wave by using the three dimensional approach, we used another approach which combined the equations of motion and the stress-free boundary condition into a system equations as follows. First, substituting the assumed solution shown in Eq. (4) into the boundary conditions below

$$\sum_{n=1}^{3} \sigma_{jk}\left(\mathbf{u}^{(n)}\right) n_k = 0 \qquad (j = 1,2,3) \tag{6}$$

gives another set of nine homogeneous equation in nine unknown components of three polarization displacement vectors $\mathbf{u}^{(n)}$ (n = 1,2,3) as follows:

$$\sum_{n=1}^{3} \left(QC_{jklm} e_m n_k + P^{(n)} C_{jklm} n_m n_k\right) \mathbf{u}^{(n)} = 0 \tag{7}$$

Referring to Eq. (7) the matrix operator T_n corresponding to each of $P^{(n)}$'s (n = 1,2,3) can be written as

$$T_n = Q^2 A + Q P^{(n)} B + (P^{(n)})^2 C - I \tag{8}$$

Another matrix operator S_n corresponding to each of $P^{(n)}$'s (n = 1,2,3) can be expressed as:

$$S_n = QB_2 + P^{(n)} C \tag{9}$$

Combining Eqs. from (5) to (7), a set of twelve homogeneous equations can be written in a matrix form. After eliminating three dependent equations a new 9×9 system matrix equations can be obtained. In order that nontrivial solution exits, it is again necessary that the determinant of the coefficient matrix of the above equations should be zero. Hence we obtain the final solution as:

$$F(Q) = \begin{vmatrix} T_1 & 0 & 0 \\ 0 & T_2 & 0 \\ 0 & 0 & T_3 \\ S_1 & S_2 & S_3 \end{vmatrix}_{9\times 9} = 0 \tag{10}$$

For such complicated algebraic F(Q) = 0, it is necessary to use an interative search to find the solution for surface wave phase velocity v = 1/Q.

An extended bisection method was used to search the slowness value of the surface waves Q (Q = 1/v) on a complex plan, so that the phase velocities of surface waves propagating along any direction in an arbitrary free surface of a half-space of a general anisotropy medium could be determined.

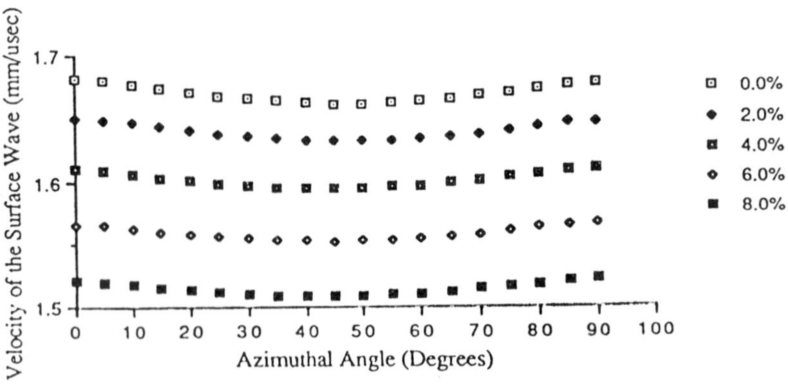

Figure 1. Surface wave sensitivity prediction to porosity content.

Surface Wave Sensitivity to Porosity

It was not possible to obtain the exact elastic constants for the individual fibers as well as the polymer thus causing problems with the utilization of the above discussed algorithm for material degradation simulation. However, since the overall woven composite material property was available, it was possible to make use of the developed porosity simulation software by using certain approximations. The cross-ply effective elastic constant was considered to be a sufficient approximation to the woven composite. The changes in the elastic constants (all nine due to orthortropic symmetry) five different levels of porosity were evaluated. Then these elastic properties were fed into the surface wave velocity estimation software and the polar surface wave velocity profiles were obtained. The polar profiles as seen from the Figure 1 illustrates the sensitivity of the surface waves to low levels of porosity.

NOVEL PROBE DESIGN CONCEPTS FOR SURFACE WAVE GENERATION

Unfortunately, conventional means of surface wave generation are not possible with these composites. These conventional means, otherwise known as the resonance technique, utilizes an obliquely incident longitudinal wave from a liquid or solid material either near or beyond a third critical angle. To overcome this problem, we have adopted a line source method. Under this alternative, a mediator is used to generate a surface wave in the composite. This mediator is coupled to a transducer whose incident angle is the surface wave critical angle, otherwise known as the third critical angle, for the given incident media/mediator combination. As the surface wave travels down the mediator, its thickness is decreased to a sharp line source which is placed on the composite. This line source is able to generate surface waves along with some other modes within the composite material.

This design employs two transducers coupled to a mediator. The transducers are fixed so that the two line sources are parallel and facing each other. In this way the spherical loss is eliminated and the simplified geometry eliminates the multiple echoes obtained in the previous design. Measurements can now be taken up to a transducer separation distance of three inches. A 1.0 MHz Harisonic Transducer is used and the separation distance is 1.5 inches. The signal to noise ratio is much better than the first design and the received signal shows no multiple echoes.

The connecting linkage holding the two transducers allows for a repeatable separation distance to be maintained between the two transducers. The next step in the design is to somehow measure accurately the surface wave travel distance over the curved surfaces. From simple geometry, the distance between the transducers can accurately be determined but unfortunately this distance does not always correspond to the surface wave travel distance over a curved surface. The assembly is also capable of conforming to different size contours, provided the top and bottom surfaces are reasonably planar. Another useful feature is that the probe can be easily held in one hand thus enabling one technician to operate the probe as well as the controls on the pulser/receiver.

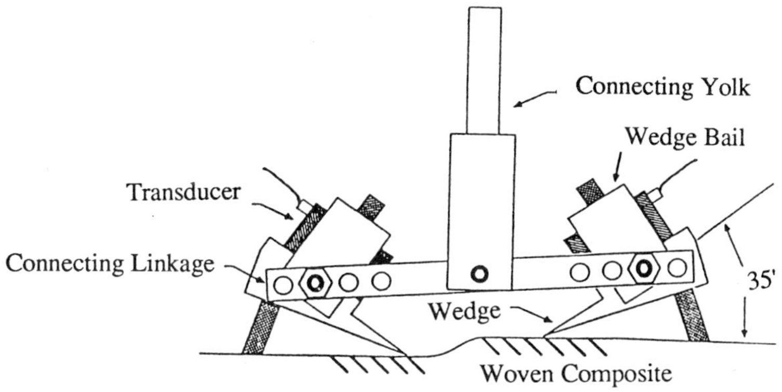

Figure 2. The new probe schematic representation.

Table 1. Surface Wave Velocity on the Control Woven Composites.

No.of the Specimen (Thickness)	Propagation Direction (Degrees)	Surface Wave Velocity (mm/msec)	
		1.0MHz	0.5MHz
2686-02 (Good) (0.4205"/10.68mm)	0 90	1.523 1.523	1.523 1.523
32664-05 Bulge (0.4301"/10.92mm)	0 90	1.449 1.485	1.414 1.449
32664-05 Base (0.2908"/7.386mm)	0 90	1.563 1.523	1.414 1.414

Table 2. Surface Waves Velocity on the Woven Composites with Different Levels of Porosity Content (PC).

No.of the Specimen (Thickness)	Propagation Direction (Degrees)	Surface Wave Velocity (mm/msec)	
		1.0MHz	0.5MHz
7A (PC Area) (0.3940"/10.01mm)	0 90	1.350 1.291	1.188 1.237
7C (PC Area) (0.4033"/10.24mm)	0 90	1.350 1.273	1.264 1.273
5A (Less PC) (0.2870"/7.290mm)	0 90	1.449 1.381	1.414 1.488
5A(More PC) (0.4850"/12.32mm)	0 90	1.188 1.150	1.237 1.080

Experimental Measurements of the Surface Wave Velocities

The results of surface wave velocity measurements on these specimens in two perpendicular directions, i.e. 0° along warp yarns direction, 90° along filling yarns direction, for both frequency 1.0 MHz and 0.5 MNz are presented in Tables 1-2.

The experimental results show that

1. The velocities of surface waves measured on the top surface area of the specimens containing high porosity is lower than that measured from the area with low porosity of a specimen.

2. The velocities of the surface waves along warp yarns (0°) and filling yarns (90°) directions on the surface area of the specimen with the less porosity are not substantially dependent on the wave propagation direction and the frequency of transducers (see results fro specimen 2686-02 and 38664-05). But, it seems that the velocities of the surface waves in the warp and filling yarns directions measured for the same frequency from the area of the specimens containing much more porosity are different (see results for specimen 7A, 7C and 5A More PC), which could be considered the anisotropic characteristics induced by the matrix pores with elliptical shape and not random distribution.

PLATE WAVE RESONANCE EXPERIMENTS

The first set of specimens used were the [0_4, +45, -45, 0_4] specimens with simulated porosity using echo spheres. Two cases of specimens with 0% and 2% porosity by volume was taken and the dispersion curves obtained for each case is shown in Figure 3 along the 0° direction. As earlier predicted from theory, the influence of porosity is to shift the curves to the left by measurable quantities.

Here, the transducer employed was a 1.0 MHz broadband, in a corner technique setup, and the dispersion curves are obtained from the minima of the reflection factors. The same specimens were examined using traditional normal incidence longitudinal waves and normal incidence shear wave mode velocity measurements. It was observed from these studies that although it is possible to measure porosity content beyond 2% with reasonable reliability, their no sensitivity as low porosity values.

From Figure 4, it can be seen that the resonance features show excellent sensitivity through out the porosity range, even below 2%, which is a significant experimental result indicating promise for low porosity evaluation. It may be noted that each individual mode behaves with a different relation to the porosity. Thus, their inter-relationships may be exploited

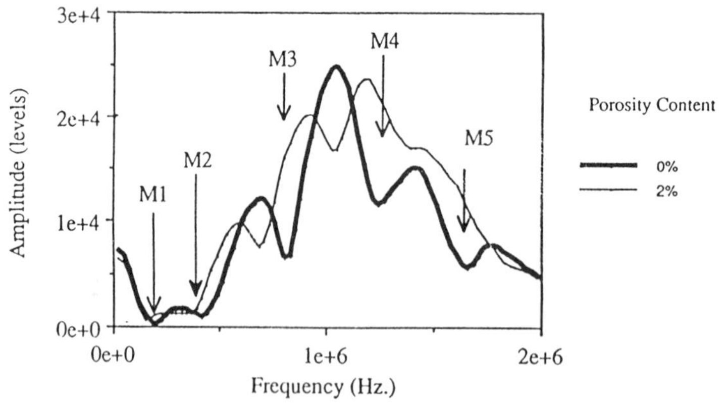

Figure 3. Experimental demonstration of sensitivity to porosity.

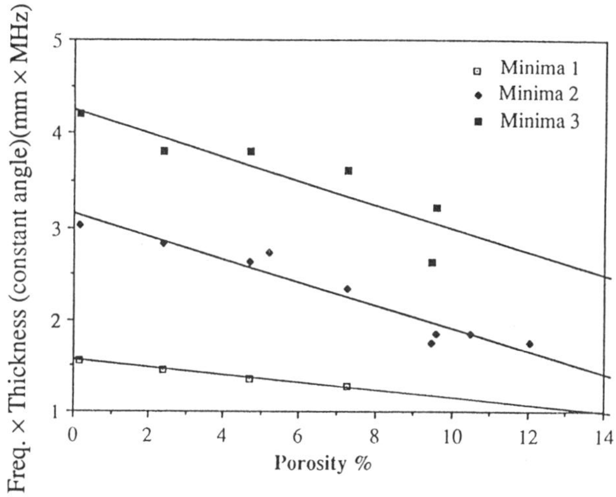

Figure 4. The Resonance shifting feature shows increased sensitivity.

to develop multi-mode experimental algorithms capable of dealing with experimental vagaries such as thickness changes. Also new features may be established using more than one mode which will serve either as complementary in some other cases even supplement any of the other primary features.

FEATURES FROM GUIDED WAVE ANALYSIS

The most common features which have been found to be useful in an NDE point of view can be listed as

- The average shifting of resonance for specific incident angles.
- The 'fd' value spacing between two modes m_1 and m_2.
- Surface wave velocity.
- Surface velocity difference between the two orthogonal direction.

CONCLUSIONS

The influence of anomalies such as spherical porosity are modeled as an influence of material degradation of the effective elastic constants.

A novel line source method has been adopted to generate surface waves in order to overcome the initial difficulties encountered due to the slow velocity of the surface waves in composites which make traditional angle beam surface wave shoes impractical for surface wave generation.

A new set of surface wave wedge probe set (Sender & Receiver) has been designed and fabricated.

New features have been defined and their sensitivity to material anomalies have been demonstrated.

REFERENCES

1. Martin, B.G., "Ultrasonic Wave Propagation in Fiber Reinforced Solids Containing Voids," J. Appl. Physics 48 (8), 1977.

2. Balasubramaniam, K., "Guided Plate Waves and Reflection Factor for NDE of Composite Materials," A PhD Thesis, Drexel University, Philadelphia, PA, March 1989.

3. Hashin, Z., "The Elastic Moduli of Heterogeneous Materials," J. Appl. Mech. 29, Transactions of the ASME 84, 143-150, 1962.

4. Rose, J.L., Huang Y., Tverdokhlebov, "Surface Waves for Anisotropic Material Characterization - A Computer Aided Evaluation System," Review of Progress in Quantitative Nondestructive Evaluation, Volume 9, pp. 1573-80, 1990. Edited by D.O. Thomson and D.E. Chimenti, Plenum Press, NY.

5. Balasubramaniam, J. and Rose, J.L., "Guided Plate Wave Potential for Damage Analysis of Composite Materials," Review of Progress in Quantitative Nondestructive Evaluation, Volume 9, pp. 1505-12, 1990. Edited by D.O. Thomson and D.E. Chimenti, Plenum Press, NY.

COMPARISON OF NON-DESTRUCTIVE TECHNIQUES FOR THE MEASUREMENT OF FIBER CONTENT IN REINFORCED ENGINEERING PLASTICS AND COMPOSITES

Peter L Warren

ICI Advanced Materials
Wilton Materials Research Centre
Middlesbrough, Cleveland, England

INTRODUCTION

The production of re-inforced engineering plastics relies on a number of non destructive (ND) instrumental techniques for monitoring product composition and quality assurance. Recent quality initiatives have meant that international standards such as ISO-9000 have to implemented. This is designed to improve process monitoring, and give customers the confidence in the manufacturing, and proof that the process is under control.

Non destructive methods of measurement are preferred as they are usually rapid and robust, they minimise sample preparation and are less operator dependent than, for instance chemical methods of analysis. They also offer the possibility of on-line measurement for rapid and automatic quality control monitoring. Ideally, assuming satisfactory and reliable performance from the on-line system, it has the potential for feed back and closed loop control. As an example, the techniques which have been used in ICI for the measurement of fiber content in a variety of materials include x-ray fluorescence, neutron and gamma-ray absorption, and ultrasonic methods. They have the common feature that they are comparative rather than absolute, hence the requirement for calibration standards.

X-RAY FLUORESCENCE SPECTROMETRY

XRF is a widely used analysis technique in ICI. Wavelength dispersive instruments (eg Siemens SRS-303) can detect and determine elements from sodium to uranium in the period table, in a concentration range from a few ppm upwards. This method has been used for the determination of additives, catalyst residues and contamination in a wide range of plastics [1]. It is also used in a qualitative mode for identifying compounds of unknown origin and investigating competitive materials.

For engineering compounds XRF can be used for measuring fillers, flame retardants, pigments and fiber-glass. The latter determination depends on monitoring one of the constituents of glass. Table 1 shows the typical composition of borosilicate glass.

Table 1. Fibre-Glass - Composition and XRF Data

Component	Content	Analyte Line	Critical Depth
SiO_2	55 %	Si Ka	33 μ
CaO	23	Ca Ka	420
Al_2O_3	14	Al Ka	30
B_2O_3	7	-	-
MgO	4	Mg Ka	25
K_2O	0.8	K Ka	310
Fe_2O_3	0.5	Fe Ka	2200
ZrO	0.5	Zr Ka	> 10000
Na_2O	0.2	Na Ka	20
SrO	0.2	Sr Ka	> 10000

Usually major elements such as silicon or calcium are chosen as the tracer element as these give high count-rates. A minor constituent eg strontium can be used but the concentration of this component in glass can vary with the source of raw materials. Another important factor in selecting the best element is the penetrating power of the fluorescent radiation. Most of the low atomic number elements have a small "critical depth" and this limits the chances of looking at a fully representative sample. Measurements can be made directly on polymer pellets but the sample must then be held in a cell with a thin film window. This window and any variation in granule size can affect the validity of calibration. The alternative form of presentation, compression moulded discs, need to be skimmed prior to analysis to remove any surface contamination or additive migration. However, they give a more reproducible surface for examination and more precise results.

NEUTRON ABSORPTION

For the determination of fiberglass in plastic compounds an alternative NDT technique to XRF has been developed. Measurements can be made directly on polymer granules (pellets approx. 3x2x2 mm) or material in the sheet form. While XRF measures the Si (or Ca,Sr) as a marker for glass, neutron absorption depends on the proportion of boron present from the borosilicate glass, to monitor this component in compounds and composites. The B^{10} isotope has a very high absorption coefficient - only the isotopes of Li and Cd approach this and these elements are not normally encountered in most compound formulations. The source of neutrons is provided by a Am^{241} / Be radio-isotope, and a sealed gas counter acts as detector. With equipment such as the Compuglass M200 (by Radiation Monitoring Devices of Boston USA) the sample cell consists of a acrylic box into which the granules are packed. The neutron beam passes though the sample (reproducible thickness about 2 cm) irradiating an area approximately 10 x 8 cm.

A log plot of change in count-rate with glass content is linear over a range of about 10 % glass. Minor differences in calibration line are noticed for change in polymer matrix, and granule size. However these are much less than matrix and surface effects that interfere with XRF analysis. NA is limited by the requirement for consistent sample packing and the need for a larger sample, about 250 g.

GAMMA RAY ABSORPTION (GRA)

As an alternative to NA, the absorption of high energy gamma rays can give quantitative information on compound composition for fillers and flame retardants as well as glass. While GRA is less specific than NA it too is compact, mobile and considerable less expensive than XRF. The equipment (eg Compuglass M300) has a very similar geometry and cell design to the neutron machine. It consists of a Cd^{109} source which decays to produce two intense lines at approx 24 and 88 KeV. Detection of the attenuated beam is by a scintillation counter. We have used the low energy band to determine glass content in a nylon matrix. A good linear calibration was achieved for sheet samples of a constant thickness over a concentration range 20 - 40 %. The high energy band is preferentially absorbed by elements of a high atomic number. Hence it is possible to monitor two components in a polymer matrix, three if information from a NA glass analyser is included. The radio-isotope has a half life of about 2.3 years. The equipment software incorporates a correction to take into account this gradual fall off in radiation flux.

NEUTRON SCATTERING

The determination of carbon fiber in a thermoset or thermoplastic material has until recently been limited to chemical separation using sulphuric acid digestion. Neutron back-scattering is one technique that is being developed and promises a faster and safer instrumental alternative to the traditional method. The technique requires weak thermal neutrons from a source, deep seated in a polyethylene moderator. In principle, this method differentiates the scattering characteristics of hydrogen atoms from those of higher atomic number (eg C,O). Thus resins such as polyester or epoxy with a significant hydrogen content can be determined quantitatively in a composite with carbon. The sample in sheet form is presented in a fixed geometry relative to the source and gas filled detector. Because beam penetration is relatively shallow it is possible to examine thin materials eg composite pre-pregs by this method. Like previously described techniques, measurements are relative rather than absolute. However, a given resin/fibre system produces a linear calibration for counts plotted against per cent resin if known materials. With a ten minute counting time a precision of about 1 % is achieved.

ULTRASONIC MEASUREMENTS

For the determination of re-inforcing fibres in polymer composites, ultrasonic measurements have been shown to be effective. Equipment (eg URA-2000 by T.E.S.T. OF San Diego) has proved to be simple to use as well as rapid and reliable. A range of fibres such as glass, carbon or polymeric (eg Kevlar) have been determined in thermosets and thermoplastics. Thus, for instance, we have determined the carbon fiber content in a PEEK matrix in the range 60-75%. This is achieved by measuring the rate of propagation of ultrasonic pulses through the sheet of material.

The time of flight must be monitored to an accuracy of 1 nano second. A simultaneous measurement of sheet thickness to the nearest micron is made. The slowness of the pulses is related to the Young's Modulus of the material, which in turn is dependent on the fiber content.

The technique is ideal for tacky pre-preg sheets where the transducers apply sufficient pressure to consolidate the sheet. Solid laminates require a coupling agent (eg water). Again the method is relative and requires standardisation against materials of known fibre/resin ratio. Measurements are valid, even with up to 3% voids, but cracks and delamination cause serious interferences.

References

[1] Warren P L, Farges O, Horton M, Humber J, J.A.A.S. 1987, 2, 245.

ULTRASONIC SCANNING SYSTEM FOR INSPECTION ON INTERNAL DEFECTS IN CFRP USING DIGITAL PEAK DETECTORS

Y. Tanaka, E. Izumi, S. Aoki and T. Shima

Hitachi Construction Machinery Co., Ltd.
650 Kandatsu-machi, Tsuchiura, Ibaraki
300 Japan

ABSTRACT

The ultrasonic pulse echo method has proved itself to be useful for nondestructive inspection (NDI). It has found practical use in the detection of internal flaws e.g. delaminations and voids in carbon fiber reinforced plastics (CFRP). Video images such as B- and C-scan based on U.S. (ultrasonic) data are useful to represent internal defects. These U.S. data provide amplitudes of echo signals and the time of flight of each echo peak. We have developed a digital peak detector which obtains this information accurately in frequencies as high as 10 kHz.

The experimental system using three DPDs easily located Teflon sheets placed in layers from the second (0.4 mm depth) to the twenty eighth layer.

INTRODUCTION

Composite materials such as carbon fiber reinforced plastics (CFRP) have become more prominent in aerospace structures. Internal defects such as delaminations and voids decrease their strength, so it is necessary to detect and represent their shape and location. The ultrasonic pulse echo method (Figure 1) has proved to be an extremely useful nondestructive inspection (NDI) technique to detect such flaws. Video images such as B- and C-scan are based on measurements of ultrasonic (U.S.) data principally the time of flight (time data) between the surface echo and reflected echo (which is equivalent to depth or thickness) and the amplitude of the reflected echo.

In Figure 1, T is the transmitted pulse signal and S and F rather than F' are the reflected echo signals from the surface and internal flaws. And if there are no flaws, the reflected echo from the bottom will be seen. G_1, G_2 and G_3 are the gates to detect the echo signals from the surface, flaws and bottom. G_2 and G_3 are triggered just as G_1 detects the surface echo signal, so they are ready to follow the surface echo and detect any flaws at the same depth. In this pulse echo method, it is important to detect correctly the time data t_{g2} and the amplitude of the highest peak echo in the gate.

Of the two procedures available to detect these U.S. data, analog is limited because when the peak holder detects the amplitude and the digital counter detects the time data the value t_{g2}' between points P and Q differs from the real value t_{g2}, even though the operation is at high frequency. If there are several echoes in G_2-gate, for example, F' is lower than F, the U.S. values become quite different and consequently, deep flaws will be more difficult to detect.

Nondestructive Characterization of Materials IV
Edited by C.O. Ruud *et al.*, Plenum Press, New York, 1991

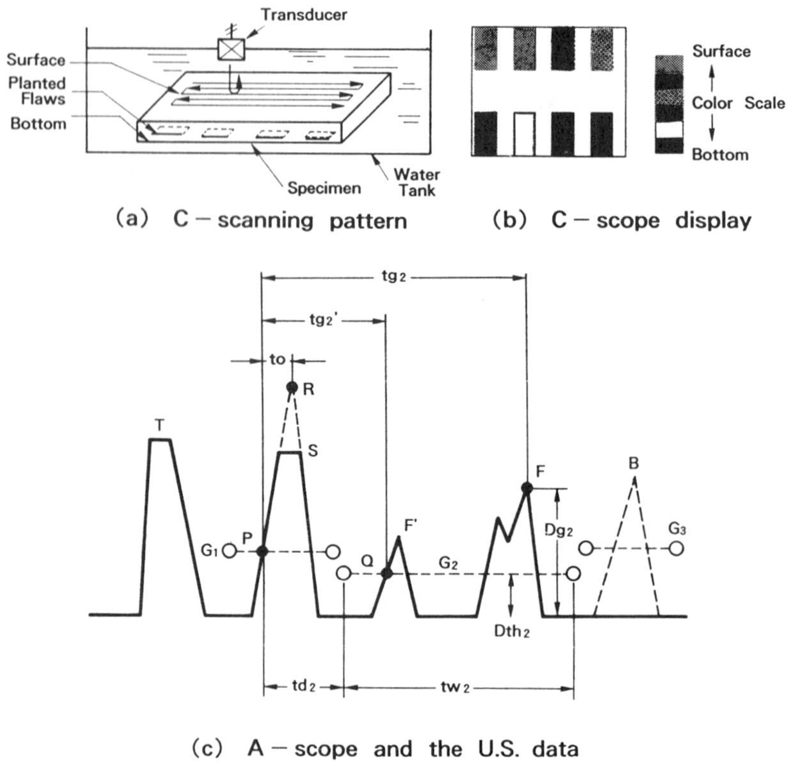

Figure 1. C-scan and ultrasonic pulse echo method

If a digital process is used, the echo wave signal is digitalized by a high speed ADC (analog to digital converter) and stored in a wave memory, the U.S. data will be picked up from the ADC data previously obtained. Accurate data result, but the procedure requires so much time that it can not operate practically over 1 kHz.

Neither procedure mentioned above can accommodate on-line inspection. To apply the pulse echo method to on-line inspection, it is necessary to detect U.S. data of frequencies as high as 10 kHz. We, therefore, developed a digital peak detector (DPD) which would detect U.S. data directly from ADC echo signal data. We succeeded in developing a prototype digital gate board with an ADC and three DPDs. World wide standard VME bus interfaces were easily controllable with a broad selection of VME-type computers etc. The result is an experimental system for ultrasonic inspection, which utilizes B- and C-scan software.

SYSTEM DESCRIPTION

General Performances

Figure 2 shows the experimental setup of our ultrasonic scanning system, and Figure 3 is a block diagram of the same. A scanning unit, ultrasonic equipment and data processing unit are all included. The procedure is displayed on the color CRT and with simple operation of the keyboard the results of C-scope are clearly seen after scanning.

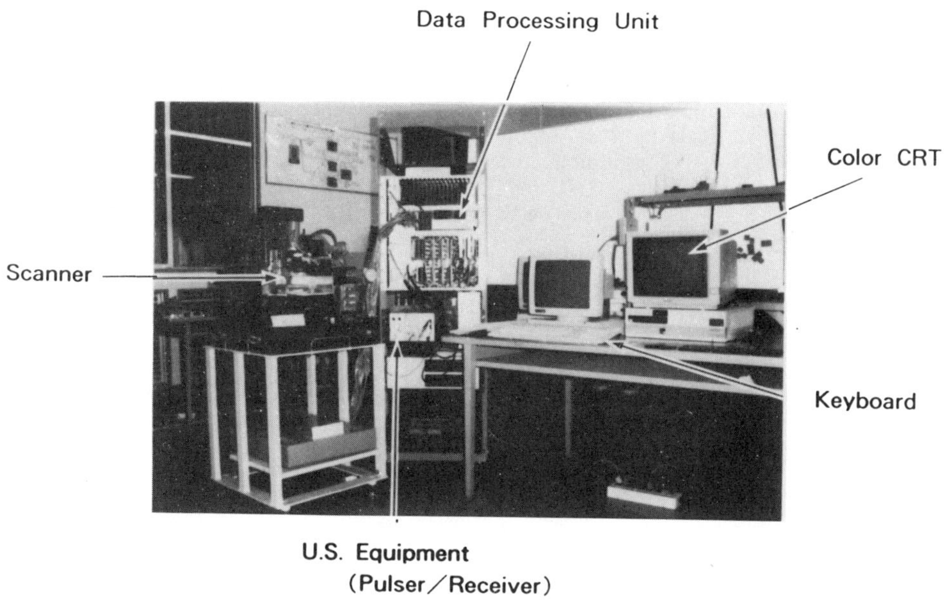

Figure 2. The experimental scanning inspection system

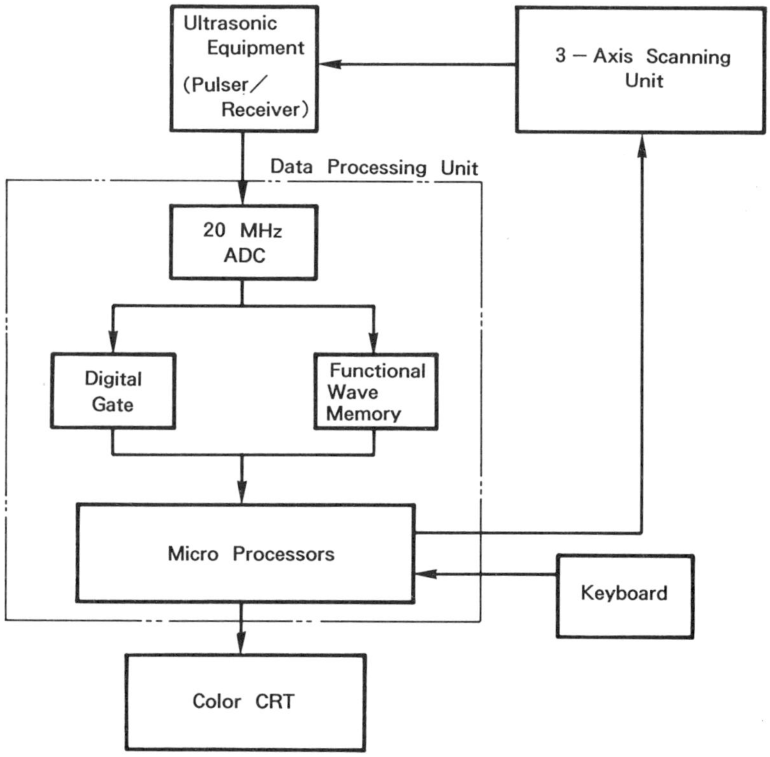

Figure 3. System block diagram

Scanning System

The scanning system consists of a three dimensional scanner and servo driver controlled by a sub-CPU board in the data processing unit. There are two operational modes, Z-scanning (vertical) and X, Y-scanning. Z-scanning is for manual focusing of the ultrasonic beam, and X, Y is for automatic scanning. The maximum scanning speed is 400 mm/sec and positioning accuracy is 0.01 mm.

Ultrasonic Equipment

A Hitachi pulser unit AT050P with pulse width 20 ns (nanosecond) and maximum amplitude 300 V is used. The basic receiver used was a Hitachi AT025R operating from 1 to 35 MHz and 60 kB amplitude with an RF signal output. We added signal conditioning circuits for detecting and filtering the analog echo signal before converting to digital data by ADC. As the most suitable probe frequency is inspect CFRP is 5 MHz and our DPD operates with a 20 MHz clock, echo signals have to be smoothed out. The receiver operates in from 1 to 3 MHz frequency and has a 60 dB amplitude.

Data Processing Unit

The data processing unit includes several boards on VME bus, so the system is easy to build because there are many VME-type computers, graphic controllers, digital I/O and other interface boards. The data processing unit itself is as follows.

1. The microcomputer consists of two CPU boards. The main-CPU board is for controlling the whole system and the processing of the U.S. data. The sub-CPU board is for the scanning units.

2. A RAM board. The U.S. data are stored on a 2 Mbyte memory.

3. A 640 Kbyte floppy disk stores the operation procedures and results.

4. The color CRT has 640 x 400 pixels on the video and displays operation procedures and results.

5. Functional wave memory for our experimental system picks up echo data for the A-scope from 20 MHz ADC data without losing peaks.

6. The digital gate board.

7. GPIB interface boards are for communication between the main- and sub- CPU boards.

Digital Gate Board and Digital Peak Detector

Figures 4, 5 and 6 are the block diagram, a photograph of the digital gate board and a block diagram of the DPD. An ADC, three DPDs and the VME bus interfaces are included in the digital gate board, so the main CPU can set the gates' parameters and produce a U.S. data read out. An external trigger mode is acceptable, so each gate starts when the trigger signal enters. Gate parameters are the delay time t_d, gate width t_w and threshold level D_{th} and the three gates' timing can be set independently or subsequently. In our system, we set the first gate to detect the surface, the other two switched on when the echo had exceeded the threshold level of the first (see Figure 1). In other words, G_1 was triggered by a series of position pulses from the scanning unit and G_2 and G_3 by the event signal E_{v1} of G_1. The minimum parameter setting is 50 ns delay time and gate width, 0.4% in the threshold level. The time of flights in the gates can be detected with 50 ns accuracy and amplitudes with 0.4% accuracy. At first, we made the digital peak detector (DPD) using TTL logic circuits. We then developed it as an LSI (Gate Array).

The DPD operation in accordance with sampling procedures is as follows:

1. The main-CPU sets the timing data (t_d and t_w) to a timing memory and the threshold level data D_{th} to a data latch through the VME bus.

2. Once the DPD is triggered, the time counter starts and a series of echo signal data is received from ADC with 20 MHz clock timing.

3. The comparator compares the incoming data with the previously obtained data. If the incoming data is larger, the time counts and incoming data are latched on to. The DPD outputs an event signal (E_v) as the first incoming data exceeds the latched data (D_{th}).

4. At the end of the gate ($t_d + t_w$), the DPD sends an Interrupt request signal S_{IRQ} to the main-CPU, after the time (T_g) and the amplitude data (D_g) have been latched.

5. These data are double buffered, so they are available on the main-CPU.

It is easy to sample the time and amplitude data using the DPD, because the digital gate board can be handled like any other external board, such as a RAM board or I/O boards etc.

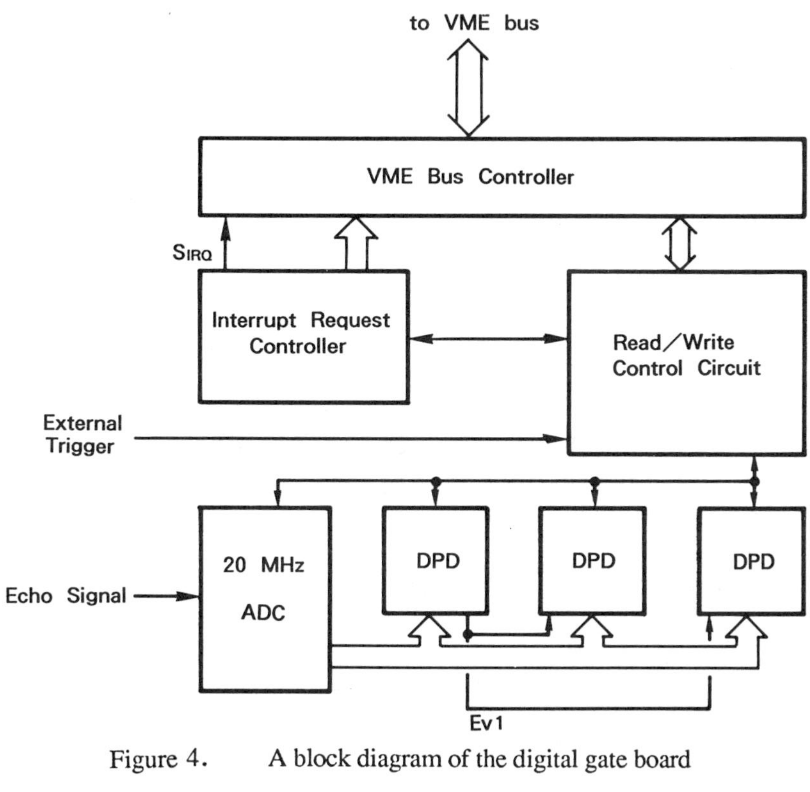

Figure 4. A block diagram of the digital gate board

Figure 5. A photograph of the digital gate board

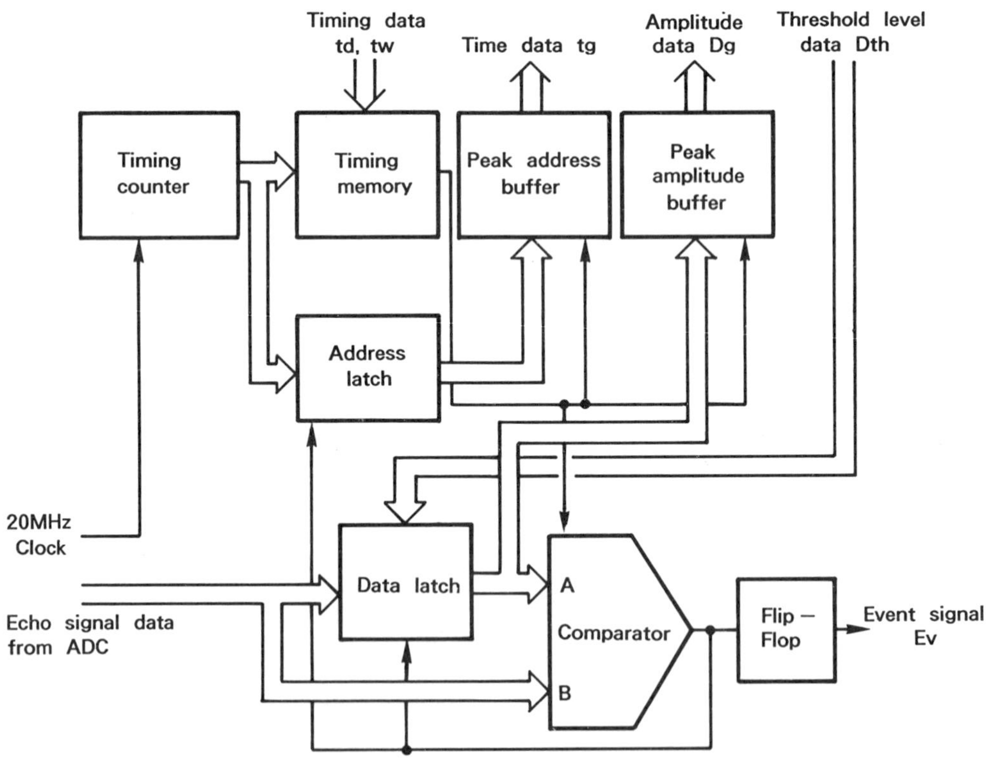

Figure 6. A block diagram of the digital peak detector

EXPERIMENTAL PROCEDURE

Data Processing of C-Scan (Software)

The C-scanning program includes data sampling procedures for the U.S. data and displaying the C-scope. The most important program in our system is the C-scanning software as this includes both the U.S. data sampling and the C-scope displaying, as follows.

1. The three gates for detecting the echoes of the surface, internal flaws and bottom are set using the keyboard while keeping an eye on the A-scope and gates' display on the color CRT. The gates' parameters of the G_2-gate are, for example, the delay time t_{d2}, the gate width t_{w2} and the threshold level D_{th2}. Signatures of the other gates are respectively names Gates 1, 2 and 3.

2. The scanning parameters are set with the scanning zone $X \times Y$ being 120×60 mm and the maximum scanning speed 400 mm/s. The scanning unit outputs a trigger pulse in each 0.3 mm position.

3. As a result of this trigger pulse, the pulser drives the probe and starts the initial gate operation.

4. The echo signal exceeds the threshold level; D_{th1} and the G_2- and G_3- gates follow.

5. The U.S. data such as the time and amplitude data are then latched.

6. The main-CPU reads out the U.S. data and stores them in the memory in each trigger pulse timing.

7. In the C-scope procedure, only the U.S. data of the G_2- and G_3-gates are used. The surface echo in the G_1-gate is sometimes saturated, so the time data between the triggered point (P in Fgure 1) and "virtual" surface point R was experimentally defined as the delay time t_o before C-scan testing. The wave velocity (V_s) in the specimen are detected as 3000 m/s. The depth of internal flaws is calculated as $(t_{g1} - t_o)*Vs/2$. The internal flaws are then displayed in different colors for different depths. If there is an echo in the G_3-gate no flaws and the area of the specimen will be displayed ultrasonically sound area as white indicating a flawless part (shown in Figure 7).

8. After the C-scope is displayed, the B-scope in any place within the specimen will be displayed when the cursor in the C-scope is moved.

RESULTS

At first we tested the processing rate of the digital peak detector and the digital gate board, and found they operated at 10 kHz frequency. In our experimental system, they operated at 4 kHz because of the limitation of the scanner speed.

Specimens

The specimens used were CRFP plates with 30 layers (0.2 mm ply). We tested two types.

Specimen-A included 8 rectangular Teflon sheets (0.05 mm thickness, 10 mm width, 15 mm length) at depth 0.4, 0.6, 0.8, 1.0, 4.2, 4.4, 4.6 and 4.8 mm. Specimen-B included 12 circular Teflon sheets (0.05 mm thickness and 2, 3, 4 and 5 mm diameter) at 0.8, 3.0 and 5.2 mm depth.

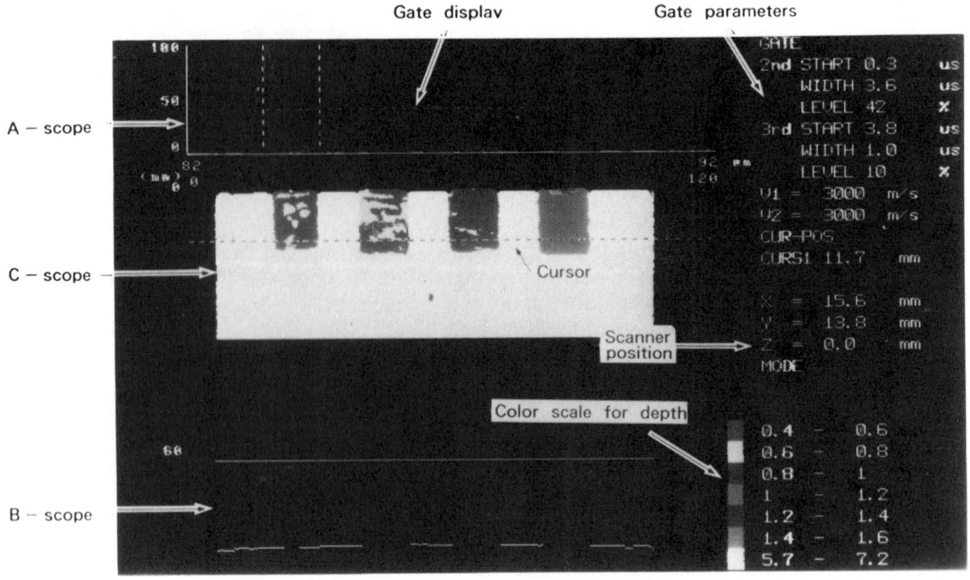

Figure 7. A graphic displays of C-scan

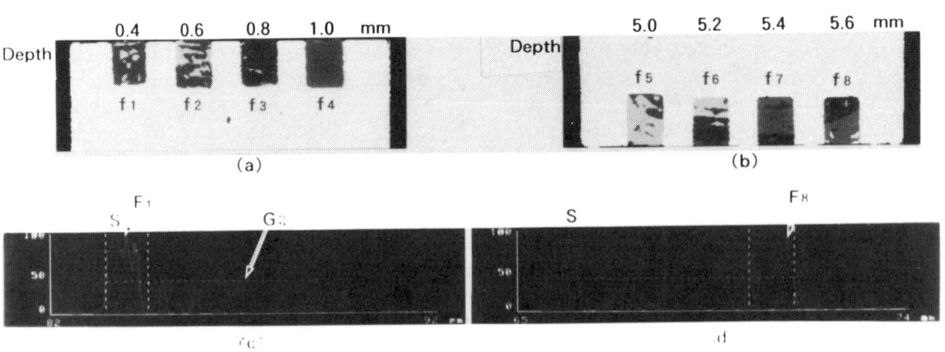

Figure 8. C-scope of specimen-A

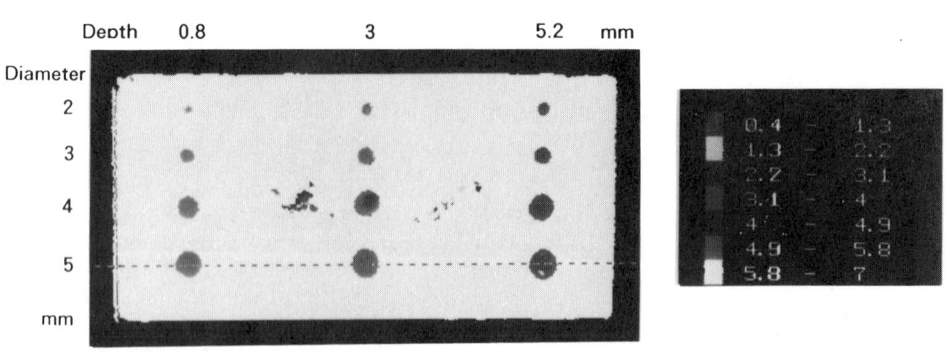

Figure 9. C-scope of specimen-B

Results

Figure 7 shows a graphic display of C-scan. The scanning parameters and A-, B-, C-scans both appear on the same color CRT. The A-scope is useful for setting the gates and checking the results of B-, C-scans. On the right, gates' parameters, positions of the scanner, color scales with depth etc. are displayed. After scanning, the C-scope (at the center) comes on the screen. The display shown is of specimen-A with the Teflon sheets under the surface at 0.4, 0.6, 0.8 and 1.0 mm depth clearly distinguishable. Moving the cursor in the C-scan shows B-scan display underneath.

Figure 8 shows a typical C-scope result for specimen-A. The benefit of using the DPD is apparent. (a) and (b) are sampled in only one time scanning and displayed after being changed to the color scale. It is not necessary to change the gate's parameters in spite of the reduction of ultrasonic waves. (c) and (d) are the A-scopes of planted Teflon sheets at 0.4 mm (f_1) and 5.6 mm (f_8) depths respectively. (c) is the A-scan of the sheet f_1 and G_2-gate. The analog gate method is not sufficient to detect the peak echo signal F_1 because echo signals have already exceeded the threshold level at the start point of the G_2-gate. The benefit of using the DPD which picks the highest peak echo out of several in the gate is very clear. Figure 9 is a C-scope of specimen-B. The circular Teflon sheets 2, 3, 4 and 5 mm diameter in the depths 0.8, 3 and 5.2 mm are all detected clearly. Some natural flaws can also be seen.

CONCLUSIONS

A computer controlled digital gate board using digital peak detectors (LSI) has been developed. The time of flight and amplitude data are accessible at high frequency as 10 kHz.

An experimental ultrasonic inspection system using digital peak detectors successfully detects planted Teflon sheets ply-by-ply from 0.4 to 5.6 mm depth and detects minimum diameter 2 mm.

ULTRASONIC CHARACTERIZATION OF COMPOSITE MICROSTRUCTURE

Ronald A. Kline
Robert Adams

School of AME
University of Oklahoma
865 Asp Avenue
Norman, Oklahoma 73019

INTRODUCTION

In this work, nondestructive methods for the quantitative evaluation of composite microstructure are described. Parameters of interest in this regard might include fiber volume fraction, porosity, degree of cure (thermoset resins) and crystallinity (thermoplastic resins). These methods rely upon the high degree of accuracy which can be obtained with ultrasonic measurements of acoustic velocities; hence the mechanical properties of the material, and well established composite micromechanics relationships between elastic properties and composite microstructure. This results in sets of nonlinear equations which can be solved for the unknown microstructural parameters.

BACKGROUND

Ultrasonic velocity measurements have been used extensively for the quantitative characterization of polymer properties. Work in this field originated with Dietz et. al.[1] who employed both velocity and attenuation measurements to study cure phenomena. Subsequent investigators have expanded on this work to incorporate critical angle measurements to determine the state of cure in completed parts[2], shear and longitudinal velocity measurements to characterize elastic moduli[3], and wide band measurements of velocity and attenuation spectra for viscoelastic property measurement throughout the cure cycle[4,5]. While most of the work to date in this area has been directed towards characterizing thermoset resins, similar considerations also apply to thermoplastics where the formation of crystallites is a primary strengthening mechanism. Results indicate that there is a linear relationship between wave speed and crystallinity for a wide range of thermoplastic materials.

Nondestructive Characterization of Materials IV
Edited by C.O. Ruud *et al.*, Plenum Press, New York, 1991

While results of this nature are useful and have been employed in commercially viable equipment, it should be pointed out that these results are principally phenomenological with correlations drawn between the measured acoustic properties and the specimen microstructure. Hence, only qualitative information is supplied about the nature of the sample. For quantitative microstructural characterization, more detailed knowledge of the mathematical relationships between the properties of the constituent materials and the resulting composite structure are required. This is domain of composite micromechanics where the basic principles of elasticity theory are applied to determine the elastic moduli of composites from the properties of the constituent media; their sizes ,shapes and distribution within the sample as well as their individual mechanical properties. This information, when coupled with direct ultrasonic measurement of elastic moduli, can be employed to precisely determine critical microstructural parameters.

A variety of techniques have been employed to mathematically relate elastic properties to specimen composition. In general,in these models the constituents are assumed to behave in an ideal fashion i.e. linearly elastic fibers and matrices, completely aligned fibers and perfectly bonded composites. In order to make the problem tractable, most of these models employ a representative volume element, or the smallest area over which the stresses are macroscopically uniform. For the laminated composites typically used in the aerospace industry, this consists of a single, cylindrical fiber surrounded by an annular matrix. The simplest approach to the problem of determining the apparent moduli in a heterogeneous solid is the mechanics of materials approach where the strains or stresses are assumed to be the same in the fiber and matrix. This leads to obvious inconsistencies at the boundary between the two components but is nonetheless helpful for estimating bounds on the apparent moduli[6]. The most important contributions in this area are attributable to Hashin and co-workers[7,8] who used the "composite cylinder assemblage" model for their representative volume element and assumed an axially symmetric displacement field. Energy methods were then applied to yield either absolute or bounded values for the pertinent moduli. Of particular significance in this model is the ability to properly account for anisotropic fiber behavior as several commercially important fibers, notably graphite and Kevlar, are anisotropic. Hashin, among others, has also employed a similar approach to study the behavior of composites reinforced with geometries other than cylindrical fibers[9]. In this study, Hashin's model of a composite reinforced with spherical inclusions (with appropriate properties) is used to represent porosity.

Reynolds and Wilkinson[10] were the first investigators to couple micromechanics models with NDE measurements. In their work, it was recognized that the presence of long, cylindrical reinforcing fibers will have a directionally dependent effect on elastic properties but the contribution from spherical pores (providing they are randomly distributed) will be directionally independent. Hence, they will have a different effect on longitudinal and shear (polarized) waves. This means each combination of porosity and fiber volume fraction is associated with a unique longitudinal velocity/shear velocity pair as illustrated in Figure 1.

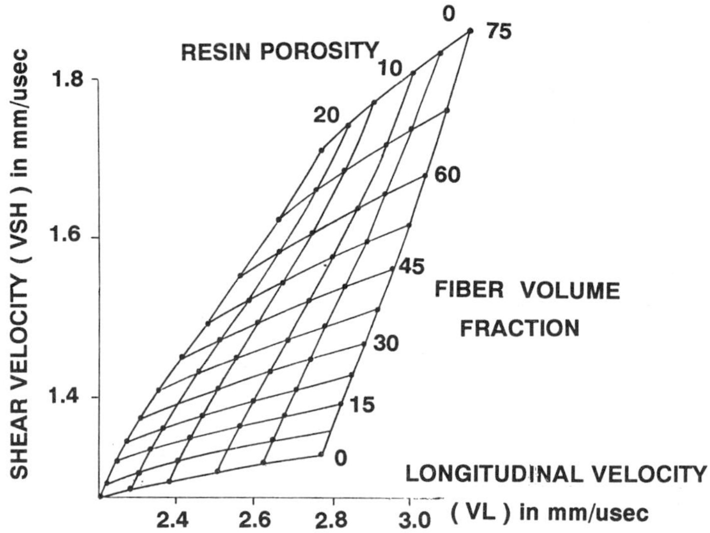

Figure 1. Variation of Shear and Longitudinal Velocities with Fiber Volume Fraction and Porosity (Unidirectional)

Reynolds and Wilkinson then measured the longitudinal velocity and a shear velocity at a given point and used graphical means of establishing the fiber volume fraction and porosity present. It should be pointed out that this approach requires precise knowledge of the properties (density, elastic moduli) of the microconstituents. Also, the Reynolds and Wilkinson approach suffers from several critical drawbacks in that it uses local, contact measurements (hence is unsuitable for scanning), employs a graphical (hence slow and inaccurate) solution technique and can only be applied to unidirectionally reinforced composites (not practical laminates) as well as the fact that no experimental data are presented to corroborate their findings. Nonetheless, their work is conceptually sound and represents an important pioneering effort in a critical discipline. Recent work in the area has attempted to address some of these shortcomings. Kline[11,12] has developed automated procedures for calculating fiber volume fraction and porosity from the measured ultrasonic data. Gruber et. al.[13] used this algorithm in a scan mode (mode conversion was used to generate the required shear waves) to map out property variations in unidirectionally reinforced panels.

THEORY

In this work we seek to extend the results of the earlier studies to the nondestructive, microstructure characterization of practical composite laminates. In particular, we wish to incorporate the effects of ply orientation and stacking sequence in our analysis.

The earlier efforts in this area are based on common micromechanics assumptions i.e. linear elasticity, perfect bonding between fiber and matrix, perfect fiber alignment, isotropic matrix, isotropic or transversely isotropic fibers, known properties of constituent materials, etc. In order to extend these results, we must employ two additional simplifying assumptions :

1) The plies are perfectly aligned with a known stacking sequence.

2) The composition of each ply in the laminate is identical.

With these assumptions the problem is tractable using similar techniques to those previously developed for unidirectionally reinforced media. Since we wish to use this method in a scan mode, mode conversion must be employed although now we must take into account the fact that pure modes will no longer be generated and energy flux deviation from the wave normal may be considerable.

The problem is posed as a coupled set of nonlinear equations. The transit time for a given wave mode will, in general, be a function of the elastic moduli of the medium, the density, the direction of propagation, the layup sequence of the composite and the path traversed in the medium say

$$TT_i = TT_i(C_{ij}, \rho, l, t)$$

Also, from composite micromechanics, the elastic moduli of the composite can be determined from the properties of the fiber and resin systems, the number of reinforcing fibers present and the porosity of the media

$$(C_{ij})_{composite} = C_{ij}[(C_{ij})_{fiber}, (C_{ij})_{resin}, V_{fiber}, V_{porosity}]$$

As we seek to determine the parameters which characterize specimen composition(here fiber volume fraction and porosity) from acoustic experiments, two independent equations are needed. Hence, two measurements of the transit time for two independent waves are required. In thin composite laminates, one choice is obvious- the through thickness longitudinal wave. The choice for the second wave, however, is less clear. Shear wave propagation with contact transducers is possible, but the need for a high viscosity coupling agent precludes scanning in most cases. Mode conversion is therefore used, but care must be exercised due to energy flux deviation from the wave normal and refraction phenomena at each ply boundary. Here we chose to use a quasilongitudnal wave generated by an obliquely incident longitudinal wave in water. This choice has the advantage of being readily isolated from other waves as it will be the first arrival in a through transmission experiment. The choice of inspection angle is based on two competing factors: localized inspection area (small angle) and the ability to resolve differences in wave speed due to the presence of reinforcing fibers (large angle). Figure 2 illustrates this problem for two oblique inspection angles in a cross-ply composite. Clearly, at 5° resolution capability for fiber volume fraction

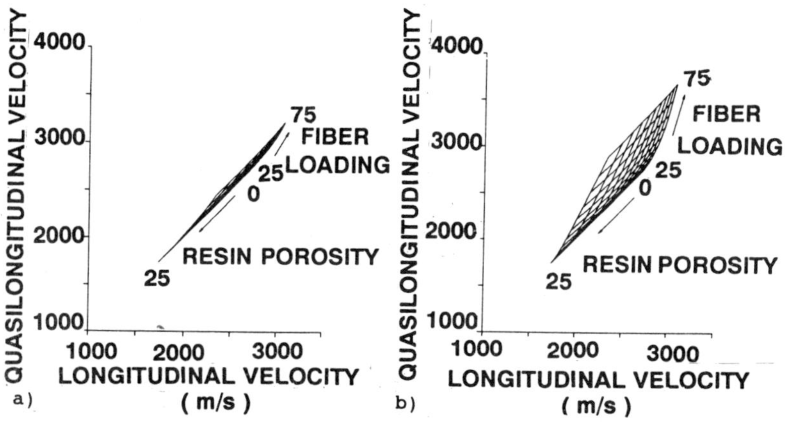

Fig. 2. Effect of Inspection Angle on Resolution
a) Normal Incidence (L) + 5° Oblique Incidence (QL)
b) Normal Incidence (L) + 7° Oblique Incidence (QL)

longitudinal wave in water. This choice has the advantage of being readily isolated from other waves as it will be the first arrival in a through transmission experiment. The choice of inspection angle is based on two competing factors: localized inspection area (small angle) and the ability to resolve differences in wave speed due to the presence of reinforcing fibers (large angle). Figure 2 illustrates this problem for two oblique inspection angles in a cross-ply composite. Clearly, at 5° resolution capability for fiber volume fraction is limited in comparison to that for the 7° inspection. However, this means that a larger volume of material is studied with each measurement.

EXPERIMENTAL PROCEDURE

Testing was conducted on a 6.0 x 12.0 x 0.066 (in) six layer (0/90/0/90/0/90) cross-ply graphite-epoxy plate supplied by Beech Aircraft. Figure 3 is a schematic of the ultrasonic testing system, showing the connection of the ultrasonic transducers to the personal computer containing the digitizer and pulser receiver boards.

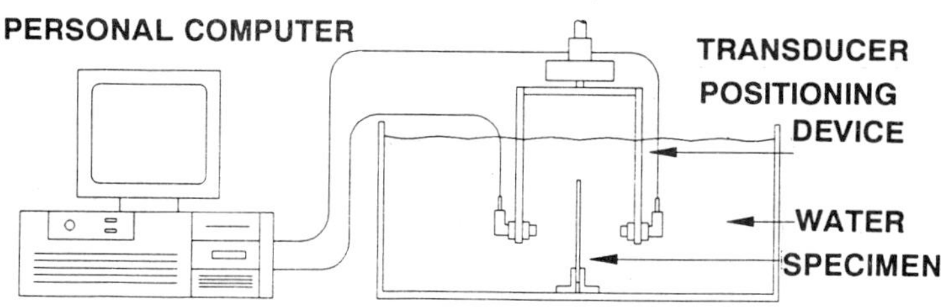

Fig 3. Schematic of ultrasonic test setup.

Two Panametrics 2.5 MHz ultrasonic transducers were utilized to generate and receive the necessary ultrasonic signals for analyzing the specimen. A special fixture was designed to control the angle of the transducers relative to the surface of the test specimen, which was mounted vertically in the scanning tank. The fixture included a RVDT to measure the relative angle of the transducers from the normal to the plate. The transducers were aligned with each test square and a longitudinal and quasilongitudinal wave speed was determined. The fiber and matrix property data from ref. 10 was used as input to the algorithm for determining fiber volume fraction and porosity at each location. After ultrasonic testing was completed, the plate was sectioned and each test square was subjected to acid digestion.

Results and Discussion

Ultrasonic velocity measurements for normal incidence (longitudinal) and oblique incidence (7° with respect to the surface normal, quasilongitudinal) were performed at several selected sites within each sample (see Figure 4). Subsequent to ultrasonic testing, the panels were sectioned along the grid lines shown (Fig. 4) and subjected to acid digestion for the determination of microstructural content. The results from a typical panel are presented in Figure 5, where fiber loading as determined both nondestructively (ultrasonic) and destructively (acid digestion) are plotted for each of the twelve sample sites in the composite panel.

Clearly, the agreement between the two approaches is quite good with most readings within +/- 2%, the inherent accuracy of the acid digestion technique. Possible improvements in the technique include, developing a transducer alignment fixture to more precisely align the transducer with the surface of the specimen and better information on the material properties of the composite constituents. In this regard, we are principally concerned with potential variations in the cure state of a sample. Alternative measures of the cure state on a local basis may be required for the successful implementation of this technique in practice. In addition, we are currently testing composite samples with systematic variations in porosity (varying amounts of hollow glass spheres) to verify this feature of the algorithm.

Conclusions

1. An NDT method for quantitative microstructure analysis in laminates has been developed.

2. Results have been presented for a typical cross-ply laminate, with good agreement between the NDT and acid digestion techniques.

3. Future work includes development of a technique for local measurement of cure state and verification of local porosity determination.

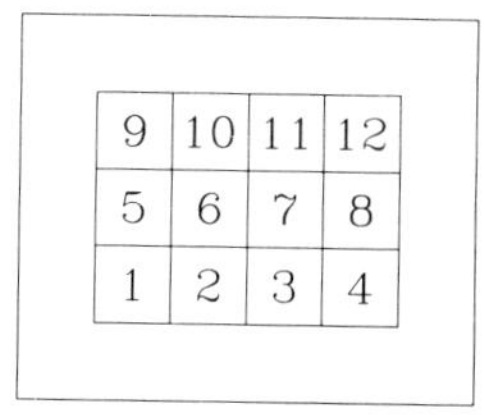

Fig. 4. Specimen Sample Location

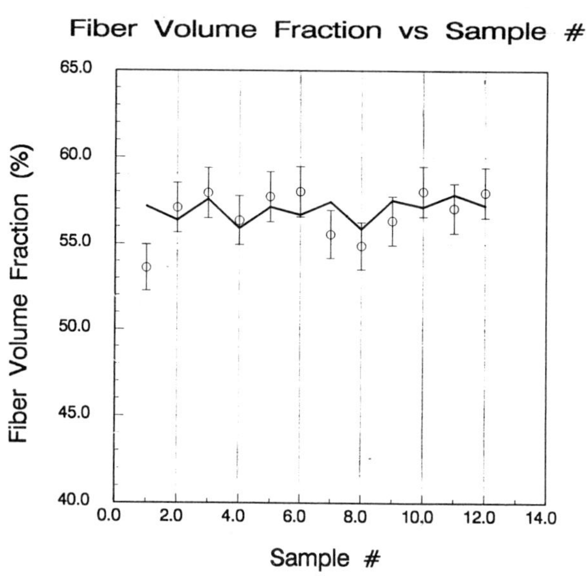

Fig. 5. Test Results.

References

1. A. G. Dietz, E. A. Hauser, F. J. McGarry and G. A. Sofer, Ultrasonic Waves as a Measure of Cure, Ind. Chem. Eng., 48:75 (1956).
2. J. H. Speake, R. G. C. Arridge and G. J. Curtis, Measurement of the Cure of Resins by an Ultrasonic Technique, J. Phys. D. Appl. Phys., 7:412, (1974).
3. A. M. Lindrose, Ultrasonic Waves and Moduli Changes in a Curing Epoxy Resin, Exp. Mech., 18:227 (1978).
4. S. I. Rokhlin, An Ultrasonic Bridge for the Study of the Properties of Thin Interface Films, J. Acoust. Soc. Am., 73:1619 (1983).
5. C. P. Hsiao and R. A. Kline, The Measurement of Viscoelastic Moduli Using an Ultrasonic Spectroscopy Technique, Proceedings IEEE Ult. Symp., 443 (1984).
6. R. M. Jones, "Mechanics of Composite Materials," Hemisphere, New York (1975).
7. Z. Hashin and B. W. Rosen, The Elastic Moduli of Fiber Reinforced Materials, J. Appl. Mech., 29:22 (1962).
8. Z. Hashin, Analysis of the Properties of Fiber Composites with Anisotropic Constituents, J. Appl. Mech., 46:543 (1979).
9. Z. Hashin, The Elastic Moduli of Heterogeneous Materials, J. Appl. Mech., 29:143 (1962).
10. W. N. Reynolds and S. J. Wilkinson, The Analysis of Fiber Reinforced Porous Composite Materials by the Measurement of Ultrasonic Wave Velocities, Ultrasonics, 16:159 (1978).
11. R. Kline, Ultrasonic Measurement of Fiber Volume Fraction and Porosity, Proceedings ENTEC (1987).
12. R. Kline, Quantitative NDE of Advanced Composites, J. Eng. Mat. Tech., 112:218 (1990).
13. J. J. Gruber, J. M. Smith and R. H. Brockelman, Ultrasonic Velocity C-Scans for Ceramic and Composite Material Characterization, Mat. Eval., 46:90 (1988).

DIRECT THERMOMECHANICAL STRESS AND FAILURE MODE ANALYSES OF A CLOTH REINFORCED CERAMIC MATRIX COMPOSITE

Joseph W. Krynicki, Dennis C. Nagle* and Robert E. Green, Jr.

Center for Nondestructive Evaluation, The Johns Hopkins University
Baltimore, Maryland 21218
*Martin Marietta Laboratories, Baltimore, Maryland 21227

INTRODUCTION

Virtually all composite systems have inherent thermomechanical stresses which arise from thermal expansion coefficient differentials between composite constituents. In the designing of any fiber reinforced composite, one needs to understand how these stresses arise in the processing of a composite and their effect on material performance in service. Since inherently brittle ceramics are often toughened through reinforcing fibers, the mechanical behavior of a ceramic matrix composite (CMC) is heavily influenced by the strength of its fiber/matrix interface, which is in turn directly related to the residual thermomechanical stresses present between the matrix and reinforcement. This study describes the use of photoelasticity to directly determine thermomechanical stresses in this engineering CMC. The use of acoustic emission (AE) and optical microscopy to characterize flexural test data in establishing fundamental failure mechanisms will also be discussed.

Thermomechanical Stresses

Most research work in the field of ceramic matrix composites indicates that in order to maximize mechanical properties for a CMC, prime attention must be paid to the different coefficients of thermal expansion of the constituents. Phillips et al. [1] studied this optimum state by altering matrices to change their thermal expansion coefficients. They observed that changes in differential thermal expansion between composite constituents correlated with mechanical property changes.

Thermal properties are also extremely important to the behavior of a composite material in a variety of structural applications. Mechanical property degradation in composite materials by thermal loading has become a common occurrence in many structures ranging from microelectronic surface mount technology to Earth orbiting satellites. In all cases, thermal loading leads to thermal fatigue within the composite system by degrading the fiber/matrix interface. Thompkins [2-6] has demonstrated the effects of thermal fatigue on aerospace materials in a variety of environments.

In 1816 Brewster discovered that a "...normally transparent isotropic substance could be made optically anisotropic by the application of mechanical stress..." [7]. This discovery led to the development of stress birefringence as a technique to evaluate stress fields within a material. Photoelastic stress analysis is very often limited to models, as opposed to real structures or engineering materials. This is simply a result of the fact that many engineering materials and structures are often not optically transparent and/or they possess microscopic multiphase systems (making analysis difficult). However, models do not adequately mimic the effects incurred by using actual composite components and therefore the model lacks the

Nondestructive Characterization of Materials IV
Edited by C.O. Ruud *et al.*, Plenum Press, New York, 1991

chemical interfaces and mechanical keying which is unique to the engineering material. Nonetheless, a number of authors have attempted to model and explain thermomechanical stresses in composite materials through the use of photoelasticity [8-10].

CMC Failure

Along with the noncatastrophic failure which is often observed during CMC flexural tests, CMC's also typically yield a "saw-tooth" behavior which is attributed to random microcracking and to the presence of strain relief mechanisms such as fiber pull-out and fracture. While the relative magnitudes of these load-deflection curve anomalies vary, authors have almost invariably shown flexural test curves that contained either a subtle "saw-tooth" behavior or a "blip". Even with the aid of an optical microscope (400X magnification), cracks were often not observed even after "blips" occurred in the load-deflection curve [11]. In agreement with previous investigators, this author has observed that during three-point bend tests of a single layer CMC, "steps" found in the stress-deflection curve did not always provide a one-to-one correlation with dye penetrant or microscopic inspection for cracks [12]. Fig. 1 shows stress-deflection curve anomalies in a single layer CMC.

Acoustic emission (the release of a transient elastic wave through a strain relief mechanism) was determined to be the the most straightforward nondestructive tool to investigate failure modes in a composite possessing inherently brittle failure mechanisms. While several authors have correlated acoustic activity with crack propagation during mechanical loading of directionaly reinforced ceramics [13-15], little work has been done to characterize failure in cloth reinforced composites.

EXPERIMENTAL PROCEDURE

Composite Manufacture and Preparation

The composite manufactured for this study was composed of a Corning class 7740 borosilicate glass powder with a Nicalon silicon carbide woven cloth as the reinforcement. The Nicalon cloth was a two dimensional woven roving comprised of equal fiber fractions in mutually perpendicular directions. All composite samples were hot-pressed at approximately 1100 ^{o}C and resulted in fiber volume fractions of approximately 7%. The use of small volume fraction, single layer specimens in part of this study allowed a more direct interpretation of experimental results. As a control for the acoustic emission monitored bend tests, a nonreinforced glass specimen was also hot-pressed. After manufacture, average theoretical densities of 95.0% and 95.5% were determined for the one layer and nonreinforced specimens, respectively.

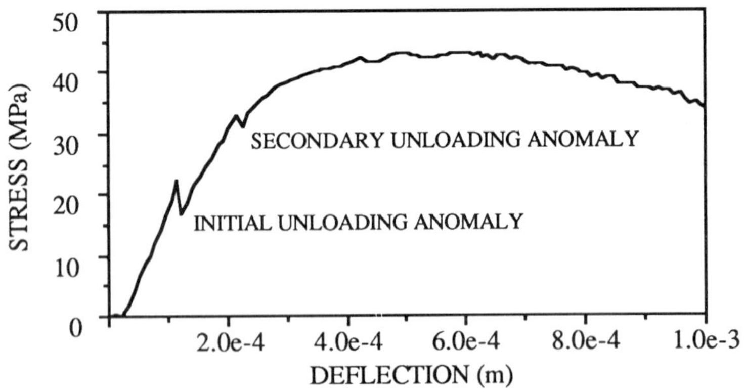

Fig. 1. Stress-deflection curve from a single layer CMC subjected to flexural loading.

Coupon specimens were cut from the hot-pressed CMC composite samples in a manner such that one of the mutually perpendicular predominant fiber directions was perpendicular to the photoelasticity viewing plane (or equivalently, the loading noses of the three-point test fixture for AE/flexural testing). Therefore, 50% of the fiber volume fraction was always perpendicular to the loading noses while the other 50% was parallel to the noses. The specimen used for photoelastic analysis was typically 4 mm wide (the width of a fiber tow) by 2.25 mm thick. Samples for flexural testing were typically 30 mm wide by 2.25 mm thick. In order to normalize surface defect effects, the specimens' faces were lightly abraded using a 500 grit, 30 micron, silicon carbide paper. Specimens used for optical inspection were polished for "through thickness" viewing using successively finer abrasives up through a 1 micron diamond paste.

A plane polariscope was used to determine principle stress directions from isoclinic fringes. The addition of quarter wave plates to the setup resulted in a circular polariscope which removed isoclinic fringes and only provided principle stress difference information.

The support span-to-depth ratio for the three-point bend tests was typically 14:1 for all specimens tested; loading noses possessed a 3.2 mm radius. An Instron screw type loading machine Type TT-CM was used to provide a constant deflection rate of 0.3 mm/min for all tests performed in this study. In order to elaborate upon load-deflection curve anomalies and corresponding acoustic emission results, additional bend tests were performed using a small test fixture designed for a light microscope (32X magnification) with a similar, specimen span-to-depth ratio.

AE monitoring was performed using a Physical Acoustics R15I Integral transducer with a detection band of 100 kHz to 450 kHz and a resonant frequency of 160 kHz. The transducer, possessing a self-contained 40 dB amplifier, was acoustically coupled through a silicone grease and then securely fastened to the sample by tape. The use of this relatively high frequency transducer allowed the filtering of extraneous background mechanical noise. The transducer output was then sent to a Physical Acoustics 3000/3104 AE monitoring system for signal amplification, digitization, storage and manipulation. Using a total gain of 60 dB (40 dB from the transducer and 20 dB from the main amplifier), a threshold of 0.2 volts was set to provide a detection level sufficiently above background noise. Signal dead time was set to 40 msec. Data was then stored at a rate which provided a one second resolution.

RESULTS AND DISCUSSION

Thermomechanical Stresses

The sketched stress trajectories compiled from the photoelastic stress analysis are shown in Fig. 2. The result of having a matrix with a slightly lower thermal expansion coefficient than that of the reinforcement (1.2×10^{-6} vs. 3.26×10^{-6}) clearly manifests itself through the existence of tensile stresses at the fiber/matrix interface. The stress trajectories illustrate how the fiber pulls away from the matrix as the composite is cooled below the glass's strain point (510 ^{o}C). (The strain point represents a temperature at which the viscosity of a glass is $10^{14.6}$ poises. At this temperature, the movements of atoms within a glass are so sluggish that residual stresses cannot be relieved.) A detailed quantitative analysis of the stresses present and the effects of these stresses on mechanical properties will be presented in a later paper.

CMC Failure

The acoustic emission monitored bend tests results shown in Fig. 3 suggested discrete failure modes which are associated with the separate highly pronounced strain relief mechanisms and AE activity changes. Since this CMC possesses an optically transparent matrix, it was possible to directly observe failure and thus elaborate upon the standard flexural test and acoustic emission data. Fig. 4, to be discussed later, shows the derivative of the maximum outer bend stress with respect to strain (instantaneous effective modulus) versus strain.

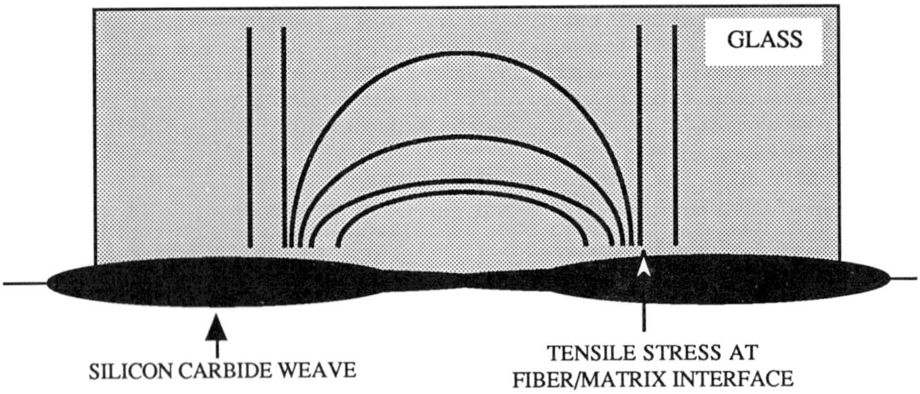

Fig. 2. Two dimensional stress trajectories sketch for thermomechanical stress build-up in a single layer ceramic matrix composite.

Utilizing the investigative tools used in this study, (flexural testing, acoustic emission and optical microscopy), the entire failure of the single layer composite can be summarized in the following three steps.

1. Crack initiation and propagation

A crack initiated and rapidly propagated from the tensile side up to the fiber/matrix interface. As one might expect, Fig. 3 revealed that the specimen released no significant acoustic emission prior to the first load-deflection curve peak since the composite was behaving in an elastic fashion. However, once the first tensile crack was initiated, a very sharp acoustic response was recorded ($\approx 10^3$ over background activity). A similar observation of acoustic emission activity was made for the loading of a nonreinforced glass specimen (Fig. 5).

2. Advanced fiber loading

Subsequent loading after the first failure yielded no further visible cracks, yet acoustic emission activity increased. It was believed that the additional loading was largely taken up by the tensile stressing of the fibers. This resulted in a complimentary compressive field being developed in the upper portion of the glass composite. The manifestations of fiber stressing due to fiber pull-out is demonstrated by the upward slope of acoustic activity between the first and second peaks of the load deflection-curve. More specifically, this acoustic activity is attributed to frictional noise.

3. Initiation and propagation of secondary crack

Upon further loading, a crack was observed to initiate at the fiber/matrix interface which was originally in axial compression. This second crack rarely originated directly opposite the tensile side crack. Thus, the fibers deflected the crack and impeded crack propagation. This second tensile crack that originated at the interface was expected since the neutral stress plane had moved up and placed the upper fiber/matrix interface in tension. The second tensile crack propagated very slowly towards the upper (compressive side) loading nose. However, it was eventually deflected laterally from the loading nose since a sizeable compressive field existed immediately in front of the nose. Initiation of the second crack also caused a sharp increase in acoustic emission. The steady increase in acoustic activity after the second peak was attributed to the propagation of the second crack and additional frictional noise generated by fiber pull-out. Fig. 6 schematically depicts the failure modes of the composite.

Since both faces of the specimen were abraded by a 500 grit cloth and the matrix was optically transparent, it was possible to observe where cracks occurred in relation to the reinforcement weave. As the specimen was further loaded, additional microcracks originated at locations near those of the first and third crack types and behaved accordingly. This additional microcracking and fiber pull-out again led to the increasing acoustic activity that was observed. The cracks on the tensile side (parallel to the loading noses) which accompanied advanced failure were observed in a regular fashion as one looked laterally from the central loading nose. Further inspection indicated that the cracks initiated along lines which delineated successive fiber tows which were parallel to the loading noses. (It is worth noting that a gap of approximately 10% of the width of a fiber tow existed between parallel fiber tows.) It is proposed that these microcracks regularly occur at the aforementioned locations due (1) the residual tensile previously discussed, (2) strain magnification in the matrix at these fiber "poor" regions. These transverse cracks (with

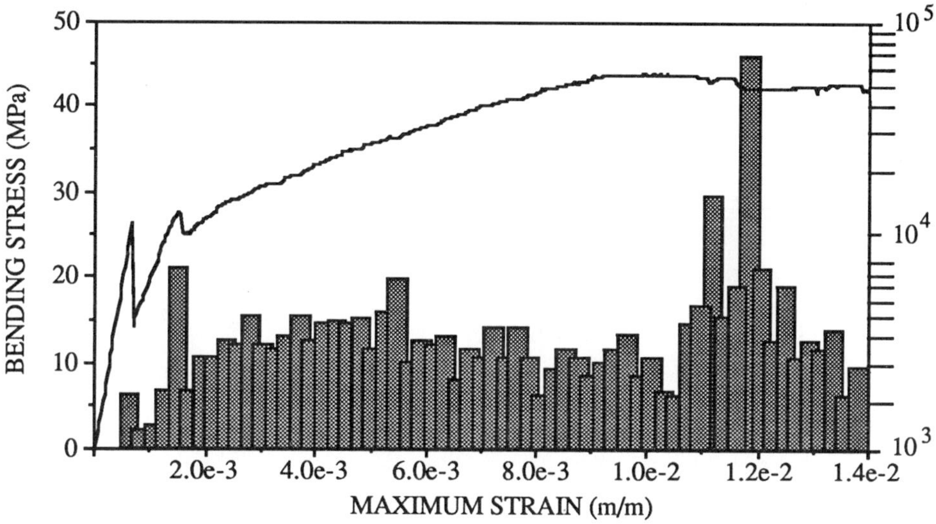

Fig. 3. Outer bending stress and total AE counts versus strain for a single layer CMC.

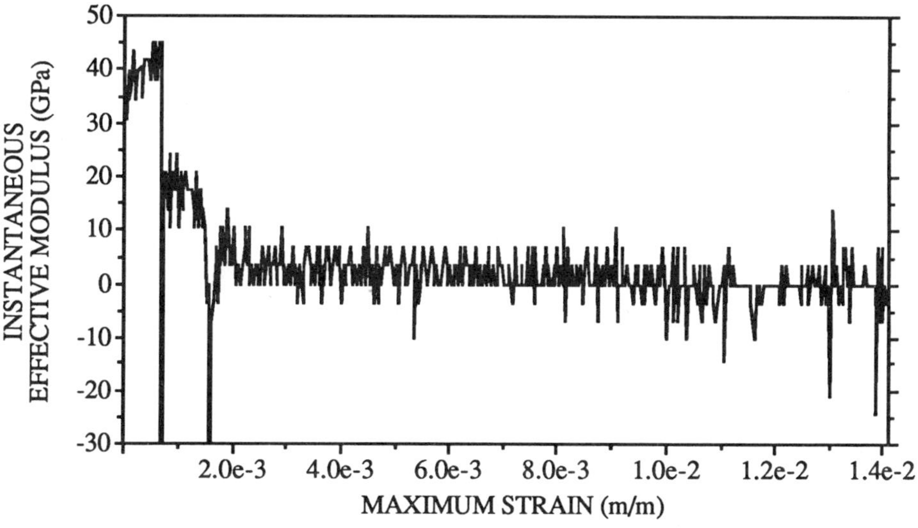

Fig. 4. Instantaneous effective modulus versus strain.

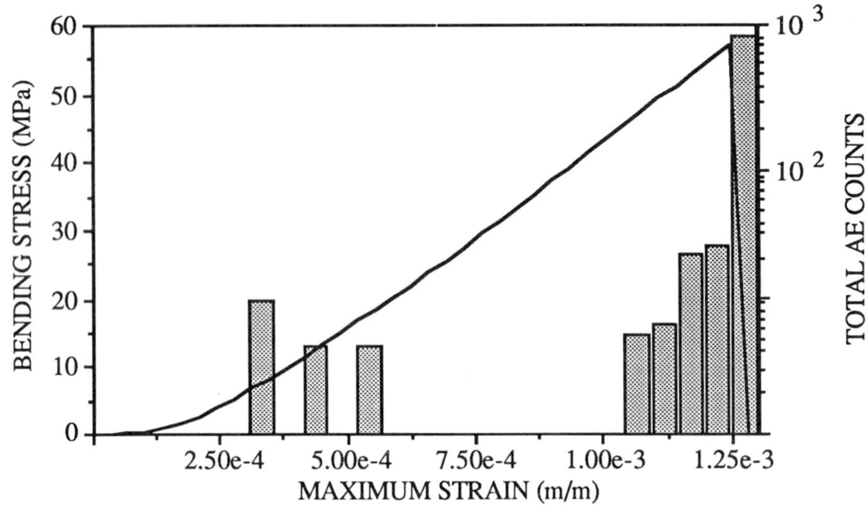

Fig.5. Stress and total AE counts versus strain for a nonreinforced glass specimen subjected to three-point bending.

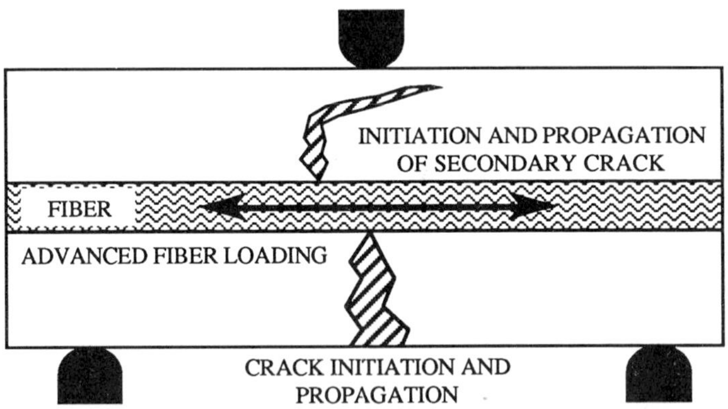

Fig. 6. Schematic representation of failure modes during three-point bend test of a single layer CMC.

reference to the principle stress direction) gave rise to major unloadings of the composite as well as the more subtle stress-strain curve slope changes. Moreover, after taking the derivative of the maximum outer face bend stress with respect to strain (yielding an instantaneous effective modulus), it was possible to correlate the calculated stress-strain curve slope changes with the more pronounced AE activity changes (Figs. 3 and 4). Even beyond the massive slope excursions which occured prior to the deflection of 1.0E-4 m, acoustic emission peaks often correlated with negative stress-deflection curve slopes which indicated strain relief areas. Lastly, as the material's ultimate strength was attained, AE activity peaked due to the multitude of failure mechanisms present.

CONCLUSIONS

Microscopic photoelasticity techniques directly revealed thermomechanical stresses which resulted from thermal expansion differentials between composite constituents. The effects of having a matrix with a thermal expansion coefficient less than that of the reinforcement was clearly shown in the stress trajectory sketches.

The results of this study have shown how stress birefringence can be applied to enhance the physical understanding and directly determine thermomechanical stresses in a cloth reinforced engineering ceramic composite material. It is believed that this technique can be used to further develop the processing of composites in general.

As shown by other authors, matrix microcracking is often not optically visible even though it still may be observed in load-deflection curve data. It is proposed that the cracks are not readily visible since the load bearing high modulus fibers prevent further opening of the already small crack openings or even close up existing cracks once the load is removed. Acoustic emission activity monitoring and optical microscopy provided great insight concerning flexural test data used to determine discrete and predictable failure mechanisms for a single layer SiC cloth reinforced glass composite. Changes in the stress-deflection curve slope were also often accompanied by AE activity changes. The cumulative results of using these three evaluative techniques provided detailed information on how CMC's fail in general and how AE may be applied in the future to nondestructively characterizing CMC structures.

ACKNOWLEDGEMENTS

This work was supported by both The Johns Hopkins University Center for Nondestructive Evaluation and Martin Marietta Laboratories. The National Science Foundation is specifically recognized for its support of the thermomechanical stress research.

REFERENCES

1. R. A. SAMBELL, D. H. BOWEN and D. C. PHILLIPS, "Carbon Fiber Composites with Ceramic and Glass Matrices (Part 1 Discontinuous Fibers)", *Journal of Materials Science*, **7** (1972) pp. 663-675.

2. G. A. DRIES and S. S. TOMPKINS, "Effects of Thermal Cycling on Graphite-Fiber-Reinforced 6061 Aluminum", *NASA Technical Paper 2612*, (National Aeronautics and Space Administration Scientific and Technical Information Branch, 1986).

3. S. S. TOMPKINS and S. L. WILLIAMS, "Effects of Thermal Cycling on Mechanical Properties of Graphite Polyamide", *J. Spacecraft*, (May-June 1984) pp. 274-280.

4. S. S. TOMPKINS, G. F. SYKES and D. E. BOWLES, "The Thermal and Mechanical Stability of Composite Materials for Space Structures", Presented at the IEEE/ASM/ASME/SME Space Technical Conference, 1985, Anaheim, California.

5. S. S. TOMPKINS, K. E. ARD and G. R. SHARP, "Thermal Expansion Behavior of Graphite/Glass and Graphite/Magnesium", Presented at the 18th International Technical Conference Society for the Advancement of Materials and Process Engineering, 1986, Seattle, Washington.

6. S. S. TOMPKINS, "Effects of Thermal Cycling on Composite Materials for Space Structures", Presented at the Space Environmental Effects on Materials Workshop, Sponsored by NASA Langley Research Center, 1988, Hampton, Virginia.

7. E. HECHT, *Optics*, Second Edition, (Addison-Wesley Publishing Company Inc., Reading, Massachusetts, 1987), p. 315.

8. I. M. DANIEL and J. DURELLI, "Shrinkage Stresses Around Rigid Inclusions", *Experimental Mechanics*, (August 1962) pp. 240-244.

9. I. M. DANIEL "Photoelastic Studies of Mechanics of Composites", *Progress in Experimental Mechanics, Durelli Anniversary Volume*, V. J. Parks, ed., (The Catholic University of America, D. C., 1975) pp. 131-166.

10. T. YOSHINO, T. OHTSUKA, and Y. NOGUCHI, "Stress Analysis by Photoelasticity and F.E.M. of Plane Woven Glass Fiber Reinforced Plastic Laminates", *Proceedings from Japan-U.S. Conference on Composite Materials*, Tokyo, 1981, pp. 47-54.

11. T. MAH and M. G. MENDIRATTA, "Room Temperature Mechanical Behavior of Fiber-Reinforced Ceramic-Matrix Composites", *Journal of the American Ceramic Society*, **68** [1] (1985) pp. C-27-C-30.

12. J. W. KRYNICKI, "Design and Behavior of a Ceramic Matrix Composite Subjected to Mechanical and Thermal Loading" Master of Science Essay, The Johns Hopkins University, Baltimore, Maryland, 1989.

13. D. J. LLOYD, K. TANGRI, "Acoustic Emission From Al_2O_3-Mo Fibre Composites", *Journal of Materials Science*, **9** (1974) pp.482-486.

14. J. LANKFORD, "Compressive Strength and Damage Mechanisms in a SiC-Fiber Reinforced Glass-Ceramic Matrix Composite", *Proceedings of the Fifth International Conference on Composite Materials, San Diego, CA*, July 29-August 1, 1985 pp. 587-602.

15. M. J. NOONE, R. L. MEHAN "Observation of Crack Propagation in Polycrystalline Ceramics and Its Relationship to Acoustic Emissions", *Proceedings of the Symposium on Fracture Mechanics of Ceramics, University Park, PA,* July 11-13 1973, **1** pp. 201-229.

THERMALLY INDUCED ACOUSTIC EMISSION

in DAMAGED COMPOSITE PLATES

John M. Liu and Paul M. Gammell

Materials Evaluation Branch (R34)
Naval Surface Warfare Center
Silver Spring, MD 20903-5000

INTRODUCTION

Thermally induced acoustic emission (AE) in brittle materials like firebricks and ceramics has long been noticed in the materials industry. Indeed, it has been suggested that the characteristics of such "clicking" sound might be related to the thermal shock response of ceramics.[1]

Recently, Norio Sato, Toshio Kurauchi, and Osami Kamigaito investigated AE generated by low level heating, in polymer-based composites that had been damaged by mechanical deformation.[2,3] They introduced the term "thermo-acoustic emission", and suggested that this may be a new tool for damage detection. The work to be reported here is motivated by our interest in the development of techniques for sensing the occurrence of damage in composites which are exposed to a cyclic thermal environment in service. We have investigated some characteristics of thermally induced AE in impact-damaged graphite/epoxy panels. The dependence of such AE on the level of impact energy is reported below.

EXPERIMENTAL SET-UP

A sketch of the experimental set-up used for the detection of thermo-A.E. is shown in Fig. 1. Since the specimen center was the impact site, heating was directed to this region using a 500 watt quartz lamp. The piezoelectric AE transducers were coupled to the specimen surface via a viscous liquid, and were separated from each other by approximately 12 cm. This separation allowed the transducers to remain relatively cool during the thermal cycle, which typically consisted of 15 seconds of heating and two minutes of cool down. The back surface temperature was sensed by an infrared

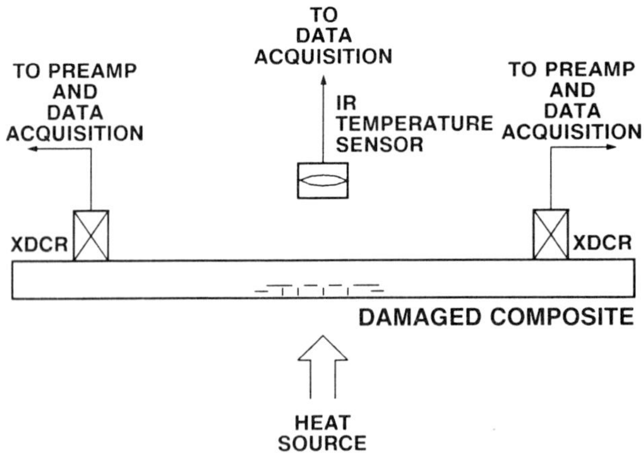

Figure 1. Sketch of experimental set-up for thermo-acoustic emission measurements.

detector. A commercially available computerized AE data logging system was used which recorded the two channel AE signals, temperature, and heat source on/off signal. AE signals originated in the centrally located damaged region, would reach the two transducers within a timing window. By means of a clock in the computer, these were logged in as "events". Signals reaching either one of the transducers with amplitude over a presetable threshold are called "hits". Ideally, there would be twice as many hits as events. Hits that originated from regions close to the specimen edges might reach only one of the transducers.

Typical system gain was 110db for recording thermo-AE with a threshold of 45db. The performance of the measurement system was checked by observing AE generated by the breaking of pencil lead at reduced amplifier gain. Rearm time for each channel had to be set at 50 ms in order not to overload the computer system when specimens with large amount of impact damage were tested.

The specimens were impacted with a hemispherical plunger of 1.27cm in diameter when they were simply supported around the edges. The height of the freefall of the plunger was changed to achieve a range of impact energies. The quasi-isotropic graphite/epoxy specimens were 0.48 cm in thickness.

RESULTS

A typical signal associated with a "hit" generated by thermo-AE in a specimen impact damaged at 7800 Jm^{-1} is shown in Fig. 2. Good signal to noise ratio is evident here. The threshold chosen for our data logging system, as indicated earlier, allowed these fairly energetic signals to be recorded. Records of the number of "hits" and the stimulating temperature versus time are shown in Fig. 3. The corresponding "events" are shown in Fig. 4. It is seen in these last two figures that the rates of increase in "hits" and "events" were large during the heating part of the thermal cycle. Furthermore, these rates appear to be related to the amount of impact damage for a fixed thermal stimulation.

The dependence of the number of "hits" and "events" for a range of impact damage levels on the number of thermal cycles in one set of specimens is summarized in Table I. It is seen that AE activity diminished as the number of cycles increased. This decrease was substantial after one cycle, in specimens with small to moderate levels of damage.

The two transducer set-up in our experiments allowed us to estimate the locations of the AE sources. In Fig. 5 are shown the locations of "events" versus time in a specimen sustaining a damage level of $4454 Jm^{-1}$. The locations of the two transducers are at the points marked -1 and +1, respectively.

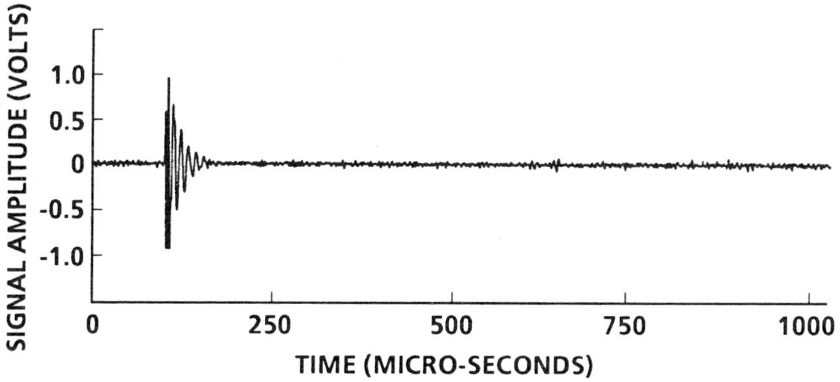

Figure 2. Thermo-AE signal generated in a graphite/epoxy panel impact-damaged at $7800 Jm^{-1}$.

The dense population of "events" in locations approximately midway between these two points was consistent with the expectation that most sources were at or close to the impact site at the center of the specimen. This figure also shows that significantly large number of sources existed closer to one of the transducers than the other. In contrast, the events in a specimen with low level of damage, shown in Fig. 6, are mostly confined to a small region midway between the transducers, and thus close to the immediate vicinity of the impact site.

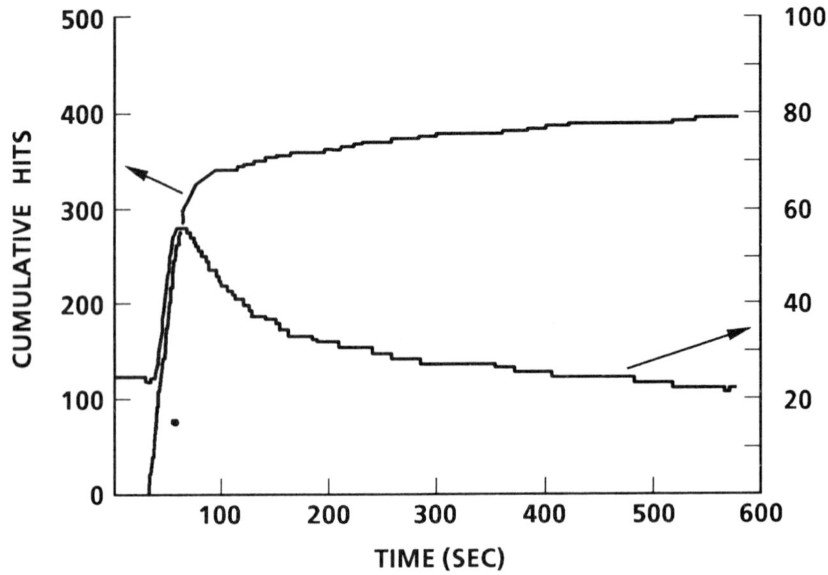

Figure 3. Number of "hits" and temperature versus time in a thermal cycle generated in a panel damaged at 4454 Jm^{-1}.

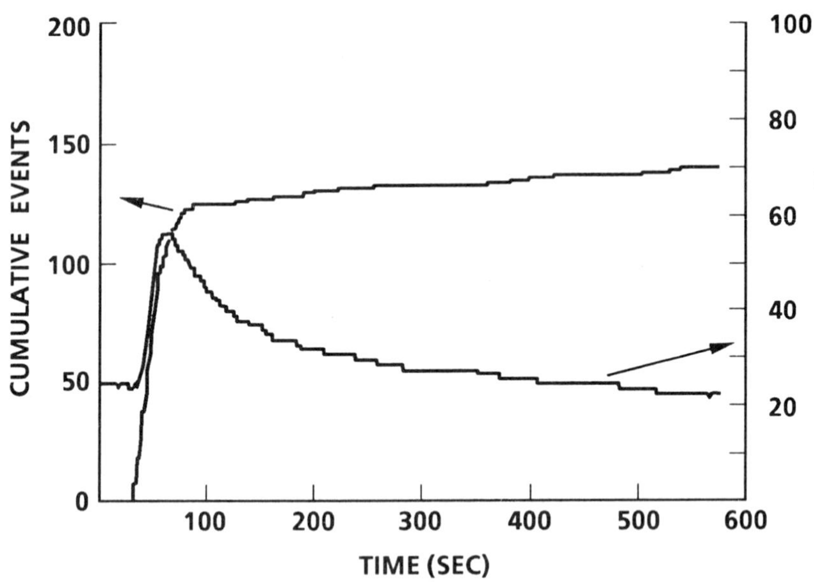

Figure 4. Number of "events" and temperature versus time in the experiment as shown in Fig. 3.

Table 1. Summary of Thermo-AE Activities for a Number of Consecutive Thermal Cycles in Quasi-Isotropic Graphite/Epoxy Panels Subjected to a Range of Impact Energies.

IMPACT ENERGY (Jm^{-1})	TOTAL HITS	TOTAL EVENTS	CYCLE
7800	1976	466	1
	1067	422	2
	787	321	3
	645	253	4
	573	229	5
6012	1350	312	1
	750	271	2
	470	166	3
4454	586	224	1
	132	56	2
	68	24	3
	67	27	4
	49	19	5
2227	193	79	1
	33	14	2
	25	11	3
1559	96	38	1
	11	4	2
	19	7	3
1113	6	3	1
	3	1	2
	5	0	3

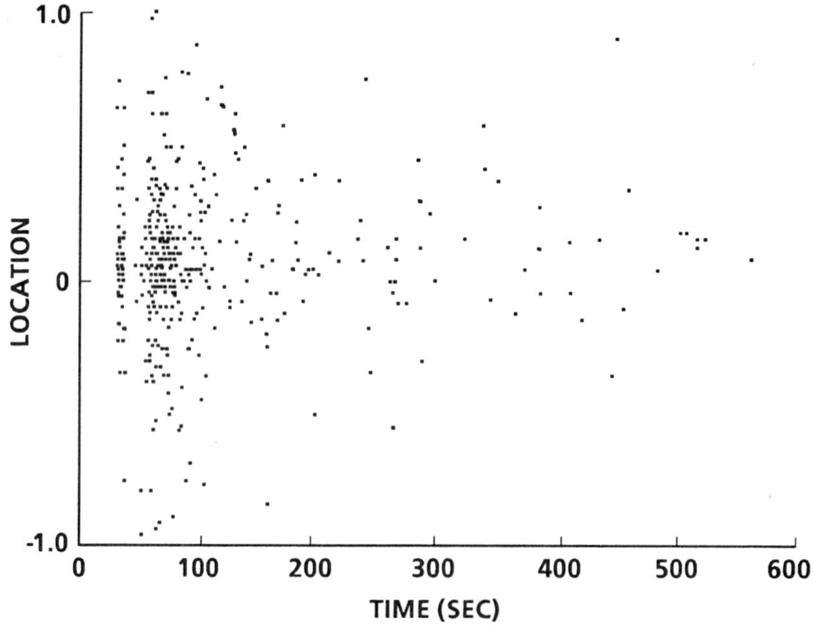

Figure 5. Locations of "events" versus time in the experiment on a panel damaged at 7800 Jm^{-1}. The positions of the transducers are at +1 and -1 along the vertical axis.

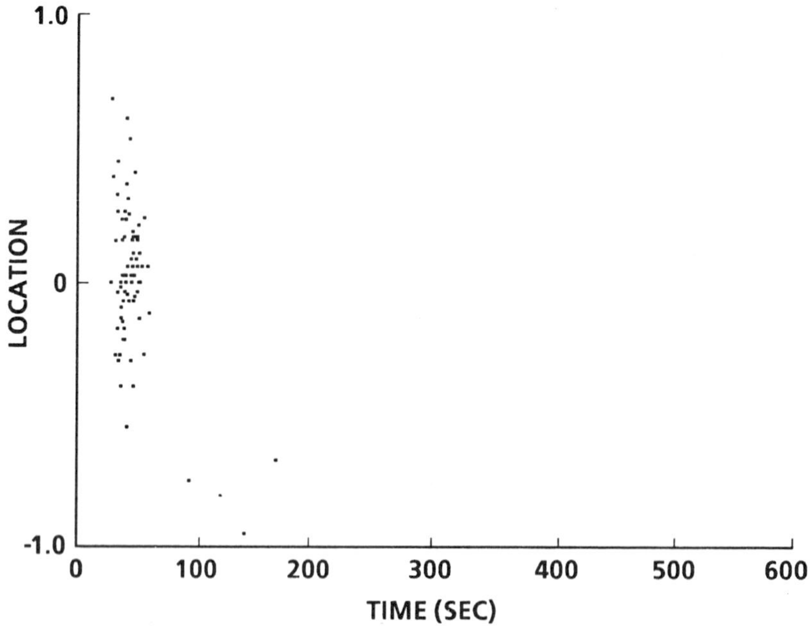

Figure 6. Locations of "events" versus time in the experiment on a panel damaged at 2227 Jm^{-1}. The positions of the transducers are at +1 and -1 along the vertical axis.

DISCUSSION

Thermally induced acoustic emission in impact-damaged graphite/epoxy panels is shown in this work to depend systematically on the energy of impact. The friction related mechanism for such AE, proposed by Norio Sato et al., appears to operate at the tip of delaminations in our impact-damaged graphite/epoxy specimens. The wider distribution of AE sources away from the center of the impact site is consistent with the observation that the regions of delamination spread farther as the energy of impact increases.

The systematic decrease of AE activity as the number of cycles increased, however, raises questions about the reliability of this thermo-AE techniques as a general NDE tool for the detection of impact damage in graphite/epoxy composites. The sensitivity of the technique also appears inadequate for characterizing composites with low levels of damage. Data in Table 1 indicate that the AE detected appeared to decrease at a smaller rate as the number of thermal cycles progressed. Whether or not it approaches a steady state, non-zero value at larger number of cycles is not known currently. It does appear, however, that if the service thermal environment of the composite is periodic, the occurrence of impact damage would produce AE in a correspondingly periodic manner. This may aid in the detection of such occurrences in the presence of random noise.

REFERENCES

1. G.C. Robinson, C.R. Reese, and E.A. La Roche, Jr., "Determination of Thermally Induced Acoustic Emission of Ceramics," Ceramics Bulletin, 53:482 (1974).

2. N. Sato, T. Kurauchi, and O. Kamigaito, "Thermo-Acoustic Emission from a Damaged Composite," SAMPE Journal, 23:48 (1987).

3. N. Sato, T. Kurauchi, and O. Kamigaito, "Detection of Damage in Composite Material by Thermo Acoustic Emission Measurement," J. Composite Materials, 22:447 (1988).

The first step is the experimental collection of inspection data (AU signals) which will be described in the following section while the second and third steps are the goal of the pattern recognition analysis. A commercial Statistical Analysis Software (SAS) package was employed for pattern recognition analysis. SAS uses a number of routines for PRA which are described briefly in the following paragraphs.

Principal Component Analysis (PCA)

The principal component analysis is a multivariate technique for examining relationships among the features. The PCA can be used to reduce the data and display them on the principal components. The PCA process calculates a correlation matrix of all features which is then used to find features that vary in a similar way(i.e. a high correlation factor) and only keep a few features from each group.

Stepwise Discriminant Analysis (SDA)

The SDA is a procedure which selects parameters that contribute the most to the separability of data. Several statistical methods are used to evaluate the contribution of a variable in the separability between groups, the most common methods are the Wilks' lambda and F statistical tests (11).

Canonical Discriminant Analysis (CDA)

Canonical discriminant analysis is a dimension-reduction technique related to principal component analysis and canonical correlation. CDA makes it possible to replace a large set of features by another set of canonical variables. The canonical space dimension is always equal to the number of groups minus one. The canonical variables are not correlated with each other and are sorted in decreasing discriminating power, so the two or three first variables can well describe the total between-group separability.

Fast Cluster Analysis (FCA)

In the FCA procedure, observations are divided into clusters such that every observation belongs to one and only one cluster. The number of clusters is equal to the number of groups. The procedure first selects cluster seeds and assigns each observation to the nearest seed. After all observations are assigned, the cluster seeds are replaced by cluster means. This step can be repeated until the changes in the cluster seeds become small or zero. Final clusters are formed by assigning each observation to the nearest seed.

Two Dimension Plots (TDP) and Frequency Tables

Finally, to visualize the results, the procedure TDP is used to project the data points on the first two-canonical-variable plane. This projection is often sufficient to see the separability between the different groups. However, clusters occasionally show overlapping areas in the projection plane that may be totally separated in multi-dimensional space. In such cases, Frequency Tables are used to provide a cross-tabulation of numbers of samples in each cluster that belong to a given group. If all samples constituting a cluster are identified as one group, then the classification is 100%. Otherwise, the classification rate is equal to the percentage of samples identified in the right group.

Table 1. Definition and Partition of the Adhesive Joint Specimens by Surface Treatment or Induced Defects.

Adhesive	Group	No. of Specimens	Surface Treatment or Defect Type
Hysol	A	5	Reference group, ASTM STD D1002-72
"	B	10	Different joint thickness
"	C	5	Same as A but no surface etching
"	E	5	Frozen for 24h at -18C
"	F	25	Teflon spray, diff. size & location
"	G	10	Voids, various size & location
FM300	A	5	Reference group, ASTM STD D1002-72
"	B	5	Same as A but no surface etching
"	C	5	Teflon spray, diff. size & location
"	D	5	Voids, various size and location

EXPERIMENTAL PROCEDURES

Test Specimens

Single-lap joint specimens (adapted from ASTM standard D1002-72) were made using aluminium adherends and either Hysol 9320 liquid adhesive or FM300 adhesive film. Controlled degradation in bond line strength was achieved by the deliberate introduction of voids in the joint, contamination of the adherend surface, modification of the standard surface preparation prior to bonding or variation of the thickness of the adhesive layer. Each group of samples consisted of at least 5 specimens.

Voids with different sizes and locations were created in the joint by removing part of the adhesive before bonding. Zero volume unbonds were created in the joint by contaminating partially or totally the surface of the aluminium adherends with teflon spray before bonding or by modifying the standard surface treatment (no surface etching). The thickness of the adhesive in the joint was controlled by a special fixture used to hold specimens during cure. The bonded specimens were then inspected using the ultrasonic C-scan method. Only specimens with voids showed detectable flaws and the actual flaw size was slightly different from the intended size due to adhesive flow during cure. A list of specimens is provided in Table 1.

Acousto-Ultrasonic Evaluation

The acousto-ultrasonic measurements were carried out using the apparatus shown schematically in Figure 2. An ultrasonic pulse generator which excited a broadband transmitting transducer with a 500 kHz centre frequency was employed to generate the input pulses. At a fixed distance away, a receiving acoustic emission transducer was used to intercept the stress waves. Both transducers were coupled to the surface using a coupling fluid. The stress waves received by the receiving transducer were amplified by 40 dB using a preamplifier and then filtered and amplified by a 0.2-2 MHz band-pass filter amplifier with the gain fixed at 20 dB. Representative waveforms of the AU signals from several measurements were digitized using a universal waveform analyzer and stored on floppy disk for further analysis in the time and frequency domains. Figure 3 illustrates a typical AU signal from a bonded specimen in the time and frequency domains.

The signals collected during the acousto-ultrasonic tests were, in a later stage, digitally processed. A discrete FFT was calculated for each signal and a large number of parameters were extracted in both the time and frequency domains to describe each signal. The choice of parameters to

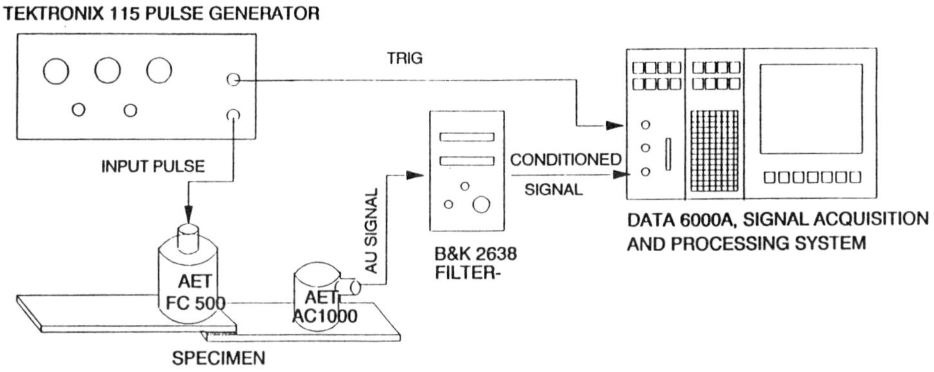

Figure 2. Components of the Acousto-Ultrasonic Measurement System.

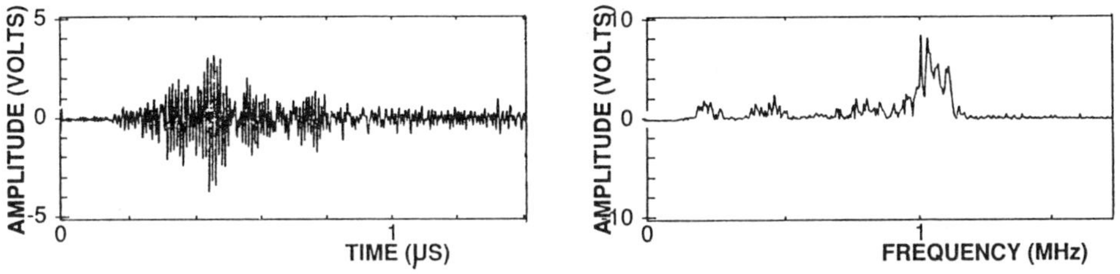

Figure 3. A Typical Acousto-Ultrasonic Signal and the Corresponding Power Spectrum from an Adhesive Bonded Joint.

Table 2. Partitioning of the Adhesive Joint Specimens by Failure Load.

Adhesive	Group	No. of Specimens	Range of Failure Load (kg)
FM300	1	4	615-695
"	2	6	790-830
"	3	7	900-985
"	4	7	1020-1060

evaluate was based on intuition and results from earlier investigations. The consistency of measurements was evaluated by comparing the parameters taken from the same sample at different times. Each specimen was then represented by the mean values of parameters from different measurements.

Mechanical Testing

After completion of the acousto-ultrasonic measurements, the test specimens were loaded in tension to failure in a universal testing machine and the failure load was measured. The shear strength was calculated by dividing the failure load by the overlap area. After failure, the fracture surfaces were examined to determine failure modes and the void or unbond areas within each specimen were measured. Specimens were divided into several groups based on their ultimate shear strength regardless to the type of defect. Table 2 gives the new grouping of FM300 specimens. Pattern recognition routines were again carried out on the new groups and the classification rate was evaluated.

Table 3. List of Features Retained for Classification by Surface Treatment or Failure load.

Feature	Surface Treatment	Failure Load
Mean Amplitude	x	x
Standard Deviation	x	x
Energy	x	
Location of the Maximum Peak	x	x
Location of the Minimum Peak	x	x
Maximum Slop	x	x
Time to First Crossing of a STD Threshold		x
Ring Down Count over a STD Threshold	x	
Area Under the Frequency Spectrum	x	x
Amplitude of the Max. Peak in Freq. Spectrum	x	x
Frequency Corresponding to Max. Peak	x	x
Partial Area in the 2nd Freq. Band		x
" " " " 6th " "	x	x
" " " " 7th " "	x	x
" " " " 8th " "	x	x
" " " " 11th " "	x	
" " " " 12th " "	x	

RESULTS AND DISCUSSION

Feature Selection and Data Reduction

The initial data set consisted of 85 observations (one AU signal for each specimen) and 36 parameters. First, the data set corresponding to each adhesive system was considered separately and grouped on the basis of surface treatment or defect (Table 1). Then, the data for FM300 samples was grouped on the basis of failure load (shear strength) for each specimen (Table 2).

After the PCA and SDA procedures were applied to the data sets, 17 features were found to be significant for group separation. A list of important features is provided in Table 3. The most significant features in the time domain were the mean amplitude, standard deviation and energy which are related to the efficiency of the joint in transmitting stress wave energy. Other features, such as the location of the maximum and minimum peaks are indicative of the stress wave propagation speed. Since the stress waves travel from one adherend across the bond line to the other adherend, the variations in bond line properties (structural and morphological) or in the adhesive-adherend interface (adhesion) affect the stress wave propagation efficiency and speed.

From the frequency domain, three global features, i.e. area under the spectrum and amplitude and frequency of the major peak in the spectrum were selected for classification as shown in Table 3. Also, three features related to partial area in different frequency bands contributed to group separation. The total area under the spectrum and the amplitude of the major peak are again measures of the energy transmitted, while the frequency corresponding to the major peak indicates the most effective frequency in the transmitted waves and can be related to the frequency response of instruments , geometry of the sample and acoustic impedance of the joint. The partial parameters are, in a similar way, a measure of energy transmission efficiency in a given frequency range. It can be seen that the more significant frequency range is 0.75 to 1.25 MHz which is the optimum range for the AE sensor employed. However, the information contained

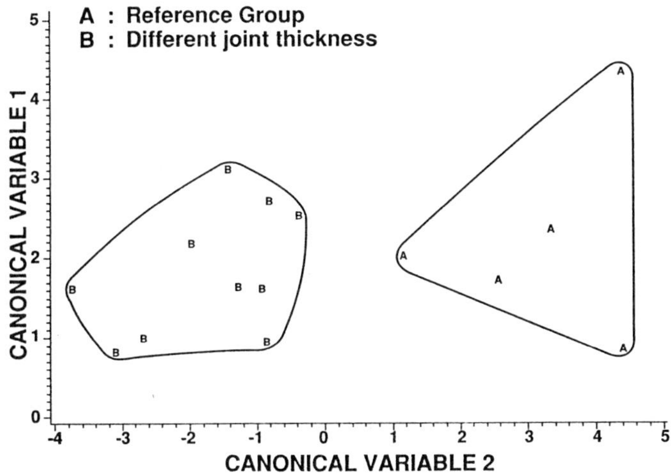

Figure 4. Classification of AU Signals by the Adhesive Thickness and the Projection of Data on Two Canonical Axis.

Table 4. Classification of Hysol Specimens by Surface Treatment

Cluster Frequency Recognition Rate (%)		Group A	C	D
	A	5 100	0 0	0 0
Group	C	0 0	5 100	0 0
	D	0 0	0 0	5 100

outside of this range is not lost due to transducer characteristics, since some partial parameters covering frequencies as low as 0.3 MHz and as high as 1.7 MHz were also retained.

The CDA procedure used with the Hysol 9320 data and the partitioning by surface condition (6 classes) produced a set of 5 canonical variables that constituted the space in which the classification was performed. In the same way, and with the data corresponding to FM300 specimens, two 3-canonical-variable spaces, one for the 4 classes by surface treatment and the other for the 4 classes by failure load were obtained.

Classification

The classification results for groups A and B of Hysol specimens which were different in terms of adhesive thickness are illustrated in Figure 4. The overall recognition rate (ORR) was 100%. Groups A, C and D of Hysol samples also classified at a 100% recognition rate, although the projection of groups C and D on a two dimensional plane did not indicate the separation. However, in canonical space these groups were quite separable as indicated using the cross-tabulation procedure (Table 4). Group C was different from group A in terms of surface pre-treatment while in group D adhesive was degraded by freezing. Neither of these groups were identified as defective by the conventional ultrasonic C-scan inspection. Groups A, E and F of Hysol specimens which contained teflon spray or voids were

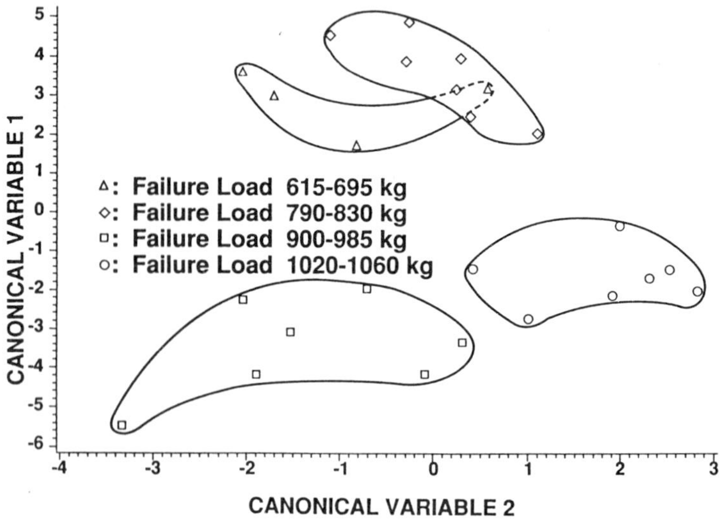

Figure 5. Classification of AU Signals by the Specimen Failure Load and the Projection of Data on Two Canonical Axis.

classified with a 87.5% recognition rate. The lower recognition rate is due to confusion between specimens with lack of adhesion created using teflon spray and those created by cutting a part of the adhesive film before bonding. The two approaches resulted in disbonds which were similar in nature. Both of these groups were quite separable from the reference group.

The classification of FM300 specimens in terms of joint surface condition or defect resulted in a 96% recognition rate. Only one sample from group C which was expected to contain an unbond area was classified in the reference group. Optical examination of fracture surfaces after mechanical tests indicated that the misclassified sample was in fact a defect free specimen.

The FM300 specimens were then loaded in tension to failure and the bond strength was calculated by dividing the failure load by the overlap area. The failure load data were divided into four ranges as shown in Table 2 regardless to the types of defect present. Pattern recognition procedures were then applied on acousto-ultrasonic signals which were stored before destructive tests. It was possible to separate the four load groups in the canonical space using the AU signals; the projection of data on two canonical axis is illustrated in Figure 5. The two groups that show an overlap were totally separated in the multi-axis canonical space.

CONCLUSIONS

The acousto-ultrasonic technique along with pattern recognition analysis can be used for the evaluation of the integrity of adhesively bonded joints. Experimental measurements of AU signals on single-lap joint specimens and their classification using the SAS pattern recognition routines indicated that it was possible to separate defective samples from well-bonded specimens with excellent recognition rates. Also, classification of AU signals by failure load regardless to pre-treatment of samples or defect resulted in positive recognition. Further investigations are required to establish the applicability of these techniques on large and complex bonded structures.

REFERENCES

1. Guyott, C.C.H., Cawley, P. and Adams, R.D., The Nondestructive Testing of Adhesively Bonded Structures: A Review, J. Adhesion, 1986, Vol. 20, pp. 129-159.
2. Adams, R.D., Cawley, P. and Guyott, C.C.H., Nondestructive Inspection of Adhesively Bonded Joints, J. Adhesion, 1987, Vol.21, pp. 279-290.
3. Fahr, A. and Tanary, S., Nondestructive Evaluation of Adhesively-Bonded Joints, Proc. 69th AGARD meeting, Brussels, 1-6 October 1989.
4. Vary, A., The Acousto-ultrasonic Approach, NASA TM89843, 1987.
5. Fahr, A., Lee, S., Tanary, S. and Haddad, Y., Estimation of Strength in Adhesively Bonded Steel Specimens by Acousto-Ultrasonic Technique, Materials Evaluation, Vol. 47, No. 2, 1989, pp.233-240.
6. Hand, D.J., Discrimination and Classification, Wiley, New York, 1981.
7. James, M., Classification Algorithms, Collins, London, 1985.
8. Afifi, A.A. and Clark, V., Computer-Aided Multivariate Analysis, Lifetime Learning Publications, Belmont, California, 1984.
9. Williams, J.H. Jr. and Lee, S., Pattern Recognition Characterization of Mechanical and Morphological Materials States via Analytical Quantitative Ultrasonics. Materials Analysis by Ultrasonics, Edited by A. Vary, Noyes Data Corp., Park Ridge, New Jersey, 1987.
10. Dickstein, P., Segal, Y., Segal, E. and Sinclair, A.N., Statistical Pattern Recognition Techniques: A Sample Problem of Ultrasonic Determination of Interfacial Weakness in Adhesive Joints, J. NDE, Vol. 8, No. 1, 1989.
11. SAS Institute Inc., SAS/STAT Guide for Personal Computers, 6th edition, Cary, 1987.

CHARACTERIZATION OF INTERPHASIAL PROPERTIES IN ADHESIVE JOINTS USING ANGLE-BEAM REFLECTION ULTRASONIC SPECTROSCOPY

W. Wang and S. I. Rokhlin

The Ohio State University
Department of Welding Engineering
Columbus, Ohio 43210

INTRODUCTION

Mechanical properties of adhesive joints depend on: (1) the bulk (cohesive) properties of the adhesive such as type of adhesive, degree of cure, porosity, et cetera and (2) the interphasial properties of the bond between the adhesive and the adherends. While several techniques have been developed for nondestructive evaluation of cohesive properties of the adhesive, the assessment of interphase properties is more complicated. It has previously been proposed [1] that an ultrasonic wave that produces shear stress on the interphase is sensitive to interphase properties. This may be achieved by utilization of interface waves [1,2] or guided waves in the bonded plates [3,4,5]. Another possibility is the use of obliquely incident ultrasonic waves for interphase evaluation, analyzing signals in the time or frequency domains [6,7,8].

The interfacial regions of a typical adhesive joint consists of a layer of anisotropic porous oxide and primer. In addition, adhesive properties in regions near the interface are also different from those of the bulk adhesive. Degradation of interphase, e.g. a moisture-caused reaction between the aluminum oxide and primer leading to debonding, often leads to failure of the adhesive joint [9]. Naturally, modeling interphase structure and assessing the interphasial properties is of importance in nondestructive evaluation of adhesive joints.

In our previous studies [10,11] we have developed a physical model for angle-beam ultrasonic evaluation of adhesive joints of aluminum alloys. The model includes layers of anisotropic porous aluminum oxide, primer, an optional weak boundary layer and adhesive. Numerical calculations of the effects of the properties of these layers have been reported [10,11]. Preliminary results obtained from samples of two extreme cases - good-bond interphases at both sides of the adhesive layer and weak-bond interphases at both sides of the adhesive layer - are in good agreement with the theoretical predictions of our model [10,11].

In this work samples with good-bond interphase at one side and controlled interphasial properties at the other side of the adhesive layer were investigated. The experiment uses an obliquely incident transverse wave and measures the spectrum of the signal reflected from the interphases. By using a nonlinear least square optimization for curve fitting the loci of minimum reflection coefficient we determine the viscoelastic properties and the thickness of the interphase.

THEORY

The physical model used in this study for ultrasonic evaluation of adhesive joints of aluminum alloys is shown in Figure 1. The adherends are considered as semi-infinite solids; this is appropriate for short ultrasonic pulses used in our experiments. Such an approach has an advantage in that interference of ultrasonic waves in the adherend can be eliminated and therefore the sensitivity of ultrasonic signals to interphasial properties increases. Anisotropy of the aluminum oxide layer is taken into consideration in our multilayer model. In addition, an optional weak boundary layer (WBL) is used to model the interface of interest.

The WBL simulates the degradation of interphase and is modeled as a thin viscoelastic boundary layer whose properties, complex bulk modulus, k, and shear modulus, μ, are calculated as:

$$\mu = \mu_\infty \left(\frac{\omega^2\tau^2}{1+\omega^2\tau^2} - \frac{\omega\tau}{1+\omega^2\tau^2}\right) \quad (1)$$

$$k = k_o + (k_\infty - k_o)\left(\frac{\omega^2\tau^2}{1+\omega^2\tau^2} - \frac{\omega\tau}{1+\omega^2\tau^2}\right) \quad (2)$$

where μ_∞ is the shear modulus of the viscoelastic material in the solid state, $\omega\tau$ is a relaxation parameter (ω is frequency and τ is relaxation time), k_o is the bulk modulus of the viscoelastic material in the liquid state, and k_∞ is the bulk modulus in the solid state. In reality, the WBL may be formed due to degradation of oxide, insufficient cure of adhesive at interface, adhesive plasticisation, contamination etc. Our model may simulate a gradual transition from viscous liquid WBL to solid with properties equal to that of adhesive. The major advantage of WBL approximation by viscoelastic material with one relaxation time is that there is only one

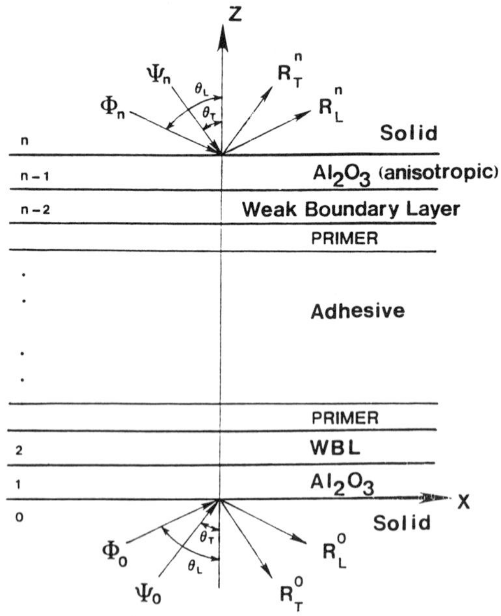

Fig. 1. The layered model of an adhesive joint used in this study.

unknown material parameter, $\omega\tau$, which must be determined from experiment. Another unknown parameter is the WBL thickness, but this can be estimated from physical surface roughness and the thickness of the oxide layer.

To calculate the reflection coefficient of an ultrasonic wave reflected from the adherend/adhesive interface, we modified a matrix algorithm [12] developed for a multilayer anisotropic system in liquid. The algorithm was extended to the case [10] of anisotropic layered system between two solids. With reference to Figure 1, the matrix equation for reflection and transmission coefficients can be written in the form:

$$\begin{bmatrix} \phi_n \\ R_L^n \\ \psi_n \\ R_t^n \end{bmatrix} = [D] \begin{bmatrix} R_L^o \\ \phi_o \\ R_t^o \\ \psi_o \end{bmatrix} \tag{3}$$

where

$$[D] = [B^{(n)}][a^{(n-1)}][a^{(n-2)}] \quad [a^{(2)}][a^{(1)}][A^{(0)}] \tag{4}$$

Isotropic and anisotropic layers are represented by 4 x 4 matrices $[a^{(i)}]$ (i=1,...,n-1). This is valid if the anisotropic layer lies in the plane of symmetry as for porous aluminum oxide. Matrices $[B^n]$ and $[A^o]$ are related to the elastic semispaces and are derived here.

In our experiments we used only a transverse wave incident from one side of the system. For this case the reflection coefficient for a reflected transverse wave is

$$R_{tt} = \frac{d_{13}d_{41} - d_{11}d_{43}}{d_{11}d_{33} - d_{13}d_{31}} \tag{5}$$

where d_{ij} (i and j are 1, 2, 3, or 4) are the elements of the matrix [D].

EXPERIMENTAL APPROACH

A novel ultrasonic goniometer was built for an obliquely-incident focused ultrasonic wave technique, as shown in Figure 2. A transducer is positioned on the arc and faces toward the center of a cylindrical reflector. Samples are translated so that the signal reflected from an internal

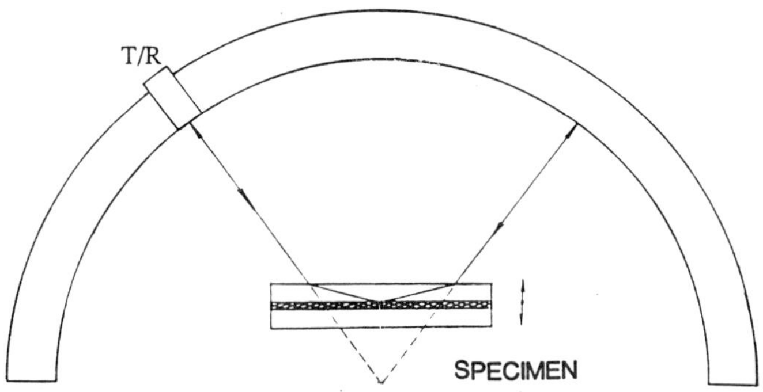

Fig. 2. Ultrasonic goniometer for measurements with the obliquely-incident ultrasonic wave focused on the interface.

interface can be reflected by the cylindrical reflector and returned to the same transducer. Such a design is distinguished by its ability to measure interface reflected signals with various incident angles using only one transducer. In addition, the sample can be moved by a translation table such that scanning a large and complex joining area is possible.

In our experimental system a computer controlled DC motor is used for simultaneous control of transducer rotation and sample translation. Angle and translation resolution and repeatability of 0.01° and 0.01 mm, respectively, are achieved in our setup. At each angle of rotation the signals received are automatically digitized and averaged by a computer controlled LeCroy 9400 125 MHz digital oscilloscope and fed to a computer through an IEEE interface. The signals are then Fourier transformed to the frequency domain and deconvolved with the spectra of signals reflected from an air-backed aluminum plate. The water tank is temperature stabilized at 29.8± 0.1°C.

Figure 3 schematically shows the type of sample used in the measurements. Each sample consists of two aluminum plates and a polymer film. A thin interphase layer made of epoxy and curing agent is applied to the interface between aluminum plates and the polymer film. The thickness of the interphase layer is in the range 20 to 50 μm (See Table 1). On one side of the polymer film the interphase layer mixture is 80% epoxy and 20% curing agent, giving completely cured properties similar to those of the polymer film. On the other side of the polymer film the interphasial layer has different percentages of curing agent for different samples. This gives the samples good bond properties on one side of the polymer film and controlled interphasial properties on the other side. Table 1 shows the percentages of curing agent used in the sample preparation and the thickness of the layers measured by a micrometer (subtracting aluminum and polymer layer thicknesses from total adhesive joint thickness).

EXPERIMENTAL RESULTS AND DISCUSSION

An example of the measured spectra is shown by the dashed line in Figure 4, where the transverse wave incident angle from the aluminum to the interphase is 40° and the sample has 20% curing agent in the controlled interphasial layer. The solid line in Figure 4 is calculated from the model using the reconstructed WBL $\omega\tau$ and thickness. The reconstruction of WBL $\omega\tau$ and thickness will be discussed in later sections. A number of sharp reflection minima, both theoretically and experimentally, appear in

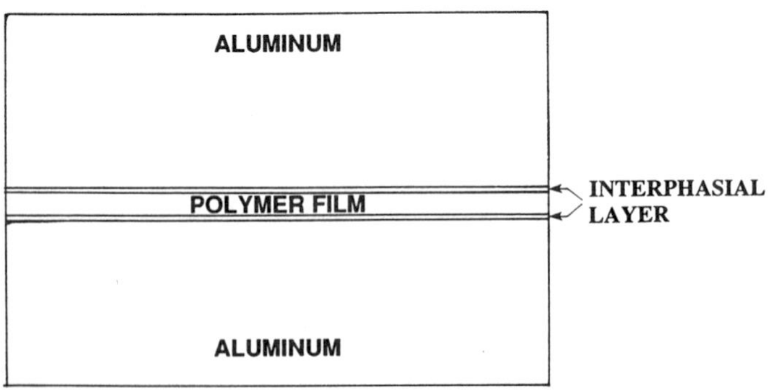

Fig. 3. Type of sample used in the measurement. Thin interphase layers made of epoxy and curing agent are applied to the polymer/aluminum plate interfaces. The percentages of curing agent are shown in Table 1.

Table 1. The properties and thicknesses of the controlled interphase layer. The percentage of curing agent shown below is to produce the interphasial region, which is on the side nearest the transducer. The other side has 20% curing agent for all samples.

percentage of curing agent in interphase (%)	0	2	5	10	20
WBL $\omega\tau$ from reconstruction	0.01	0.8	3	10	100
thickness measured by subtracting Al and polymer thicknesses from total thickness (μm)	32	29	45	53	23
thickness from reconstruction (μm)	32	34	45	47	23

the figure. Similar results were also found in the experimental results from other samples. However, each sample shows a unique spectrum pattern of its own. It was noted that the minimum-reflection frequencies and the amplitudes at these frequencies change with the percentage of curing agent in the sample. An example of these changes can be seen by comparing Figure 4 to Figure 5. In the latter, the spectrum is measured at the same incident angle (i.e., 40°) but from a sample with 0% curing agent in the controlled interphase layer. As can be seen in Figures 4 and 5, the minimum-reflection frequency and the depth of the reflection minima change with change of the percentage of curing agent in the controlled interphase. For example, a decrease of minimum-reflection frequency as high as 1.9 MHz, e.g., from 8.1 MHz in a sample with 20% curing agent to 6.2 MHz in a sample of 0% curing agent, is observed in Figures 4 and 5, and whereas almost all

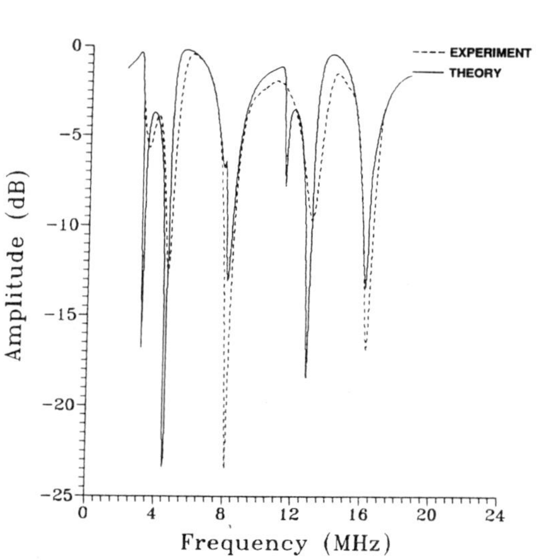

Fig. 4. Measured and calculated reflected spectra from interface at 40° incidence from aluminum (incident angle in water is 18°). The sample is prepared with 20% curing agent in the epoxy interphase layer on one side of the polymer film; the other side is completely cured.

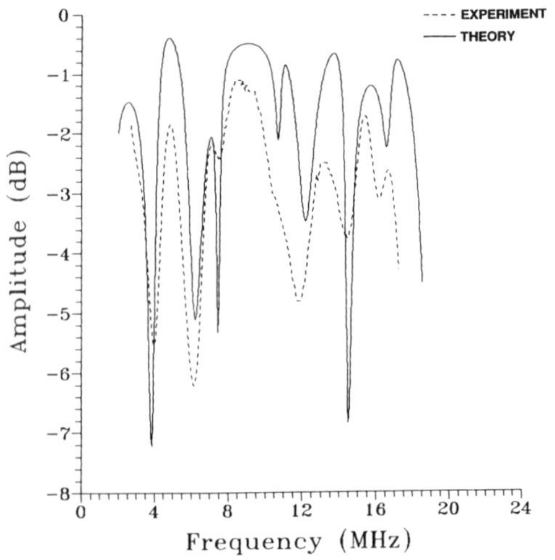

Fig. 5. Measured and calculated reflected spectra from interface at 40° incidence from aluminum (incident angle in water is 18°). The sample is prepared with 0% curing agent in the epoxy interphase layer on one side of the polymer film; the other side is completely cured.

the minima in Figure 4 have depth larger than 10 dB, the depths in Figure 5 are all less than 7 dB. These distinct features are used in later sections for the reconstruction of the WBL $\omega\tau$ and thickness.

By measuring spectra at various incident angles, loci of minimum reflection coefficient were obtained. For example, the loci of the minimum reflection coefficient for a sample with 20% curing agent in the controlled interphasial layer are shown in Figure 6, where the experimental data are shown as discrete points and the dotted lines are those calculated from the model using the reconstructed WBL $\omega\tau$ and thickness. Five branches of reflection minima are found in the frequency range 3 to 17 MHz and almost all these minima have depths greater than 3 dB. The loci of minimum reflection coefficient for other samples were obtained in a similar fashion but each has its own pattern. For example, loci for the sample with 0% curing agent are shown in Figure 7. Six branches of minima are observed. In contrast to Figure 6 almost all the minima in Figure 7 have depths less than 3 dB and the frequencies of the minima are quite different from those in Figure 6.

DETERMINATION OF INTERPHASIAL PROPERTIES

To determine the properties of the interphasial layer, a nonlinear least square approach is used. Minimum-reflection frequencies are determined by using Equation (5). Interphasial properties are calculated from equations (1) and (2), where weak-boundary-layer thickness and $\omega\tau$ are considered as free variables for curve fitting. The fitting is done for the incident angle range 16° to 25° from water to aluminum plate (32° to 60° from aluminum to interphase). The results are shown in Table 1.

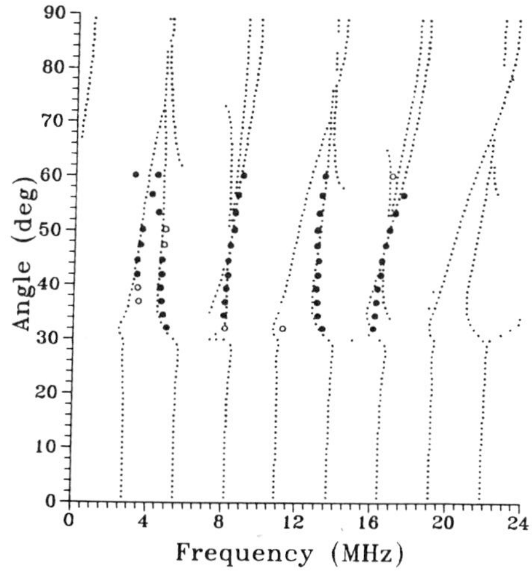

Fig. 6. Measured and calculated loci of minimum reflection coefficient at 40° incidence from aluminum (incident angle in water is 18°). The sample is prepared with 20% curing agent in the epoxy interphase layer on one side of the polymer film; the other side is completely cured. (• depth greater than 3dB, o depth less than 3dB)

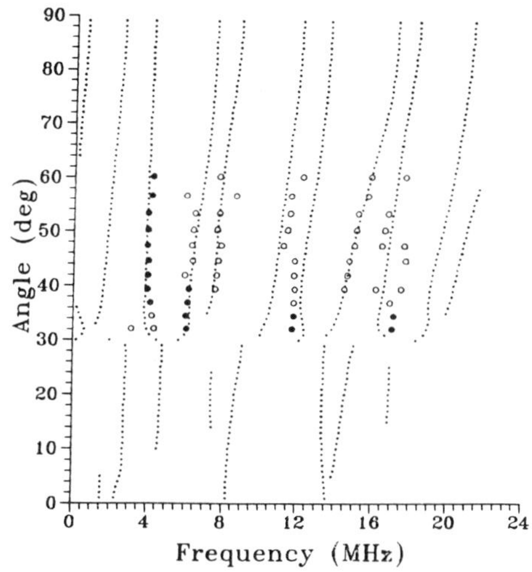

Fig. 7. Measured and calculated loci of minimum reflection coefficient at 40° incidence from aluminum (incident angle in water is 18°). The sample is prepared with 0% curing agent in the interphase layer on one side of the polymer film; the other side is completely cured. (• depth greater than 3dB, o depth less than 3dB)

As can be seen in Table 1, the reconstructed interphase thicknesses are in good agreement with those measured by subtracting the thicknesses of aluminum and polymer from total adhesive thicknesses. From the reconstructed WBL thickness and $\omega\tau$ the theoretical spectra and loci of minimum reflection coefficient are calculated. Examples of comparison of the calculated results with the experimental data can be seen in Figures 4 to 7. The agreement between the calculated results and the experimental data is satisfactory.

CONCLUSION

In this work, samples with good-bond interphase on one side of an adhesive layer and various interphasial properties on the other side of the layer have been studied. The samples are characterized by angle-beam reflection ultrasonic spectroscopy. The interphasial properties are determined from the loci of minimum reflection coefficient by a nonlinear least square curve fitting algorithm. The calculated results are in good agreement with the experimental data.

ACKNOWLEDGEMENT

This work was sponsored by the Center for Advanced NDE, operated by the Ames Laboratory, USDOE, for the Air Force Wright Aeronautical Laboratories/Materials Laboratory under contract #SC-80-148B.

REFERENCES

1. S. I. Rokhlin, M. Hefets and M. Rosen, J. Appl. Phys. 52, 2847-2851 (1981).
2. S. I. Rokhlin, J. Acoust. Soc. Am. 73, 1619-1623 (1983).
3. G. A. Alers and R. B. Thompson, 1976 Ultrasonic Symposium Proceedings, IEEE, 138-142 (1976).
4. S. I. Rokhlin, in Review of Progress in Quantitative NDE, D. O. Thompson and D. E. Chimenti, eds., Vol. 5B, Plenum Press, N.Y., 1301-1308 (1986).
5. C. M. Teller, K. J. Diercks, Y. Bar-Cohen and N. N. Shah, in Review of Progress in Quantitative NDE, D. O. Thompson and D. E. Chimenti, eds., Vol. 7B, Plenum, N.Y., 935-942 (1988).
6. S. I. Rokhlin and D. Marom, J. Acoust. Soc. of Am., 80, 585-590 (1986).
7. A. Pilarski and J. L. Rose, J. Appl. Phys. 63, 300-307 (1987).
8. P. B. Nagy and L. Adler, in Review of Progress in Quantitative NDE, D. O. Thompson and D. E. Chimenti, eds. Vol. 8B, Plenum Press, N.Y., 1417-1424 (1989).
9. J. D. Venables, in Adhesive Joints, K. L. Mittal, ed., Plenum Press, N.Y., 453-467 (1984).
10. S. I. Rokhlin, W. Wang, Y. J. Wang and J. L. Hammill, Proceedings of the 1989 ASNT Fall Conference, NDE of Adhesive Bonds and Bondlines Track, Oct. 9-13, 1989, Valley Forge, PA, 41-50 (1989).
11. S. I. Rokhlin, W. Wang and Y. J. Wang, in Review of Progress in Quantitative NDE, D. O. Thompson and D. E. Chimenti, eds. Vol. 9B, Plenum Press, N.Y., 1231-1238 (1990).
12. F. I. Solyanik, Sov. Phys. Acoust., 23, 533-536 (1977).

CHARACTERIZATION OF CMOS PROCESS VARIATIONS BY MEASURING SUBTHRESHOLD CURRENT

Aleksandra Pavasović, Andreas G. Andreou and Charles R. Westgate

Department of Electrical and Computer Engineering
The Johns Hopkins University
Baltimore, MD 21218

1 INTRODUCTION

Analog integrated circuit performance depends on the matching characteristics of individual components; transistors, resistors or capacitors. For example, in a current mirror, two identically biased transistors that are designed to have the same geometry, are said to be matched if their drain currents are equal. Transistor mismatch is one of the more acute problems in the advancement of analog VLSI collective computational systems [1].

Perfect transistor matching cannot be achieved because of fabrication process variations. Local mobility fluctuations, oxide granularity and thickness, local variations in the numbers of oxide and surface charges, Gaussian distribution of dopants in the substrate and variations in transistor length and width all contribute to the transistor mismatch. Despite the importance of transistor matching in analog circuits, there are very few published studies in this area [2,3,4,5].

In this paper we report on an experimental method to characterize the effects due to fabrication process imperfections and of the material quality, on the matching properties of MOS transistors operating in the subthreshold region. A non-destructive measurement technique and the experimental setup are described, that provide for accurate and repeatable measurements. We have studied large, dense arrays of p and n–channel MOS transistors in standard p–well and n–well CMOS technologies available through foundry services. More than 150,000 transistors with areas ranging from 9 to 400 μm^2 have been tested.

2 SUBTHRESHOLD REGION OF MOSFET OPERATION

Subthreshold MOS transistor current first attracted attention as a leakage current in the early seventies [6]. In the early eighties, Eric Vittoz [7] introduced subthreshold conduction as a principal mode of operation in micropower integrated circuits. Through their work which was aimed at the electronic watch industry, it was demonstrated convincingly that subthreshold current was predictable and its origin could be identified with specific physical process in the device.

MOS transistor is biased in the subthreshold (also referred to as weak inversion or cut–off) region when its gate to source voltage is below the threshold voltage of the device V_t. The channel is then weakly inverted and carriers diffuse from source to the drain much the same way as in the base region of a bipolar transistor. Assuming pure subthreshold conduction, drain voltage larger than a few hundreds of milivolts, zero substrate bias and transistors having square geometry, the simplified expression for the transistor current [9] is:

$$I_D = I_0 exp\left(\frac{V_{GS}}{\kappa U_T}\right) \quad (1)$$

where:

κ is the gate efficiency, which is approximately constant for a given technology,

I_0 is the subthreshold current factor, directly proportional to the oxide capacitance, carrier mobility in silicon, and κ. I_0 is inversely proportional to the doping concentration.

The above simple equation relating subthreshold current in the terminals of the transistor will be the basis of our characterization technique. Since I_0 is directly related to the silicon substrate properties measurement of this quantity over large number of transistors in arrays can provide valuable information on the quality of the fabrication process and its effects on MOS transistor matching properties.

3 TESTING METHODS

Most studies of MOS transistor mismatch were done either in the linear [3,4] or the saturation [5] region of above threshold operation. There are two reasons for doing so: First, the analysis was done to aid design of circuits for above threshold operation. Second, the measurements done in the subthreshold region were believed to be imprecise and/or buried in noise, and thus not suitable for statistical analysis. In our study we not only disproved the latter reason, but also demonstrated that the subthreshold region of transistor operation is superior for the analysis of transistor mismatch [9]. We have shown that:

1. The resolution of threshold voltage variation measurements is maximum in the subthreshold region of operation, due to exponential dependence of subthreshold drain current on gate to source voltage.

2. In the subthreshold region, the relative difference between two currents is independent of nominal V_{GS}. (This is also a consequence of the exponential dependence of I_D on V_{GS}.) Thus, analysis of measured data is made easier, and a direct comparison of results obtained from different chips is made possible without the necessity of taking more than one data point per transistor and extracting parameters.

3. The precision of subthreshold current measurements is good. A thorough investigation of the experimental apparatus demonstrated the regularity of transistor transfer characteristics in the subthreshold region. The signal to noise level ratio was found to be large (> 1000). Leakage currents were less than 1pA.

We have designed and fabricated transistor arrays that consist of 1024 (32×32) devices. In some arrays all the transistors have equal nominal dimensions ($4\mu m \times 4\mu m$ or $16\mu m \times 16\mu m$), while others consist of four blocks, each composed of transistors of different sizes. To ensure valid statistical analysis, no single–size transistor block in any array is smaller than 256 transistors. All previous studies of device mismatch in CMOS processes are based on smaller sample sizes. For example, Pelgrom[4] in his study used six transistors of the same size in each module (the number of modules per chip is not reported), Buehler[2] used test structures ranging from 63 to 223 devices, while Tanner[10] used a 6×9 multiplier array.

On our test chips, transistors are placed with a minimum separation allowed by VLSI design rules (Fig. 1). Each array is designed as dense as possible by placing substrate contacts only at its edges. The circuitry for selecting individual transistors in the arrays was included on the chips. It consists of two 5–to–32 decoders per array, one for selecting rows (gates) and the other for columns (drains). Decoders are interfaced to the arrays through a switching system which grounds all unselected rows/columns, while the selected row/column is connected to a bonding pad where the desired gate/drain voltage is applied.

The data acquisition system (Fig. 2) consists of a Macintosh SE computer with 4 Mbytes of RAM and a 40 Mbyte hard disk. The Macintosh HyperCard environment and its programing language HyperTalk are used for all data acquisition tasks. A IOtech MacSCSI488 bus controller is used to interface between the Macintosh SCI and the Keithley IEEE488 data buses. The measuring device in the system, is the Keithley 617 programmable electrometer. In cases where more than 2 controllable power supplies are needed, the Keithley 617 internal power supply is also used.

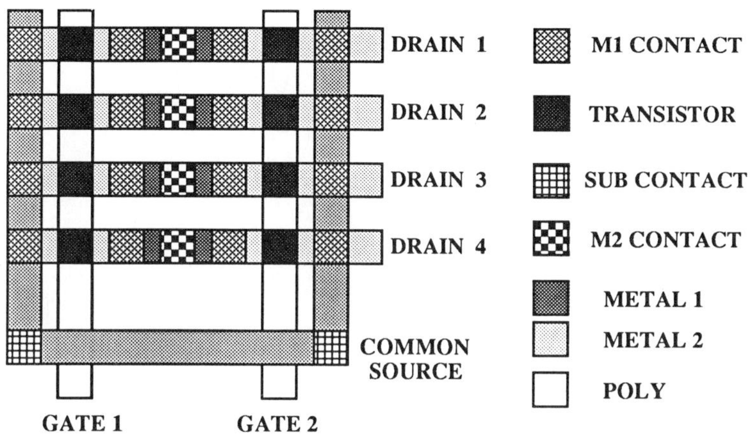

Figure 1. Layout of a corner of a $4\mu m \times 4\mu m$ n–channel transistor array

A Keithley 556 Measurement & Control System is also employed. Two cards are used: one with two 16-bit D/A converters and the other with 32 digital I/O channels (used as outputs). The D/A converters provide variable analog input test voltages. One of the digital output channels is used to control the switch connecting the non-controllable power supply to the test chip, while the rest are inputs to the decoders, located on the test chips themselves. The last component of the data acquisition system is the chip interface box, used to connect the instruments to the test chip. This interface box was custom designed. The leakage current at any pin of the chip holder is, in the worst case, smaller than 300 fA (typically about 30 fA).

Several measurements were performed to check the accuracy and repeatability of the data obtained from our experimental setup. First, transistor arrays were measured twice under the same conditions (equal nominal drain currents). The nominal drain current was 100 pA.

A typical distribution of relative errors between two measurements is shown in Fig. 3. The mean (or expected value) of the relative *absolute* difference between two values of current for the same transistor is:

$$\overline{Da} = \overline{\left| \frac{(ID1)_{ij} - (ID2)_{ij}}{ID_m} \right|} = 0.0029$$

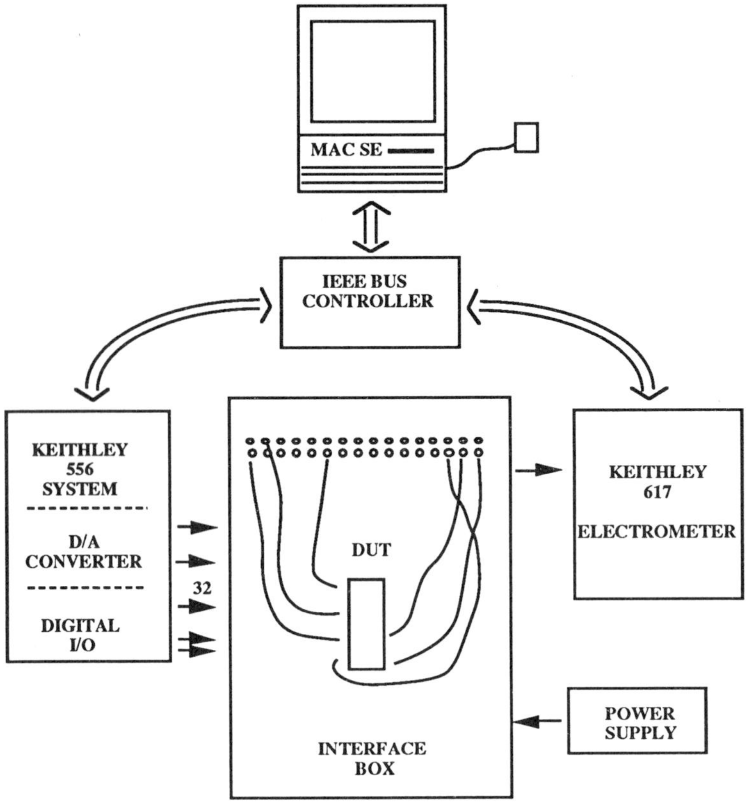

Figure 2. Configuration of the data acquisition system

where $(ID1)_{ij}$ and $(ID2)_{ij}$ are first and second measurement values of drain current of transistor in row i and column j . The corresponding median absolute difference is 0.0024, and the estimated standard deviation is 0.0032. It is important to point out that the standard deviation is calculated on

$$\frac{(ID1)_{ij} - (ID2)_{ij}}{ID_m}$$

which approximately satisfies the normal distribution law. As the value of interest is the relative absolute difference

$$Da = \left|\frac{(ID1)_{ij} - (ID2)_{ij}}{ID_m}\right|$$

we can conclude that more than two thirds of relative absolute differences are smaller than 0.3% of the nominal current value. The corresponding standard deviation (assuming $\kappa = 1.6$) for the threshold voltage is:

$$\sigma(\Delta V_{th}) = \kappa U_T \ln\left[\sigma\left(\frac{I_{D_1}}{I_{D_2}}\right)\right] \approx 0.1mV$$

Thus, the error made in estimating the threshold voltage for a particular transistor is less than ±0.1mV in 67% of the cases. This result compares favorably with the value reported in [4] ($\sigma(\Delta V_{th}) = 0.15$ mV).

In all tests conducted the normalized standard deviations were approximately equal for the two measurements (in the case above they were equal:

$$\frac{\sigma(I_{DA})}{I_{D_{NOM}}} = \frac{\sigma(I_{DB})}{I_{D_{NOM}}} = 0.128$$

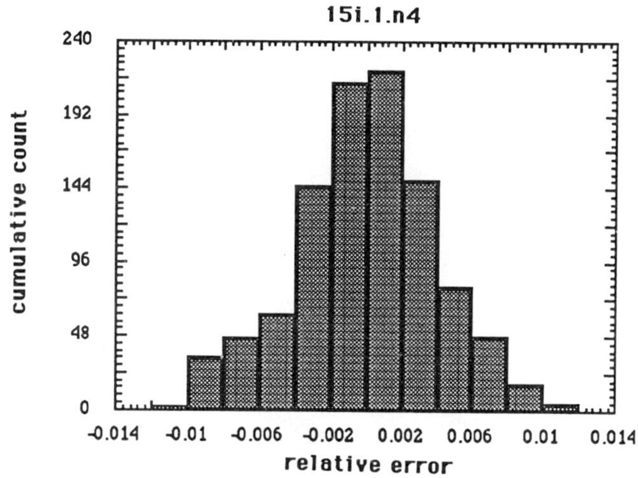

Figure 3. Distribution of relative differences between two measurements of $4\mu m \times 4\mu m$ n-channel transistor arrays

Thus, the standard deviation of the relative error is almost two orders of magnitude smaller than the quantity under investigation.

4 RESULTS

In addition to the random variations in the transistor drain currents, two quasi-deterministic effects on transistor matching were observed in our measurements. Transistors on the borders of the arrays behave differently than the transistors near the center. We refer to this effect as the "edge effect". In this paper we concentrate on a second spatial variation in transistor currents which we call the "striation effect". By *quasi-deterministic* we mean that the effects could have been predicted if the details of the fabrication process were available, if the exact position of the tested chip on the die was known and if influences of all circuits surrounding the test structure were characterized. Since none of these conditions are satisfied, we can not treat these effects as purely deterministic.

In one p-well process studied, the transistor currents follow a spatial, sinusoidal oscillation of slowly varying frequency. Figure 4 shows the density plots of several arrays demonstrating this effect. Each transistor is presented as one square (32×32 squares in all). The current value is coded by shades of gray; the minimum and maximum values represented by black and white, respectively. Figures 4a and b show the density plots of an array of $16\mu m \times 16\mu m$ n-channel transistors, the second one has edge effect removed to show the striation effect more clearly. The corresponding plots for $4 \times 4\mu m^2$ transistors are presented in Fig. 4c and d. The striation effect is relatively small. It shows clearly in the array with larger transistors where the random variations are small; however, it is not evident in the array with smaller transistors where the random variations are larger. We can regard this striation effect as a deterministic signal buried in "white noise". Its peak amplitude is approximately 6% of the average current value, while the average noise amplitude (equal to the standard deviation) is approximately 2.0% for larger transistors and 7.2% for smaller ones. The "signal-to-noise" ratios in these two cases result in the signal being noticeable in the first case and not noticeable in the second. It should be pointed out that the presence of this "signal" is undesirable since it affects the circuit performance in the same way as random mismatch. However, its effect on circuit performance is even worse than random mismatch since it does not decrease with increasing transistor area. Its nature is not known at the time of circuit design and thus, since, although it is deterministic, its effects cannot be eliminated.

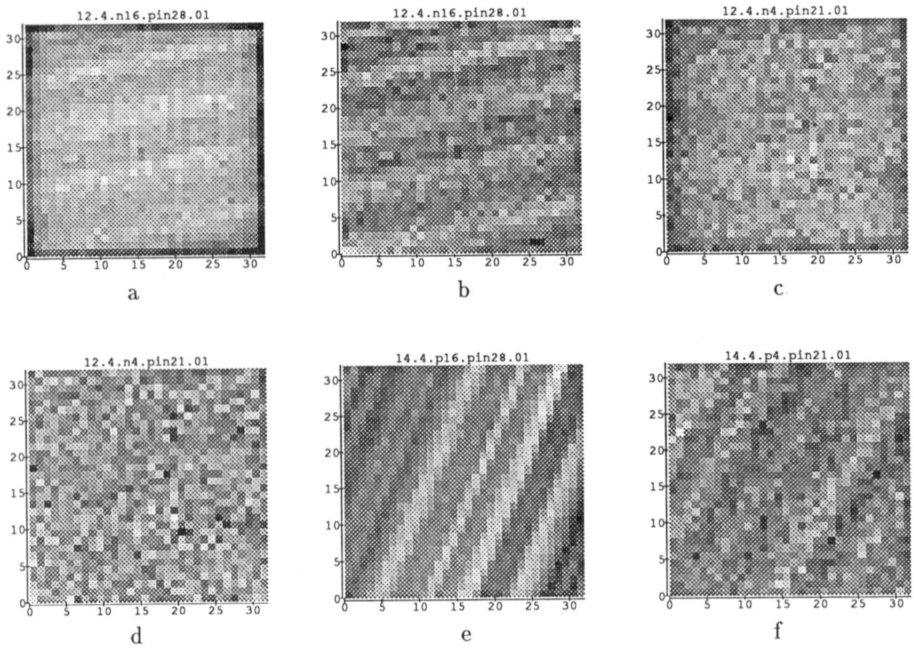

Figure 4. Density plots for several arrays on chips fabricated in p–well technology: a. n–channel $16\mu m \times 16\mu m$ array; b. same array with edge effect removed; c. n–channel $4\mu m \times 4\mu m$ array; d. same array with edge effect removed; e. p–channel $16\mu m \times 16\mu m$ array; f. p–channel $4\mu m \times 4\mu m$ array.

Density plots of arrays of p–channel transistors are presented in Fig. 4e and f. The striation effect is striking. The peak amplitude of the sinusoidal variation is up to 30% of the average current, overcoming any random variations for the larger devices. The central part of Fig. 4e was rotated so that the striation effect is perpendicular to the x axis; the result is presented in Fig. 5a. Column averages from this plot, presented in Fig. 5b, show the extent of the striation effect. The period of spatial oscillations is in the range of 100 to 200 μm.

The cause of this irregularity is not fully understood yet, but a probably cause is the threshold adjustment ion implantation process. In most modern CMOS processes, n^+ polysilicon is used for the gate electrode, requiring smaller threshold voltage adjustment for n–channel devices than for the p–channel devices. Thus, larger ion implantation dose is needed for the p–devices, and any irregularity in the implantation process would affect the p–channel transistors significantly more than the n–channel, which is evident from our data.

The adverse effects of these striations cannot be avoided in the circuit design stage, especially if the area for each circuit is limited. It would be better to modify the fabrication process such that the striation effect is reduced much below the level of the random noise. Fortunately, it seems that the state of today's technology is advanced enough to achieve this goal.

5 CONCLUSIONS

We have described a novel method for characterizing irregularities in CMOS fabrication process by testing transistor mismatch in subthreshold region. Our method is accurate, offers high sensitivity, and is automated, thus could be used for on-line monitoring of the unifor-

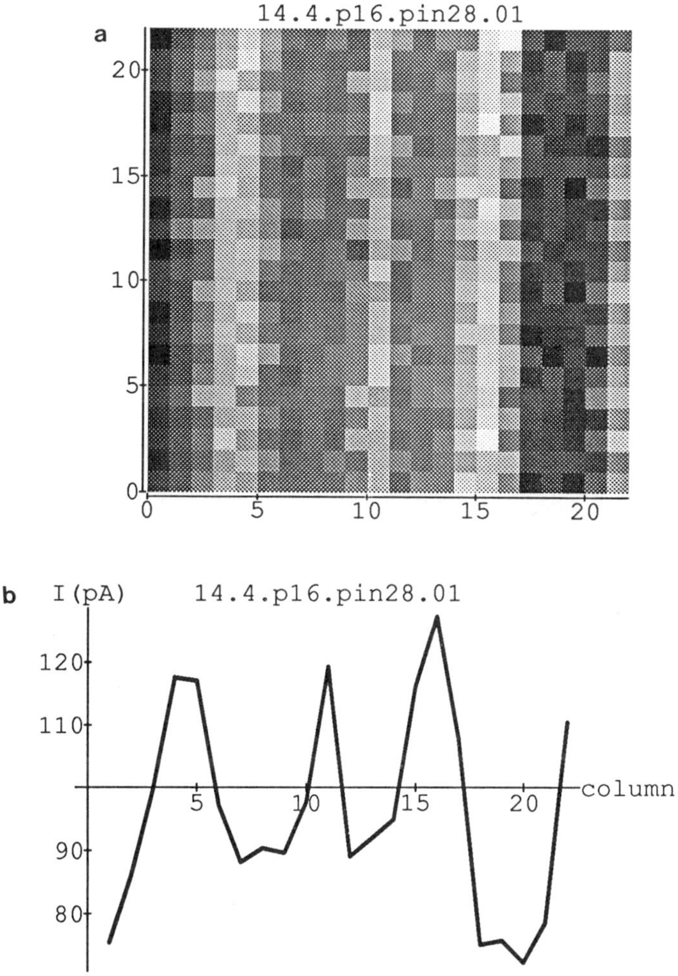

Figure 5. (a) Density plot for the central part of the array in Fig. 4e rotated by 25°; (b) corresponding column averages.

mity in the doping process. Our measurements have shown the existence of two effects, not reported until now in the open literature, which are related to the details of the fabrication process. These result in severe mismatching of transistor characteristics. Previous studies do not report similar effects because small test samples were used and devices under test were not in proximity. Statistical analysis of small sample sizes cannot discriminate between the various underlying statistical distributions, and thus the assumption that data are normally distributed can be neither confirmed nor refuted.

6 Acknowledgement

The authors would like to thank Robert Jenkins of the Applied Physics Laboratory for providing financial support and encouragement. Thanks also go to P.O. Pouliquen for designing the custom measuring box and to Kwabena Boahen for usefull discussions.

References

[1] C. A. Mead: *"Analog VLSI and Neural Systems"*, Addison–Wesley Publishing Company,1989.

[2] Buehler M.G., Blaes B.R. Sayah H.R. and Lieneweg U.: *"Parameter Distributions for Complex VLSI circuits"*, Proceedings of the Decennial Caltech Conference on VLSI, MIT Press, March 1989.

[3] Lakshmukumar K.R., Hadaway R.A. and Copeland M.A.: *"Characterization and Modeling of Mismatch in MOS Transistors for Precision Analog Design"*, IEEE J. of Solid-State Circuits, Vol. SC-21, No. 6, Dec. 1986.

[4] Pelgrom M.J., Kuinmaijer A.C.J. and Welbers A.P.G.: *"Matching Properties of MOS Transistors"*, IEEE J. of Solid-State Circuits, Vol. SC-24, No. 5, Oct. 1989.

[5] Shyu J.B., Temes G.C. and Krummenacher F.: *"Random Error Effects in Matched MOS Capacitors and Current Sources"*, IEEE J. of Solid-State Circuits, Vol. SC-19, No. 6, Dec. 1984.

[6] Gosney W. M.: *"Subthreshold Drain Leakage Currents in MOS Field-Effect Transistors"*, IEEE Transactions on Electron Devices, Vol. ED-19, No. 2, Feb. 1972.

[7] Vittoz E.: "Micropower Techniques", in *Design of MOS VLSI Circuits for telecomunications*, Eds. Y.Tsividis, P.Antognetti, 1985., pp. 104-144.

[8] Pavasović A: *"Subthreshold Region MOSFET Mismatch Analysis and Modeling for Analog VLSI Systems"*, Ph.D. dissertation, The Johns Hopkins University, 1990 (unpublished).

[9] Tanner J.: *"Integrated Optical Motion Detection"*, Ph.D. disertation, California Institute of Technology, 1986.

The choice of the operating frequency, for a given material ($\sqrt{\varepsilon'_r}$) is a compromise between the highest resolution (inversely proportional to the wavelength) and microwave penetration (proportional wavelength). For the first trials, we chose a frequency of 94 GHz (λ_o = 3,19 mm) and for instance with a ceramic material: $\varepsilon_r = 9$ and $tg\delta = 2{,}6.\ 10^{-2}$, we get a resolution of 1 mm and a penetration depth of 13 mm.

PRINCIPLES OF THE EXPERIMENTAL MILLIMETER N.D.T. SENSOR

We designed a 94 GHz FM-CW Radar sensor (frequency modulated-continuous waves) associated with a focusing ellipsoïdal reflector antenna of diameter 30 cm and focal length 55 cm[1] (Figure 1). The material under test is located at the second focal point of the ellipsoïdal reflector antenna; where the focal spot diameter is equal to

$$4/\sqrt{\varepsilon_r}\ \text{mm (at - 3 dB points)}$$

and ε_r is the dielectric constant of the sample. At this location, there are quasi-plane waves[2] and any parallel plane layer theory can be applied to rely the reflexion coefficient to the material characteristics. The transmitted signal from the source is a 15 mW continuous wave oscillator output, linearly swept over 1 GHz bandwidth.

The receiver detects the combination of a fraction of this signal and the delayed signal reflected by the sample under test. The resulting intermediate frequency is then filtered and amplified before processing. Because the FM waveform is periodic, the actual output spectrum is a line spectrum with components at harmonics of the sweep frequency (25 KHz). Because the FM waveform is periodic, the actual output spectrum is a line spectrum with components at harmonics of the sweep frequency (25 KHz). The amplitude of the nth component which satisfied the relationship,

$$n = 2\,\frac{L\,\Delta F}{C}$$

where ΔF = bandwith of the oscillator, C = speed of the light, and L = range from the antenna to the sample, and is related to the energy reflected by the sample and so to its physical characteristics.

The 94 GHz FM-CW radar sensor is mounted on a platform with four mechanical translators driven by step-by-step motors (Figure 2). Three of them realize an horizontal X-Y scanning across the sample in a 1 m × 1 m range, with 100 μm × 10 μm resolution and 20 cm/s maximum speed. The fourth translator executes a vertical Z movement in order to focus the beam either at the surface or inside the material. The Z movement can vary in the range of 0.6 m and with 10 μm resolution.

A desktop computer drives the scanning platform, performs signal acquisition and image processing. Thus we observe on the screen of the computer an X-Y colored image related to the material under test. This image is a true image of the normal incidence reflectivity variations of the sample.

EXPERIMENTAL RESULTS

Many experimental trials have been made on various dielectric materials to detect in depth defects or inhomogeneity distributions in composite materials. For example, in Figure 3, we see a composite material mapping. The composite material is made of carbon powder embedded in epoxy resin matrix. The black areas are mainly the inhomogeneous distribution of the carbon in the mixture. Figures 4 and 5 are examples of in depth defects detection. In Figure 4, we see glue defects between slices in sonic absorbing material. Figure 5 shows two metal grains (less than 1 mm in diameter) embedded in a piece of advanced ceramic which is 50 mm in diameter.

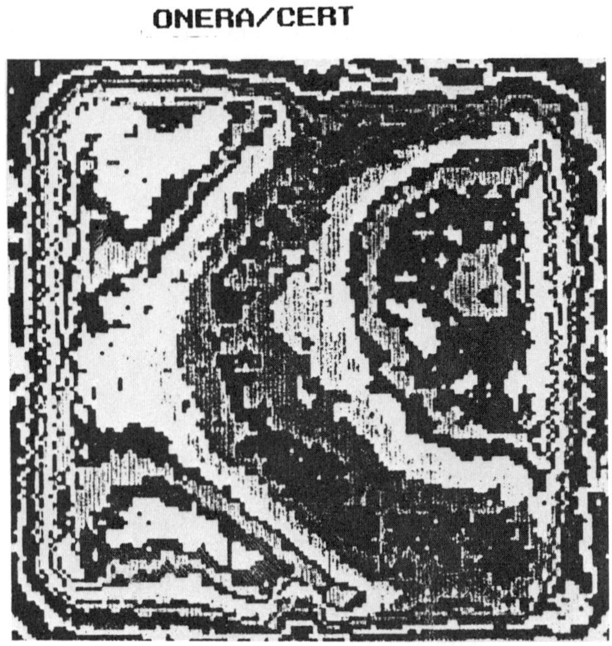

Figure 3. Mapping of a Composite Material

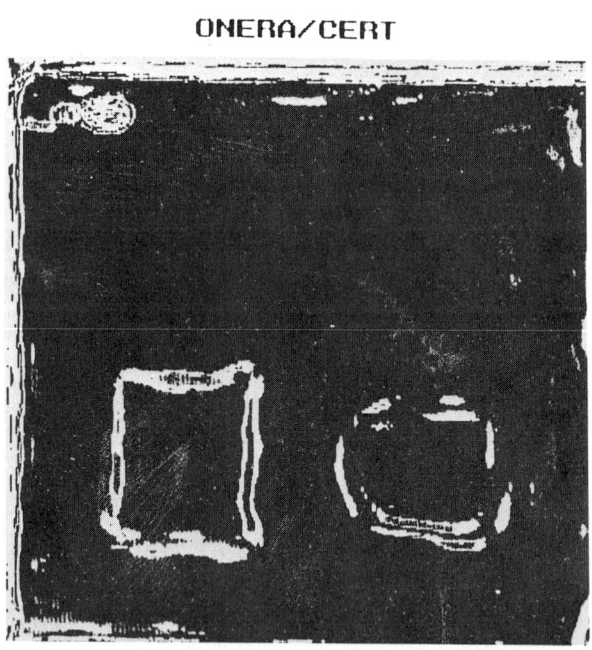

Figure 4. Glue Defects Inside Sonic Absorbing Material

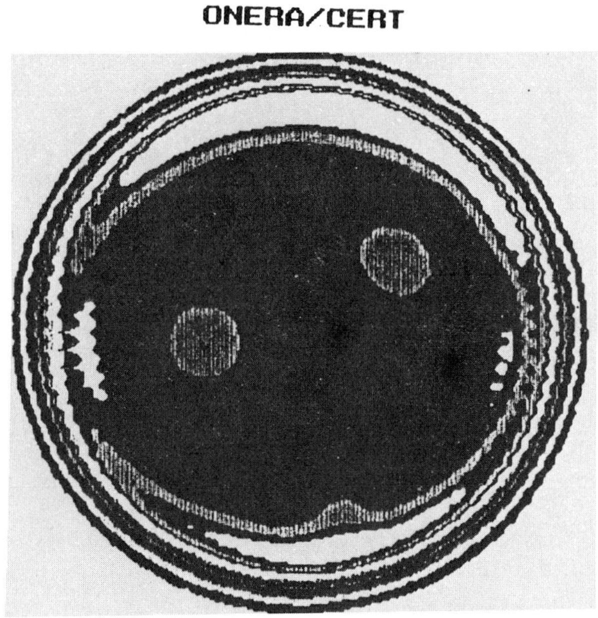

Figure 5. Two Metal Grains Embedded Inside Advanced Ceramic

CONCLUSION

A microwave N.D.T. platform was built to test composite materials. The first observed results are very encouraging. It appears that microwave N.D.T. provided data for the detection of in depth defects and gives information about the inhomogeneity of composite materials.

The application of this technique to other kinds of samples - foreign bodies, air bubbles, porosity, resin defect, inhomogeneity of heterogeneous materials built for absorbing material - is presently under consideration.

The platform performance may be improved by using other frequency ranges and wave polarizations.

REFERENCES

1. S. Roques and A. Prious, "Millimeter Waves Non Destructif Testing," 8th International Technical Transfer Conference, Berlin, Germany (September 1989).

2. J. Musil and F. Zacek, Studies in Electrical and Electronic Engineering, Elsevier (1986).

ULTRASONIC CHARACTERIZATION OF ATTENUATIVE MATERIALS BY MEANS OF A CORRELATOR SYSTEM

Eric A. Lindgren and M. Rosen

Center for Nondestructive Evaluation
The Johns Hopkins University
Baltimore, MD 21218

INTRODUCTION

The new demands of material performance, such as higher strength per weight and high operating temperatures, have caused an increase in the use of polymers and composites in daily engineering practices. When these materials are nondestructively evaluated by means of ultrasound, a problem that frequently occurs is the high attenuation or the absorption of the energy of the sound wave by the material being tested.

A possible solution to this problem proposed in this paper is the use of a correlation system. This method has the possibility of having, in principle, an infinite signal to noise ratio enhancement (SNRE). The correlator used in this study has a SNRE value of 70 dB. This high value cannot be duplicated by conventional pulse-echo systems as there are limits on the ability to increase the power of the emitted signal without losing range or resolution[1]. Signal averaging systems can reach this SNRE value, but only after a prohibitively large number of averages[2].

In addition to its ability to penetrate highly attenuative samples, the correlation method is also capable of measuring the impulse response of the material that is being tested[3]. This response can be transformed into the frequency domain. The frequency response can then be used to further characterize the material.

The correlator has been used to penetrate many different highly attenuative samples. These include polytetrafluoroethylene (teflon), neoprene rubber, C-ρ rubber, carbon-carbon composites with different lay-ups and different weaves, concrete, wood and foam insulation material.

It is also possible that the correlation method can be used to measure the travel time of an ultrasonic wave in very thin materials if the proper experimental conditions are met.

Currently, these conditions are not available, however, they should be implemented in later models of the correlator.

THEORY

Correlation occurs when there is a systematic connection between two items. The mathematical definition of correlation is given by the following equation:

$$R(\tau) = \lim_{T \to \infty} \frac{1}{T} \int_0^T x(t)\,y(t+\tau)\,dt \tag{1}$$

where the variables are defined as:

$R(\tau)$ = the correlation function

$x(t)$ = a waveform

$y(t)$ = another waveform

τ = time delay between $x(t)$ and $y(t)$

T = the limit of the integration

Any common phase and frequency information between the two waveforms would cause the correlator function to have a non-zero value. There are two types of correlation functions: one, when $x(t) = y(t)$, i.e. the two waveforms being correlated are from the same source; the second, when $x(t) \neq y(t)$ or the two waveforms are not the same. The first method is called autocorrelation and the second cross-correlation. The method used with the correlator in this work is actually a combination of the two. The two waveforms are identical, but they do not originate from the same source. This feature is especially important in order to derive that the correlator output is proportional to the impulse response of the material being tested. Before this derivation is shown, the ability of the correlator to test highly attenuative materials will be discussed.

A. Signal-to-Noise Ratio Enhancement

The correlation scheme has a great advantage over the conventional pulse-echo methods of ultrasonic detection. The pulse-echo systems are limited by the following expression[4]:

$$\frac{\text{Maximum range}}{\text{Range resolution}} \leq \frac{\text{Burst interval}}{\text{Burst width}} = \frac{\text{Peak power}}{\text{Average power}} \tag{2}$$

This means that the ratio between the maximum range and the range resolution is limited by the peak power that can be applied to the transducers before electric breakdown by depolarization of the transducer occurs[5].

Therefore, when highly attenuating materials are investigated using the pulse-echo method, the amount of average power that is applied to the material is desired to be large to increase the size of the returning echoes and make them detectable by the receiving transducer. However, since the peak power of a transducer has a limit, the increase in the average power will cause a deterioration of either the range or the resolution of the test system[1]. This limitation is not present in the correlation system as it uses large time-bandwidth transmit signals[6]. The large time bandwidth is determined by the length of the maximal-length sequence which is used to generate the noise input signal[7]. These sequences will be discussed later. Thus, the resolution of the correlation system is determined by only the bandwidth of the received signal[5] and the average power delivered by the transducers can be larger than conventional methods.

In addition to increasing the power that is used to test a material, the correlator has the ability to arbitrarily increase the signal-to-noise ratio enhancement (SNRE). SNRE is defined by the following equation[1,8]:

$$\mathrm{SNRE} = \frac{\mathrm{SNR}_{out}}{\mathrm{SNR}_{in}} = a\frac{B_{in}}{B_{out}} = aBT \qquad (3)$$

where: a = on-off ratio of the transmitted signal
B_{in} = Bandwidth of the input transmitted noise
B_{out} = Bandwidth of the output signal from the correlator (= 1/T)
T = integration time of the correlator

This expression is a result of the thermal noise power of amplifiers being proportional to their bandwidths[8]. Therefore, the signal-to-noise power at the output of an ultrasonic system divided by the signal-to-noise ratio of the input signal into the material being tested can be shown to equal the input signal bandwidth divided by the output signal bandwidth.

Equation 3 cannot be readily used in conjunction with pulse-echo systems as the bandwidth of the transmitted pulse and the bandwidth of the output signal from the detector are approximately the same. In correlation systems, however, the integration time is the inverse of the output bandwidth. Thus, the SNRE can theoretically be made arbitrarily large by either increasing the integration time or by increasing the input bandwidth. In the system used in this study, the integration time was fixed by the time constant of the low pass filter that functioned as the integrator at 1 second. The input bandwidth was determined by the rate of sweep through the maximal-length sequence. This sweep rate was set by an internal clock at a rate of 10 MHz. Thus, the SNRE is 70 dB for the correlator.

B. Impulse Response

When a linear system is tested with the correlation function, the resulting correlation is the impulse response of the system if the input signal into the system is white noise. The output from the system is correlated with the white noise signal to give the correlation output. This can be shown in the following derivation[3].

If x(t) is taken as the white noise input to the system, then the output, y(t) is:

$$y(t) = h(t) * x(t) \qquad (4)$$

where h(t) is the impulse response of the system and * is the convolution operation. This can be written as:

$$y(t) = \int_{-\infty}^{\infty} h(v)x(t-\tau)\, dv \qquad (5)$$

Substituting in the values of y(t) into the correlation equation (1):

$$R_{xy}(\tau) = \lim_{T \to \infty} \frac{1}{T}\int_0^T x(t) \int_{-\infty}^{\infty} h(v)x(t-v-\tau)\, dvdt \qquad (6)$$

Reversing the order of integration gives:

$$R_{xy}(\tau) = \int_{-\infty}^{\infty} h(v) \lim_{T \to \infty} \frac{1}{T}\int_{0}^{T} x(t)x(t-v-\tau)\, dtdv \qquad (7)$$

The definition of autocorrelation of the input signal of x(t) at time $(\tau - v)$ is:

$$R_{xx}(\tau-v) = \lim_{T \to \infty} \frac{1}{T}\int_{0}^{T} x(t)x(t+(\tau-v))\, dt \qquad (8)$$

This autocorrelation can be substituted into equation (7):

$$R_{xy}(\tau) = \int_{-\infty}^{\infty} h(v)R_{xx}(\tau-v)\, dv \qquad (9)$$

This can be written as:

$$R_{xy}(\tau) = h(\tau)*R_{xx}(\tau) \qquad (10)$$

and since x(t) is white noise, $R_{XX}(\tau)$ is equal to the delta function, $\delta(\tau)$ multiplied by a constant, K. The output is now:

$$R_{xy}(\tau) = Kh(\tau) \qquad (11)$$

This demonstrates that the correlation output is proportional to the impulse response of the system tested.

In the correlator used in this project, the input signal is not white noise but a pseudorandom signal of a broad frequency range. The broad frequency is important as this allows inspection of a sample due to interactions at these various frequencies as well as amplitude variations[9]. The study of frequency response was started in the 1960's[10] and specific responses have been characterized to come from specific materials subjected to certain test conditions[4]. Thus, any signal processing done on the correlator output has the advantage that it can give frequency information over a wide bandwidth.

The signal used in the present correlator system was generated by a maximal-length sequence. The m-sequence, as it is commonly called, comes from circuitry that includes an n-bit shift register, an exclusive-or gate, and a clock which causes the shift register to shift to the right at certain times[3]. The method to produce the m-sequence is described in the literature[11,12]. The m-sequence is desirable as the resulting bit pattern approximates white noise that has a broad bandwidth.

The output of this broad-banded signal is performed on a continuous wave basis. When continuous wave transmission is used, it has the advantage that self noise is not introduced into the output of the correlation system, as the self noise is a result of breaking the m-sequence into short bursts[13]. This is especially important when the goal is to investigate characteristics of highly attenuating media. While this dictates the use of two transducers, one as a transmitter and one as a receiver, this is not seen as a serious experimental limitation. Earlier work has been done on a pulsed wave basis where self noise was minimized by means of increased correlation times[5] or specially paired Golay[14] codes to replace the m-sequence. While these methods seem to be successful, it is desirable that when testing highly attenuating materials, any possibility of self noise should be eliminated. Also, when the testing is done on a continuous basis, the amount of power that can be delivered to the sample is limited by its ability to dissipate that power and not on the length of the pulses.

Another problem encountered in previous correlation schemes is the use of a delay between the signal used to test the sample and the reference signal. Early correlators used a mechanical delay line[1,8], and a later design used a digital delay line[5]. By using a mechanical system, it becomes difficult to produce the continuous linearly varying delay necessary for the correlation function output to be proportional to the impulse response of the system tested[15]. If bucket-brigade delay lines are used, the useful delay is approximately 0.2 ms to 10 ms, thus, eliminating any impulses arriving after this time.

The correlator used in this study uses another approach to generate this delay without using a delay unit. An initial delay is introduced between two separate, but equal, noise generators. Special circuits[3] cause the correlator to cycle through the m-sequence of one generator one clock rate slower than in the other generator. Thus, the change in delay between the two generators will be 1.0×10^{-7} seconds per second. This is called the sweep rate of the correlator. It is important to note that the inverse of the sweep rate is the input bandwidth of the correlator which determines the amount of SNRE that is possible. It is also important to note that the clock rate determines the time resolution of the correlator, in this case it is 100 nanoseconds.

Figure 1 shows a block diagram of the correlator. The clock controls the sweep across the test and reference pseudorandom noise generators. The reference noise generator has the delay circuitry which creates the continuous linearly varying delay between the two generators. The test signal, which is five volts peak to peak, is sent through the sample and is pre-amplified before being correlated with the reference signal. The propagation delay circuit is a convenience circuit which allows the elimination of output before the arrival of the test signal which traveled through the specimen. Thus, output which would yield a straight line (multiplied by zero) can be eliminated which decreases the amount of time to make measurements.

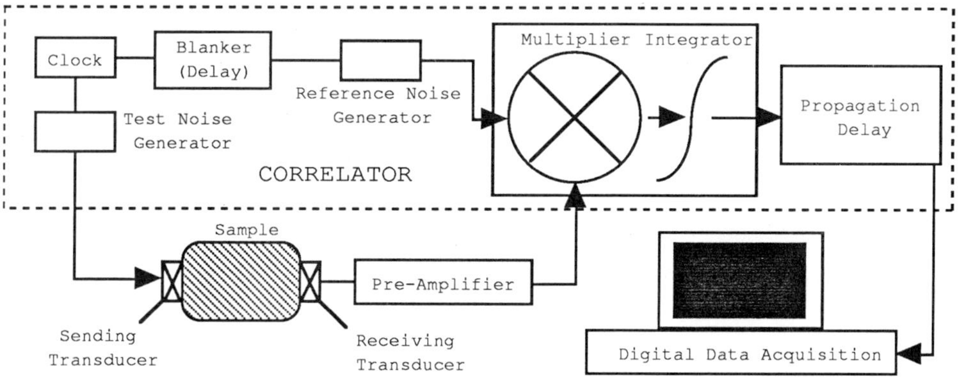

Figure 1. Block diagram of the correlator and the experimental setup used to make measurements.

RESULTS

Many different highly attenuative materials have been tested using the correlation system. These include polytetrafluoroethylene (teflon), neoprene rubber, c-ρ rubber, carbon-carbon composites, concrete, wood, and foam insulation material. Figure 2 shows the correlation output from a piece of concrete with a 15 inch travel path tested in through transmission with commercial 1.0 MHz transducers. The acquisition time for this sample was three minutes and 20 seconds. Figure 3 shows the result from the same specimen. This time the propagation delay control was set at 85.1 microseconds, thus reducing the acquisition time to 51 seconds. This demonstrates the correlator's ability to reduce the acquisition time when gathering data from samples with a long travel time.

Figure 4 shows the result of a through transmission test of a 0.5 inch piece of foam insulation material used to insulate liquid fuel cells. This test was carried out using 0.5 MHz transducers. The purpose of this test is to demonstrate the possibility of using the correlator to examine actual fuel cells and to detect delamination between the foam insulation and the steel casing for the fuel. Figure 5 shows the geometry of the sample. The steel casing is only 0.015 inches thick, requiring a frequency of 5.0 MHz in order for the wavelength of the ultrasonic wave to be small enough to resolve its presence. If the incident sound wave can resolve the steel, then the reflected wave should have a phase inversion if there is an air gap between the steel and the insulation. If there is no air gap, then there should be no phase inversion. It is hoped that the correlator will be able to detect any airgaps by using this property of the sound wave.

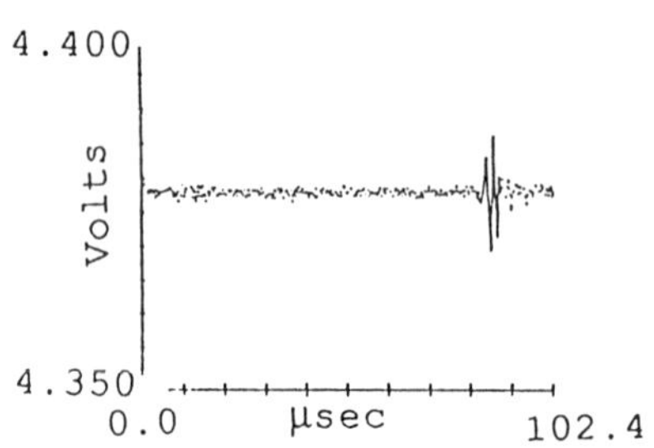

Figure 2. Correlation output from 15 inches of concrete.

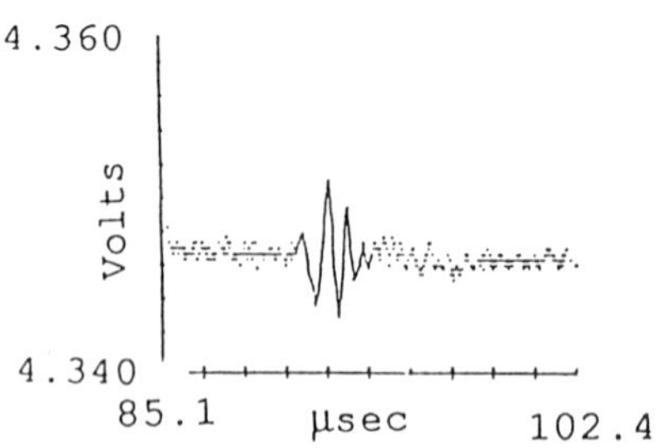

Figure 3. Correlator output from 15 inches of concrete with the propagation delay set to 85.1 μsec.

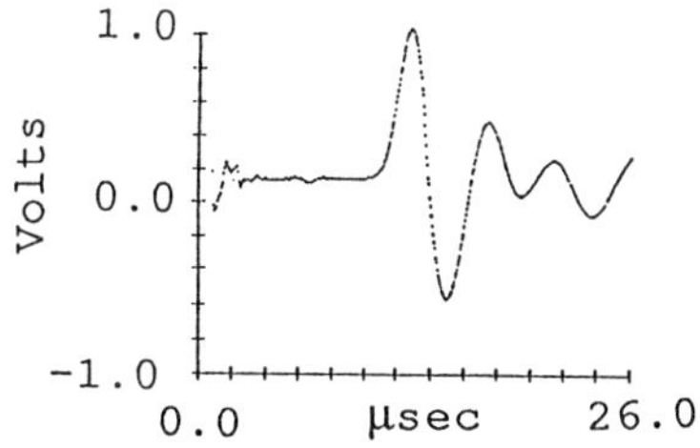

Figure 4. Correlation output from 0.5 inches of liquid fuel cell insulation material.

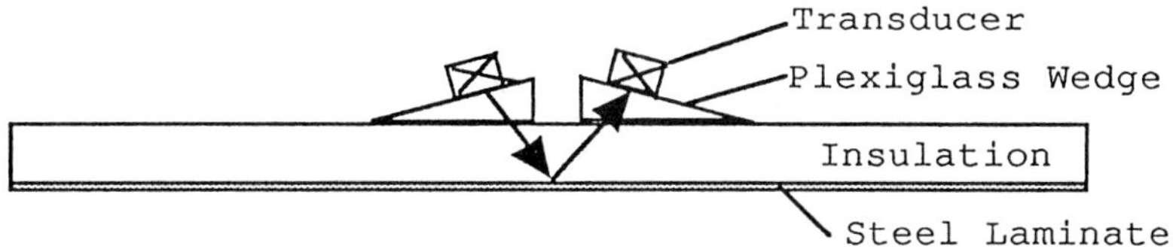

Figure 5. The sample geometry and the experimental setup to test the liquid fuel cell.

A complete comparison of the correlator to pulsed signal averaging methods has been performed[2]. The results show that the correlator has an increased SNRE as it is able to transmit more power into the sample than the signal averaging method. The limit of the transmitted power for pulsed systems is the breakdown voltage of the transducer. For the correlator, which sends its signal on a continuous wave basis, this is not true. The low amplitude of the transmitted signal does not approach the breakdown voltage of the transducer. Instead, the limit on the amount of power that can be transmitted is the maximum power that can be dissipated by the sample. Therefore, the total power is usually much higher than in pulsed systems, resulting in stronger signals.

Both systems use an increase in the acquisition time to obtain a SNRE. The method that is used is different between the two systems. The pulsed system has a fixed time window in which it collects data. The SNRE increases as the acquisition time increases as the signals are added together and the random noise is averaged out and the true signal, which is not random, is summed and emerges from the noise floor. The correlator has a fixed SNRE which is determined by its sweep rate which, in turn, determines the input bandwidth. The time window of the data increases as the acquisition time increases, showing the penetration of the ultrasound into the sample. Thus, for the same amount of acquisition time, the two systems should obtain the same amount of SNRE. This is not usually the case as the propagation delay control on the correlator can greatly reduce the amount of acquisition time, especially when inspecting a localized region within a sample. If the travel time is known for an ultrasonic arrival, then the propagation delay control can be set to only observe a limited time window around this arrival time. Also, the acquisition time for signal averagers depends on the speed of the digitizer and the averager. For example, the Data Precision 8000 takes 32000 averages in 100 minutes. This still only gives a SNRE of 45 dB while the correlator has an instant SNRE of 70 dB.

Another feature of the correlator which has not been utilized is the absence of an excitation pulse which occurs when a sample is tested by a pulse method. A small d.c. offset occurs when a test is started, but the magnitude of this offset can be controlled. If a faster clock were placed in the correlator to increase its time resolution, then it is possible that sound wave velocities could be determined for samples with very short travel times.

SUMMARY

The use of the correlation system has demonstrated its ability to test highly attenuative materials. The SNRE of the correlator used in this study was 70 dB. This proved useful when testing such materials as concrete and foam insulation. The source of the 70 dB SNRE was shown to be the sweep rate of the correlator which determines the input bandwidth of the system. It has also been demonstrated that the correlator can measure the impulse response of the material tested as the test waveforms are pseudorandom white noise. A comparison to pulsed signal averaging methods show that the correlator has a larger SNRE due to its increase in transmitted power and also that the correlator can decrease the acquisition time due to the propagation delay feature.

REFERENCES

1. Bilgutay, N.M., Furgason, E.S. and Newhouse, V.L., IEEE Trans. Sonics Ultrasonics, SU-23, 5, (1976), 329.
2. Kishoni, D. and Pietsch, B., "Ultrasonic Correlator versus Signal Averager as a Signal to Noise Enhancement Instrument", Proceedings of the 35th International Instrumentation Symposium, 1989
3. Shyn, Hung-Fa, Goncharoff, Vladimir and Jacobs, John, Intl. Ad. Nondestructive Testing, 10, (1984), 43.
4. Adler, L. and Lewis, D.K., Mat. Eval., 35, (1977), 51.
5. Lee, B.B. and Furgason, E.S., IEEE Trans. Sonics Ultrasonics, SU-29, (1982), 359.
6. Furgason, E.S., Newhouse, V.L. and Sanice, J.,"Digital Techniques to Improve Flaw Detection by Ultrasound Systems," EPRI Final Project Report 1395-4, (1986).
7. Lam, F.K. and Hui, M.S., Ultrasonics, 20, (1982), 107.
8. Furgason, E.S., Newhouse, V.L., Bilgutay, N.M.and Cooper, G.R., Ultrasonics, 13, (1975), 11.
9. Brown, A.F., Ultrasonics, 11, (1973), 202.
10. Gerike, G.R., J. Acous. Soc. Am., 35, (1963), 364
11. Haykin, S., Communication Systems, John Wiley and Sons, New York, N.Y., 1978.
12. Forney, G.D.,Jr., Spread Spectrum Techniques, IEEE Press, New York, N.Y., 1976.
13. Elias, C., IEEE Trans. Sonics Ultrasonics, SU-27, (1980),1.
14. Lee, B.B. and Furgason, E.S., Ultrasonics, 21, (1983), 153.
15. "Ultrasonic Correlator for Nondestructive Characterization of Materials," NASA SBIR Proposal No. 84-1-04-10-0332, (1984).

ULTRASONIC LEAKY WAVES FOR NONDESTRUCTIVE INTERFACE CHARACTERIZATION

T.M. Hsieh and M. Rosen*

Industrial Quality, Inc.
Gaithersburg, MD 20879

INTRODUCTION

There has been a great deal of emphasis recently in the development of composite materials, specifically metal and ceramic matrix composites. These types of composite materials offer the advantages of being lighter, stiffer, stronger, and more resistant to creep and corrosion. However, because of the physical and chemical differences of the matrix and reinforcing agents the interface is plagued by chemical reactions and a high level of residual stress. This impedes the ability of the interface to bear and transfer load and results in fracture upon subsequent stress. Thus, the need for nondestructive characterization of interfaces is critical to the development of these high technology composite materials.

Feedback from a nondestructive interface characterization technique is critical to the further development and refinement of materials processing procedures as well as for nondestructive testing of components. The unique propagation characteristics of leaky interface waves appear to be ideally suited for these purposes. While bulk moduli measurements usually yield relatively poor average values of modulus for fiber reinforced composites, leaky waves offer the potential for microscopic measurement of these physical parameters at localized regions of interest. References to the existence of acoustic waves that travel along an interface have sporadically appeared in the literature since they were first predicted by Richard Stoneley in 1924 [1-5]. Many of these presented theoretical predictions of wave propagation between two ideal media with

*Johns Hopkins University, Department of Materials Science and Engineering, Baltimore, MD 21218

Nondestructive Characterization of Materials IV
Edited by C.O. Ruud *et al.*, Plenum Press, New York, 1991

only a few papers citing experimental observations. Interface waves are elastic waves which propagate along the boundary between two solid half-spaces with both longitudinal and shear components of particle motion. It has been shown that they are sensitive to near bond line conditions including microstructural variations and degree of contact. Renewed interest in these types of waves has developed because of their potential applications for characterizing composite interfaces.

On a macroscopic scale, nondestructive characterization of interfaces is important to insure structural integrity. It is clear that nondestructive interface characterization is also necessary for the accurate modeling of the micro and macrobehavior of composites. The sensitivity of interface waves to microstructural variations and bond quality at a planar interface was reported by the authors in a previous publication[6]. It was demonstrated that both Stoneley and leaky waves show good quantitative agreement to the analytically predicted velocities and are highly sensitive to near bond-line conditions. In this study, remote techniques for the generation and detection of leaky waves at a planar interface were developed and utilized to evaluate the potential of leaky waves for nondestructive evaluation of composite interfaces.

THEORY

The boundary condition matrix equation for the planar interface case has been examined in detail by Lee and Corbly[7] and Simmons and Krasicka[8]. In order to calculate theoretical leaky waves in the Al/SiC system, experimentally determined values for density and velocity were substituted into the determinantal boundary equation. The following analytical results were obtained utilizing computer models developed for this program at the National Institute for Standards and Technology[9] (NIST).

All 16 branches of the characteristic matrix equation can produce solutions, however, not all solutions may physically exist. Leaky waves were calculated from the characteristic wave equation by locating the approximate real and pure imaginary roots and then using Newton's method for computing the real and complex roots. Solutions to the boundary matrix equation produces two types of modes. Decomposable modes which are a superposition of longitudinal and shear plane waves and non-decomposable modes. The decomposable modes represent certain reflection/refraction situations in which one or more output wave is missing. Table I shows the 10 analytical interface modes for the SiC/Al system. There are six non-decomposable modes and four decomposable modes that appear. A double index notation was used, referring to s, t modes where $s, t = 2p + q$. Both p and q may be 0 if the negative principal square root value is chosen or 1 if the positive principal square root value for the longitudinal (p) and shear (q) components. Therefore, s and t range from 0 to 3. The first index refers to the matrix material, aluminum in this case and the second index to the reinforcing material, silicon carbide. The model also assumes an interphase layer between the matrix and reinforcing material which can be anisotropic, elastic, viscoelastic or anelastic.

In order to better visualize these analytically determined leaky waves, Figure 1 shows the energy flow and particle displacement for a non-decomposable type wave. The ellipses represent the shape of the particle orbit at a number of points with the arrowheads indicating the sense of the particle motion. The arrow from the center of the ellipse shows the energy flow or Poynting vector averaged over one period. Although the ellipse sizes and arrow lengths are arbitrarily scaled for graphing purposes, the sizes within the figure is consistent. Figure 1 which plots the energy flow for the non-decomposable 1,3 mode shows most of the acoustic energy propagating through the silicon carbide and leaking into the aluminum.

EXPERIMENT

An aluminum hemisphere 4 inches (10.16 cm) in diameter was machined to enable determination of the leakage angle, since as with all high modulus fiber reinforcement in relatively low modulus matrix composites the leakage is into the metal matrix. Two aluminum shims approximately the same diameter as the transducer face were also machined from the same block of 2024 aluminum to focus the wave and to allow

TABLE 1. Theoretical leaky waves calculated for a planar silicon carbide/aluminum interface.

Nondecomposable Leaky Waves in Al/Sic

Type	Velocity: Real part, imaginary part	Asymtotic Leakage Angle in Al
0,3	5562,1113	21.0
1,3	8643, 2458	49.0
2,3	4730, 231	49.2
1,1	4619, 3416	48.3
3,1	10481, 2454	-73.8
3,0	5562, 1113	-57.9

Decomposable Leaky Waves in Al/SiC

Type	Velocity	Longitudinal Wave Angle in Al	Longitudinal Wave Angle in SiC	Shear Wave Angle in Al	Shear Wave Angle in SiC
0,3	10030	50.9	15.8	72.0	58.8
2,3	10077	-51.3	16.8	72.0	59.0
3,2	11531	-56.7	33.2	-74.4	-63.2
2,1	16377	-67.3	-53.0	79.0	71.5

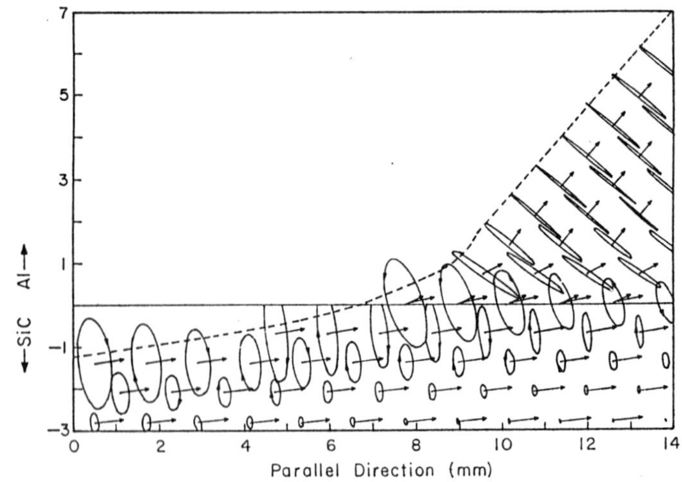

Figure 1. Representation of the displacement field of a non-decomposable leaky wave propagating on an Al/SiC interface.

the flat transducer faces to conform to the curvature of the hemisphere. They were attached to the transducers by a solid wax couplant.

A sintered SiC plate with dimensions 4" x 4" x 3/4" (10.16 x 10.16 x 1.91 cm) was placed under the aluminum hemisphere to form a planar interface. The velocities of both materials used for the calculation of leaky wave existence were determined by conventional pulse-echo ultrasonic techniques.

Leaky waves were generated on the SiC/Al interface through a mode conversion from incident longitudinal waves. A pair of 1 MHz commercial longitudinal transducers, 0.5 inches in diameter, were mounted on the aluminum shims with a solid wax couplant. These in turn were coupled to the aluminum hemisphere by silicone grease. Additionally, a pinducer was mounted at the far end of the specimen normal to the SiC/Al interface. The specimen was fixed to a rotation stage which allowed precise determination of the incident and reflected angles.

Longitudinal waves were generated by the transducers into the aluminum shims. This focused wave was introduced into the hemisphere and allowed to reflect off the interface. Detection was accomplished by the second longitudinal transducer mounted on the opposite side of the hemisphere. Simultaneous monitoring of energy propagating along the interface was accomplished by the pinducer. The 1.5 mm diameter pinducer utilized a 1 mm thick PZT 5A element. These transducers are nominally 4 MHz and extremely broadband. Measurements were conducted with no bonding between the surfaces and with the surfaces joined by epoxy. Travel times were measured at 5° increments from zero to 85° with the transducers placed symmetrically about the normal axis of the hemisphere. Degrees are measured counter-clockwise from the right interface edge arbitrarily designated 0° to the normal at 90° for the received signal. The incident wave angle designation is a mirror reflection of this with the normal also at 90°. When a signal was detected by the pinducer, increments of the sending and receiving transducers was decreased to 1°.

RESULTS

When the aluminum was bonded to the silicon carbide by a thin layer of adhesive, two distinct leaky modes were evident at 1 MHz. Although no effort was made to quantitatively measure the amplitude of the detected signal at the second transducer, the appearance of the reflected longitudinal wave even at angles of leaky wave generation indicates that only a fraction of the energy is mode converted into a leaky wave. Several factors including beam spread and difficulties determining exact reflection angle made amplitude measurements impractical. The measured travel times of the two waves at approximately 23° and 55° represent a combination of velocities which follows a law of mixtures type argument. Each travel time is composed of a longitudinal wave which passes through the aluminum shim and then propagates through the aluminum hemisphere a distance equal to the radius of the hemisphere. At the interface, part of this wave is mode converted to a leaky wave which travels the length of the radius at the interface and is detected by the pinducer. The leaky wave generated by mode conversion from approximately 53° to 57° traveled at a velocity of 10064 ± 82 m/sec. The second leaky wave generated from approximately 21° to 25° traveled at a velocity of 5224 ± 119 m/sec.

The velocity of the wave generated by mode conversion at $55 \pm 2^{\circ}$ implies that it is largely affected by the silicon

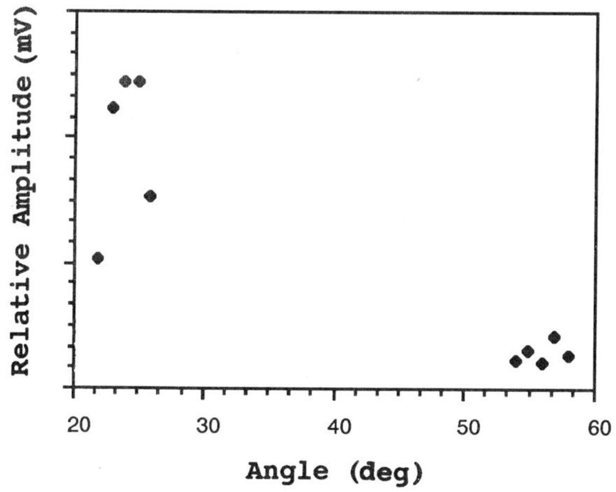

Figure 2. Relative amplitudes of the two leaky waves.

carbide whose longitudinal velocity is 11965 m/sec. As this wave propagates along the interface, an examination of the displacement field normal to the interface would show that the majority of the energy is propagated through the denser silicon carbide as predicted by analytical studies. Energy is leaked away from the interface through the lighter medium, aluminum in this case, at a fairly steep angle (~55°) indicating a relatively attenuative leaky wave. The second wave has a much shallower leakage angle approximately 23 ± 2° and is less attenuative than the faster wave. This is corroborated by Figure 2 which plots the relative amplitudes of the leaky waves as measured by the pinducer. Since more energy is radiated away from the 55° leaky wave, its relative displacement amplitude measured by the pinducer parallel to the interface is much less than that of the 23° wave. Therefore the steeper leaky wave appears as the more attenuative wave. This leaky wave travels at a much slower velocity of approximately 5224 m/sec. This wave appears to be largely affected by the aluminum, since its leaky wave velocity is closer to that of the longitudinal wave velocity of the aluminum. The leakage is also into the aluminum, but at a much steeper angle.

The appearance of these interface waves demonstrates the effectiveness of remotely generating leaky waves by mode conversion through the bulk of one of the materials. The critical angles of generation are consistent with the angles of leakage for these waves so that any modes that exist can be generated by varying the incident angle of the transducer. Remote generation and detection of these waves is an important feature for ultrasonic interface wave characterization. Conventional ultrasonic interface wave techniques require direct access to the interface or surfaces comprising the interface to measure travel times. Utilizing the particular propagation characteristics of leaky waves demonstrated by these experiments, interface studies may be accomplished through either bulk material without direct contact with the interface itself.

The two experimentally detected waves correlate to the analytically predicted leaky modes in a planar silicon carbide/aluminum interface. The first leaky wave experimentally measured corresponds to the non-decomposable (0,3) leaky mode predicted to leak at an angle of 21.0^o and to travel with a velocity of 5562 m/sec. The experimentally determined wave had a leakage angle of 23 ± 2° and traveled at a velocity of 5224 ± 119 m/sec. The (1,3), (2,3), and the (1,1) analytical modes leak at angles of 49.0^o, 49.2^o, and 48.3^o respectively. Although the experimental measurement of the leaky waves was not able to distinguish those three individual modes, the fastest mode was observed. The (1,3) mode has a leakage angle of 49.0^o and a predicted velocity of 8643 m/sec. This corresponds to the appearance of the second experimentally observed leaky wave at 55 ± 2° which travelled with a velocity of 10064 ± 82 m/sec. The three analytical modes are grouped within one degree of each other and therefore could not be distinguished experimentally by leakage angle. Again, only the fastest of these three modes the (1,3) was detected by the pinducer.

No attempt was made to measure any of the decomposable leaky modes. This was due to the potential difficulty generating and detecting these modes.

Direct measurement of the leakage at both generation angles was also accomplished. Figure 3a shows the measured relative amplitudes of the leakage of the first interface wave to appear at 1 MHz. At a specific incident angle of generation of leaky waves determined by a pinducer signal at the interface, the pinducer was scanned along the reflected face of the aluminum hemisphere. This figure shows the relative amplitudes of the leakage for an interface generated at an incident angle of 23°. The reflected longitudinal wave was not plotted. A line drawn from the center of the hemisphere which corresponds to the point of generation of the leaky wave through the maximum leakage point shows the angle of leakage to be consistent with the incident angle. Note that by earlier definition the center of the hemisphere corresponds to the 90° designation at the surface of the hemisphere. This line represents the angle of maximum acoustic energy flow of the leakage away from the interface.

Figure 3b shows the direct leakage measurements of an interface wave generated from an incident angle of 55°. Its relative leakage amplitude maximum is larger than that of the leakage from the 23° wave. This substantiates the contention that the steeper leaky wave is the more attenuative of the two interface waves. The line drawn from the point of generation of the wave on the interface through the maximum leakage points again shows that the angle of maximum energy flow away from the interface is consistent with the incident angle of generation.

Utilizing these types of measurements, the quality of an interfacial bond may be examined by nondestructive acoustic means. The measurement of a specific interface wave velocity for known propagation distances will allow the bond along that interface to be qualitatively determined. By decomposing the measured interface wave velocity into its leaky wave and/or Rayleigh wave components through a law of mixtures type argument, the contribution of the Rayleigh wave can be determined and directly related to the amount of disbonding. In the same manner, the percentage contribution of the leaky wave can be directly related to the percentage by bonding.

Remote generation and detection of leakage and leaky waves demonstrated by these experiments offers even greater possibilities for nondestructive acoustic evaluation of the interfacial structure. While direct measurement of leaky waves offers several benefits as describe previously, these results allow considerable insight into the structure and properties of the planar silicon carbide/aluminum interface. Each leaky wave propagates through both materials and contains certain information about the elastic properties of the interface depending on the mode. Remote generation by reflection from a surface generated bulk wave allows the

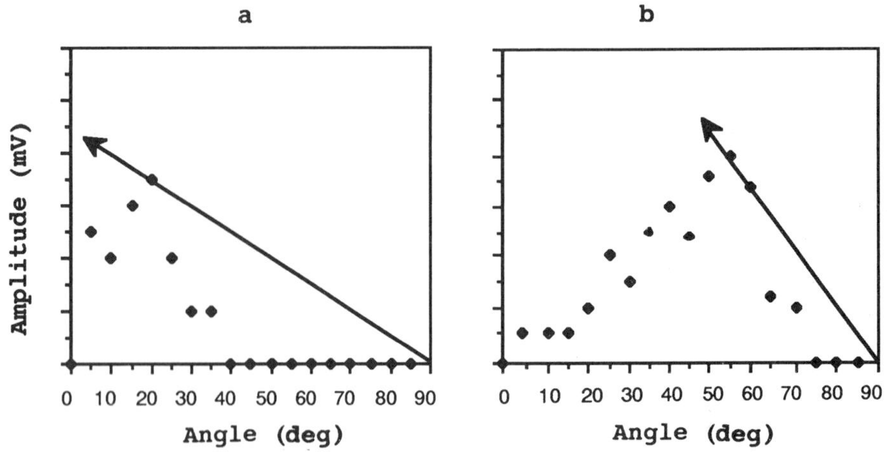

Figure 3. Direct measurement of leakage of both interface waves.

potential for all modes to be produced. Additionally, detection of the leakage at the surface of either material is crucial for nondestructive interface characterization. By measuring the leakage or absence of leakage of a certain leaky mode, the presence and location of disbonding can be determined. Leakage measurements also offer insight into the elastic character of both of the materials comprising the interface as well as the interface itself if that structure is different (i.e. heat affected zone, etc.). Additional work needs to be done to in order to detect the other non-decomposible modes, as well as the four decomposible modes.

CONCLUSIONS

1. Two leaky modes were generated by mode conversion from a bulk wave incident on a bonded silicon carbide/aluminum interface at a critical angle. One wave was generated at ~23° and travelled with a velocity of ~5224 m/sec, while the second leaky wave was generated at ~55° and travelled with a velocity of 10064 m/sec.

2. Theoretical analysis by NIST computer modeling predicts four decomposable modes and six non-decomposable modes for a planar silicon carbide/aluminum interface. Two of these have been experimentally verified.

3. No interface waves were generated on the unbonded silicon carbide plate and the aluminum hemisphere.

4. Relative amplitude measurements of the leakage energy show that maximum leakage energy of the steeper wave was greater than that of the shallower wave. Direct measurement of the leaky wave amplitude at the interface confirmed this fact.

The steeper wave was more attenuative than the shallower leaky wave.

5. Measurement at the surface of the aluminum hemisphere of the leakage energy was accomplished. These measurements confirmed that the leakage angle was the same as the incident generation angle. It was concluded that this direct measurement of leakage is a valuable method for separating modes and will contribute to nondestructive interface characterization of composites.

ACKNOWLEDGEMENTS

This work was supported by the Office of Naval Research under contract #N00014-87-K-0341, with Dr. Steven Fishman as the program director. The authors would also like to thank Dr. Ewa Krasicka and Dr. Johns Simmons for their contributions to this research.

REFERENCES

1. Stoneley, R., Elastic Waves at the Surface of Separation of Two Solids Proc. Roy Soc., 106:416 (1924).
2. Sezawa, K., Kanai, K., The Range of Existence of Stoneley-waves and Some Related Problems, Bull. Earthquake Res. Inst., 17:1 (1939).
3. Claus, R.O., Palmer, C.H.,Optical Measurement of Ultrasonic Waves on Interfaces Between Bonded Solids, IEEE Trans Sonics Ultrason., SU-27:97 (1980).
4. Owen, T.E., Surface Wave Phenomena in Ultrasonics, Prog. Appl. Mater. Res., 6:69 (1964).
5. Yamaguchi, R., Sato, Y., Stoneley Wave- Its Velocity, Orbit and the Distribution of Amplitude, Bull. Res. Earthquake Inst., 33:549 (1955).
6. Hsieh, T.M., Rosen, M., Lindgren, E.L., The Effect of Interfacial Properties on Stonely Wave Propagation, Ultrasonics, (in press 1990)
7. Lee, D.A., Corbly, D.M., Use of Interface Waves for Nondestructive Inspection, IEEE Trans Sonics Ultrason., SU-24:206 (1977).
8. Krasicka, E., Simmons, J., Propagation and Dispersion of Radial-Axial Leaky Modes in Infinitely Clad Rods, to be published.
9. Computer models were developed by Dr. John Simmons and Dr. Eva Krasicka of NIST.

AN ULTRASONIC TECHNIQUE FOR THE IDENTIFICATION OF CRACK SUSCEPTIBLE MICROSTRUCTURES IN NICKEL-COPPER ALLOY K-500 FORGINGS

M.E. Natishan and C.A. Lebowitz

David Taylor Research Center
Bethesda, Maryland

INTRODUCTION

Centerline cracking has recently been observed in nickel-copper alloy K-500 large forged bars and components. These cracks were identified to be intergranular and completely internal, with varying degrees of continuity. The cracks appeared as transverse or longitudinal in orientation and were flat and tightly closed. They initiated at the center of the forging and arrested just short of the outer diameter. The extensive nature of the cracks, and quantity of cracks observed indicated that there was a problem inherent in this material that needed to be resolved to decrease the expense of using this alloy. The cause of the intergranular cracking has been attributed to a grain boundary graphite content greater than 15%. It is currently believed that these graphite precipitates weaken the grain boundaries such that the stresses produced during the forging and heat treating process are sufficient to cause intergranular cracking. Identification of material susceptibility to cracking prior to finish machining of K-500 components can reduce the cost of using this alloy.

This program focused on identifying an NDE method which could differentiate between crack susceptible and nonsusceptible K-500. It is believed that the continuous nature of the grain boundary graphite would effect the attenutive nature of the grain boundaries, making ultrasonic inspection a possible method for detection of crack susceptibe material. To determine whether UT was a feasible technique for this application, a simple heat treat study was conducted to see if UT could differentiate between various microstructural conditions in K-500. The results of this study are described below.

EXPERIMENTAL PROCEDURE

Nickel-copper alloy K-500 is an FCC solid solution of nickel and copper, strengthened by a fine dispersion of gamma' precipitates (Ni_3Al, Ti, Fe) [1]. Due to its high strength, toughness and corrosion resistance it is commonly used in marine applications. Table 1 summarizes the chemistry and properties of K-500.

Table 1. Chemistries and properties of bars used in this study.

Element	Alloy K-500[1]	Bar A	Bar B
Ni	63.0-70.0	65.9	66.3
Cu	27.0-33.0	30.2	28.5
Si	0.5 max.	0.18	0.17
P	0.020 max.	<0.004	<0.004
Pb	0.006 max.	<0.004	<0.004
Sn	0.006 max.	<0.004	<0.004
C	0.25 max.	0.159	0.134
Fe	2.0 max.	0.65	1.03
Ti	0.35-0.85	0.43	0.5
Al	2.30-3.15	2.81	2.7
Mn	1.5 max.	0.77	0.78
Properties			
0.2% Y.S.	80-100 ksi	92 ksi	97 ksi
U.T.S.	130-160 ksi	146 ksi	158 ksi
% Elongation	20	31	27

Two large diameter alloy K-500 bars were obtained and cut into one inch cubic specimens. Using a Krautkramer Branson USD-10 Digital Ultrasonic Flaw Detector with a Aerotech 5 MHz, 0.5 inch diameter gamma transducer, each cube was examined in the three orthogonal directions. The measurement consisted of recording the instrument gain required to peak the ultrasonic signal to 80% full screen height (FSH). This was done to gain information concerning effects of anisotropy on the UT signal and to get a baseline signal for the as-recieved microstructures. The specimens were also polished in the three orthogonal directions, etched and examined using an optical microscope at magnifications up to 1000X to characterize the as-recieved grain size and distribution as well as the extent of any grain boundary precipitation. Scanning Auger electron spectroscopy was used to identify and measure the grain boundary precipitate content.

A preliminary heat treat study was conducted to determine the temperature regions within which various precipitation and grain growth/recrystallization phenomena take place in alloy K-500. Figure 1 shows the temperature regions identified. After the initial characterization, the blocks were heat treated for a one hour hold time at temperatures within this temperature range (1000°F to 2300°F) to produce various microstructures of interest. Heat treat temperatures were chosen to isolate the various microstructural features having an effect on the ultrasonic signal.

After the heat treatments, the cubes were again evaluated ultrasonically and metallographically in the three orthogonal directions to characterize the changes produced by the heat treatments. The average change in gain for each specimen/heat treatment was calculated by subtracting the gain recorded before heat treatment from the gain recorded after heat treatment for each of the orthogonal directions and then averaging the three values. The microstructural changes were then compared to the average change in gain to determine if it would be feasible to use ultrasonic inspection to differentiate between the various microstructural conditions produced.

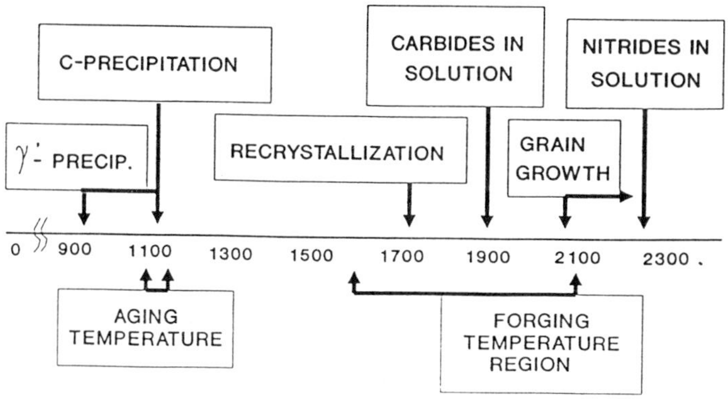

Figure 1. Temperature line showing critical temperature regions for precipitation and grain growth events.

RESULTS

The as-received microstructures of the two bars differed in average grain size, grain size distribution and in the distribution of inclusions and precipitates. Bar A had a uniform grain size distribution with an average grain size of 30 microns, as shown in figure 2a. The distribution of particles was also quite uniform and the grain boundary graphite content was approximately 2% to 5%. Bar B had a much larger average grain size of 120 microns as shown in figure 2b. Most of the particles were in the vicinity of grain boundaries and the grain boundary graphite content was measured at approximately 15% to 20%. The grain size in the two transverse bar directions were similar and finer than the grain size in the longitudinal direction for both bars.

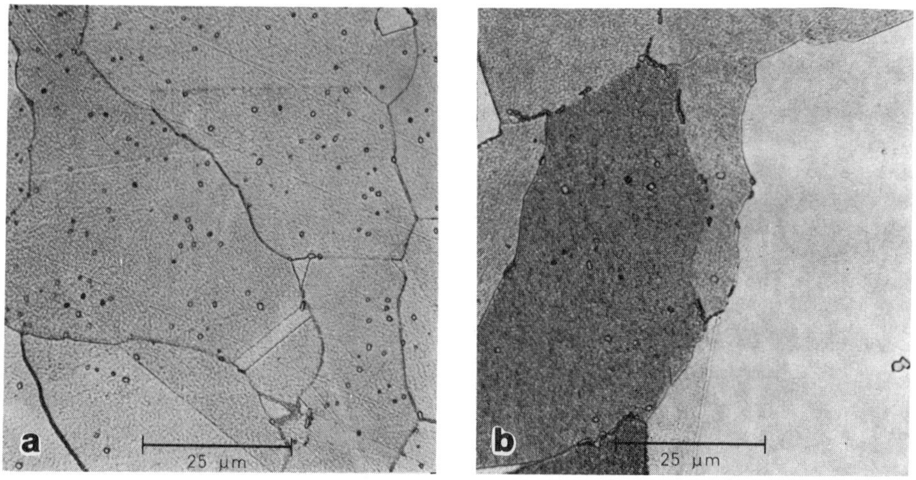

Figure 2. Optical photomicrographs showing microstructures typical of specimens from a) Bar A, and b) Bar B.

Typical UT signals for the two materials in the as-received condition are shown in figure 3. As expected Bar B, with the larger grain size, was more attenutive than Bar A and required more energy to obtain a peak signal of 80% full screen height. The directionality of the as-received grain size was apparent in the UT signal as the energy required to peak the signal to 80% FSH was lower in the two transverse bar directions (consistant with the grain size) than in the longitudinal direction.

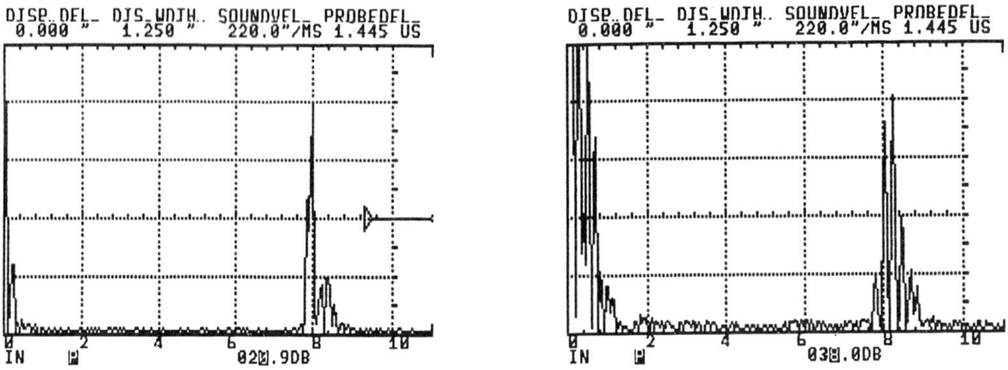

(a) (b)

Figure 3. Typical ultrasonic technique signals for the two bars studied, in the as-received condition. a) UT signal from a Bar A specimen. b) UT signal from a Bar B specimen.

While 33 different heat treatments were conducted on specimens from each bar only six different microstructural conditions were identified, with many of the specimens overlapping. Since only the average change in gain was considered for each specimen, and since there was more than one specimen exhibiting the same microstructural condition after heat treatment, only the trends in the UT signal changes will be discussed for each microstructural condition.

The first category of microstructural change produced by heat treatment was actually no change at all. This gave a baseline for comparing the trends in UT variation shown by the samples which contained microstructural changes. As expected for the specimens from both bars, there was no change in the energy required to obtain a peak signal of 80% FSH when no change occurred in the microstructure, Table 2.

The second category of microstructural change noted was a decrease in the amount of grain boundary graphite. The absolute decrease for both specimens was approximately 2%, however, this is a much larger relative change in Bar A specimens which only contained 2% to 5% to begin with. The gain required to peak the signal to 80% FSH decreased along with the decrease in grain boundary graphite content. This decrease was larger for Bar A than for Bar B, which approximates the relative change in microstructure.

Table 2. Summary of the microstructural changes resulting from the various heat treatments and the corresponding changes in the UT signals.

Microstructural Category	Microstructural Change	Change in the UT Signal Bar A	Bar B
1	No change	0	0
2	2% decrease in grain boundary graphite content	-2 dB	-1 dB
3	4% decrease in grain boundary graphite content	-3 dB	-2 dB
4	Realignment of carbides on grain boundaries, slight decrease in g.b. graphite	0	---
5	Recrystallization (33% decrease in grain size	-2 dB	-2 dB
6	Excessive grain growth	20 dB	26 dB

Category 3 was a larger absolute decrease in the amount of grain boundary graphite content (-4%) again resulting in a greater relative decrease in the Bar A specimens with the lower as-received graphite content than in the Bar B specimens. The gain required for 80% FSH resulting from this microstructure decreased with Bar A showing a slightly greater decrease consistant with the greater relative change in graphite content.

Category 4 was only identified in Bar A specimens heat treated at higher temperatures. This category was identified as a change in the carbide distribution from a uniform distribution in the as-recieved specimens to primarily a grain boundary distribution of carbides. While this was suspected of effecting the attenutive nature of the grain boundaries, the carbides themselves were discrete (not continuous as were the graphite precipitates) and the realignment of carbides along the grain boundaries occurred in conjunction with a slight decrease in the grain boundary graphite content. Thus there was no change noted in the UT signal resulting from this microstructural change.

Category 5 was identified as a decrease in grain size caused by heat treating in the recrystallization temperature region. The relative decrease for specimens from both materials was approximately 33% with a larger absolute decrease in Bar B specimens where the grain size was larger to begin with. The change in the UT signal for both materials was consistant with the relative change, with both showing a decrease of approximately 2 dB in the gain required to get a peak signal of 80% full screen height.

Category 6 was defined as extensive grain growth due to heat treatment well above the grain growth temperature region. This extensive grain growth resulted in a large increase in the attenuative nature of the specimens and a large increase in the amount of energy required to obtain 80% full screen height for the peak signal.

SUMMARY AND DISCUSSION

Six categories of microstructural change were identified out of a possible 33 heat treat conditions. These were; 1) no change in microstructure, 2) a 2% decrease in grain boundary graphite content, 3) a 4% decrease in grain boundary graphite content, 4) a change in the carbide distribution (for Bar A only), 5) a 33% decrease in grain size, and 6) an excessive (order of magnitude) increase in the grain size. The trends noted in the UT signals corresponding to these changes were consistent for all of the catergories defined. Decreases in grain boundary graphite content resulted in a decrease in the instrument gain required to peak the signal to 80% FSH. This indicates that relatively continuous grain boundary graphite precipitates cause an increase in the attenutive nature of the grain boundaries. Slight decreases in grain size resulted in a decrease in the attenutive nature of the grain boundaries decreasing the amount of energy required to peak the signal to 80% FSH. This was expected since a decrease in grain size relative to the ultrasonic frequency results in a decrease in the scattering of the ultrasound [2].

The fact that no change in the UT signal was noted with the alignment of carbides along the grain boundary of Bar A specimens was inconclusive. As suggested above this result may be explained by the fact that the carbides were not continuous along the grain boundaries as was the graphite. It may also be that the small decrease noted in the amount of grain boundary graphite offset the effects of the increase in the grain boundary carbide content.

CONCLUSIONS

This experiment was successful in creating microstructures in which the various features expected to effect the UT signal were isolated. Trends in the UT signal changes indicate the UT is a technique which can successfully differentiate between these various microstructural features. Grain size has a very large effect on the UT signal while the effects of grain boundary graphite were much more subtle. To be a viable method for differentiating between crack susceptible versus non-crack susceptible alloy K-500 forgings development of a UT procedure which is more sensitive to the effects of the grain boundary graphite would be necessary. Work is progressing to quantify the effects of the various microstructural features using artificial neural networks and in developing an acceptance criteria consistant with the metallurgical criteria.

REFERENCES

[1] Monel Alloys, Huntington Alloys Division of International Nickel Company, Product Pamphlet, 1972

[2] Krautkramer, J., and H. Krautkramer, Ultrasonic Testing of Materials, Third Edition, Springer-Verlag, New York, 1983, pp107-108

ULTRASONIC CHARACTERIZATION OF TENSILE OVERLOAD EFFECTS AND STRESS INTENSITY SHIELDING

D. K. Rehbein, O. Buck and R. B. Thompson

Ames Laboratory - USDOE
Materials Science and Engineering Department
Iowa State University
Ames, Iowa 50011

INTRODUCTION

The growth of a fatigue crack is modified by the development of contacts between the crack faces[1,2], creating shielding and thus canceling a portion of the applied load. These contacts develop through a number of mechanisms, including plastic deformation, sliding of the faces with respect to each other and the creation and collection of debris such as oxide particles[3]. Compressive stresses are created on either side of the partially contacting crack faces resulting in opening loads that must be overcome in order to apply a driving force to the crack tip for growth. In this way, the crack tip is shielded from a portion of the applied load, thus creating the need for modification[1] of the applied stress intensity range from $\Delta K = K_{Imax} - K_{Imin}$ to $\Delta K = K_{Imax} - K_{Ish}$. Determination of the contact size and density in the region of closure from ultrasonic transmission and diffraction experiments[4] has allowed estimation of the magnitude of K_{Ish} on a crack grown under constant ΔK conditions. The calculation has since[5] been extended to fatigue cracks grown with a tensile overload block. The calculation was also successful in predicting the growth rate of the crack after reinitiation had occurred. This paper reports the results of attempts to define the amount of retardation remaining before reinitiation of crack growth in terms of the parameters used by the distributed spring model. In addition, this paper reports on the acoustic response of a crack grown using a decreasing ΔK and the subsequent implications for determination of the threshold stress intensity factor.

ULTRASONIC EXPERIMENTS AND MODELING

The experimental configuration for the ultrasonic interrogation of the fatigue cracks under consideration is shown in Fig. 1. The fatigue crack in a modified compact tension specimen is illuminated by a normally incident longitudinal wave focussed in the plane of the crack. The longitudinal wave transmitted is detected by a receiver placed coaxially on the opposite side of the sample. The frequency spectrum of this received signal is deconvolved with that observed in a reference experiment where the beam is transmitted through the uncracked ligament. In this way, most of the influence of the measurement system and

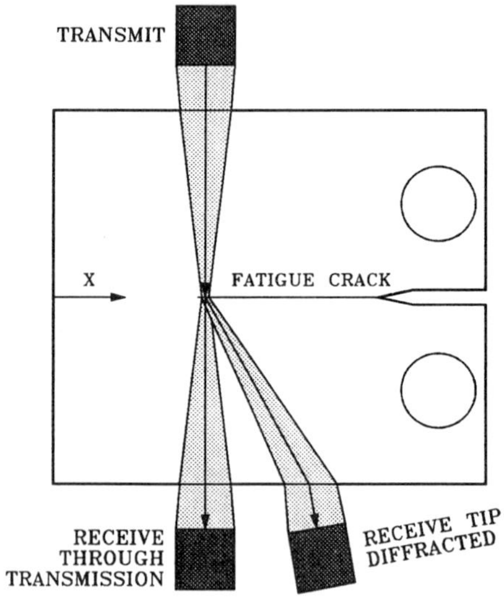

Fig. 1 Arrangement for measurement of acoustic response of fatigue crack.

material microstructure is eliminated with the information obtained being directly characteristic of the magnitude of the crack closure.

The plane of a partially contacting fatigue crack is viewed as consisting of contacting asperities separated by crack-like voids which act as scatterers of an elastic wave. This scattering can be modeled based on the electromechanical reciprocity theorem[6] applied to the conditions of normal incidence and utilizing[7] an averaged crack opening displacement which describes the contact at the crack surface in terms of a spring constant κ which is inversely related to the crack opening displacement.

This development has been used to obtain descriptions of through transmission results from a variety of samples[8-11]. Fig. 2 presents a comparison between experimental through transmission results (left) and model predictions (right) for a fatigue crack grown under constant ΔK conditions. For this case, κ is described using an exponentially decaying function,

$$\kappa(x) = \kappa_0 e^{-\beta x} \tag{1}$$

in the closure zone near the crack tip where κ_0 is the spring constant at the crack tip and β describes the decay of the spring constant in the crack tip closure region.

One interesting feature of both the experimental and model results in Fig. 2 is the form of the cross-over of the data in the vicinity of the crack tip. As seen in the figure, the transmission response for the lowest frequency begins to decrease earliest and persists longest with this phenomena reversing as frequency increases. This effect is caused by the variation of the radii of the focussed ultrasonic beam with frequency. It is well known that the spot size of the effective

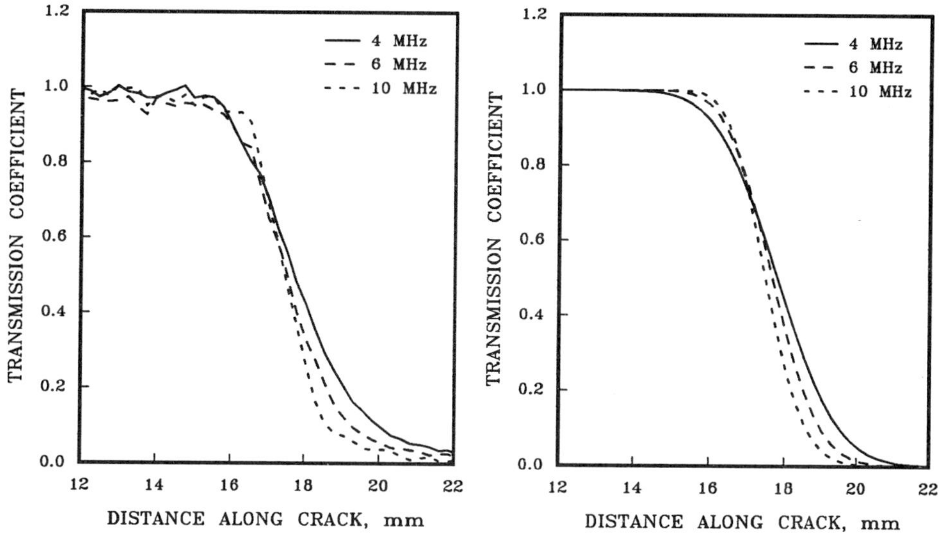

Fig. 2 Experimental acoustic response (left) and distributed spring model prediction (right) of crack grown with constant ΔK.

beam at a given frequency becomes smaller as the frequency increases for the focussed transducers being used. Thus, the beam impinges on the crack tip earlier for low frequencies than for high frequencies as the beam approaches the crack from the uncracked ligament causing the low frequency transmission response to decrease first. A portion of the low frequency spot also stays in the uncracked ligament longer for the lower frequency, resulting in a higher transmission response at that position. When the beam is centered on the tip of an ideal crack, the transmission response from all frequencies will be identical and equal to 50% transmission since the effective beam centers are coincident.

Changes in this observed behavior are apparent in Fig. 2 due to crack closure, causing the position of the "crossover" in the frequency response to occur at a transmission coefficient of greater than 0.5 and making an accurate determination of the position of the crack tip more difficult. Further changes in this behavior have been observed for the case of a crack currently in the retarded condition[12] and are important for detection of the "tight" crack formed during tensile overloading.

CHANGES IN ACOUSTIC RESPONSE DURING RETARDATION

It has been observed[12] that during application of a tensile overload a section of a tightly closed crack is created with the observation made that changes in the frequency response of the crack with the ultrasonic beam is evident. Further investigation of this effect has shown that these changes are consistent at least during the first half of the retardation period and may be indicative of the amount of retardation that remains. Figure 3 shows the acoustic response of a crack in 2024-T351 aluminum that has undergone a five cycle tensile overload at twice the ΔK of the preceding growth. The response curves shown in Fig. 3 were taken at various intervals during fatigue cycling subsequent to the overload application as shown for each set of response curves. Subsequent cycling was accomplished at the same ΔK as was used prior to the overload.

The results show the appearance of a trend in the frequency response as cycling occurs through the retardation period. Immediately after the overload application and after 8,000 cycles of subsequent cycling, no "cross-over" of the data at the various frequencies is seen as would be expected. Beginning with the data at 25,000 cycles, a "cross-over" in the 4 and 6 MHz data is observed with the cross-over appearing to move to lower values of the transmission magnitude as additional cycling is accomplished. Only at the highest number of cycles, 85,000, does the 10 MHz data begin to approach the expected behavior.

It is speculated that the section of tightly closed crack created during the overload application has altered the frequency response from that normally observed. The method by which this alteration occurs is not clear at this time. It can also be observed that the curves appear to become steeper as additional fatigue cycles are applied. This steepening is difficult to quantify from the experimental results but will be discussed further in a later section.

CRACK TIP POSITION DETERMINATION

Determination of the position of the crack tip can be achieved by a number of experimental methods, i.e. optical measurements, clip gauge opening, strain gauge measurements, etc. Each of these measurements contains some degree of error due to the curvature of the crack front as the crack changes from the plane stress regions on the edges of the compact tension specimen to the plane strain region in the center. The ultrasonic response of the crack conceptually should

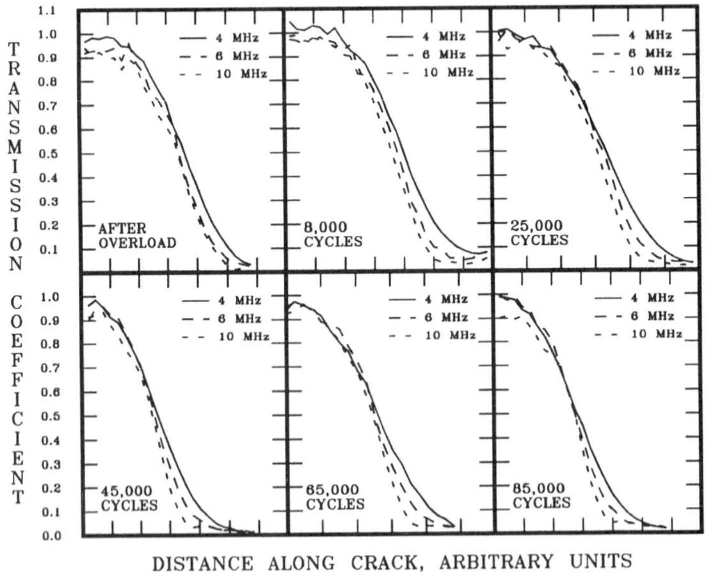

Fig. 3 Acoustic response of overloaded crack at intervals during retardation period.

provide the most accurate determination of the crack tip position since the focussed beam is entirely within the plane strain region and should detect the furthest extent of the crack front. This determination is however essentially impossible directly from the ultrasonic response of the crack due to the variable effects of closure on the ultrasonic response.

The distributed spring model as described earlier must have a value for the crack tip position in order for Eq. (3) to be used. It is possible using iterative computing techniques to obtain a best-fit set of values for the crack tip position and κ_0 and β by comparison with the experimental data. Two examples of determination of the crack tip position by this method are shown in Fig. 4. Also included for comparison are measurements of the crack tip position by optical measurement, clip gauge opening and fractography after destruction of the sample.

The crack tip positions as determined by the spring model were established using a non-linear regression technique comparing the experimental response and the model predictions from a large set of input values of the crack tip position and κ_0 and β. It is seen from Fig. 4 that for each of the cracks shown, the model determination of the crack tip position yielded a more accurate position that either the optical or clip gauge opening measurements thus validating that the distributed spring model is capable of an accurate prediction of the crack tip position.

OVERLOAD CRACK ANALYSIS

The iterative procedure used earlier for determination of the crack tip position was also applied to estimate the best values for the crack tip position as well as κ_0 and β for the acoustic response curves shown in Fig. 3. In addition, the computations were done on a second set of response curves that were obtained with the sample under a static load of 65% of the K_{max} that was used during the growth of the original crack. This static load was intended to fully open the crack so that precise determination of the crack tip position could be made experimentally since the crack response would then be that of an ideal crack. The results of these computations are shown in Table 1. Also included are the results for the acoustic response of the crack immediately before application of the overload. The difference between the computed crack tip positions from the two response curves at each fatigue interval is also given.

The difference in the crack tip positions shows a rather confusing set of data. If indeed the higher static load opened the crack to approximate an ideal crack, then this difference should have been a positive number throughout as it was for the acoustic response data taken just before the overload (-0). Instead the differences for the responses early in the retardation period show negative values with essentially no difference at 45,000 cycles. It is possible that these results are within the computational uncertainty of the technique based on the inescapable experimental error. The opinion of the authors is that the only significant differences are those shown just before the overload and those for 65,000 and 85,000 cycles. At these last two fatigue intervals, the difference is approximating the length of tightly closed crack created during the overload. It is possible for the responses early in the retardation that the response curve shows the original crack tip that has been blunted by the overload application in accordance with earlier speculation[12].

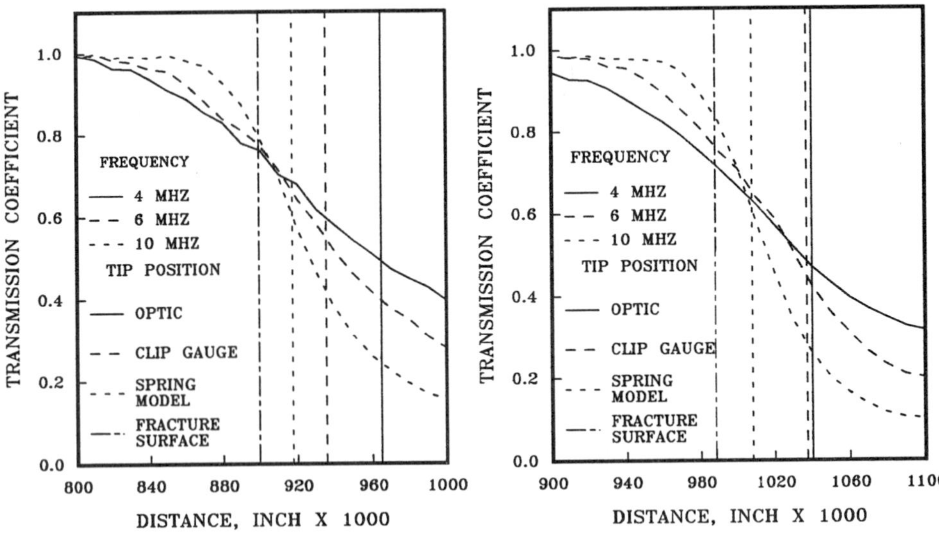

Fig. 4 Acoustic response and crack tip position measurements for cracks grown with constant ΔK.

The κ_0 values shown do not exhibit a consistent trend at either static load setting except that the values appear to peak at 45,000 cycles for both static load settings. The computational uncertainty of these values is unknown at present, precluding any conclusions at this time from this apparent trend.

The one consistent trend shown is in the values for the exponential decay constant, β, at the lower static load setting. These values show almost a monotonic increase from just after the overload (+0) to 85,000 cycles. This trend agrees with the earlier observation that the response curves in Fig. 3 appeared to steepen as additional fatigue was applied during the retardation period. Whether this trend is actually an indicator of the amount of retardation remaining will have to be tested in further investigation. Unfortunately, this specimen became corroded before reinitiation of the crack occurred. In order for growth to reoccur, the values for κ_0 and β should return to values close to those just before the overload.

LOAD SPECTRUM AND SHIELDING

A contribution to the shielding stress intensity factor is made by each contacting asperity in the closure region of the crack with the magnitude of the contribution dependent on the load carried by the individual asperity. Residual stresses in the closure region have been calculated[4], based on contact pressure calculations[13], that agree well with corresponding results using x-ray diffraction[14]. Integration over all of the contacting asperities, then, yields the total shielding stress intensity factor, K_{Ish}. For a crack grown with a constant ΔK, the shielding stress intensity factor was found to be 40% of the applied ΔK[4].

A crack was grown using a continuously decreasing applied stress intensity factor. The crack propagation data from this crack is shown in Fig. 5, indicating that the growth changed from the Paris regime to the threshold regime. Acoustic transmission data was taken at crack

TABLE 1. Values of crack tip position, κ_0 and β for overloaded crack at various fatigue intervals.

FATIGUE CYCLES RELATIVE TO OVERLOAD BLOCK (cm)	DIFFERENCE IN CRACK TIP POSITIONS	κ_0 (dyne/cm^2)		β (cm^{-1})	
		LOW LOAD	HIGH LOAD	LOW LOAD	HIGH LOAD
-0	0.08	7E+13	4E+14	8	40
+0	-0.04	2E+14	3E+14	16	34
8000	-0.04	2E+14	5E+14	16	30
25000	-0.03	3E+14	7E+14	18	38
45000	0.01	7E+14	5E+14	24	30
65000	0.16	3E+14	6E+14	20	32
85000	0.09	2E+14	2E+14	26	30

lengths corresponding to points A and B in the figure. The response at point A was similar to that shown in the left side of Fig. 2.

The acoustic response of the crack at point B was remarkably different. At large distances away from the crack tip (6 to 10 mm), the transmission remained at a relatively constant, high value of 0.2, indicating considerable closure. This indicates to us that the closure region is extended with respect to that in the Paris regime and the resulting K_{Ish} is about 45% larger[15]. A conversion of da/dN versus ΔK as in Fig. 5 to da/dN versus ΔK_{eff}, where $\Delta K_{eff} = K_{Imax} - K_{Ish}$, provides a new correlation for the data in Fig. 5. This new correlation has a shape close to that in the Paris regime throughout the entire correlation. Thus it appears that the "threshold" behavior has been artificially created by the extensive closure found, in agreement with earlier observations[16].

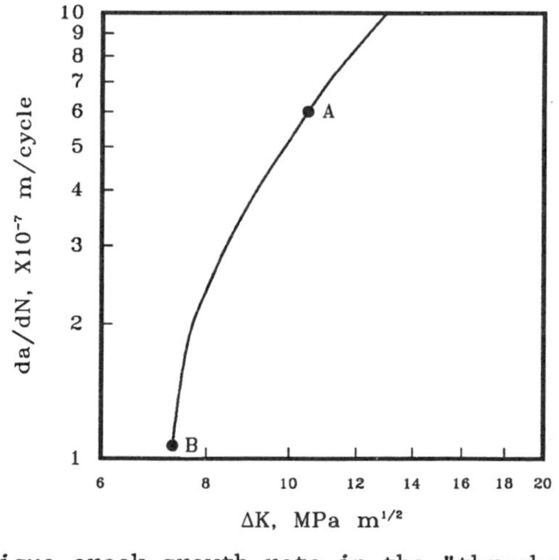

Fig. 5 Fatigue crack growth rate in the "threshold" regime.

SUMMARY

A possible method of determination of the growth state of a crack based on the ultrasonic response of the crack in the unloaded condition is given. This method uses the frequency dependence of the through transmission coefficient in the crack tip region and the parameters determined using the distributed spring model to determine whether the crack is capable of growth under the loading conditions specified. The effect of shielding on the growth rate of cracks grown using nonconstant loading has been calculated with the results in agreement with previous work.

ACKNOWLEDGEMENT

Ames Laboratory is operated for the U. S. Department of Energy by Iowa State University under contract W-7405-ENG-82. This work was supported by the Office of Basic Energy Sciences.

REFERENCES

1. W. Elber, in Damage Tolerance in Aircraft Structures, ASTM STP 486 (Am. Soc. Test. Mat., Philadelphia, PA), 230 (1971).
2. C. Q. Bowles and J. Schijve, in Fatigue Mechanisms-Advances in Quantitative Measurement of Physical Damage, ASTM STP 811 (Am. Soc. Test, Mat., Philadelphia, PA), 400 (1983).
3. S. Suresh and R. O. Ritchie, Scripta Met. 17, 595 (1983).
4. O. Buck, D. K. Rehbein and R. B. Thompson, Eng. Frac. Mech. 28, 413 (1987).
5. D. K. Rehbein, L. Van Wyk, R. B. Thompson and O. Buck, in Review of Progress in Quantitative NDE 8B, edited by D. O. Thompson and D. E. Chimenti, Plenum Press, New York and London, 1787 (1989).
6. B. A. Auld, Wave Motion 1, 3 (1979).
7. J.-M. Baik and R. B. Thompson, J. Nondes. Eval. 4, 177 (1984).
8. R. B. Thompson, C. J. Fiedler and O. Buck, in Nondestructive Methods for Materials Property Determination, edited by C. O. Ruud and R. E. Green, Plenum Press, New York and London, 161 (1984).
9. R. B. Thompson and C. J. Fiedler, in Review of Progress in Quantitative NDE 3A, edited by D. O. Thompson and D. E. Chimenti, Plenum Press, New York and London, 207 (1984).
10. O. Buck, R. B. Thompson and D. K. Rehbein, J. Nondes. Eval. 4, 203 (1984).
11. D. K. Rehbein, R. B. Thompson and O. Buck, in Review of Progress in Quantitative NDE 4A, edited by D. O. Thompson and D. E. Chimenti, Plenum Press, New York and London, 61 (1985).
12. D. K. Rehbein, L. Van Wyk, R. B. Thompson and O. Buck, in Review of Progress in Quantitative NDE 9B, edited by D. O. Thompson and D. E. Chimenti, Plenum Press, New York and London, 1589 (1990).
13. K. Kendall and D. Tabor, Proc. Roy. Soc. London A232, 231 (1971).
14. E. Welsch, D. Eifler, B. Scholtes and B. Maucherach, in Residual Stresses in Science and Technology, DGM Informationsgesellschaft Verlag, Oberursel, 785 (1987).
15. O. Buck, D. K. Rehbein, and R. B. Thompson, in Effects of Load and Thermal Histories on Mechanical Behavior of Materials, edited by P. K. Liaw and T. Nicholas, Met. Soc. of AIME, Warrendale, 49 (1987).
16. L. M. Van Wyk, A Study on Ultrasonic Detection and Characterization of Partially Closed Fatigue Cracks, M.S. Thesis, Iowa State University, Ames, IA (1989).

INTERROGATION OF RESIDUAL STRESSES OF MACHINED SURFACE BY AN X-RAY DIFFRACTION TECHNIQUE

Y. C. Shin[1], S. J. Oh[1] and C. O. Ruud[2]

[1]School of Mechanical Engineering
Purdue University, West Lafayette, IN

[2]Department of Industrial and Management Systems Engineering
The Pennsylvania State University, University Park, PA

ABSTRACT

The surface quality of a machined product is a very important consideration in determining process parameters. Due to the inherent nature of the machining process, there always exists a certain amount of residual stress in the material after machining. This paper presents the results of the surface interrogation of machined workpieces by using an x-ray diffraction technique. The machining tests were performed with 4140 hot rolled steel at two speeds. During the tests, tool wear and surface finish were measured and compared to the measured residual stresses. The results show that the x-ray diffraction technique provides valuable information regarding the machining process and if used as an in-process monitoring method could provide a sensing technique for intelligent processing of materials.

INTRODUCTION

In recent years, automation has become one of the primary thrusts for industries to cope with ever increasing demands for higher productivity and quality. In machining, with the rapid increase in the number of CNC machines and the advent of technologies, there has also been a vast amount of effort made in pursuing flexible automation and even unmanned machining. One of the critical prerequisites in implementing automation is the development of robust and accurate in-process sensing methods for the monitoring of processes to be controlled.

Over the last few decades, this thrust has led to the development of several different sensing techniques for machining applications [1]. While there have been many successful cases with in-process monitoring of process conditions such as tool wear, torque, vibration, etc., measurement of workpiece quality has usually been performed off-line due to the lack of suitable in-process sensing methods. Here workpiece quality comprises both geometric accuracy and surface integrity. Surface integrity includes surface roughness, residual stress, crack existence and density, chemical and physical property changes of subsurface. This paper deals with the

Nondestructive Characterization of Materials IV
Edited by C.O. Ruud *et al.*, Plenum Press, New York, 1991

residual stress aspect of surface integrity and presents an investigation on the possibility of using an x-ray diffraction technique for in-process monitoring of workpiece quality and tool wear.

During a machining process, chips are generated as a result of a large amount of plastic deformation, which consequently induce residual stress on the surface. It is generally known that residual stresses affect the functional quality of the workpiece [2-5]. The static strength of a material can be lowered due to the existence of residual stresses [4] and tensile residual stresses usually shorten fatigue life [5]. Residual stresses can also contribute to stress corrosion for certain materials under a corrosive environment [6,7]. Therefore, when designing a machining process, one must not only use the maximum metal removal rate or tool life as a criterion, but also take surface integrity including residual stresses into consideration for the production of reliable products.

During machining, the development of residual stresses can be affected by many parameters, including feed, speed, depth of cut, tool geometry, and tool and workpiece material properties. Dependent parameters such as tool wear and temperature are also related to the development of residual stresses [6,8]. Leskovar and Peklenik [8] showed that higher machining speeds tend to induce higher tensile stresses. Higher surface temperature also was observed to result in higher residual stresses and different tool materials can create completely different patterns of residual stresses [6,9]. However, it is difficult to design optimum process conditions incorporating specifications of residual stresses since the relationships between cutting conditions and residual stresses are generally unknown or very sensitive to the composition and microstructural condition of the workpiece, even if known for a particular set of conditions. For example, prior deformation, preferred crystallographic orientation or heat treat condition of the workpiece can all affect the machining. Often these characteristics are seen as variability in machinability from piece-to-piece, batch-to-batch, or heat-to-heat. These workpiece variabilities cause different wear rates for cutting tools and different workpiece quality, and lead to the more conservative setting of machining conditions, than those reqired if those properties were known or measurable.

Measurement Techniques of Residual Stress

There have been many different techniques proposed to date for the measurement of residual stresses. One of the earlier techniques is called "deflection technique" which measures the deformation of the piece sectioned out from the workpiece [9,10] after machining. This method yields macroscopic average stress. Apart from being a useful laboratory technique, this method requires careful preparation of the specimen and obviously cannot be used for in-process measurement.

Many different non-destructive techniques have been applied to the measurement of surface structure and residual stress which have promise for on-line application. These include x-ray diffraction, ultrasonic, eddy current and electrochemical techniques [11,12]. Electrical and ultrasonic techniques have been investigated extensively as a means of nondestructive characterization of metallic components. Es-Said and Morris [13] described electrical conductivity measurements for the detection of compositional changes in cast aluminum, and Buxbaumn and Green [14] described the applications of ultrasonic techniques to the characterization of titanium weldments.

Zankl et. al [15] used an x-ray diffraction analysis to determine the cold work during milling. Brinksmeier et. al [6] conducted a rather broad investigation of residual stresses resulting from several types of machining processes including turning and recently Matsumoto et. al [16] applied an x-ray diffraction technique to the measurement of residual stress on machined surfaces. Using x-ray diffraction techniques and a slope etching method for stress depth distribution, Tonshoff and Brinksmeier concluded that tensile residual stresses were dominant in turning [17].

X-ray diffraction techniques have been used often as a tool for the study of residual stresses of machined parts. There is little doubt that this is possible on an off-line basis, however virtually all of the applications of x-ray diffraction to machining studies have been with technologies that are too slow and cumbersome for on-line monitoring. However, Ruud [18], has developed a fiber optic x-ray detector that provides the capability for bringing x-ray diffraction measurements on-line. The present study was performed by using a computer controlled x-ray diffraction measurement system with a fiber optic detector, and hence applying measurement times suitable for on-line application. Moreover, while most of other work has been concerned only with stress measurements under a specific set of machining conditions, the current study presents measurements of residual stresses as machining progresses and machining variables such as tool wear changes.

EXPERIMENTAL PROCEDURE

Machining experiments were designed and performed on a 30-HP LeBlond lathe. The work material chosen for the present investigation was hot rolled 4140 steel alloy bar which is known to have good machinability. The workpieces, about 20 inches long and 4 inches in diameter, were supported by a three jaw chuck and a live center of the tailstock. The bars were premachined to a fine surface finish after clamping so that uniform depth of cut could be maintained throughout the machining tests. Machining tests were carried out by using uncoated positive carbide inserts at two different machining speeds (650 fpm and 700 fpm). Depth of cut was set to 0.025 in and feed was 0.007 in/rev. Three directional cutting forces were measured using a three component piezo-electric dynamometer, and were collected and processed by a Tektronix 2630 Fourier Analyzer, however the results are not presented in this paper. A schematic diagram for the machining experimental set-up is depicted in Figure 1.

During testing machining was stopped and tool wear was measured using a tool microscope at every 2 inch interval along a bar. Surface roughness was measured after machining was completed using a SurfAnalyzer 4000 with a contact stylus. Residual stresses were measured on the bars after machining was complete at nine positions each, on two axial traverses ninety azimuthal degrees apart. The measurements were made at these eighteen positions in three directions to provide sufficient information to calculate in-plane principle stresses. The three directions were axial (0°), circumferential (90°) and oblique (45°).

The single exposure technique (SET) for residual stress measurement was selected to provide rapid measurements consistent with good precision and accuracy, and to be compatible with the possibility of on line measurement [19,20]. However, a slower and more elaborate technique of

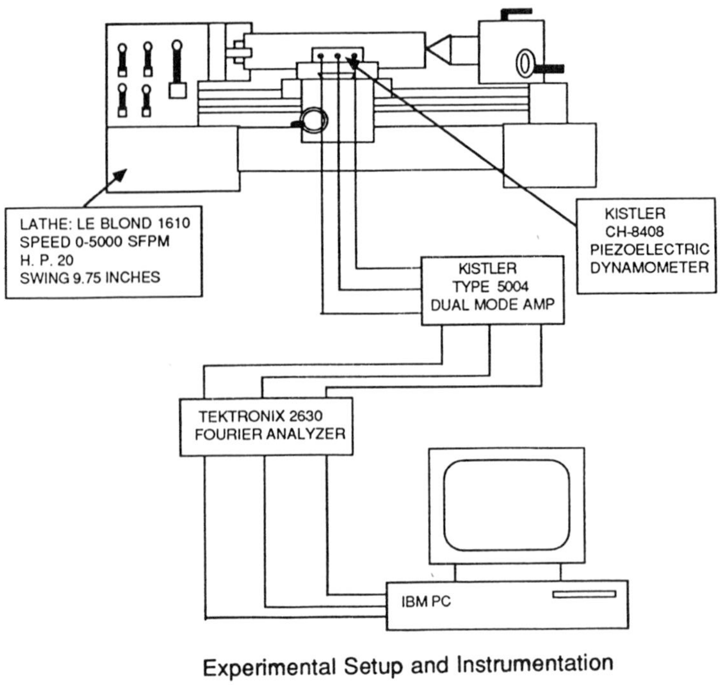

Figure 1. Schematic Diagram of Experimental Lathe Machining Arrangement.

stress measurement, sine-square-psi, was used to confirm the validity of use of the SET with respect to out of surface plane stresses and plastically induced preferred orientation [21]. The SET of residual stress measurement was performed using a fiber optic based position sensitive scintilation detector (PSSD) [20]. The combination of the SET and PSSD provides for the speed, compactness, and adaptability of x-ray residual stress measurement necessary for potential in-process (on-line) measurement of machining induced residual stresses in the workpiece. The data collecting times for each measurement was 1 to 2 seconds and the measuring instrument was at a distance of about 1.5 inches from the workpiece. However, the time could be reduced to 0.1 to 0.4 seconds by using a modern x-ray source. The source which was available is over 30 years old and limited in power.

The aforementioned x-ray residual stress techniques, i.e. SET and sine-square-psi, both assume the condition of plane stress. This is a reasonable assumption considering that the x-rays measure stresses only to a depth of about 0.0005 inches into the workpiece. Thus, when stresses were measured in three directions, i.e. axial (0°), circumferential (90°) and oblique (45°), the data could be used to calculate the surface plane principle stresses. The principle stresses are designated sigma 1 (the major stress) and sigma 2 (the minor stress).

After machining was completed, the workpiece surface quality was measured in terms of surface roughness and residual stress. The workpieces were divided into nine regions and measurements were obtained from each region since there was a substantial amount of tool wear during the cutting. The high tool wear rate was due to the high speed

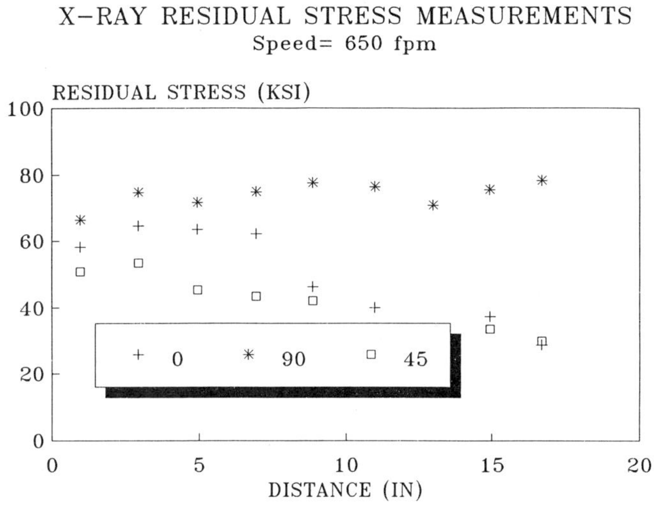

Figure 2. Average residual stress plotted versus distance along the axis of a 4140 steel bar machined at 650 fpm, showing axial (0°), circumferential (90°) and oblique (45°) stresses.

intentionally used to generate rapid wear. The workpiece surface quality in each region was expected to reflect the different levels of tool wear.

RESULTS AND DISCUSSION

Figure 2 shows the measured three components of residual stress, axial (0°), circumferential (90°) and oblique (45°), for the 4140 steel bar machined at the lower speed (650 fpm). It can be seen that axial stresses decrease relatively sharply as cutting progresses while the hoop stresses show a slight increase. Figure 3 presents similar results for higher machining speed (700 fpm). Comparison of the plots indicate that the stresses follow the same trend for both machining speeds but the circumferential stress is lower for the higher speed. This phenomenon is contradictory to the results presented by Leskovar and Peklenik [8]. However, it must be pointed out that the machining speed range they used was much lower compared to that used in the present investigation; therefore, direct comparison cannot be made. Nevertheless it indicates that simple extrapolation of results obtained at low speeds, to high speed machining may not be valid. It is believed that the chip morphology in the speed range used for the present study may be substantially different than for low cutting speeds.

The remainder of the data presented herein is concerned with the bar machined at 650 fpm only, but the results from the bar machined at the higher speed are nearly identical. Figures 4 and 5 present the workpiece surface roughness (arithmatic average) and flank wear measurements made on the cutting tool insert with respect to cutting length. As expected, both quantities increase almost linearly with increasing length of cut although surface roughness reaches its maximum and levels off after a certain amount of machining. This is due to the fact surface roughness depends on

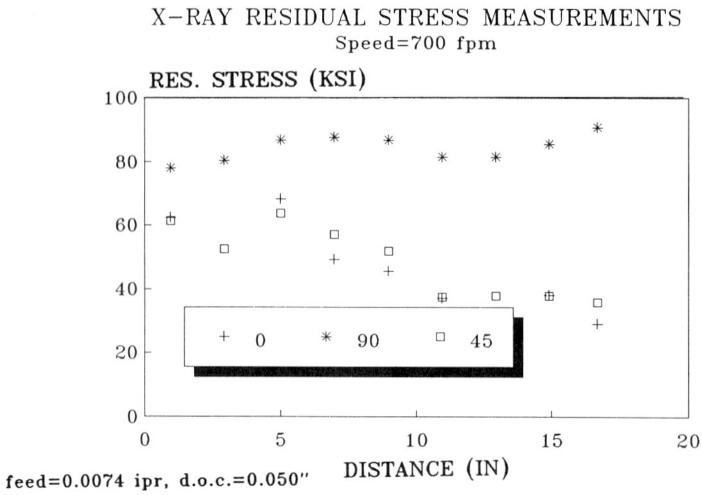

Figure 3. Average residual stresses plotted versus distance along the axis of a 4140 steel bar machined at 700 fpm, showing axial (0°), circumferential (90°) and oblique (45°) stresses.

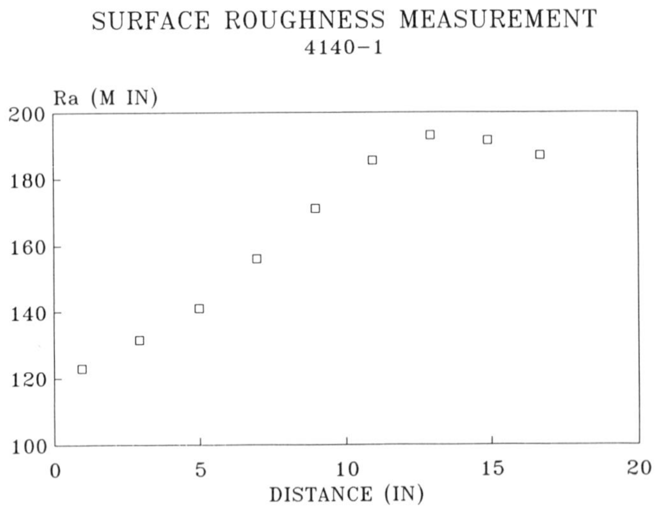

Figure 4. Surface roughness plotted versus distance along the axis of a 4140 steel bar machined at 650 fpm.

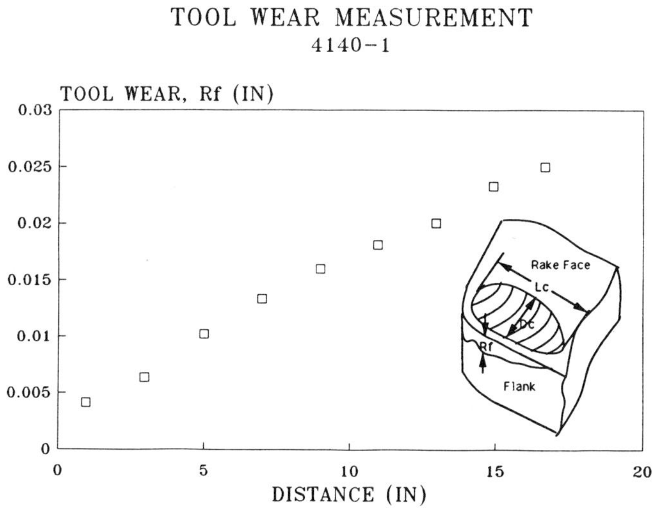

Figure 5. Tool flank wear plotted versus distance along the axis of a 4140 steel bar machined at 656 fpm.

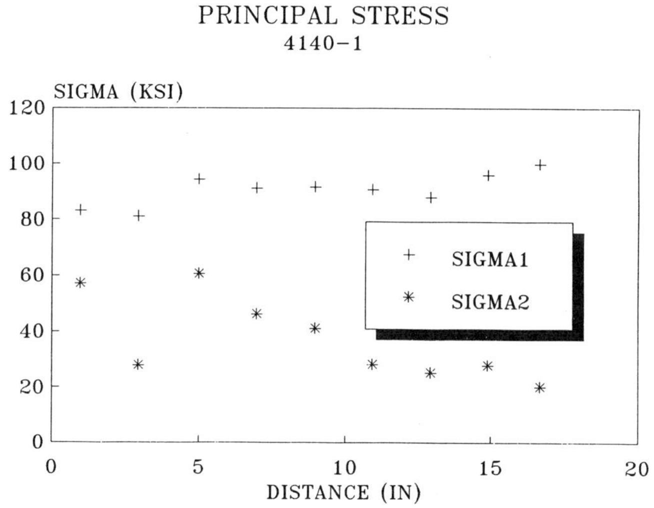

Figure 6. Sigma 2 principal stresses vs. cutting distance for the 4140 steel bar machined at 650 fpm.

other machining conditions beside tool geometry and wear. One possible explanation is the reduction of vibration, because of the increased stiffness of the workpiece as the tool approaches the end of the workpiece towards the chuck.

The two surface plane principal stresses were calculated from the measured residual stresses and shown in Figure 6. It is interesting to note that the minor principal stress sigma 2, which is near the axial direction, is more sensitive to the tool wear. Hence the minor principal stress was used to investigate correlations to other quantities. Figure 7 shows the minor principle stress (sigma 2) plotted against the surface quality for the 4140 steel bar machined at 650 fpm and indicates a linear correlation of 0.90. Figure 8 shows the minor principle stress (sigma 2) plotted against the flank wear of the tool insert and also indicates a nearly linear correlation of 0.87. These results indicate that through x-ray diffraction residual stress measurements information about machining conditions can be determined. As stated earlier, the success of automation relies upon the capability of reliable and robust in-process monitoring of important machining parameters and conditions, and x-ray diffraction shows promise as a monitoring technique.

CONCLUSIONS

It has been demonstrated that x-ray diffraction residual stress measurement technology shows promise for in-process monitoring of machining conditions.

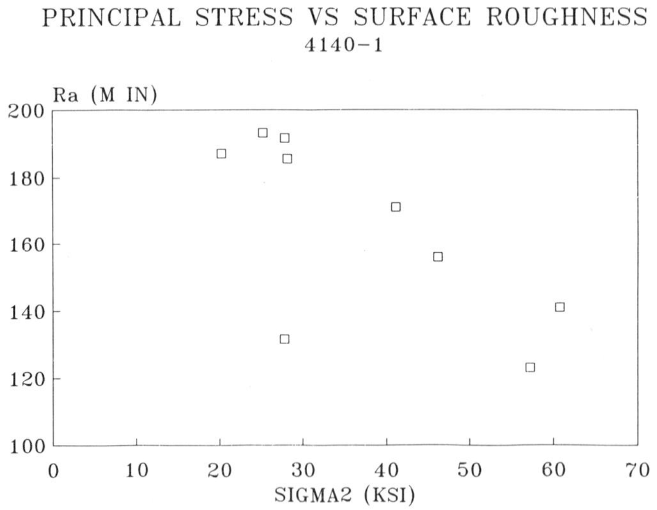

Figure 7. Sigma 2 principal stress vs. surface roughness for the 4140 steel bar machined at 650 fpm.

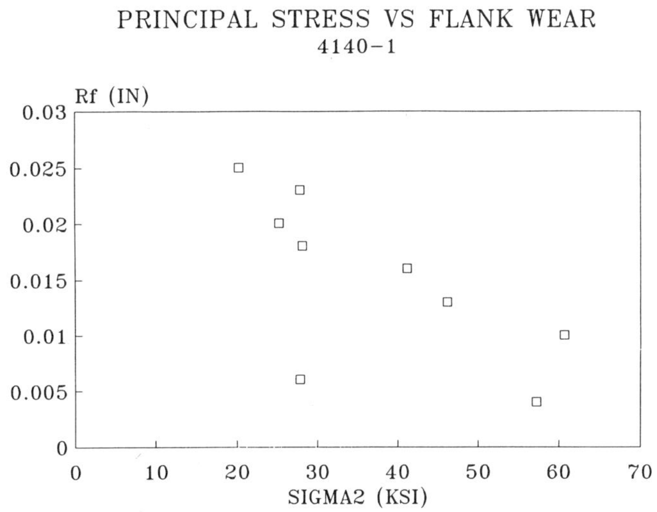

Figure 8. Sigma 2 principal stress vs. flank wear for the 4140 steel bar machined at 650 fpm.

-- Residual stresses correlate well with tool wear and surface roughness.

-- For the present experimental study, it was shown that the minor principle stress, which was near the axial direction of the bar, was more sensitive to tool wear change than the major principle stress.

REFERENCES

1. J. Tlusty and G. C. Andres, A Critical Review of Sensors for Unmanned Machining, Annals of the CIRP, Vol 32/2, pp. 563-572, (1983).
2. N. Skalli and R. Geslot, Effect of Grinding Conditions on Fatigue Behavior of 42 CO 4 Grade Steel, Annals of the CIRP, 32/1, pp. 481, (1983).
3. H. K. Tonshoff and F. Hetz, Influence of Thermal and Mechanical Impacts on Fatigue in Precision Grinding, "Advances in Surface Treatments," Pergamon Press, Oxford, (1984).
4. N. H. Polakowshi, Effect of Residual Stress on Yielding and Strain-Agening of Carbon Steel, J. Iron Steel Inst., pp. 369-376, (1952).
5. L. P. Tarasov, W. S. Hyler and H. R. Letner, Effect of Grinding Conditions and Resultant Residual Stresses on the Fatigue Strength of Hardened Steel, Proc. Am. Soc. Test. Mat., 57, pp. 601-622, (1957).
6. E. Brinksmeier et. al, Residual Stresses - Measurement and Causes in Machining Processes Annals of the CIRP, Vol. 31/2, pp. 491-510, (1982).

7. C. O. Ruud, D. J. Snoha, D. P. Ivkovich, and A. R. McIlree, "Residual Stresses in Inconel 600 U-Bend Heat Exchanger Tubes," J. of Press Vess. Tech., 112, pp 169-174, May (1990).
8. P. Leskovar and J. Peklenik, Influences Affecting Surface Integrity in the Cutting Process, Annals of the CIRP, Vol. 31/1, pp. 447-450, (1982).
9. E. Brinksmeier, X-ray Stress Measurement - A Tool for the Study and Layout of Machining Processes, Annals of the CIRP, Vol 34/1, pp. 485-490, (1985).
10. R. G. Treuting and W. T. Read, A Mechanical Determination of Biaxial Stress in Sheet Materials, J.Appl. Phys. 22, pp. 130-134, (1951).
11. C. O. Ruud, G. H. Pennington, E. M. Brauss and S. D. Weedman, "Simultaneous Residual Stress and Retained Austenite Measurement by X-Ray Diffraction," Nondestructive Characterization of Materials III, pp. 406-412, Springer-Verlag, Berlin (1989).
12. M. Field, J. F. Kahles and J. T. Cammett, A Review of Measuring Methods for Surface Integrity, Annals of the CIRP, Vol. 21/2, pp. 219-238, (1972).
13. O. S. Es-said and J. G. Morris, A Method to Calculate the Resistivity of Strip Cast 300 Series Aluminum Alloys Based on their Nominal Solute Content, NDC of Mat. II, Plenum Press, pp. 271-280, (1987).
14. S. R. Buxbaumn and R. E. Green, Ultrasonic Characterization of Titanium 6211 Weld Materials, Nondestructive Methods for Mat. Prop. Det., Plenum Pres, pp. 271-288, (1984).
15. F. Zankl, D. G. Barkow and A. O. Schmidt, X-Ray Diffraction as a Gage for Measuring Cold Work Produced in Milling, Trans. of ASME, pp. 307-318, (1947).
16. Y. Matsumoto, M. M. Barash and C. R. Liu, Residual Stress in the Machined Surface of Hardened Steel, ASME PED-Vol. 12, pp. 193-204, (1984).
17. H. K. Tonshoff and E. Brinksmeier, Optimization of Computer Controlled X-Ray Stress Analysis, Annals of the CIRP, 30, pp. 509-13, (1983).
18. C. O. Ruud, Position-Sensitive Detector Improves X-Ray Powder Diffraction, Ind. R&D, pp. 84-86, (1983).
19. SAE, Residual Stress Measurement by X-Ray Diffraction - SAE J784a, Soc. of Auto. Engr., 400 Commonwealth Dr., Warrendale, PA (1971).
20. C. O. Ruud, "Application of a Position Sensitive Sintilation Detector to Nondestructive X-Ray Diffraction Characterization of Metallic Components," Nondest. Methods for Mat. Prop. Meas., Plenum Press, pp 21-38, (1984).
21. C. M. Brackman, "Residual Stresses in Cubic Materials with Orthorhombic or Monoclinic Specimen Symmetry: Influence of Texture on Psi Splitting and Non-linear Behavior," J. Appl Cys., 16, 325-340, June (1983).

X-RAY DIFFRACTION STUDY OF RESIDUAL STRESSES IN METAL MATRIX COMPOSITE-JACKETED STEEL CYLINDERS SUBJECTED TO INTERNAL PRESSURE

S. L. Lee, M. Doxbeck, and G. Capsimalis

U.S. Army ARDEC
Close Combat Armaments Center
Benet Laboratories
Watervliet, NY 12189-4050

ABSTRACT

The study of aluminum/silicon carbide metal matrix composite (MMC)-jacketed steel structural components was made because of their light weight and high stiffness. Steel 'liner' cylinders were wrapped with MMC 'jackets' with all-hoop layup and were put through various degrees of hydraulic autofrettage and thermal soak. In this paper, we summarize the results from our x-ray diffraction residual stress measurements on a number of cylinders using a position-sensitive scintillation detection (PSSD) system. Our experimental results are compared with theoretical predictions from a model based on the elastic-plastic analysis of a thick cylinder subjected to internal pressure. Interpretation of the interference effect caused by the MMC jacket on the steel liner is discussed.

INTRODUCTION

Jacketed steel cylinders have been designed and fabricated to obtain the same pressure containment as steel cylinders, but with much less weight. The manufacturing and testing of the jacketed cylinder program has been summarized in a Benet Laboratories technical report [1]. From the material properties for steel and MMC listed in Table 1 [1], the steel has a density of 0.28 lb/in.^{3}, and the metal matrix jacket (continuous silicon carbide wrapped in the hoop direction and sprayed with 6061 aluminum alloy matrix) has a density of 0.11 lb/in.^{3}. The ultimate tensile stress (UTS) is 178 Ksi for the steel and 217 Ksi for the MMC. The reason for using the MMC is due to the reduction in weight (2.6 times) and the comparable UTS.

In the present paper, we report x-ray diffraction measurements of the residual stress using a PSSD system. The cylinders were subjected to various pressure cycles and then autofrettaged and thermal soaked. The autofrettage process can induce favorable residual stress distribution in the cylinders and improve fatigue and fracture behavior. Because applied and residual stresses add algebraically, the autofrettage procedure has found applications in increasing the elastic operating pressure of a system. However, cylinders subjected to heating can lose residual stress. In high temperature applications, a thermal soak process is

Nondestructive Characterization of Materials IV
Edited by C.O. Ruud *et al.*, Plenum Press, New York, 1991

applied to release some of the residual stress generated by autofrettage to obtain thermally stable components. Our experimental results are compared with predictions from a model based on the elastic and plastic analyses of a thick-walled composite tube subjected to internal pressure. The effect of the presence of the MMC on the steel liner is discussed.

Table 1. Comparison of Material Properties of Steel and Aluminum/Silicon Carbide MMC Composites

Property	Steel	MMC
Weight density (lb/in.^{3})	0.283	0.11
Ultimate tensile strength (Ksi) (MMC with hoop layup)	178	217
Yield strength (Ksi)	155	-
Young's modulus (msi)- (MMC with hoop layup)		
E11	30	32
E22, E33		19
Poisson's ratio- (MMC with hoop layup)		
V12, V13	0.28	0.25
V21, V31		0.14
V23, V32		0.36

EXPERIMENTAL METHOD

A D-1000-A Denver X-Ray Instruments Model stress analyzer design was used based on a Ruud-Barrett PSSD. The instrumentation and experimental methods are described in References [2] and [3]. The instrument utilizes a miniature x-ray tube that fits inside a cylinder with a 3.70-inch interior diameter. The operating principles for this detection system consist of converting incident x-rays into light through cadmium-zinc sulfide scintillation coating; transporting this light to a detector package through fiber optics bundles; amplifying the light intensity through an image intensifier; and converting it into an electronic signal through a 512-element reticon photodiode array. The instrument uses a chromium target tube and is capable of exploiting x-ray sources of 100 watts or less and providing a very short data collection time interval. The characteristic x-ray radiation from the chromium tube does not fluoresce any major alloys in steel. The chromium K-α reflects from the 211 plane of the body-centered-cubic (BCC) carbon steel and the 211 plane of the BCC martensitic steel at 2θ = 156.41 degrees. The divergent x-ray beam was collimated by a square slit, the irradiated area at the specimen was approximately $(1/16)^2$ in.^{2}.

An IBM AT computer was connected to the system for data acquisition control, storage, and analysis. The instrumentation allows both single exposure and $\sin^2\psi$ methods of residual stress determination. In the single-exposure technique, residual stress is determined from two x-ray diffraction patterns obtained simultaneously in a single exposure. In the $\sin^2\psi$ technique, stress is determined from diffraction patterns obtained at multiple angles of inclination. The diffraction peaks were noise- and gain-corrected. Typical data collection time for a steel sample was 2 to 3 seconds. In addition, the software used a number of algorithms to correct electronic and mechanical hardware fluctuations, x-ray focusing errors, and effects due to polarization and absorption. A parabolic fit taking 50 percent of the upper part of the diffraction peak was used in making the final peak determination.

In this work, stress distribution measurements were made in the steel liner using the single-exposure technique. The bore and interface stresses were further confirmed by $\sin^2\psi$ measurements. In the latter method, we used four β angles at 12, 20, 27, and 33 degrees, giving eight data points for the linear regression fit. The β angle is the angle between the specimen surface normal and the incident x-ray beam. In addition to the system software, SYMPHONY and FREELANCE software were used for data analysis and graphics.

CYLINDER PROCESSING HISTORY

The geometry of the jacketed cylinders is shown in Fig. 1. The cylinders are ASTM A723 steel forging approximately 1 foot long, with a bore diameter of 4.5 inches, a steel liner thickness of 0.4 inch, and a MMC jacket thickness of 0.2 inch. This geometry gives a 2.25-inch steel liner inside radius, a 2.65-inch steel/MMC interface radius, and a 2.85-inch MMC outside radius. As a result, the steel liner has 64 percent of the total volume and 82 percent of the total weight, while the MMC has 36 percent of the total volume and 18 percent of the total weight. The total component weight savings of the compound cylinder compared to an all-steel cylinder is 22 percent. The liner was wrapped directly with the silicon carbide fibers in the circumferential direction, and the 6061 aluminum alloy matrix was applied by plasma spray. The monofilament reinforcing silicon carbide fibers were 0.0056 inch in diameter and occupied 47 percent of the total MMC volume. The steel cylinders were manufactured by Benet Laboratories, while the MMC wrapping was done by Textron Specialty Materials Division, Lowell, MA. Interlayers were plated on the liners consisting of 0.001 to 0.002 inch of nickel, silver, and aluminum/10 percent silver alloy. The purpose of the interlayers was to create a better bond and to reduce galvanic corrosion between the jacket and the liner.

The cylinders were pressurized in increments over various pressure ranges before autofrettage, followed by hydraulic autofrettage and thermal soak operations. Table 2 gives the processing history of these cylinders.

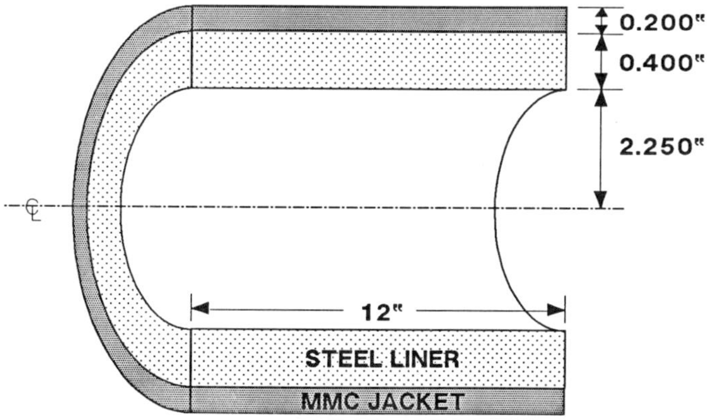

Fig. 1. Jacketed cylinder geometry.

Table 2. Cylinder Processing History

Cylinder No.	3AE2	3AE5	3AA	3AN
Pressure history (Ksi)	14-33	14-33	16-35	16-35
Autofrettage pressure (Ksi)	37.1	37.1	38.6	39.3
Thermal soak (4 hrs)	700°F	600°F		

SPECIMEN PREPARATION

To measure the surface residual stress through the cross section, normal to the cylinder axis, 1-inch thick rings were cut from the middle section of the cylinders. One sample each was obtained from 3AA and 3AN and two were cut from 3AE, for a total of four rings. The rings were cut at Benet Laboratories on an Agie DEM315 travelling wire electrostatic discharge machine, which can produce strain-free samples with minimum mechanical and thermal damage. The ring surface was polished manually with sandpaper and then electropolished to remove surface effects due to sanding, machining, and oxidation. It was then washed with acetone, alcohol, and distilled water. The electropolishing setup was as follows: anode - specimen to be polished; cathode - lead; solution - 50% H_3PO_4 (85%), 25% H_2SO_4 (98%), 25% H_2O; current density - 30 to 90 amp/dm²; temperature - room; agitation - no external agitation.

CALIBRATION AND VERIFICATION PROCEDURE

A four-point bend experiment was formed to determine the elastic stress constant for the steel BCC 211 plane and to test the x-ray method. A 1-inch by 6-inch by 3/16-inch thick steel sample was used for this purpose. An electrical strain gauge was applied to the specimen adjacent to where the x-ray measurements were made. Four-point bending generated tensile and compressive stresses between +100 Ksi and -100 Ksi. From the slope of the calibration curve, we determined the elastic modulus, E, to be 30.7 msi (211.8 GPA). The linear regression correlation coefficient was better than 0.95. To verify the x-ray method, four-point bend stress measurements using x-ray method versus strain gauge method are plotted in Fig. 2. The agreement is very good, with a linear regression correlation coefficient of 0.97.

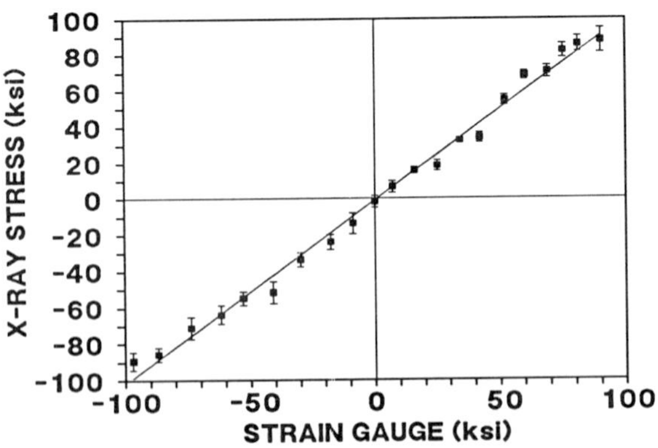

Fig. 2. Four-point bend verification.

EXPERIMENTAL RESULTS

The hoop residual stress distributions in the steel liner as functions of the radial distance from the center of the cylinder are given in Figs. 3 (3AE02), 4 (3AE05), 5 (3AA), and 6 (3AN). In these figures, each experimental point represents the average of five measurements taken at two-second intervals. The error bars represent the dispersion in the measurements and do not include alignment, focusing errors, and effects due to surface irregularities. The solid curves are discussed later.

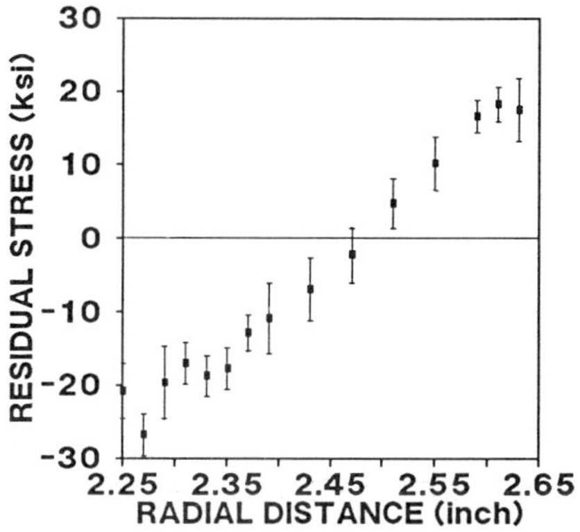

Fig. 3. The measured 3AE02 hoop residual stress distribution.

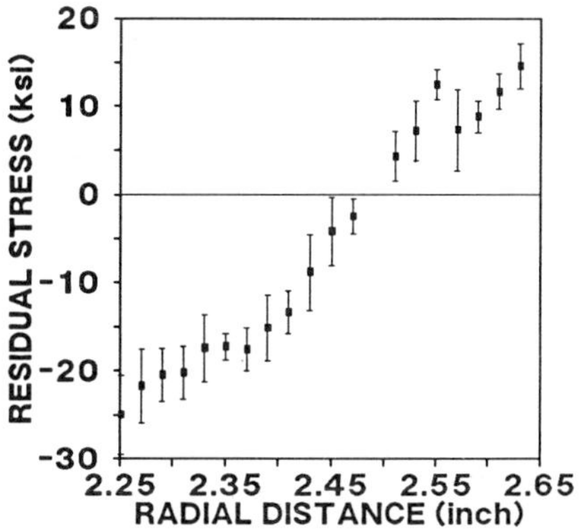

Fig. 4. The measured 3AE05 hoop residual stress distribution.

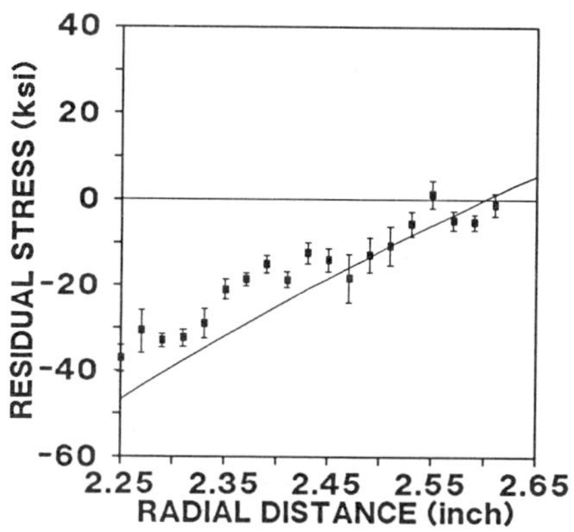

Fig. 5. The measured 3AA hoop residual stress distribution with a line superimposed showing the predicted stress distribution.

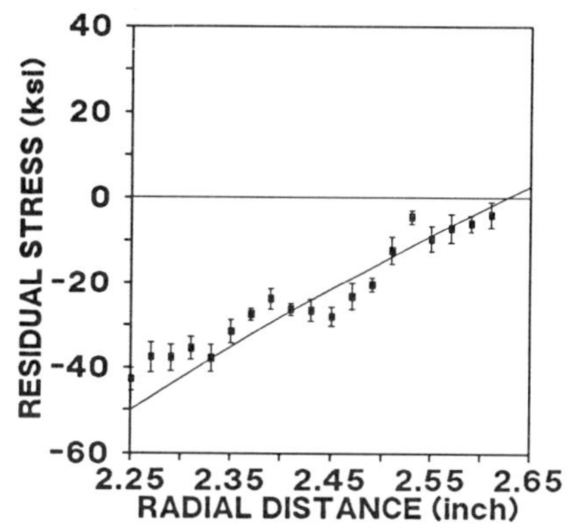

Fig. 6. The measured 3AN hoop residual stress distribution with a line superimposed showing the predicted stress distribution.

From our results, we see that residual stress has been induced at the bore of the cylinders, diminishing and changing to tensile stress toward the outer edge of the steel liner. In cylinders 3AE02 and 3AE05, which were autofrettaged and thermal soaked, both compressive and tensile stresses were observed in the steel liner. In cylinders 3AA and 3AN, which were only autofrettaged, the residual stresses in the steel liner were mostly compressive. The characteristics of the distribution curves are similar to the distributions for autofrettaged steel cylinders in the literature [4].

Plots of strain versus $\sin^2\psi$ from the bore are given in Fig. 7 (3AE05, 3AN) and of steel/MMC interface are given in Fig. 8 (3AE05, 3AN). From the slope of the (D-DP)/DP versus $\sin^2\psi$ plot, where DP is the perpendicular d spacing, compressive stresses at the bore are -17 Ksi (3AE05), -35 Ksi (3AN), and tensile stresses at the interface are 14 Ksi (3AE05), 0.5 Ksi (3AN). We used our previously determined Young's modulus for the steel 211 plane. The correlation coefficients for the linear regression fit for all the curves were better than 0.95 with the exception of 3AN interface which was 0.46.

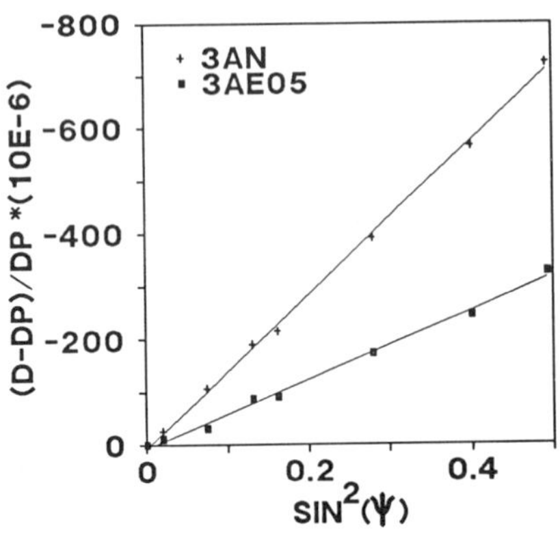

Fig. 7. $\sin^2\psi$ determination of stresses at the cylinder bore.

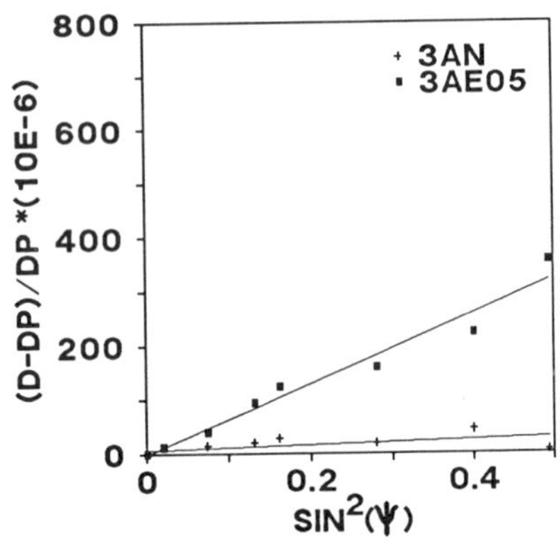

Fig. 8. $\sin^2\psi$ determination of stresses at the interface.

The autofrettage pressure for cylinders 3AA and 3AN were 38.6 Ksi and 39.3 Ksi, respectively. Comparing Figs. 5 and 6, compressive residual stresses 5 to 10 Ksi less have been generated in 3AA than in 3AN. Comparing Figs. 3 and 4 with Fig. 6, compressive stresses approximately 15 to 20 Ksi less have been generated in 3AE02 and 3AE05 than in 3AN. The theory in the next section predicts less compressive residual stresses for smaller autofrettage pressures. Moreover, less stress was observed in 3AE02 and 3AE05 compared to 3AA and 3AN for the following reasons: (1) less autofrettage pressure in 3AE02 and 3AE05 (37.1 Ksi), (2) thermal relaxation in 3AE02 and 3AE05 causing residual stress relief [5,6], and (3) possible gap formation between the liner and jacket due to the thermal process. We did not observe a difference in the residual stresses in 3AE02 and 3AE05, which were thermal soaked at 700°F and 600°F, respectively.

THEORETICAL CONSIDERATION

Figure 9 shows how the compound cylinder problem can be decomposed into two parts: (1) a steel cylinder with internal applied pressure p and a pressure q acting on the steel by the MMC, and (2) the MMC with a pressure q acting on the MMC by the steel cylinder. We compare our experimental results with two theoretical models.

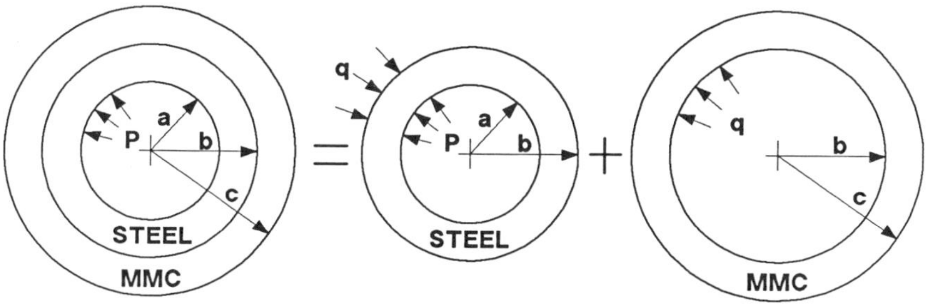

a = 2.25 inch   b = 2.65 inch   c = 2.85 inch

Fig. 9. Decomposition of composite/steel cylinder.

Superimposed on the experimental distribution data in Figs. 5 and 6 are the analytical predictions obtained from an elastic-plastic deformation autofrettage model [7]. The calculations based on this model utilized the elastic constants in Table 1 and assumed the MMC to be elastic. For the autofrettage pressure range of 37.1 Ksi to 39.7 Ksi, the steel cylinder can be considered fully plastic. Experimental results compare well with theoretical curves to within +10 Ksi.

The theoretical distribution of residual stresses based on Tresca's yield criterion was reported by Davidson et al. [8]. We obtained theoretical curves by assuming an all-steel monoblock 100 percent overstrain autofrettage model. In this model, the cylinders are considered fully plastic and the hoop residual stress distribution is given by:

$$\sigma_t = \sigma_y \left(\frac{-\log w}{(w^2-1)}\left(1 + \frac{b^2}{r^2}\right) + \left(1-\log \frac{b}{r}\right)\right) \tag{1}$$

where σ_t = hoop stress, σ_y = yield strength, w = b/a, and a,b,r = bore, outside, and variable radius.

Figure 10 gives the theoretical predictions based on this simple monoblock autofrettage model assuming: (1) the cylinder consists of the steel liner only and extends to a radius of 2.65 inches (top curve), and (2) the steel extends to the whole extent of the cylinder at 2.85 inches radius (bottom curve). We used 155 Ksi as the yield strength for steel.

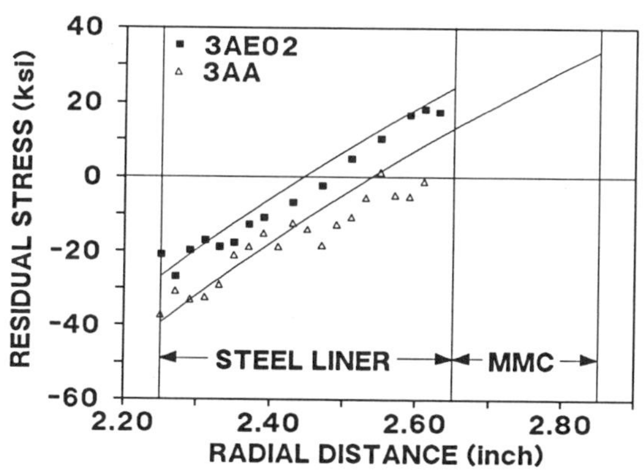

Fig. 10. Monoblock plastic deformation calculations with 3AA and 3AE02 hoop residual stress distributions superimposed.

The experimental stress distribution results for cylinders 3AA and 3AE02 are superimposed on the analytical curves. The 3AE02 data points fall closely to the top monoblock curve, while the 3AA data points fall more closely to the lower monoblock curve.

DISCUSSION

As indicated in Fig. 10, in an all-steel plastic deformation autofrettage model, the compressive stress at the bore is balanced by the tensile stress at the outer part of the cylinder. A comparison of the zero crossings in our stress distribution curves indicates that in an MMC steel cylinder, tensile stress has been transferred into the MMC. From a consideration of the area under the stress curve in cylinders 3AE02 and 3AE05 (Figs. 3 and 4), approximately 40 percent of the tensile residual stress has been transferred. In cylinders 3AA and 3AN (Figs. 5 and 6), almost 100 percent of the tensile stress has been transferred into the MMC.

In the MMC, if we assume that the residual stress is approximately uniform, and the net compressive stress in the steel cylinder is balanced by the tensile stress in the MMC, then we can estimate the average tensile stress from an area consideration. The results are as follows: 6 Ksi (3AE02, 3AE05), 30 Ksi (3AA), 42 Ksi (3AN).

In the autofrettage process, the bore is plastically enlarged and the MMC is elastically enlarged due to the internal hydraulic pressure. As this internal pressure is released, the outer MMC attempts to contract elastically to its original dimension. The steel near the bore, which has been deformed the largest, resists this action. This results in compressive hoop residual stress near the bore and changes to tensile stress near the outer surface. Since our 3AA and 3AN results compare favorably with theoretical predictions [7], it appears that the model can be used to describe the autofrettaged compound cylinder.

In the thermal soak process, thermal stresses are added to the residual stresses and exceed the lower yield strength at elevated temperatures, permitting thermal relaxation in steel and aluminum. This causes the loss of bore expansion and compressive residual stresses in autofrettaged cylinders. The fact that 3AE02 was thermal soaked for four hours at 700°F and the close comparison of 3AE02 data with the top monoblock curve in Fig. 10 suggest possible gap formation at the steel/MMC interface. If this should occur, the liner is dissociated from the jacket, and the monoblock model is adequate to interpret the 3AE02 data. The effect of thermal relaxation on autofrettage in a compound cylinder should be investigated further both experimentally and theoretically.

ACKNOWLEDGEMENTS

The authors wish to thank Peter Chen for sharing his theoretical results, John Underwood for the helpful discussions, Gary Cunningham for providing the compound cylinders and the material constants, and Robert Messier and John Hoffman for preparing the rings.

REFERENCES

1. R.G. Hasenbein, E. Hyland, and G. Cunningham, Laboratory Tests of Metal Matrix Composite-Jacketed Cylinders, U.S. Army ARDEC Technical Report, Benet Laboratories, Watervliet, NY, in press.

2. C.O. Ruud, P.S. DiMascio, and D.J. Snoha, A Miniature Instrument for Residual Stress Measurement, in: "Advances in X-Ray Analysis," Plenum Press, New York (1974), p. 27.

3. S.L. Lee, M. Doxbeck, and G. Capsimalis, Experimental Methods in Residual Stress Measurement Using a Position-Sensitive Scintillation Detection System, U.S. Army ARDEC Technical Report, Benet Laboratories, Watervliet, NY, in press.

4. L. Pintschovius, E. Macherauch, and B. Scholtes, Determination of Residual Stresses in Autofrettaged Steel Tubes by Neutron and X-Ray Diffraction, in: "Residual Stresses in Science and Technology," Vol. 1, DGM Informationsgesellschaft mbH, Federal Republic of Germany (1987), pp. 159-165.

5. V.C.D. Dawson and J.W. Jackson, Investigation of the Relaxation of Residual Stresses in Autofrettaged Cylinders, Journal of Basic Engineering, March 1969.

6. J. Throop, J. Underwood, and G. Leger, Thermal Relaxation in Autofrettaged Cylinders, in: "Residual Stress and Stress Relaxation," E. Kula and V. Weiss, eds., Plenum Press, New York (1982), p. 205.

7. P.C.T. Chen, Elastic-Plastic Analysis of a Thick-Walled Composite Tube Subjected to Internal Pressure, U.S. Army ARDEC Technical Report ARCCB-TR-89027, Benet Laboratories, Watervliet, NY, October 1989.

8. T.E. Davidson, D.P. Kendall, and A.N. Reiner, Residual Stresses in Thick-Walled Cylinders Resulting from Mechanically Induced Overstrain, Experimental Mechanics, November 1963.

CRYSTALLOGRAPHIC TEXTURE ANALYSIS AS A NEW METHOD FOR MATERIAL CHARACTERIZATION

J. R. Hirsch

Alcoa Laboratories
Alcoa Center, Pennsylvania 15069

ABSTRACT

For the major processes of metal production, i.e., casting (solidification), rolling (plastic deformation) and annealing (recrystallization) of FCC metals, some of the evolving typical textures are presented and examples of characteristic effects of important material and process parameters on microstructure and related textural features are discussed. It is emphasized that texture analysis can be applied to investigate these processes and their underlying physical mechanisms and so better understand, control and improve related fabricating processes.

INTRODUCTION

Almost all processes of metals production, i.e., casting, deformation and annealing, result in characteristic textures and the important role of textures in material science has been more and more recognized in recent years. So far, however, most applications of texture research mainly make use of the correlation between texture and anisotropy to improve material properties like plastic anisotropy of sheet material (Bunge, 1988).

However, in recent years textures have found a much broader field of application in physical metallurgy by the investigation of the origin of characteristic textures; the basic mechanisms involved are analyzed like the deformation or recrystallization mechanisms in metals and alloys or in rocks in geology (Kallend, Gottstein, 1987). Still, the majority of current activities in the broad field of texture research deal with the questions of how textures develop in the various steps of materials processing and what the physical reasons are for the formation of the different characteristic texture peaks. However, in recent years with the help of fast automatic measuring equipment and modern computer supported evaluation methods (Bunge, 1986 and 1987) the quality and quantity of texture studies have rapidly increased. Thus, a wider, more systematic overview of the great variety of textures and their changes with various processing parameters has been obtained and a quite deep understanding of the underlying mechanisms of the various texture formation processes has been achieved (e.g., the geometry and degree of deformation, the effects of alloy element and content, annealing time and temperature, etc.).

With the knowledge about the origin of characteristic textures and the basic principles of their formation and correlation to microstructure, texture data themselves could be used in turn to draw conclusions about these processes. Some properties of texture data are of special advantage for this purpose, like the statistical nature of X-ray texture data. Also, their magnifying effect for certain events is very useful, for example,

Nondestructive Characterization of Materials IV
Edited by C.O. Ruud *et al.*, Plenum Press, New York, 1991

the nucleation process of recrystallization. In many cases, the influence of certain processing parameters like the effect of solute atoms or of particles on recrystallization is most difficult to observe directly. Making use of these advantages, applied texture research can become a new, powerful tool to analyze the microstructure and the underlying physical mechanism as a natural extension of metallography.

It must be noted that, of course, only those mechanisms can be observed and analyzed which lead to notable and characteristic texture changes. However, with the help of modern measuring techniques and evolution methods, an extremely accurate and detailed analysis of even minor effects is possible. Here, the question of reproducibility of texture effects must be considered carefully, which for spot check experiments and processing on a laboratory scale is often a problem. But for industrially-fabricated material the reproducibility is usually so good that slight variations in production parameters can cause notable and clearly distinguishable texture changes.

In the following, some examples for the results of fundamental texture research will be presented which illustrate the analysis of the determining mechanisms and aspects of their application for material characterization will be discussed.

CASTING AND SOLIDIFICATION

The cast ingot can show a characteristic texture which is caused by the effect of growth selection for the fastest growing crystallites during solidification. For example, in FCC metals often a <100> direction is preferred so that a clear <100> fiber texture is developed in uniaxial solidification, Figure 1a (Hirsch et al., 1987a). Since the direction of this fiber axis indicates the solidification direction in a cast ingot, this effect might be used to check solidification geometry. In the case of rapid solidification of melt spun metal cast on a fast rotating wheel, where this effect has been observed (Eucken et al., 1988) the information can be used to determine the important value of the rate of solidification. Simple geometrical considerations making use of the observed tilting angle α of the fiber axis taken from a pole figure (Figure 1b) are sufficient to estimate this value (which otherwise would require enormous efforts for local temperature measurements). With v_R being the velocity of the metal strip on the rotating wheel, the solidification speed v_S and the time t_d required for the complete solidification of the strip thickness d can be estimated as:

$$v_S = v_R \cdot \sin\alpha \quad \text{and} \quad t_d = \frac{d}{v_R \cdot \tan\alpha} \quad \text{(see Figure 2)}$$

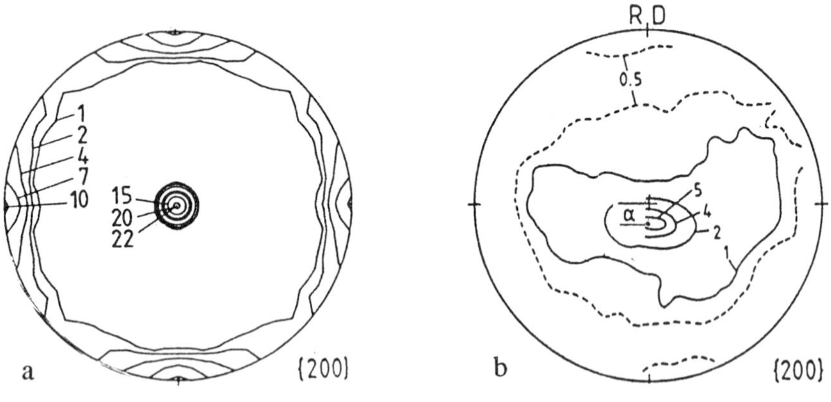

a b

Figure 1. {200} pole figures of solidification textures:
a) Texture of directionally solidified pure aluminum with a pronounced <100> fiber axis parallel to the solidification direction (= center of the pole figure, Hirsch et al., 1987a).
b) Textures and schematic solidification structure in a melt spun Cu alloy (RD see Figure 2) (Eucken et al., 1988)

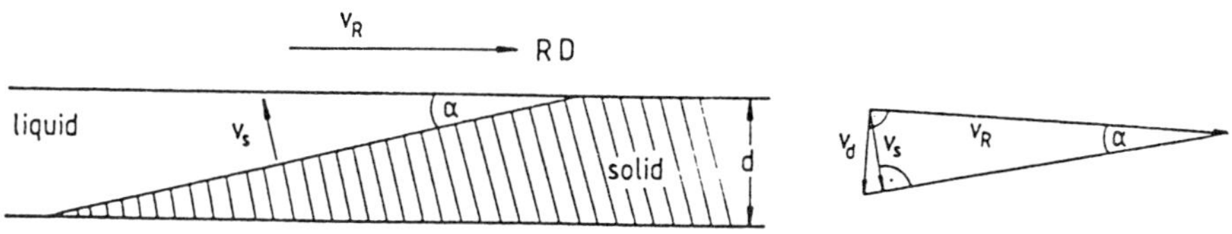

Figure 2. Schematic plot of the solidification geometry in melt spinning. (RD = rotation direction of melt spin wheel)

Furthermore, the texture can also reveal information about other effects of the solidification process, e.g., nucleation on the wheel surface and microstructural changes in different layers of the strip (Eucken at al., 1988).

PLASTIC DEFORMATION

Geometry of Deformation

Deformation in metals occurs by glide on a few crystallographic slip systems which can cause quite distinct textures. Depending on the geometry of deformation, characteristically different textures can develop. In turn, from the texture type as observed experimentally, these geometrical conditions of the deformation history can be deduced (which is an important application in geology research). For example, the pole figure of the cross section of an extruded Al-Li bar (Figure 3) (Tempus et al., 1987) shows a double fiber texture (<111> and <100>) characteristic for uniaxial deformation in the left side of the deformed area (a) which represents mostly an equiaxed deformation geometry. The ratio of intensity of these two fiber texture components depends on material and process parameters (like stacking fault energy and deformation temperature) and therefore can be used to determine their influences. In the right side of the deformed region (b), however, a typical plane strain (rolling-type) texture is developed. So, this case demonstrates how the texture type reveals information about the geometry of deformation and materials flow.

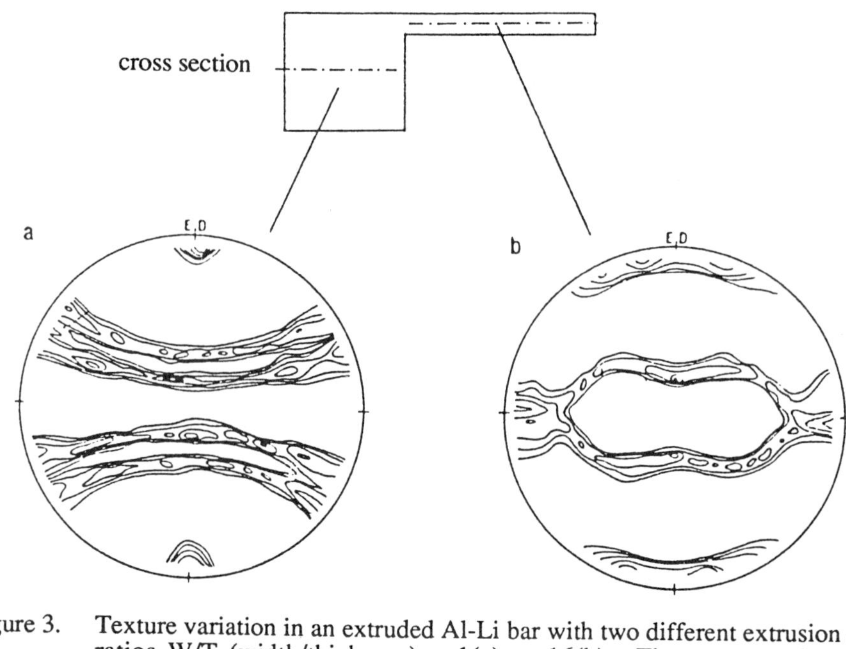

Figure 3. Texture variation in an extruded Al-Li bar with two different extrusion aspect ratios W/T (width/thickness) = 1(a) or 16(b). The cross section of the extrusion is plotted on top. The corresponding {111} pole figures indicate the different geometries of deformation (ED = extrusion direction).

Another example concerning the geometry of deformation is the important effect of surface friction in rolling which can be analyzed with the help of textures. In order to estimate the depth of the influenced zone the height and variation of the characteristic texture peak for surface friction in Al (e.g., the ND rotated cube (001) <110>) can be attributed to the depth of shear deformation. In Figure 4, the orientation density distribution along the τ-fiber, i.e., for orientations with a common <110> axis parallel to the (long) transverse direction TD is plotted. (ϕ = <110> axis rotation angle from $Cube_{ND}$). On an inhomogeneously rolled Al sheet (ratio of roll contact length ℓ_c to sheet thickness t: $\ell_c/t<1$) different depth of the textures measured are indicated by the value of S (surface S = 1, at mid-thickness S = 0). The intensity of the stable rolling texture orientation {112} <111> "C" (at $\phi \approx 30°$) is strong at the mid-thickness (S ≈ 0) but it continuously decreases with increasing S (i.e., closer to the surface) at the expense of the surface shear texture $Cube_{ND}$ {001}<100> ($\phi = 0°$, locations of C and $Cube_{ND}$ are indicated on top line of Figure 4). So the friction conditions responsible for the development of this zone can be quantified and measures to minimize it can be analyzed.

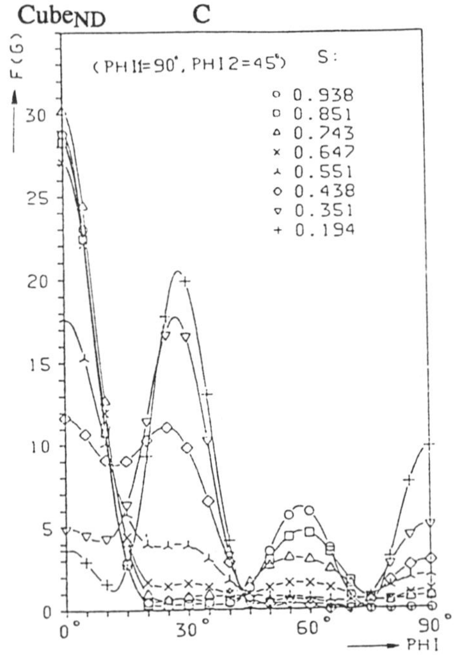

Figure 4. The variation of texture peak intensities f(g) at {112}<111> = C and {001}<110> = $Cube_{ND}$ orientations, characteristic for shear deformation in inhomogeneously rolled Al at different depths from sheet surface (S=1: sheet surface, S=0: mid-thickness T/2).

Mechanisms of Deformation

In general, texture intensity is a function of the degree of deformation, so in a comparative study it can be used to estimate this parameter. However, characteristic texture changes can occur when the mechanisms of deformation are changed, e.g., for FCC metals with low stacking fault energies, SFE. During rolling deformation medium to high SFE material (Copper, Aluminum) form the typical pure metal (or copper) - type rolling texture which consists of a characteristic orientation tube (called "β-fiber") composed of three texture components {112}<111> "C" (Euler angle $\varphi_1, \phi, \varphi_2$ = 90°, 35°, 45°), {123}<634> "S" (59°, 37°, 63°) and {011}<211> "B" (35°, 45°, 90°). For low SFE material (α-brass, silver) the latter component is dominant forming the characteristic brass-type rolling texture together with a minor component {011}<100> "G" (0°, 45°, 90°) (Hirsch et al., 1988). Besides the mostly continuous increase of component intensities with strain, in the case of low SFE material, a characteristic change in texture, namely the transition from the copper to the brass-type rolling texture, occurs which is associated with characteristic changes in deformation mechanisms (mechanical twinning and shear band formation). Characteristic texture components occur at different strain levels and the texture transition depends on the parameters, SFE and strain, and temperature in a rather complex manner. This has been analyzed in great detail by Hirsch et al. (1988). Based on these data, textures can be used to identify the parameters. For example, in Figure 5 the texture transition is plotted in the form of the variation of volume fractions for the main texture components for various homogeneous copper alloys for a constant strain of 95% rolling reduction. In this plot the scale of alloy content was chosen in such a way that the different curves for the same texture component are closest to each other, thus providing a common measure for the SFE.

Certain extractions of the ODF can be used to compare different texture effects, like strain levels or different materials and can help to analyze the associated deformation mechanisms. For FCC rolling textures the tube of orientations formed in the 3-D orientation space can be characterized by the line of maximum intensity (called "skeleton line") and its exact orientation which can be described by a (limited) fiber texture* called "β-fiber," Hirsch et al. 1988). Figure 6 shows typical skeleton line and β-fiber plots (f(g) and $\varphi_{1,\phi}$ vs. φ_2, respectively) for 95% cold rolled Al and Cu alloys. The characteristic variation in maximum intensity distribution along the skeleton line (Figure 6a) show the texture transition from the copper type to the brass type texture with decreasing SFE. This and especially the corresponding exact β-fiber orientations (Figure 6b) can be compared to theoretical predictions, like the Taylor Model (included in Figure 6b) which reveals details about active slip systems.

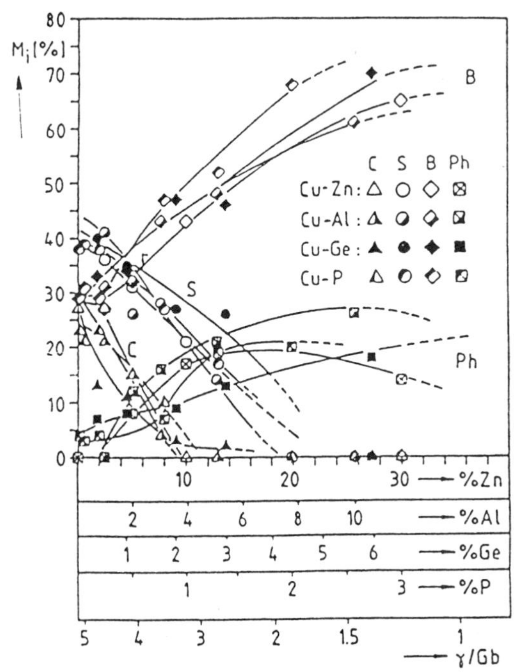

Name	=	{hkl} <uvw> (φ_1, ϕ, φ_2)
C	=	{112} <11$\bar{1}$> (90°, 35°, 45°)
S	=	{123} <63$\bar{4}$> (59°, 37°, 63°)
B	=	{011} <21$\bar{1}$> (35°, 45°, 90°)
Ph	=	random (phon)

Figure 5. Volume fractions of main texture components in various Cu alloys (rolled to 95% reduction) as function of the alloy content and normalized stacking fault energy SFE (Hirsch et al., 1988).

* The ideal β-fiber is a ±36° rotation around a <110> axis located at an angle α = 60° from the normal direction of the sheet towards the rolling direction.

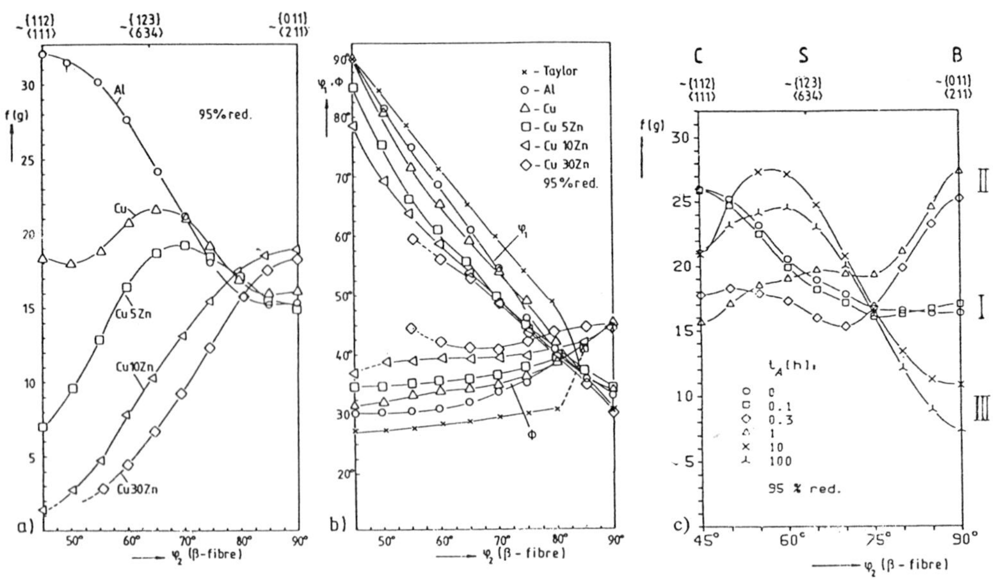

Figure 6. Skeleton line (= maximum intensity) plots extracted from ODFs (locations of the main texture components are indicated on the top line)

a) Skeleton intensities of various FCC materials rolled to 95% reduction and

b) Their corresponding exact orientation φ_1 and ϕ vs. φ_2. (Inserted are the theoretically predicted lines of the Taylor model.)

c) Skeleton line plots of 95% rolled Al 1.7% Cu with different precipitation states (Engler et al., 1989):

I = underaged (0 and 0.1 hr).
II = peak aged (0.3 and 1 hr).
III = overaged (10 and 100 hr).
All at 300°C

Influence of Other Material and Processing Parameters

Increasing the deformation temperature in low SFE alloys strongly reduces mechanical twinning and so shifts the texture transition. Above a critical temperature, dynamic recrystallization can occur which dramatically changes the evoling microstructure and texture (Hatherly et al., 1986). In Al alloys, increasing the deformation temperature leads to a brass type rolling texture, however, with a different appearance than the low SFE texture (Hirsch, 1990). So, the actual temperature of deformation can significantly change the microstructure which can also be identified by texture effects.

In technical alloys, precipitates can affect the texture evolution. In general, they reduce texture sharpness by disturbing the homogeneous material flow but they can also create distinctive orientation effects. In a detailed ODF analysis, systematic effects on the rolling textures have been detected. For example, excluding all other influences (e.g., of different initial textures), the final rolling texture in an Al 1.7% Cu alloy was found to clearly depend on the precipitation state which was varied systematically by different annealing treatments before the final cold rolling (Engler et al., 1989). Figure 6c shows these differences in the form of the variation in intensity along the skeleton line. Here the ~{112}<111> C-orientation is strongest in the underaged condition, the ~{123}<634> S-orientation is dominant in the overaged condition and in the peak aged samples the {011}<211> B-orientation is strongest. (The effects of particles are even stronger on the subsequent annealing textures.)

Some of these texture effects are relatively small (the pole figures all look very similar) but they are clearly visible in the ODF data. So using ODFs is essential, if one wants in turn derive the effect of precipitation from texture data. If the problem becomes more complex, e.g., if several alloy elements are present which are contributing differently, being partly in solid solution or producing bimodal particle distributions, also then textures might help to separate the effects and to estimate the relative importance of the different contributions.

It should be mentioned that for medium and low SFE materials, e.g., Cu and Cu-alloys, the texture development has good reproducibility. But, especially for high SFE material like Al, the initial texture has a very strong influence (Hirsch et al., 1986a). So, depending on the preceding processing and the initial structure, different intensities of the stable rolling texture components can be observed. But this strong influence of the initial state, which can be a problem for a direct comparison of deformation textures of differently prepared materials, on the other hand can be useful as it allows for the detection of variations from preceding processing steps (Hirsch, 1987b).

ANNEALING AND RECRYSTALLIZATION

As recrystallization is a mechanism of large angle boundary motion it is strongly correlated with changes in textures which, therefore, in turn can be used to analyze the corresponding underlying mechanisms (Hirsch, 1986b). Since it is closely related to the rolling texture, it is evident that the recrystallization texture also depends on the parameters of deformation. However, another structure element--the grain boundary--plays a decisive role. The creation and motion of large angle grain boundaries and the varying influences of alloy elements and other parameters on their mobilities also affect texture formation. Here, effects are often strongly magnified since in many cases totally different texture components can occur. Different nucleation mechanisms for recrystallization, provided by different (local) texture and structure evolution during deformation and further growth effects, provide strong orientation selection effects so that only very few sharp texture components determine most of the final recrystallization texture. Figure 7 shows some typical examples. In highly rolled and annealed FCC material with high SFE the cube orientation {001}<100> is the prominent texture component (Figure 7a,b). In technical pure Al in some cases, the R orientation {123}<634> occurs as a second recrystallization texture component (Figure 7a). It is similar to the major deformation texture component S but its intensity and exact position can vary due to preferred nucleation and growth processes, thus indicating specific recrystallization mechanisms (Hirsch, 1986b).

A decrease in SFE increases the frequency for recrystallization twinning which can very well be observed in the textures of medium SFE material (Cu) in the form of twins of the cube (Figure 7b). A further decrease in SFE (by adding other alloy elements to Cu) significantly reduces the amount of cube, mainly due to the increase in recrystallization twins. Furthermore, changes in deformation texture and grain boundary mobility cause more complex texture changes. In low SFE materials, a new and different texture component becomes dominant (a brass-type recrystallization texture, Figure 7c). In other cases when recrystallization is determined by nucleation events (for example in shear bands) and growth processes are less effective, a Goss texture occurs (as in CuP, Figure 7d) (Hirsch, 1986b). Also, in Al alloys, small amounts of alloy elements can cause drastic changes in the recrystallization behavior. Particles being present before deformation strongly influence recrystallization and so the final recrystallization texture. Texture studies showed how these effects can help to analyze the changes in recrystallization mechanism (Hirsch, 1988 and 1990). For many cases, texture analysis provides a useful method to identify the underlying mechanism of recrystallization, their statistical relevance and microstructure effects for various metals and alloys and processing parameters.

SUMMARY AND OUTLOOK

The crystallographic texture is an essential feature of the materials microstructure and consequently can be very sensitive measure for the evolution of microstructure and the

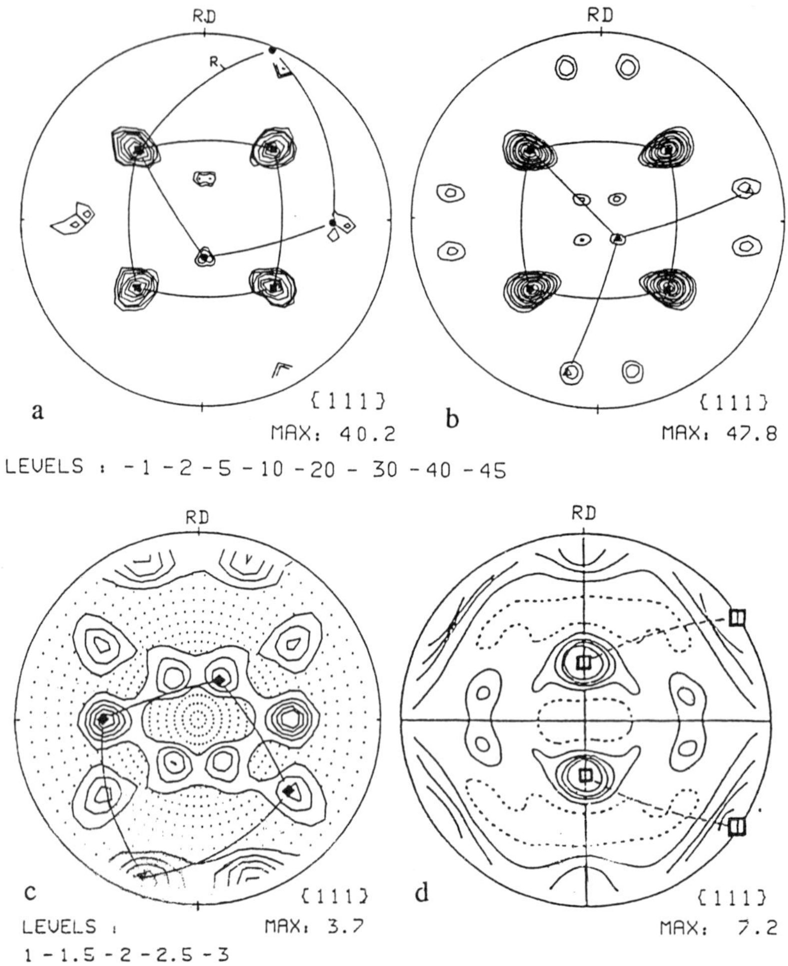

Figure 7. Some typical recrystallization textures of rolled FCC metals (95% red):
a) Pure aluminum (Cube {001}<100> and R {123}<634> texture).
b) Pure copper (Cube {001}<100> and its twins {122}<221>).
c) 70/30 brass (BsR {236}<385>).
d) Cu 1% P (Goss {011}<100>).

influence of many important parameters in metal processing. In all major processing steps of material production, i.e., casting (solidification), plastic deformation and annealing (recrystallization), the corresponding mechanisms are controlled--or at least affected--by the crystalline structure of metals. This correlation with the crystallographic orientation in many cases leads to the formation of typical textures which in turn can be used to analyze the underlying mechanisms.

For a number of cases in solidification, deformation and annealing of typical FCC materials, it has been demonstrated how systematic texture investigation can be applied to investigate material and process parameters and provide new information about microstructures and related mechanisms. The increasing activities in texture research will further broaden the experience on texture formation and the understanding of the underlying physical processes. Systematic investigations analyzing the complex dependences of textures on the different parameters currently carried out in fundamental as well as in applied research will further open this new field of material characterization. The introduction of on-line texture measuring facilities (Kopinek, 1988) is a significant step to introduce the use of texture data to control industrial production processes. Its major purpose now is the continuous check of material anisotropy, but it actually is--and with further knowledge about the interdependencies will also be more used as--a most suitable method for on-line microstructure control.

So besides their important role in affecting the anisotropy of the mechanical properties, texture data can also be applied to characterize the microstructure and investigate the different processes of material production and help to better understand, control and improve related fabricating processes. Texture investigation and the analysis of characteristic texture variations with process parameters are already well established as a useful method for materials characterization in fundamental research. It is so far developed that it now can be applied also as a new investigative tool for process and quality control in industrial practice.

ACKNOWLEDGMENT

Many fruitful discussions with texture experts at the Alcoa Laboratories and my former colleagues at the Iustitut für Allgemeine Metallkunde and Metallphysik der RWTH Aachen, especially with Professor K. Lücke are gratefully acknowledged. Special thanks to C. L. Schall for careful preparation of this manuscript.

REFERENCES

H. J. Bunge, 1984, Proc. ICOTOM 7, Noordwijkerhout, The Netherlands, p. 467.
H. J. Bunge (editor), 1986, "Experimental Techniques of Texture Analysis," DGM Inf. Ges. Oberursel, Federal Republic of Germany.
H. J. Bunge (editor), 1987, "Theoretical Techniques of Texture Analysis," DGM Inf. Ges. Oberursel, Federal Republic of Germany.
H. J. Bunge (editor) (1988), "Directional Properties of Materials," DGM Inf. Ges. Oberursel, Federal Republic of Germany.
O. Engler, J. R. Hirsch and K. Lücke, (1989) Acta Metall., Vol. 37, No. 10, p. 2743.
S. Eucken, J. R. Hirsch and E. Hornbogen, 1988, Texture and Microstructure (Wassermann Memorial Edition), Vols. 8 and 9, p. 415.
M. Hatherly, A. S. Malin, C. N. Carmichael, H. J. Humphreys, J. R. Hirsch (1986) Acta Metall 34, p. 2247.
J. R. Hirsch, W. Mao and K. Lücke, 1986a, Proc. Aluminium Technology '86, 3, The Institute of Metals, London, England, B 70.
J. R. Hirsch, 1986b, Seventh RISO Int. Symposium, Denmark, p. 361.
J. R. Hirsch, E. Nes and K. Lücke, 1987a, Acta Metall., 35, pp. 427-438.
J. R. Hirsch, 1987b, Proc. ICOTOM 8, Santa Fe, New Mexico, p. 1011.
J. R. Hirsch (1988), "Untersuchung der primären Rekristallisation und ihrer Mechanismen mit Hilfe der Texturanalyse," Habilitation-thesis RWTH-Aachen, Germany.
J. R. Hirsch and K. Lücke, 1988, Overview 76, Acta Metall., Vol. 36, p. 2863.

J. R. Hirsch, 1990, in Proc. "Recrystallization '90," Wollongong, Australia, p. 759.
J. S. Kallend and G. Gottstein (editors), 1987, "Textures of Materials," Proc. ICOTOM 8, Santa Fe, New Mexico.
H. J. Kopinek, 1988, in "Directional Properties of Materials," DGM Inf. Ges. Oberursel, Federal Republic of Germany, p. 251.
G. Tempus, G. Scharf and W. Calles, 1987, Journal de Physique, C3-187 (Proc. 4th Al-Li Conference, Paris, France).

ANGULAR DEPENDENT ULTRASONIC WAVE VELOCITIES IN THIN ZIRCALOY PLATE

S. S. Lee*, B. Y. Ahn*, H. J. Kim†, and Y. C. Kim†

* Korea Standard Research Institute
P.O.Box 3, Taedok Science Town, Taejon, Korea

† Korea Atomic Energy Research Institute
P.O.Box 7, Cheong Ryang, Seoul, Korea

INTRODUCTION

Zirconium alloys which are extensively used in nuclear industry have texture from manufacturing processes, and the presence of such a texture is one of the main causes of various dimensional instabilities due to irradiation creep and growth as well as yielding characteristics.[1] Therefore quantitative determination of texture is important not only for the estimation of texture–related macroscopic properties of the material but also for the control of the manufacturing process. Traditionally texture has been determined by X–ray or neutron diffraction.

From the recently developed several subsequent advances,[2-6] interests for the quantitative characterization of texture by ultrasonic method have been increased. However, the technique has not been applied experimentally to hexagonal crystallite aggregates. As the theoretical expressions for the angular dependence of the longitudinal and shear wave velocities in hexagonal crystallite aggregates have been derived,[7] there is a motive to apply the technique to the Zirconium based alloy.

In this paper, angular dependent velocities of SH_0 (fundamental horizontally polarized shear mode) and S_0 (fundamental symmetric Lamb mode) of the plate were obtained and compared by two experimental techniques. The one was obtained from ultrasonic wave velocity measurement. The other was the calculated angular dependent ultrasonic wave velocities using the ODC's from neutron diffraction. Considered were Voigt and Reuss approximation for the averaging.

THEORETICAL BACKGROUND

Let x_1°, x_2°, x_3° be a set of rectangular coordinate axes fixed in the hexagonal crystal and X_1, X_2, X_3 be fixed in the sample parallel to rolling, transverse, and normal direction respectively. Elastic compliance and stiffness of polycrystalline materials can be estimated by averaging those of the single crystal weighed by the orientation distribution function over the whole orientation space. The 4th rank tensor of compliance $S^\circ_{i_1i_2i_3i_4}$ and stiffness $C^\circ_{i_1i_2i_3i_4}$ for the single crystal averaged in this way may be expressed as

$$\begin{Bmatrix} S^R_{i_1i_2i_3i_4} \\ C^V_{i_1i_2i_3i_4} \end{Bmatrix} = \bar{a}(i_1i_2i_3i_4;j_1j_2j_3j_4) \begin{Bmatrix} S^\circ_{j_1j_2j_3j_4} \\ C^\circ_{j_1j_2j_3j_4} \end{Bmatrix} \tag{1}$$

where $S^R_{i_1i_2i_3i_4}$ and $C^V_{i_1i_2i_3i_4}$ are averaged quantities in Reuss and Voigt approximation respectively. Following the procedure which is explained in detail in reference 2 with the orientation distribution coefficients from the neutron diffraction measurements, we are able to get 9 independent stiffness constants C_{11}, C_{12}, C_{13}, C_{22}, C_{23}, C_{33}, C_{44}, C_{55}, and C_{66} in Voigt averaging scheme and Reuss averaging scheme.

The angular dependent ultrasonic wave velocities propagating in the X_1X_2 plane at an angle θ to the rolling direction are given by

$$\rho v_{SH}^2 = \frac{1}{2}\{(n_1^2C_{11}+n_2^2C_{22}+C_{66}) \\ -[(n_1^2C_{11}-n_2^2C_{22}-(n_1^2-n_2^2)C_{66})^2+4n_1^2n_2^2(C_{12}+C_{66})^2]^{\frac{1}{2}}\} \tag{2}$$

$$\rho v_L^2 = \frac{1}{2}\{(n_1^2C_{11}+n_2^2C_{22}+C_{66}) \\ +[(n_1^2C_{11}-n_2^2C_{22}-(n_1^2-n_2^2)C_{66})^2+4n_1^2n_2^2(C_{12}+C_{66})^2]^{\frac{1}{2}}\} \tag{3}$$

where $n_1=\cos\theta$ and $n_2=\sin\theta$. Here v_{SH} and v_L are the velocities of shear horizontal and longitudinal waves. On the other hand Sayers[7] derived the analytical expressions for the angular dependence as follows.

$$\delta v_{SH}(\theta) \equiv v_{SH}(\theta)-v_S^0 = \frac{2\sqrt{10}\pi^2(7a_2+2a_3)}{315\rho v_S^0} W_{200} \\ + \frac{2\sqrt{2}\pi^2 a_1}{105\rho v_S^0}(W_{400}-\sqrt{70}W_{440}\cos 4\theta) \tag{4}$$

$$\delta v_L(\theta) \equiv v_L(\theta) - v_L^0 = \frac{4\sqrt{10}\pi^2 a_3}{105\rho v_L^0}\left(W_{200} - \sqrt{6}\, W_{220}\cos 2\theta\right)$$

$$+ \frac{2\sqrt{2}\pi^2 a_1}{35\rho v_L^0}\left(W_{400} - \frac{2\sqrt{10}}{3} W_{420}\cos 2\theta + \frac{\sqrt{70}}{3} W_{440}\cos 4\theta\right) \qquad (5)$$

where W_{lmn} are the orientation distribution coefficients and a_1, a_2, and a_3 are the anisotropy factors which were given in reference 7.

The absolute velocities and anisotropy of the SH_0 plate mode are identical to those for plane SH wave and the absolute velocities of the S_0 mode propagating along symmetry axes are reduced from the longitudinal plane wave velocities.[8] Therefore equation (3) can be modified by considering plane stress condition.

$$\rho v_{S_0}^2 = \frac{1}{2}\{(n_1^2\hat{C}_{11} + n_2^2\hat{C}_{22} + \hat{C}_{66})$$

$$+ [(n_1^2\hat{C}_{11} - n_2^2\hat{C}_{22} - (n_1^2 - n_2^2)\hat{C}_{66})^2 + 4n_1^2 n_2^2(\hat{C}_{12} + \hat{C}_{66})^2]^{\frac{1}{2}}\} \qquad (6)$$

where $\hat{C}_{11} = C_{11} - C_{13}^2/C_{33}$, $\hat{C}_{12} = C_{12} - C_{13}C_{23}/C_{33}$, $\hat{C}_{22} = C_{22} - C_{23}^2/C_{33}$, $\hat{C}_{66} = C_{66}$.

Through the examination of equation (4), W_{440} for hexagonal crystallite aggregates could be calculated as follows.

$$W_{440} = \frac{105\rho v_S^0}{8\sqrt{35}\pi^2 a_1}\left[v_{SH_0}(45^\circ) - v_{SH_0}(0^\circ)\right] \qquad (7)$$

EXPERIMENTAL PROCEDURE

Samples

Measurements were made on a sample of Zircaloy–4 thin plate which was supplied for fabrication of spacer pads of nuclear fuel rods. The thickness of the plate was 0.8 mm. For the ultrasonic measurements, the sample had lateral dimension of about 46 cm and were examined as–received condition. The sample for the neutron diffraction measurements was a cube made of stack of thin pieces with 1.5 by 1.5 cm area cut from the edge of the plate.

Ultrasonic Measurements

The ultrasonic velocities were measured with electromagnetic acoustic transducers (EMAT's) at a nominal frequency of 500 kHz for SH_0 waves and 800 kHz for S_0 waves. The SH_0 waves were excited and detected by periodic permanent magnets EMAT's,[9] while the S_0 waves were excited and detected by meander coil EMAT's.[10] The propagation path was approximately 16 cm for the SH_0 measurements and 23 cm for the S_0 measurements. The transducers were driven by a high power pulse. After reception and amplification of the waveform, the arrival time of a zero–crossing of a particular cycle near the center of the burst was monitored with a 125 MHz digital waveform analyzer. Typically, 200 averages were employed.

Two separate measurement procedures were employed. The angular dependence of the wave velocities was first measured by separating the transmitter and receiver by a rigid spacer. The change in arrival time was then recorded as a function of propagation direction. The relative velocity measurement uncertainty was less than ±0.01 pct. In the second measurement procedure, the absolute velocities were determined by removing the rigid spacers and recording transducer separation vs. wave arrival time, with separations typically varied over 4 cm. The velocities of both modes were measured at angles of 0°, 45°, 90°, and 135°. The absolute velocity measurement uncertainty was about ±0.1 pct.

Neutron Diffraction Measurements

The measurement was made by the neutron diffractometer (λ=1.204Å) installed on the horizontal beam port of TRIGA–III reactor at KAERI. Using Two Circle Goniometer, pole figures for 5 crystallographic planes were taken by Schultz reflection method. Each first quadrant complete pole figure was measured at 19 x 19 angular positions in steps of 5 degree from 0° to 90° around the two circles. Background counting for correction was made at positions 3° – 6° away from the Bragg reflection. The parallelism of one side of the sample with the rolling direction was confirmed by exact symmetry shown on the $(10\bar{1}0)$ pole figure which was measured both on the first and the second quadrant.

RESULTS AND DISCUSSIONS

The measured pole figures for the crystallographic planes $(10\bar{1}0)$, (0002), $(10\bar{1}1)$, $(10\bar{1}2)$, and $(11\bar{2}0)$ were analyzed with the computer program TXSHTH developed at KAERI which employed Bunge's series expansion method for the calculation of

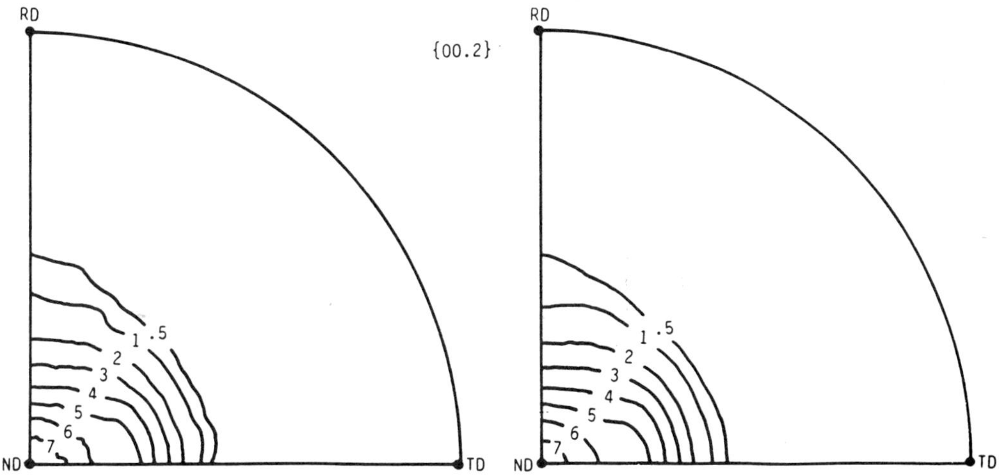

(a) measured (b) calculated

Fig.1 (0002) Pole Figure.

orientation distribution function. The above pole figures provided orientation distribution coefficients up to l= 22 using the least square method. Fig. 1 shows the measured (0002) pole figure and the calculated or "reconstructed" one from the ODC's. In Table 1 are listed relevant ODC's for the determination of the orientation-averaged elastic constants and consequently anisotropic ultrasonic velocities. Here, ODC's in Roe's notation W_{lmn} are converted from those of Bunge's $C_l^{\mu\nu}$.

Figure 2 presents a plot of the angular dependence of SH_0 wave velocities in a form defined as the relative variation, $[v(\theta)-v(0^\circ)]/v(0^\circ)$. The data points were obtained from the relative ultrasonic measurements. The solid line was obtained from Sayers formula, equation(4), for Voigt approximation. The dashed line was obtained from equation(2) for Reuss approximation. Figure 3 presents a plot of the

Table 1. ODF coefficients from neutron diffraction

W_{lmn}	W_{200}	W_{220}	W_{400}	W_{420}	W_{440}
values	0.02799	0.00037	0.01451	0.00196	0.00022

angular dependence of S_0 wave velocities. Here data points were obtained from the ultrasonic measurements, but the solid and dashed lines for Voigt and Reuss approximation were obtained from equation(6) with C_{11}, C_{12}, and C_{22} replaced by $\hat{C}_{11}$, $\hat{C}_{12}$, and $\hat{C}_{22}$ for the plate wave correction.

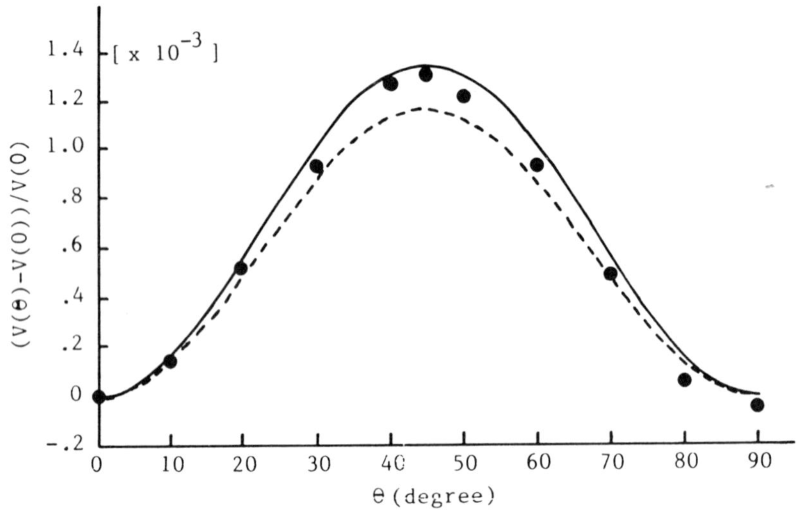

Fig.2 Relative variation $[v(\theta)-v(0^\circ)]/v(0^\circ)$ of SH_0 wave velocities ———, Voigt; - - - - , Reuss; •, experiment.

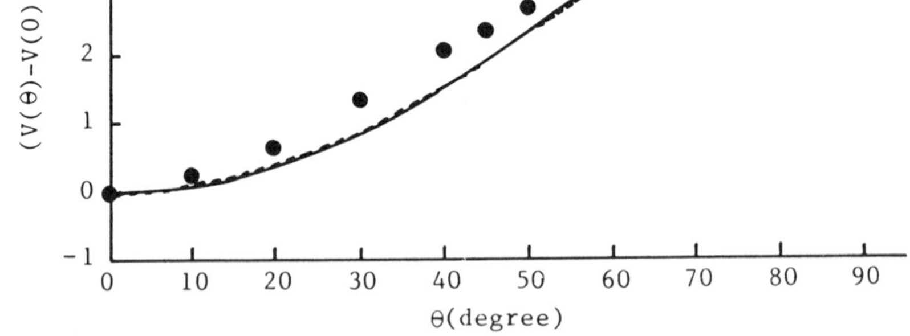

Fig.3 Relative variation $[v(\theta)-v(0^\circ)]/v(0^\circ)$ of S_0 wave velocities ———, Voigt; - - - - , Reuss; •, experiment.

Table 2. Measurements of absolute ultrasonic velocities

(Unit: $\times 10^5$cm/sec)

Angles	0°		90°	45°		135°
SH_0	2.3135		2.3130	2.3160		2.3153
Avg		2.3133			2.3157	
S_0	4.0908		4.1019	4.0961		4.0961
Avg		—			4.0961	

The agreement between the angular variations of wave velocities from ultrasonic measurements and the calculated angular variations of wave velocities from neutron diffraction is excellent.

From the absolute velocity measurements, the velocities at angles of 0°, 45°, 90°, and 135° are presented in Table 2. In the rows denoted "Avg", the averages of the velocities expected to be equal by symmetry are presented. As mentioned before, the error in the absolute measurements was about 10 times larger than that in the relative measurements.

The prediction of W_{440}, based on equation(7), from the ultrasonic data is presented in Table 3. Also included is the prediction of W_{440} from the neutron diffraction measurements there. In the ultrasonic prediction of W_{440}, the relative velocities were utilized to reduce the error. $v_{SH_0}(0°)$ was first factored out of the expression in equation(7). Absolute value for $v_{SH_0}(0°)$ from Table 2 were then combined with relative value for $v_{SH_0}(45°)$ from Figure 2 to make the prediction.

Table 3. Ultrasonic and neutron estimates of W_{440}

Method	Ultrasonic	Neutron
W_{440}	0.00020	0.00022

CONCLUSIONS

The agreement between the prediction of texture from the ultrasonic velocity measurements and that for the neutron diffraction is excellent to an accuracy of

$|\Delta W_{440}| \sim 2x10^{-5}$ in this particular experimental condition. This experiment may serve as a test not only for the accuracy of texture determination by ultrasonic method but also evaluation of anisotropic physical properties from ODF.

REFERENCES

1) C.H.Woo, J. Nuc. Mat., **131**, 105 (1985)
2) H.J.Bunge, Texture Analysis in Materials Science (Butterworth, London, 1982)
3) R.J.Roe, J. Appl. Phys., **36**, 2024 (1965)
4) R.J.Roe, J. Appl. Phys., **37**, 2069 (1966)
5) C.M.Sayers, J. Phys. D.: Appl. Phys., **15**, 2157 (1982)
6) R.B.Thompson, J.F.Smith, S.S.Lee, and G.C.Johnson, Metall. Trans., **20A**, 2431 (1989)
7) C.M.Sayers, Ultrasonics, **24**, 289 (1986)
8) R.B.Thompson, S.S.Lee, and J.F.Smith, Ultrasonics, **25**, 133 (1987)
9) C.F.Vasile and R.B.Thompson, J. Appl. Phys., **65**, 324 (1979)
10) R.B.Thompson, IEEE Trans. on Sonics and Ultrasonics, **SU–20**, 340 (1973)

AN AUTOMATIC INSTRUMENT FOR THE ULTRASONIC MEASUREMENT OF TEXTURE

Emmanuel P. Papadakis[a], R. Bruce Thompson[a], Delwyn D. Bluhm[b], George A. Alers[c], Kaveh Forouraghi[a], *Harold D. Skank[b], and Samuel J. Wormley[a]

[a] Center for Nondestructive Evaluation, Iowa State University, Ames, IA 50011
[b] Ames Laboratory, USDOE, Iowa State University Ames, IA 50011
[c] Magnasonics, Inc., Albuquerque, NM 87108

INTRODUCTION

Texture in sheet metal must be controlled in the rolling process to assure the fabrication properties desired in later manufacturing. Drawability is one of the required engineering properties in a family of applications including the manufacture of beverage cans, propane tanks, and automotive parts.

Currently-in-place technology includes several measures of drawability of sheet metal including average plastic strain ratio $\bar{r}$, planar anisotropy Δr, limiting dome height, limiting drawing ratio, and earing height of cups. All of these are destructive, time-consuming, and prone to inaccuracies because they require cutting of specimens from the sheet, plastic deformation by precise mechanisms, and accurate metrology of less than ideal shapes. Empirical correlations exist for the case of steel between $\bar{r}$ and Δr, on one hand, and the average value of Young's modulus $\bar{E}$ and its anisotropy ΔE in the plane of the plate, on the other hand. An instrument exists to provide $\bar{E}$ and ΔE (in ferromagnetic metals) by measuring the resonance frequencies of coupons cut from the plate so that $\bar{r}$ and Δr can be predicted from the empirical relationship. While somewhat faster and easier, this method is still destructive, time-consuming, and prone to errors.

Improvements can be effected by modern electronic methods. It is now possible to measure ultrasonic velocity rapidly and to make calculations based upon it at computer speeds. Since ultrasonic velocity depends functionally upon Young's modulus, the velocity of a convenient elastic wave mode propagating at various directions in a plate can be used to calculate $\bar{E}$ and ΔE. Then, for ferrous materials, $\bar{r}$ and Δr should be obtainable from the aforementioned empirical relationships. (Correlations with $\bar{E}$ and ΔE have not yet been established for other materials and for other drawability parameters.) The S_o Lamb wave mode measured by EMAT technology is an appropriate candidate for the ultrasonic technique. With this mode, three of the texture Orientation Distribution Coefficients (ODCs), namely W_{400}, W_{420}, and W_{440}, can also be calculated.

In the present work, EMAT technology was chosen to measure the velocity of S_o Lamb waves at ultrasonic frequencies in steel and aluminum sheet. An industrial grade instrument was built to measure wave travel times automatically and to make all the necessary calculations of velocities, moduli, etc., to arrive at the parameters of interest for

Nondestructive Characterization of Materials IV
Edited by C.O. Ruud *et al.*, Plenum Press, New York, 1991

drawability. Specifically, $\bar{E}$, $\bar{r}$, Δr, W_{400}, W_{420}, and W_{440} are outputs along with the travel times, velocities, and moduli at 0°, 45°, and 90° to the rolling direction. (The value of $\bar{r}$ and Δr are available for steel only.) The work included demonstrations at steel and aluminum suppliers' and users' facilities. The instrument was shipped to several locations where measurements were made on material supplied by the industrial attendees to show agreement with their fundamental tensile measurements.

In this chapter, the technical details of the measurement method and the instrument configuration will be given. Further, the testing to assure the survivability of the instrument in industrial environments will be reported. The field trial will be outlined, and the technical results of measurements on $\bar{r}$ will be given for 25 steel sheets from two major industrial partners cooperating in this test.

BACKGROUND

Foundations of Ultrasonic Predictions of $\bar{r}$

General: The possibility of relating ultrasonic measurements to r-values and other mechanical properties dates back at least to the late 1960's. At that time, four independent lines of scientific study were beginning which are now being merged to form an important new technology for the sheet steel industry. These four developments are briefly reviewed in this section.

Relationship of Elastic and Plastic Anisotropy: The practical utilization of relationships between elastic and plastic anisotropy dates back to the demonstrations by Liu and Alers [1], then at the Ford Motor Company Scientific Research Laboratories, that measurements of Young's modulus on vibrating bar coupons cut from textured sheets at various angles with respect to the rolling direction could be used to obtain important qualitative information regarding the texture state of the material. Papadakis et al. demonstrated that similar information could be obtained from ultrasonic velocity measurements. [2,3] Motivated by the work of Liu and Alers and the notion that the texture sensed by Young's modulus anisotropy would also control plastic anisotropy, Stickels (Ford) and Mould (U. S. Steel) demonstrated a correlation between the anisotropies of Young's modulus and the plastic strain ratio [4], whose definition is illustrated in Fig. 1. For r measured in the rolling direction $\theta = 0°$ and at 45° and 90° with respect to it, the average value $\bar{r}$ is given by

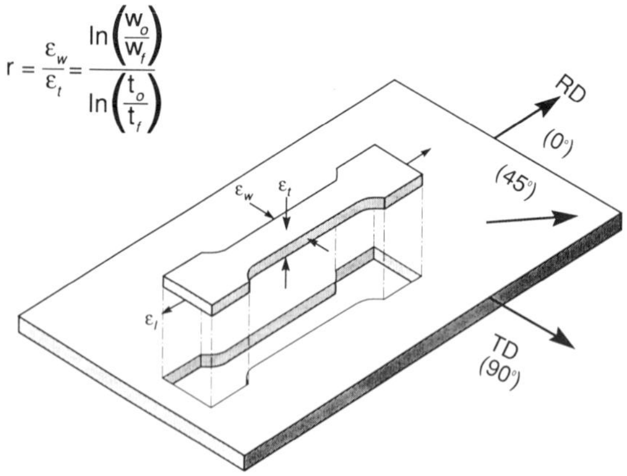

Fig. 1. Definition of plastic strain ratio, r.

$$\bar{r} = [r(0°)+r(90°)+2r(45°)]/4 \tag{1}$$

while the planar anisotropy Δr is given by

$$\Delta r = [r(0°)+r(90°)-2r(45°)]/2. \tag{2}$$

Likewise, $\bar{E}$ is given by

$$\bar{E} = [E(0°)+E(90°)+2E(45°)]/4 \tag{3}$$

and ΔE is given by

$$\Delta E = [E(0°)+E(90°)-2E(45°)]/2. \tag{4}$$

The correlation between $\bar{r}$ and $\bar{E}$ and Δr and ΔE were refined on a larger number of samples by Mould and Johnson [5]. This correlation became the basis for the Module-$\bar{r}$, marketed by Tinius Olson. In that instrument, Young's modulus is inferred from the extensional resonance frequency of coupons cut from the sheet. Operational characteristics of the Module -$\bar{r}$ include the facts that a) it makes a destructive measurement, b) it does not work on sheets less than a minimum thickness, c) it only works on ferrous alloys, and d) it does not take into account coating thicknesses.

Quantitative Texture Analysis: The scientific understanding of these correlations was strengthened by research in quantitative texture analysis that also had its origin in the mid-1960's. Roe, in the United States [6], and Bunge, in East Germany [7], developed a mathematical formalism to describe the preferred grain orientation of a material in terms of a probability distribution or orientation distribution function (ODF). Let $w(\xi,\psi,\phi)$ be the probability that a crystallite has an orientation with respect to the plate coordinates defined by the Euler angles ξ,ψ and ϕ. It was found to be convenient to represent this ODF as a sum of generalized spherical harmonics, much as one can represent a function by a Fourier series.

$$w(\xi,\psi,\phi) = \sum_{l=0}^{\omega} \sum_{m=-l}^{l} \sum_{n=-l}^{l} W_{lmn} Z_{lmn}(\xi) e^{im\psi} e^{-in\phi} \tag{5}$$

The weights of the spherical harmonics W_{lmn}, known as orientation distribution coefficients (ODCs), provide a convenient set of parameters to describe the texture. Using this formalism, Davies, Goodwill and Kallend established the physical basis [8] for the correlation of Mould and Johnson. In particular, they noted that because of particular properties of a representation of the plastic anisotropy in terms of the ODC's in steel, "when making predictions of plastic properties the analysis can be truncated at the fourth order and still give rise to good approximate values." It was also deduced that "The coefficient, W_{400}, sets the overall value of the average strain ratio, $\bar{r}$; W_{420} controls the tendency to form two ears during deep drawing while W_{440} controls the tendency to form four ears."

Couplant-Free Measurements with EMAT's: The ability to make the measurement nondestructively, while the sheet is moving, depends on sensors which can determine elastic anisotropy without any contact with the sheet. This capability is provided by measurement of the anisotropy of the elastic wave speed using electromagnetic-acoustic transducers (EMATs). EMATs consist of a coil of wire that inductively couples to dynamic currents and fields in the metal sheet and either a permanent or pulsed electromagnet that establishes a bias magnetic field [9] [10]. Through a combination of Lorentz force and magnetostrictive interactions, these fields can excite and detect ultrasonic waves in metal plates. When the coil is wound in a particular fashion, one can selectively couple to guided modes propagating in the plane of the plate. The fact that no couplant is required allows very precise measurements of velocity to be made and opens the way for on-line monitoring.

Guided Wave Propagation in Anisotropic Plates: When ultrasonic waves propagate in plates, the motion of the material is different from the purely compressional (longitudinal) or shear (transverse) deformations that occur in unbounded plates (plane wave motion). Instead, the forms of deformation are coupled so that their superposition can satisfy the stress-free conditions at the surfaces of the plate. The resulting solutions, known as guided or plate modes, travel with speeds which depend on the product of frequency and plate thickness as well as the physical properties of the material [11]. It was shown as early at 1973 that the presence of texture induced anisotropy in the plate wave speeds could be sensed by EMATs [12].

Recent Advances in r-Value Predictions

Measurement Theory: Based on these foundations, rapid progress has been made towards an r-value sensor in the last five years. In 1985, Thompson, Smith, and Lee proposed the concept of the EMAT approach [13]. In 1987, the correct equations relating the measured velocities to the ODC's W_{400}, W_{420}, and W_{440} were reported by the same authors [14]. That theory assumed that the measurement frequency was sufficiently low that the wavelength was large with respect to the plate thickness. Corrections were subsequently developed to allow data obtained on thicker plates to be interpreted [15]. Hirao et. al. have recently reported a correction for the effects of coating thickness on the velocity measurement [16].

Correlation with $\bar{r}$: Based on the demonstration that ultrasonic techniques could measure elastic anisotropy, the correlation between elastic and plastic anisotropy demonstrated by Stickles and Mould, and the interpretation of the correlation by Davies et al., several groups have demonstrated correlations between ultrasonic measurements and plastic strain ratios [16-20]. Although based on the same underlying ideas, these have taken somewhat different forms, and the preferred approach has not fully emerged. Spies and Schneider have used shear waves polarized in the plane of the plate and have developed purely empirical, alloy specific correlations [17] with $\bar{r}$. Other workers have employed longitudinal Lamb waves (S_o mode). Cassier et al [18] and Sakata et al [19] have determined the ODC's from the ultrasonic velocity data. They then use these as input to theories predicting anisotropy plastic behavior (i.e. $\bar{r}$ and Δr). Thompson and Clark [20] and Hirao et al [16], have examined correlation approaches. Included was the demonstration of two empirical correlations on a variety of samples obtained from different vendors. In the first, $\bar{r}$ was correlated with the average S_o mode velocity in the plane of the sheet, $[V(0°)+V(90°)+2V(45°)]/4$, where V refers to the measured velocity and the angle with respect to the rolling direction is indicated in parentheses. In the second, the velocity data was corrected in an approximate sense to estimate the Young's modulus that would be measured at the corresponding angles. This approximate correction assumed the relationship between Young's modulus and the S_o mode velocity in the textured plate to be the same as in an untextured plate, i.e. the effects of texture on this relationship were viewed as a second order effect. The Young's moduli so obtained were then used as input to the Mould-Johnson [5] correlation. In each case, corrections were made for sheet [15] and coating [16] thicknesses as discussed above. Both of these correlations led to good predictions of $\bar{r}$. [20]

Predictions of Earing

Prediction of other formability parameters, and parameters in materials other than steel, has not been investigated nearly as fully. However, some preliminary results in aluminum suggest that the ODCs W_{420} and W_{440} may correlate with the degree of two-fold and four-fold earing [21,22].

Instrument Development

General: The demonstration of these successful correlations has led to several projects aimed at the development of automatic instrumentation for $\bar{r}$ prediction. The essential features of these projects are 1) the mode type selected, 2) the type of EMAT utilized, 3) the configuration of EMAT's utilized to make a velocity measurement, 4) the manner in which the propagation direction of the waves is changed and 5) the signal processing utilized to determine velocities and plastic strain ratios. Breadboards have been developed by the Center for Nondestructive Evaluation (CNDE) at Iowa State University (ISU) [23], Institute for Nondestructive Testing (IZFP) in West Germany [24], Sumitomo in Japan [25], and the National Institute of Standards (NIST) [26]. An industrial quality prototype developed with funding from the Center for Advanced Technology Development (CATD) at ISU will be reported in the next section.

Detailed review of the design criteria used in the construction of these instruments is not possible given the length limitations on this chapter. However, the motivation for the design of the ISU-CATD industrial quality prototype follows. Some of these ideas are in the literature [27].

The basic measurement strategy is as follows. The long wavelength limit of the S_0 Lamb mode phase velocity is first determined for waves propagating in the plane of the plate at 0°, 45° and 90° with respect to the rolling direction. The S_0 mode was chosen over the SH_0 mode because a) the ODC predictions appear to be less sensitive to metallurgically related variations in the texture free moduli than those based on the SH_0 mode [15] and b) information on W_{420} is obtained (as distinct from SH_0 measurements).

As shown in Fig. 2, the absolute velocity is inferred from the relative arrival times of signals propagating from a pair of transmitters to single receiver. By driving the transmitters in series, it is also possible to reduce sensitivity to lift-off. The velocity is computed using a Fourier transform, phase-slope technique [27]. This involves

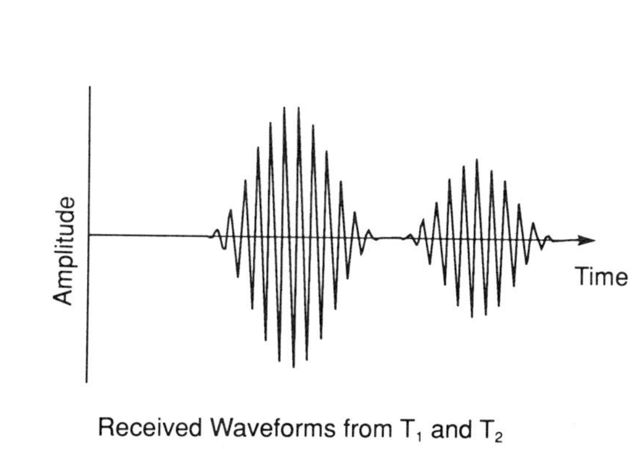

Fig. 2. Dual transmitter EMAT configuration for velocity measurement.

taking the ratio of the Fourier transforms of the two received signals, plotting the phase of that ratio versus frequency, and determining the slope. The phase velocity is given by

$$V_p = \Delta d/[(\tau_1 - \tau_2) + (\text{slope}/2\pi)] \quad (6)$$

where it has been assumed that the data has been corrected for dispersion effects. Here Δd is the separation of the two transmitting EMATs and τ_1 and τ_2 are additional delays imposed in the signal processing to roughly align the waveforms. This preshifting minimizes difficulties associated with phase folding. Analysis reveals that, when the regression is forced through the origin, much greater precision is obtained than when a simple linear regression is used, and that the results agree with the theoretical limits demonstrated by Cramer-Rao. [27] Time precisions of less than a nsec can be realized at digital sampling rates of 100 nsec.

Given the measured velocities, and after correction for dispersion and coating thickness, the ODC's are computed [14]. Good agreement with independent measurements by X-ray or neutron diffraction have been obtained [15]. The ODC's, along with independently determined values of the single crystal elastic moduli (or texture free isotropic moduli of the polycrystal), allow computation of anisotropic values of Young's modulus. It should be noted that the Young's moduli values computed in this way are sometimes significantly different from, and more accurate than, those obtained using the approximate estimation, as described in Ref. 20 [28]. The Young's moduli, in turn, serve as inputs to the correlation, previously developed by Mould and Johnson, [5] for the prediction of $\bar{r}$ and Δr in steel.

INDUSTRIAL INSTRUMENT

The Program. On the basis of the research instrumentation described above, an industrial quality automatic instrument was designed and built. As a preliminary step, a market study was carried out to assess the need for such an instrument. On the basis of the evidence gained in this study, the instrument was designed with user-friendly and ergonomically efficient features to implement the technology being embodied.

The Steps

The development program was executed in the steps discussed below.

a. Market. The initial market assessment was done by the Iowa Small Business Development Center (SBDC) through telephone contacts after consultation with CNDE personnel concerning the relevant markets. The results were positive and emphasized the desire in the marketplace for an on-line instrument and, to a lesser degree, a static instrument.

b. Configuration. The configuration of the instrument was planned to permit either static or dynamic measurements providing the sensors could be positioned correctly. Dynamic, on-line sensors were considered to be beyond the scope of the project, so a system for off-line quality assurance and incoming inspection was decided upon. A block diagram of the system is shown in Fig. 3. It was decided to make the system automatic such that an entire measurement, computation, and output sequence would be initiated by a single "Start" command. For ease of set-up and clarity of output display, a touch-screen video panel was incorporated for input/output. The capability of hard-copy output and data transfer to a mainframe was also decided upon. Other features such as caster mounting, EMI immunity, and transcontinental transportability were incorporated. Specific requirements on the probes were formulated. Structurally, they were to be integral units of nine (9) EMATs with two series-connected transmitters plus one receiver in each of three directions (0°, 45°, and 90° with respect to the rolling direction). A pulsed electromagnet system was specified for ferrous materials to eliminate magnetic clamping and the pick-up of chips in a factory. Non-ferromagnetic materials are interrogated by EMATs with permanent magnets. The instrument senses the probe type and makes the correct internal connections for steel versus aluminum, for instance.

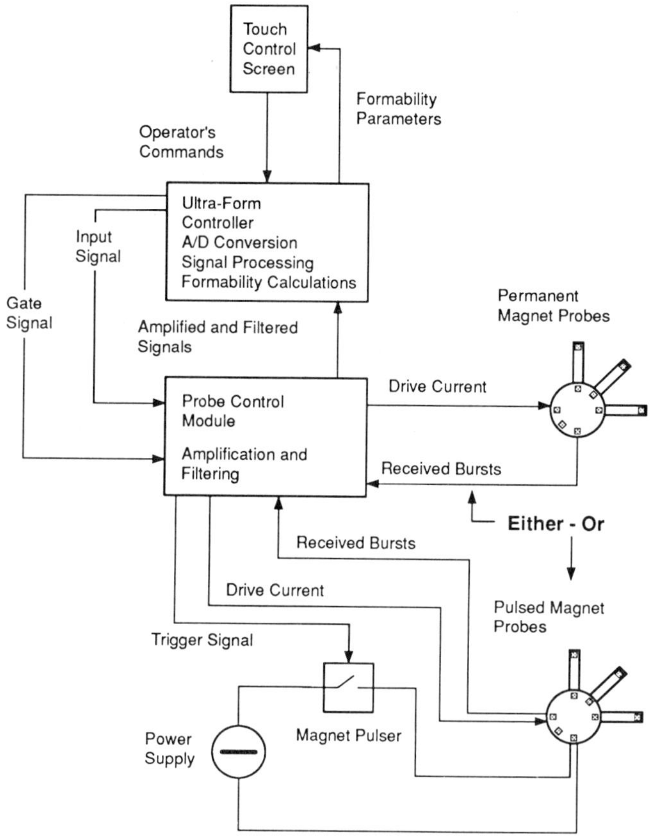

Fig. 3. Block diagram of instrument.

c. Design. With the configuration of the instrument specified, circuit and mechanical design was initiated. The Ames Laboratory electronic-computer-aided-design (ECAD) system for circuit boards was employed. Specifications for the analog portion to drive and receive from the probes and specifications for the probes as a compatible set were drawn up.

d. Fabrication. This is the period to build and fabricate the instrument included "populating" the circuit boards, building the circuit board containment and interconnections ("back plane" etc.) for the Ames Lab controller/computer subsystem of the instrument, and building the framework, power net, cooling means, electromagnetic interference (EMI) protection, and so on in the instrument case. One populated multi-layer circuit board is shown in Fig. 4. As hardware was finished, it was checked out. This procedure required some computer programming contemporaneously.

e. Check-Out. As elements were completed, the internal computers (actually 3 for different functions) were programmed to run the subsystems. All hardware and software had to be completed before the computer-controlled components could be tested realistically.

Check-out was performed when the system was complete. A number of difficulties were encountered and rectified. The check-out showed that certain assumptions had to be rethought and that further theory had to be developed and incorporated into algorithms before the instrument would function. The completed instrument is shown in Fig. 5. The final instrument obtains $\bar{r}$ in steel in 8 seconds and the Ws for aluminum in 5 seconds. The measurement of time delay is accurate to better than 1 ns.

Fig. 4. A populated circuit board.

f. Durability. The test for durability and accuracy revealed various difficulties. A temperature and vibration test patterned after the AT&T test for central office equipment was applied to the instrument with a structural fatigue cycling system in the civil engineering laboratory. This system applied 1g of acceleration at frequencies up to 100 Hz to the 755 lb instrument (gross shipping weight). Many design flaws from suppliers and one flaw of our own were detected; most of these had the characteristic of being heavy components not adequately restrained for cantilevered flexure under g-loading. Had these inadequacies not been detected and rectified, the field trial (next paragraph) which involved transportation hazards would have resulted in failure.

g. Field Trial. The field trial was arranged by the SBDC as a dynamic marketing exercise. A prerequisite was an agreement with each host corporation to share data. ISU would measure $\bar{r}$ on any specimen of steel or the Ws on any specimen of aluminum providing the host company determined tensile $\bar{r}$ (and if possible Module-$\bar{r}$® values of $\bar{r}$) on the steel or earing percentages on the aluminum. The ISU measurements were to be performed with the instrument at the metal manufacturers' or users' facilities with audiences of their choosing. This trial tour was planned for three users and three manufacturers of each aluminum and steel sheet, but was pared back to four locations. One was an aluminum manufacturer which was also a large-scale user of material for deep drawing, one was a steel user (an automobile company) who invited a competitor and several suppliers, and two were steel companies. One of the latter invited a third steel company to participate. The results of the tests at the automobile company and at one steel company are given in Tables 1 and 2. Excellent agreement (most points within ±0.05 in $\bar{r}$) is seen between the values of $\bar{r}$ from tensile tests and from the texture instrument. The results are presented graphically in Fig. 6. Further work remains to be done on aluminum and other alloys.

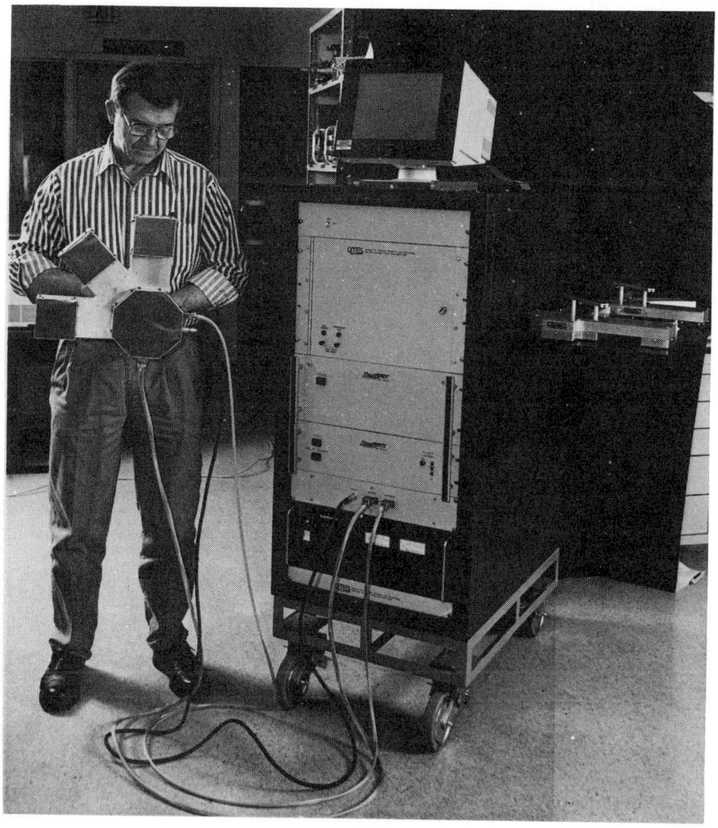

Fig. 5. The completed instrument with one probe in use.

CONCLUSIONS

1. Using special-purpose ad hoc funding [29,30], it has been shown that an industrial quality automatic instrument could be constructed to measure formability parameters of sheet metal.

2. The resulting instrument has accuracy equal to or better than the best laboratory breadboard test set.

3. The speed of the resulting instrument (5-8 sec) is compatible with off-line QC and with blanking lines in press shops and with reroll lines in steel plants.

4. Further work on probe supporting mechanisms should permit the instrument to be used on moving sheet in a rolling mill.

ACKNOWLEDGEMENTS

Funding for development of this instrument was provided by the United States Department of Commerce through the Center for Advanced Technology Development at Iowa State University under Grant ITA 87-02. Engineering expertise was provided by the Facilities Services Department

Table 1. Results of Ford Demonstration

Sample	Base Steel Gauge	Coating Thickness	TEST RESULTS Tensile $\bar{r}$	 Test Instrument $\bar{r}$
UB FNDR E7UB 160 15 RU	0.0314"	None	TBD	1.54
Lot 0886 VD-HD GAL	0.030"	0.00075"	1.89	1.84
Sumitomo MS 8056 coating	0.030"	0.0010"	1.91	1.91
Nippon BH Excellite coating	0.030"	0.0010"	1.17	1.37
Colled Rolled AKDQ	0.029"	None	1.77	1.68
VEN 023 ETTB GAL	0.039"	0.0010"	TBD	1.15
WSS1, WSS2 GAL	0.036"	0.0010"	TBD	1.66
E7UB VEN 001	0.037"	0.0010"	TBD	1.67

Conditions:
VD-HD = Vacuum Degassed - Hot Dipped 40g/40g both sides
GAL = Galvanized
TBD = "To be determined" by Ford Motor Company (promised)

Table 2. Results of Wheeling-Pittsburgh Demonstration

Sample	Base Steel Gauge	Condition*	Coating Thickness	TEST RESULTS Tensile $\bar{r}$	 Test Instrument $\bar{r}$
1	0.1147"	HR		.91	.92
2	0.1180"	HR		.91	.89
3	0.1184"	HR		.91	.95
4	0.0480"	CR		1.81	1.84
5	0.0862"	CR		1.49	1.52
6	0.0412"	CR		1.65	1.65
7	0.0725"	CR		1.55	1.55
8	0.0296"	CR, GAL	0.0015"	1.50	1.57
9	0.0598"	CR, GAL	0.0015"	1.17	1.06
10	0.0624"	CR, GAL	0.0015"	1.53	1.61
11	0.0148"	FH, GAL	0.0010"	**	.896
12	0.0110"	FH		**	.890
13	0.0236"	CR		1.62	1.57
14	0.0185"	CR		1.11	1.18
15	0.0104"	E.Sn	0.000012"	1.59	1.64
16	0.0085"	E.Sn	0.000015"	**	0.912
17	0.0064"	E.Sn	0.000012"	**	1.24

*Conditions:
CR = Cold rolled and annealed
HR = Hot rolled
FH = Cold rolled, not annealed, full hard
GAL = Galvanized
E.Sn = Electroplated Tin

**Inadequate elongation for tensile $\bar{r}$ determination

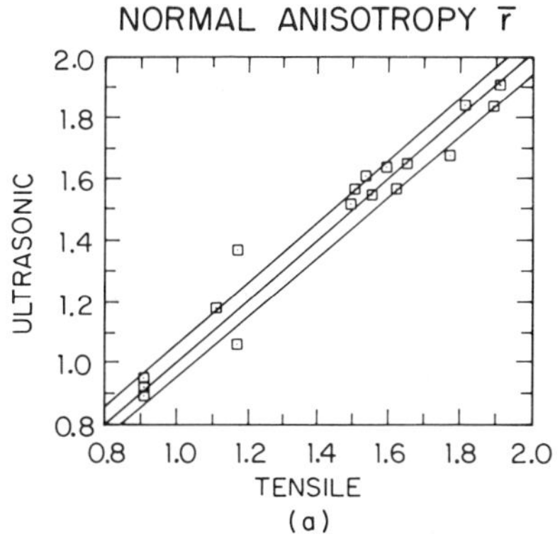

Fig. 6. Graph comparing tensile values of $\bar{r}$ with values obtained with this instrument by ultrasonics.

of the Ames Laboratory, USDOE. The authors are particularly grateful to J. H. Erickson and D. Birlingmair and their staffs. Technology transfer from the laboratory into Ames Laboratory was provided by staff from the Center for Nondestructive Evaluation who also provided necessary help on unresolved technical issues. J. S. Griffiths and R. M. Parker of the Small Business Development Center lent invaluable assistance as did W. Kuo of Industrial Engineering, M. Porter of Civil and Structural Engineering, D. Maclauchlan and his staff at Magnasonics, Inc., and C. Lazo of Lazo Transportation, Inc.

REFERENCES

1. Y. C. Liu and G. A. Alers, The anisotropy of young's modulus in cold-rolled sheets of binary Cu-Zn alloys, Trans. Met. Soc. AIME 236: 489 (1966).
2. E. P. Papadakis, Elastic wave velocities in cube-textured copper sheet, Trans. Met. Soc. AIME 236: 1609 (1966).
3. E. P. Papadakis, Elastic wave velocities in various alloy strips, Metall. Trans. 2: 575 (1971).
4. C. A. Stickels and P. R. Mould, The use of Young's modulus for predicting the plastic-strain ratio of low-carbon steel sheets, Met. Trans. 1: 1303 (1970).
5. P. R. Mould and T. E. Johnson, Rapid assessment of drawability of cold-rolled low-carbon steel sheets, Sheet Metal Industries 328 (1973).
6. R. J. Roe, Inversion of pole figures for materials having elastic anisotropy, J. Appl. Phys. 37: 2069 (1966).
7. H. J. Bunge, "Mathematische Methoden der Texturanalyse", Akademie-Verlag, Berlin, (1969).
8. G. T. Davies, D. J. Goodwill and R. S. Kallend, Elastic and plastic anisotropy in sheets of cubic metals, Met. Trans. 3: 1677 (1972).
9. G. A. Alers, G. Huebschen, B. W. Maxfield, W. Repplinger, J. Salzburger, R. B. Thompson and A. Wilbrand, Electromagnetic acoustic transducers in: "Nondestructive Testing Handbook" ASNT, Columbus, Ohio, in press.
10. R. B. Thompson, Physical principles of measurements with EMAT transducers, in: Ultrasonic measurement methods, R. N. Thurston and A. D. Pierce, eds., Academic Press, N.Y., (1990), p. 157.

11. Y. Li and R. B. Thompson, Influence of anisotropy on the dispersion characteristics of guided ultrasonic plate modes, J. Acoust. Soc. Amer 87: 1911 (1990).
12. R. B. Thompson, Electromagnetic, Noncontact Transducers, in: 1973 Ultrasonics Symposium Proceedings, J. de Klerk, ed., IEEE, New York (1973), p. 385.
13. R. B. Thompson, J. F. Smith and S. S. Lee, Inference of stress and texture from the angular dependance of ultrasonic plate mode velocities, in: "NDE of microstructure for process control", H. N. G. Wadley, ed., ASM, Metals Park, OH (1985), p. 7.
14. R. B. Thompson, S. S. Lee and J. F. Smith, Relative anisotropies of plane waves and guided modes in thin orthorhombic plates: implication for texture characterization, Ultrasonics 25: 133 (1987).
15. R. B. Thompson, S. S. Lee, J. F. Smith and G. C. Johnson, A comparison of ultrasonic and x-ray determinations of texture in thin Cu and Al plates, Met. Trans. 20A: 243 (1989).
16. M. Hirao, H. Fukuoka, K. Fujisawa and R. Murayama, Ultrasonic characterization texture in zinc-coated steel sheets, in: "Elastic waves and ultrasonic nondestructive evaluation" North-Holland, Amsterdam, (1990), p. 319.
17. M. Spies and E. Schneider, Nondestructive analysis of the deep-drawing behavior of rolling sheets with ultrasonic techniques, in: "Nondestructive characterization of materials", Springer-Verlag, Berlin, (1989), p. 296.
18. O. Cassier, C. Donadille and B. Bacroix, Lankford coefficient evaluation in steel sheets by an ultrasonic method, ibid, p. 303.
19. K. Sakata, D. Daniel, J. J. Jonas and J. F. Bussiere, Acoustoelastic determination of the higher order ODF coefficients up to l = 12 and their use for the on-line prediction of r-value, Met. Trans. (in press).
20. R. B. Thompson and A. V. Clark, Jr., Ultrasonic characterization of texture: basic science and technology transfer, in: "Intelligent processing of materials", H. N. G. Wadley, ed. (TMS, in press).
21. A. V. Clark, Jr., A. Govada, R. B. Thompson, J. F. Smith, G. V. Blessing, P. P. Delsanto, and R. B. Mignona, The use of ultrasonics for texture monitoring in aluminum alloys, in "Revies of Progress in Quantitative Nondestructive Evaluation" Vol. 6, D. O. Thompson and D. E. Chimenti, Eds., Plenum Press, NY (1987), p. 1515.
22. W. Y. Lu, J. G. Morris, and Q. Gu, Ultrasonic measurement of the earing behavior of aluminum plate, in "Review of Progress in Quantitative Nondestructive Evaluation", D. O. Thompson and D. E. Chimenti, Eds., Plenum Press, Vol. 10, NY, in press.
23. S. J. Wormley, R. B. Thompson and Y. Li, Analysis of a semi-automatic system for the ultrasonic measurement of texture, in: "Review of progress in quantitative nondestructive evaluation", Vol. 7B, D. O. Thompson and D. E. Chimenti, eds., Plenum Press, NY (1987), p. 1639.
24. E. Schneider, R. Herzer, D. Bruche and A. Wilbrand, Towards the automated ultrasonic characterization of deep drawability of steel sheets, these proceedings.
25. R. Murayama, K. Fujisana, H. Fukuoka, M. Hirao and S. Yonehara, Nondestructive evaluation of materials properties with EMAT in "1989 Ultrasonics Symposium Proceedings", IEEE, NY (1989), p. 1159.
26. A. V. Clark, R. B. Thompson, G. V. Blessing and D. Matlock, Ultrasonic measurement of formability in thin ferritic steel sheets, in "Review of Progress in Quantitative Nondestructive Evaluation," Vol. 8, D. O. Thompson and D. E. Chimenti, Eds. (Plenum Press, NY, 1989) p. 1031.
27. S. J. Wormley, K. Forouraghi, Y. Li, R. B. Thompson, and E. P. Papadakis, Application of a fourier transform-phase-slope technique to the design of an instrument for the ultrasonic measurement of texture and stress, in: "Review of progress in quantitative nondestructive evaluation," Vol. 9A, D. O. Thompson and D. E. Chimenti, eds., Plenum, New York, (1990), p. 951.
28. Y. Li and R. B. Thompson, unpublished results.
29. E. P. Papadakis, Future growth of nondestructive evaluation, Trans. IEEE, SU-23: 284 (1976).
30. E. P. Papadakis, Future growth of nondestructive evaluation: an update, Mater. Eval., 41: 1130 (1983).

ULTRASONIC PREDICTION OF ANISOTROPY IN COPPER AND COPPER ALLOY SHEETS

C.J. Yu, J. C. Conway, Jr., J. Hirsch*, C. O. Ruud and K. Kozaczek

The Pennsylvania State University
University Park, PA 16802

* Current Address : Alcoa Technical Center Pittsburgh, PA 15069

INTRODUCTION

It is well recognized that the presence of preferred crystallographic orientation (texture) results in anisotropy of material properties [1-4]. To predict the plastic deformation response of materials where the texture character dominants, it is noted that a non-intrusive sensor system for on-line, in-process texture monitoring will meet the future demand of high quality in the final product, and total automation in manufacturing processes [5-8]. Several non-destructive techniques can be applied to extract quantitative information concerning anisotropic properties in materials. Ultrasonic techniques are one of them. Previous work has indicated the possibility of inferring texture from ultrasonic velocity measurements [9-11]. Mathematical formalisms have also been developed to relate ultrasonic velocity to texture-induced anisotropy [12-14]. However, few investigations have been performed regarding the nature of metallurgical parameters that affect the obtained ultrasonic velocity information. An understanding of these interdependent relationships will be necessary when trying to use ultrasonic velocity information for the purpose of materials characterization. It is the objective of this investigation to evaluate the effectiveness of ultrasonic velocity measurements as a means for prediction of material anisotropy.

THEORY

Texture

The texture of a polystalline material is defined by the orientation distribution function (ODF) of its crystallites irrespective of the size, the shape, and the arrangement of the crystal within the material [15-18]. Mathematical formalisms have been developed to predict the elastic and plastic anisotropies by describing the three dimensional crystalline orientations in a polystalline aggregate. This prediction is based on the fact that both the elastic and plastic anisotropies are influenced by the series expansion coefficients of the ODF. The ODF expresses the probability of a crystallite having an orientation described by Euler angles ψ, θ, and η, and is given by a series of generalized spherical harmonics in the form:

$$w(\varphi,\varepsilon,\eta) = \sum_{\ell=0}^{\infty} \sum_{m=-\ell}^{\ell} \sum_{n=-\ell}^{\ell} W_{\ell mn} Z_{\ell mn}(\varepsilon)\exp(-im\varphi)\exp(-im\eta)$$

Nondestructive Characterization of Materials IV
Edited by C.O. Ruud *et al.*, Plenum Press, New York, 1991

where $W_{\ell mn}$ are the series coefficients, $Z_{\ell mn}(\varepsilon)$ is the generalized associated Legendre function, and $\varepsilon = \cos\theta$.

For cubic crystallites having orthotropic symmetry, it has been shown [15,18] that a fourth order ODF can give a good approximation of elastic moduli. For the particular case under consideration, the Euler reference axes are rotated to the sheet coordinates, i.e. rolling, transverse, and normal directions. The mean value of the modulus in the plane of the sheet at an angle ϕ to the rolling direction becomes

$$\gamma_m(\phi) = 4\pi^2 \sum_{\ell=0}^{4} \sum_{m=-\ell}^{\ell} \sum_{n=-\ell}^{\ell} G_{\ell mn} W_{\ell mn} \cos(m\phi)$$

where $\gamma_m(\phi)$ refers to the mean value of the modulus in the plane of the sheet, and $G_{\ell mn}$ are the corresponding series coefficients of generalized harmonics obtained in $\gamma_m(\phi)$.

By using the Voigt average method which determines the elastic stiffness by assuming that all grains are subjected to the same strain, the corresponding Voigt average modulus can be expressed as

$$E_v(\phi) = \frac{(C_{11}+2C_{12})[C_{11}-C_{12}-3\gamma_m(\phi)]}{C_{11}+C_{12}-C\gamma_m(\phi)}$$

where $E_v(\phi)$ is the Voigt average elastic modulus, C_{ij} are the cubic single crystal constants and $C = C_{11}-C_{12}-2C_{44}$.

Determination of W_{4m0} coefficients

To determine W_{4m0} coefficients of the ODF, ultrasonic Lamb wave velocity and X-ray diffraction measurement techniques were employed.

When the proper guided mode wave propagation is considered, it was shown by Thompson et. al. [19] that these three W_{4m0} (m=0,2,4) coefficients can be expressed in terms of the fundamental symmetrical Lamb mode(S_0) velocity measured in three directions (rolling, 45 degree, and transverse directions in the plane of sheet). This velocity asymptotically approaches a constant when the product of the applied resonance frequency and sheet thickness decreases to zero.

A method was presented by Roe [20-21] by which the ODF of crystallites in anisotropic polycrystalline material can be derived from a set of functions (pole figures) evaluated from X-ray diffraction measurements. The essence of this method is to expand those observed plane-normal distributions (pole figures) into a series of generalized spherical harmonics, the coefficients of which can be determined by numerical integration of experimental diffraction data. Since the ODF can also be similarly expanded into a series of generalized spherical harmonics, the coefficients W_{4m0} can then be obtained as linear combinations of those coefficients determined from pole figures.

SPECIMENS

C110 copper and 68:32 brass specimens, 12 inches long, 6 inches wide and 0.01 inches thick were supplied with the long axis parallel to the rolling direction in the strip. Two different sets of samples were provided, each representing different annealed grain sizes with a small grain size of 5 μm and a large grain size of 20 μm. Each set of

Table 1. Test Matrix

set	sample	condition 1	condition 2	condition 3	condition 4
1	C110/A	Annealed	C.W. 15%	C.W. 45%	C.W. 75%
2	C110/B	Annealed	C.W. 15%	C.W. 45%	C.W. 75%
3	32%Zn Brass/A	Annealed	C.W. 15%	C.W. 45%	C.W. 75%
4	32%Zn Brass/B	Annealed	C.W. 15%	C.W. 45%	C.W. 75%

C.W.= Cold Work
Grain Size A=5 μm
Grain Size B=20 μm

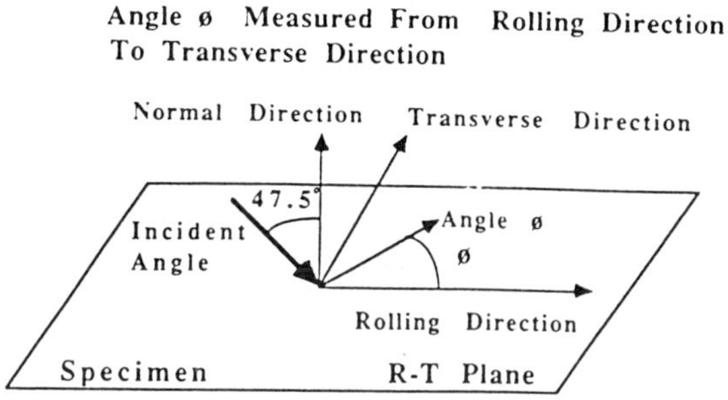

Fig. 1 Definition of sheet coordinates.

samples contained four different degrees of cold roll: annealed, 15% cold work, 45% cold work, and 75% cold work. All specimens were finish rolled to a surface roughness of approximately 5-9 μ inch. The test matrix is shown in Table 1.

EXPERIMENTAL TECHNIQUES

Both ultrasonic and X-ray diffraction techniques were employed to predict the fourth order ODF. A brief description is given below:

Ultrasound

Longitudinal waves with a frequency of 2.25 MHz were transmitted into the specimen at an incident angle of 47.5 degrees with respect to the normal as shown in Fig. 1. At this critical angle, the incident longitudinal wave mode was converted to the fundamental symmetric Lamb wave mode(S_0 mode) in the specimen. In-plane velocity measurements were taken every 5 degrees from the rolling direction to the transverse direction at two distances. However, only velocities measured in the rolling, 45 degree and transverse directions will be used in evaluating the ODF W_{4m0} coefficients. Details concerning the experimental setup can be found in [21].

X-ray diffraction

X-ray pole figures were measured in the reflection mode (Alpha max =85 degrees) using Cu K-alpha radiation on a fully automatic texture goniometer [22]. For each specimen, 4 incomplete pole figures {111}, {200}, {220}, and {113} were measured from which the ODF was calculated by the series expansion method [16,23]

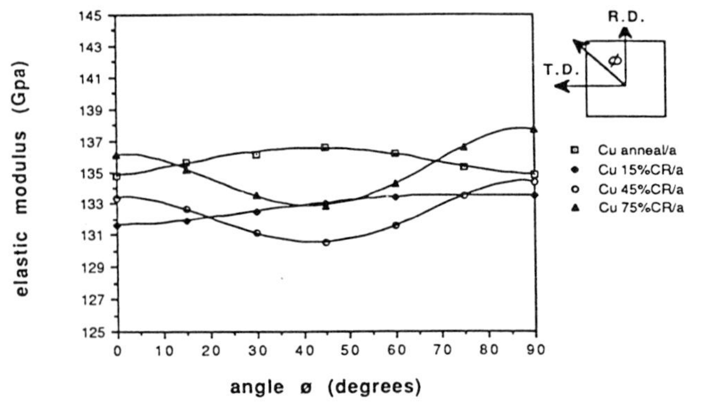

Fig. 2 Calcualted elastic moduli from Lamb wave velocity measurements in copper specimens at 5 μm grain size.

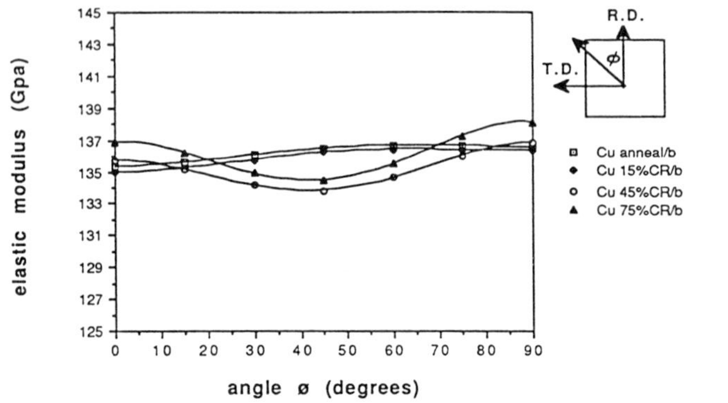

Fig. 3 Calculated elastic moduli from Lamb wave velocity measurements in copper specimens at 20 μm grain size.

with a degree of expansion ℓ_{max}=4. From these 4th order series expansion coefficients, the orientation dependence of the elastic modulus was calculated as the Voigt mean value over the texture [16].

RESULTS AND DISCUSSIONS

There have been few investigations dealing with the interdependent relationships among microstructure, texture, and ultrasonic wave propagation. In this paper, we will restrict ourselves to a comparison of the texture-induced anisotropy evaluated from X-ray diffraction and ultrasonic Lamb wave velocity measurements.

Effect of Cold Work in Copper (Texture Characteristics)

The 4th order ODF were calculated from the measurements of both techniques in copper specimens containing two grain sizes, and the elastic moduli were evaluated based on these two techniques. Comparison of Fig. (2), (3), (4), and (5) showed that the general distribution of the corresponding profiles was similar. The annealed specimen developed the anticipated annealing texture which has been attributed to the existence of cube texture development in annealed copper [1,24]. The concave downward trend of elastic modulus as a function of ø angle measured from the rolling direction in the annealed copper was predicted by both techniques. The 15% cold rolled specimens showed early development of deformation texture [25]. The effect of a combined annealing texture and rolling texture presented a weak but slight increase of elastic modulus with increase in ø

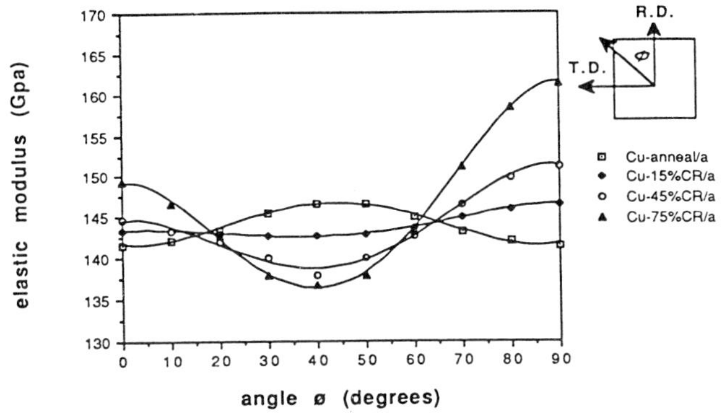

Fig. 4 Calculated elastic moduli from x-ray diffraction measurmeents in copper specimens at 5 μm grain size.

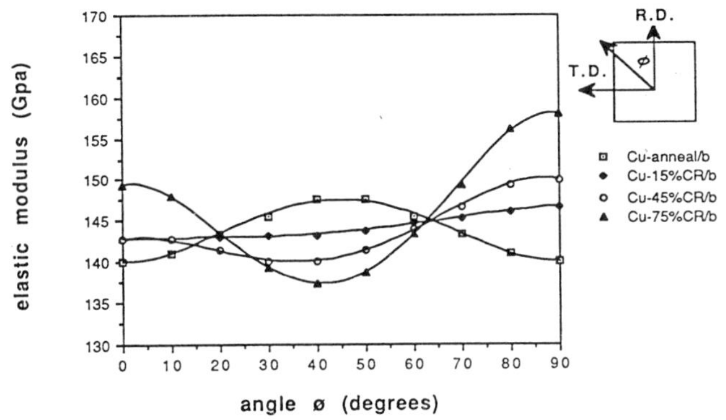

Fig. 5 Calculated elastic moduli from x-ray diffration measurements in copper specimens at 20 μm grain size.

angle from the rolling to the transverse direction. After 45% of cold work, the characteristic texture which developed in the deformed copper is called copper type texture [1,3,26,27]. The elastic modulus profile exhibited a decrease from the rolling direction to the 45 degree direction followed by an increase toward the transverse direction. This was the main characteristic of the elastic anisotropy in highly cold rolled copper.

Effect of Grain Size in Copper (Texture Severity)

Though the similarity of the trend for the corresponding elastic modulus profiles was noticed, the profiles calculated from X-ray diffraction measurements were shown to be much more severe and higher in absolute magnitude than those evaluated from ultrasonic velocity measurements. The comparison of Figs. (2), (3), (4), and (5) indicated that the elastic modulus profiles evaluated from X-ray diffraction measurements did not appear sensitive enough to sense the difference in grain sizes whereas the Lamb wave velocity measurements did sense the difference in grain size. The effects of grain size on the texture development of high stacking fault energy material (like copper and aluminum) were studied by Leffers and his co-workers [28-30]. It was observed that more severe texture developed in the small grain size specimens, and this agrees with the larger variation of anisotropy predicted by Lamb wave velocity measurements in the small grain size copper. Therefore it is suspected that the reflection type of X-ray diffraction techniques employed here most likely involved only the near surface portion of specimen, and cannot be representative for the development of bulk texture [19].

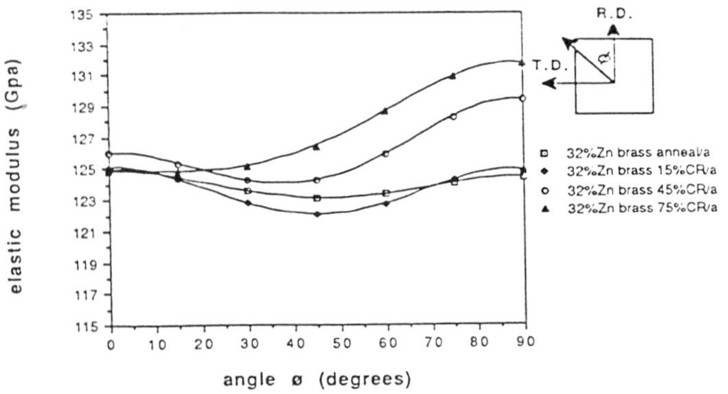

Fig. 6 Calculated elastic moduli from Lamb wave velocity measurmeents in 68:32 brass specimens at 5 μm grain size.

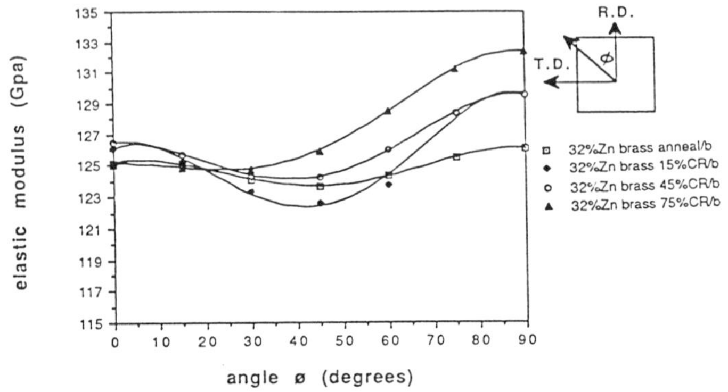

Fig. 7 Calculated elastic moduli from Lamb wave velocity measurements in 68:32 brass specimens at 20 μmgrain size.

Effect of Cold Work in Brass (Texture Characteristics)

The effect of cold work in 68:32 brass is presented in Fig. (6), (7), (8), and (9). Elastic modulus profiles predicted from both techniques in the annealed brass showed that the annealing texture in brass differs from that measured in annealed copper. It was noted that the α-brass annealing texture contained in the annealed brass material would cause this trend of anisotropy which is evidenced by a concave upward form of elastic modulus as function of propagation angle from the rolling direction [1,32]. Initial deformation up to 15% cold reduction showed early development of a copper type texture [25] which is similar in elastic modulus profile to that obtained in highly cold rolled copper specimens. Further plastic deformation caused the transition of copper type texture to brass type texture [1,30,31,33] which is evidenced as an increase in modulus from the rolling to the transverse direction. Changes in deformation mechanisms from those observed in deformed copper are responsible for this transition. Detail discussions can be found elsewhere [31,32].

Effect of Grain Size in Brass (Texture Severity)

Unlike the copper specimens, comparison of Fig. (6), (7), (8), and (9) indicated that the calculated elastic modulus profiles evaluated from both techniques in 68:32 brass specimens were barely influenced by grain size. It was also noticed that almost no difference in the predicted elastic modulus exists for the highly cold rolled brass specimens. The effect of grain size on the rolling texture of cold rolled brass has been discussed by Öztürk et. al. [34] who found only small deviation of texture in brass when

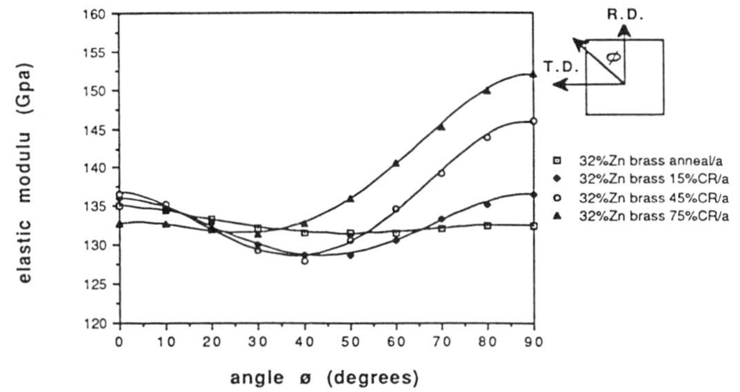

Fig. 8 Calculated elastic moduli from x-ray diffraction measurements in 68:32 brass specimens at 5 μm grain size.

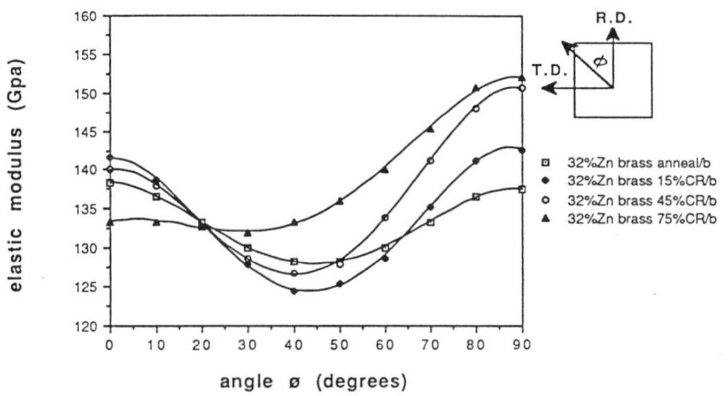

Fig. 9 Calculate elastic moduli from x-ray diffraction measurements in 68:32 brass specimens at 20 μm grain size.

the grain size was changed from 20 μm to 200 μm. Both techniques employed here can predict fairly similar results when grain size effect is not significant.

CONCLUSIONS

Evaluation of the effectiveness of Lamb wave velocity measurements as a means for nondestructive characterization of anisotropy in copper and brass sheets was conducted. Velocity as a function of propagation angle from the rolling direction to the transverse direction was measured in a suite of specimens containing two grain sizes, two compositions, and four cold rolling reductions. The elastic moduli, calculated from 4th order orientation distribution functions determined from Lamb velocity measurements, were compared with those from X-ray diffraction measurements. Good agreement was obtained, and results indicated the possibility of on-line, in-process prediction of properties in copper and copper alloy sheets when ultrasound is coupled to X-ray diffraction measurements.

ACKNOWLEDGEMENT

The research reported in this paper is supported by National Science Foundation under the contract number DMC 8615863. The authors would like to acknowledge the X-ray work conducted in Achen University, Germany.

REFERENCES

1. H. Hu Texture Vol. 1 p.233 (1974)
2. R. Penelle Textures of Materials VII eds. G. Gottstein and K. Leucke p.129 (1978)
3. G.A. Alers and Y.C. Liu Trans. TMS-AIME Vol. 236 p.482 (1966)
4. H.J. Bunge, R. Ebert, and F. Gunther Phys. Stat Sol. 31 p.565 (1969)
5. J.F. Bussiere Materials Evaluation 44, p.560 (1986)
6. H.J. Kopineck, H. Otten Textures and Microstructures 7 p.97 (1987)
7. W.G. Fricke, J.T. Ioannov ICOTOM 8, eds. J.S. Kallend and G. Gottstein (Warrendale, PA: TMS 1988) p.257
8. K.Kozaczek, C.O. Ruud, J.C. Conway Jr., and C.J. Yu Review of QNDE, Vol.9 eds. by D.O. Thompson and D.E. Chimenti p.1813 (1990)
9. R.B. Thompson, J.F. Smith and S.S. Lee NDE of Microstructure for Process Control H.N.G. Wadley, eds. ASM p.73 (1985)
10. M. Hirao, K. Aoki, and H. Fukuoka J. Acoust. Soc. Am. 81(5) May p.1434 (1987)
11. C.M. Sayers J. Phys. D 15 p.2157 (1982)
12. R.B. Thompson, S.S. Lee, and J.F. Smith J Acoust. Soc. Am. 80(3) Sep. p.921 (1986)
13. P.P. Delsanto, R.B. Mignogna, and A.V. Clark,Jr., Nondestructive Characterization of Materials II p.535 (1987)
14. D.R. Allen and C.M. Sayers Ultrasonics 21 No. 5 p.217 (1983)
15. H.J. Bunge in " Texturs in Non-Ferrous Metals and Alloys " eds. H.D. Merchant and J.G. Morries p.145 (1984)
16. H.J. Bunge in " Quantative Texture Analysis " eds. H.J. Bunge and C.Esling DGM Inf.Ges. Oberursel p.1 (1986)
17. R.-J. Roe J. Appl. Phys. 37 p.2069 (1966)
18. G.J. Davies, D.J. Goodwill and J.S. Kallend Metall. Trans. Vol. 3 Jun. p.1627 (1972)
19. R.B. Thompson, J.F. Smith S.S. Lee and G.C. Johnson Metall. Trans A Vol. 20A Nov. p.2431 (1989)
20. R.-J. Roe J. Appl. Phys. Vol. 36 No. 6 Jun. p.2024 (1965)
21. C.J. Yu, J.C. Conway,Jr., C.O. Ruud, and K.Kozaczek Review of QNDE, Vol.9 eds. by D.O. Thompson and D.E. Chimenti p.1993 (1990)
22. Hirsch, G. Burmeister, L. Hoenen and K. Luecke Experimental Techniques of Texture Analysis eds. H.J. Bunge DGM Inf.Ges. Oberursel p.63 (1983)
23. J. Jura and J. Pospiech Texture 3 p.1 (1978)
24. G.J. Davies, J.S. Kallend and T. Ruberg Metal Science Vol. 9 (1975)
25. T. Leffers, D.J. Jensen and B. Major ICOTOM 8, eds. J.S. Kallend and G. Gottstein (Warrendale, PA: TMS 1988) p.461
26. J.S. Kallend and G.J. Davies Texture Vol.1 p.51 (1972)
27. J. Hirsch and K. Leucke Acta Metall. Vol.36 No.11 p.2863 (1988)
28. T. Leffers, D.J. Jensen and F.J. Humphreys ICOTOM 8, eds. J.S. Kallend and G. Gottstein(Warrendale, PA: TMS 1988) p.449
29 D.J. Jensen, N hansen, and F.J. Humphreys ICOTOM 8, eds. J.S. Kallend and G. Gottstein(Warrendale, PA: TMS 1988) P.431
30. N. Hansen, B. Bay, D.J. Jensen and T. Leffers Strength of Metals and Alloys (ICSMA7) eds. H.J. McQueen, J.-P. Bailon, J.I. Dickson, J.J. Jonas and M.G. Akben p.317 (1985)
31. A Merlini and P.A. Beck Trans AIME Feb. J. Metals p. 385 (1955)
32. W.B. Hutchinson, B. J. Duggan and M. Hatherly Metals Technique P.398 (1980)
33. J. Tobisch and A. Mucklich Texture Vol.1 p.211 (1974)
34. T. Öztürk, J.S. Kallend and G.J. Davies ICOTOM 6 eds. Nagashima p. 507 (1982)

DETERMINATION OF TEXTURE IN PLATES OF HCP METALS ZIRCONIUM AND TITANIUM BY ULTRASOUND AND NEUTRON DIFFRACTION

Yan Li and R. Bruce Thompson
Ames Laboratory
Iowa State University
Ames, Iowa 50011

John H. Root and Tom M. Holden
Atomic Energy of Canada Limited
Chalk River Nuclear Laboratories
Chalk River, Ontario, Canada K0J 1J0

INTRODUCTION

The demand for nondestructive characterization of texture in hexagonal close packed (hcp) materials has been increasing rapidly in the last few years, particularly in aerospace and nuclear industries for quality control purposes. Texture study has traditionally been done by X-ray or neutron diffraction techniques. These diffraction techniques are generally slow or destructive. Recent advances in ultrasonics have shown that texture can also be determined by fast and nondestructive ultrasonic techniques [1,2,3]. An ultrasonic texture measurement instrument is currently being built at Iowa State University which can calculate texture parameters of cubic polycrystallites in a few seconds [4].

The texture of a polycrystalline material is generally described by a set of orientation distribution coefficients (ODC's) or $W_{\ell mn}$ in Roe's notation. The detailed definitions of these coefficients can be found in Refs. 5 and 6. In principle, only $W_{\ell mn}$ for ℓ up to 4 can be inferred ultrasonically. For materials of cubic polycrystallites, there are only three nonzero and independent coefficients: W_{400}, W_{420}, and W_{440}. For materials of hexagonal polycrystallites, there are two extra independent coefficients: W_{200} and W_{220}.

Until recently, most of the study in ultrasonic characterization of texture has been concentrated on polycrystals of cubic crystallites. In the determination of texture in plates of cubic materials, long wave length limit of S_0 Lamb waves and SH_0 waves are generally used and the velocities are measured for propagation directions of 0°, 45°, and 90° with respect to the rolling direction [1,2,3]. These conventional techniques are sufficient and have been proved to be successful for cubic polycrystallites. Typically, EMAT's (electro-magnetic acoustic transducers) are used in texture study because their couplant free operation enhances the potential for industrial applications.

ULTRASONIC TECHNIQUES AND SAMPLES

The key to the extension of the conventional techniques to texture characterization in hexagonal materials is the relations between the elastic constants and the texture parameters $W_{\ell mn}$. These relations have been established and published recently in Ref. 7. Once we have the relations, the following equations, describing the information provided by the conventional ultrasonic techniques, can be obtained:

$$
\begin{aligned}
&BW_{400} - \sqrt{5}A_3W_{200} = \frac{105\sqrt{2}}{16\pi^2}\,\rho[V^2_{SH_0}(0^0) + V^2_{SH_0}(45^0) - 2C^o_{44}/\rho] \\
&W_{440} = \frac{3\sqrt{35}}{16\pi^2 B}\,\rho[V^2_{SH_0}(45^0) - V^2_{SH_0}(0^0)] \\
&[3+8(C^o_{12}/C^o_{11})+8(C^o_{12}/C^o_{11})^2]BW_{400} + 2\sqrt{5}\{[1-2(C^o_{12}/C^o_{11})^2]A_1 - (C^o_{12}/C^o_{11})A_2\}W_{200} \\
&\qquad = \frac{105\sqrt{2}}{32\pi^2}\,\rho[V^2_{S_0}(0^0) + V^2_{S_0}(90^0) + 2V^2_{S_0}(45^0) - 4(C^o_{11} - C^{o^2}_{12}/C^o_{11})/\rho] \qquad (1) \\
&(1+2C^o_{12}/C^o_{11})BW_{420} + \sqrt{3}[A_1+(C^o_{12}/C^o_{11})A_2]W_{220} = \frac{21\sqrt{5}}{32\pi^2}\,\rho[V^2_{S_0}(90^0) - V^2_{S_0}(0^0)] \\
&W_{440} = \frac{3\sqrt{35}}{32\pi^2 B}\,\rho[V^2_{S_0}(0^0) + V^2_{S_0}(90^0) - 2V^2_{S_0}(45^0)]
\end{aligned}
$$

where A_1, A_2, A_3, and B are elastic anisotropy constants and C^o_{ij} are isotropic elastic constants of the materials. These constants are related to the elastic constants of single crystals [7] and they are averaging method dependent. The elastic constants for Voigt, Hill, and Reuss averaging method and material densities of Ti and Zr are listed in Table I. Throughout this paper, we will use the Hill averaging method, unless otherwise specified.

In Eqs. (1), one sees clearly that W_{200} and W_{400} are coupled in the first and the third equations and W_{420} and W_{220} are coupled in the fourth equation. These couplings are due to the nonzero terms W_{200} and W_{220} in the hexagonal class. W_{440} is still uncoupled, computable directly from the second or the fifth equations.

Obviously, the conventional techniques cannot solve for all five W's. New ultrasonic techniques need to be developed. The one technique that we are going to discuss in this paper utilizes SH_1 modes. The dispersion equations for SH_1 waves are:

Propagating at 0 degrees:	$C^o_{44}W = C_{66}K + C_{44}$	
Propagating at 45 degrees:	$C^o_{44}W = \frac{1}{4}(C_{11}-2C_{12}+C_{22})K + \frac{1}{2}(C_{44}+C_{55})$	(2)
Propagating at 90 degrees:	$C^o_{44}W = C_{66}K + C_{55}$	

where $K=(\frac{b}{\pi}k)^2$ is the square of dimensionless wave number with b being the plate thickness, and $W=(\frac{2bf}{V_t})^2$ is the square of dimensionless frequency with $V_t=\sqrt{C^o_{44}/\rho}$ being the normalization shear wave velocity.

Note that all three equations in Eqs. (2) are linear equations. The slopes of the solutions for 0° and 90° propagation are identical, indicating a constant separation distance between the two solutions, and the intercept of 45° propagation solution is exactly half way between the other two. Using the relations in Ref. 7, Eqs. (2) can be expressed in terms of ODC's.

There are two measurement methods that are used in our texture study: zero-crossing method for determining phase velocity and Fourier-phase method for obtaining wave vectors. Each method can be further subdivided into two measurement schemes: absolute and relative measurements.

In the zero-crossing absolute velocity measurement, one changes the transducer separation distance and keeps track of the time delays of a specific zero crossing within the waveform. In the zero-crossing relative measurement, one fixes transducer separation distance but changes the wave propagation direction and keeps track of the time delays of a specific zero crossing within the waveform. The phase velocity is associated with the shifts of the arrival time with angle or distance.

In the Fourier-phase absolute measurement, one takes waveforms at two different transducer separation distances and uses the deconvolution technique to obtain the wave

Table I. Elastic isotropy and anisotropy constants in GPa and densities in g/cm^3

Mat'l	method	C^o_{11}	C^o_{12}	C^o_{44}	A_1	A_2	A_3	B	ρ
	Voigt	163.93	75.53	44.20	-62.00	-145.00	207.00	23.00	
Ti	Hill	162.86	76.07	43.40	-61.80	-141.81	203.61	16.69	4.510
	Reuss	161.78	76.60	42.59	-61.60	-138.61	200.21	10.38	
	Voigt	145.68	72.16	36.76	-55.80	-6.60	62.40	42.40	
Zr	Hill	145.18	72.28	36.45	-49.05	-5.22	54.27	38.99	6.505
	Reuss	144.68	72.39	36.14	-42.30	-3.85	46.14	35.59	

number information. In the Fourier-phase relative measurement, one takes waveforms at different propagation directions for a fixed transducer separation distance and uses the deconvolution technique to obtain the information on the change of wave numbers.

When using the zero-crossing method, one generally measures the propagation velocity of a nondispersive or weakly dispersive wave. When using the Fourier-phase method, one can measure either the propagation velocity or the wave number of a wave which can be either nondispersive or strongly dispersive. In real applications, the Fourier-phase method is favorable because it can be easily automated.

We used two samples in our study. The first one is a 1.86 mm thick titanium sample and the second one is a 0.89 mm thick zirconium plate. Both samples were purchased at a local vender. These samples are of commercial purity although it is unclear if second phase exists in either of these samples. We did notice some apparent thickness variations within the zirconium sample.

EXPERIMENTAL DATA

Five sets of neutron pole figures have been obtained for the Ti and the Zr samples. $W_{\ell mn}$ for ℓ up to 10 have been computed based on these pole figures. The $(10\bar{1}0)$ and (0002) pole figures are shown in Figs. 1 and 2 and the $W_{\ell mn}$ for ℓ up to 4 for the two samples are given in Table II. The errors in Table II correspond to the last two digits of $W_{\ell mn}$.

The measured ultrasonic data for the two samples are given in Table III and Table IV. In these tables, the zero-crossing measurements are a combination of absolute measurements at 0° and relative measurements at other angles. The relative measurements in these tables are given with respect to the rolling direction (0°). Each entry in these two tables is the mean value averaged from three repetitions of independent measurements. For the Ti sample, measurements were made using S_{0lim}, SH_0 and SH_1 modes. The values for the S_{0lim} and SH_0 modes are velocities and the values for the SH_1 modes are wave numbers. We were not successful in obtaining useful SH_1 waveforms for the Zr plate hence the values in Table IV are all velocities. The errors for the absolute velocity measurements are about ±0.005 mm/μs for S_{0lim} and ±0.003 mm/μs for SH_0. The errors for the relative measurements are about an order of magnitude smaller. The measurement errors for the dimensionless wave numbers are about ±0.003 in Table III.

Theoretically, when the plate is orthotropic, the velocities or wave numbers for propagation in 45° and 135° directions are identical. The entries in the 45° and 135° columns are generally very close and can be considered to be the same if the measurement error is taken into account. Further, SH_0 velocities for propagation in 0° and 90° directions should also be the same for orthotropic media. This is affirmed by the data in Tables III and IV.

FORWARD COMPARISONS

Using the $W_{\ell mn}$ obtained from neutron diffraction, we can calculate the elastic constants from the relations given in Ref. 7. Then dispersion curves for wave propagating in the two

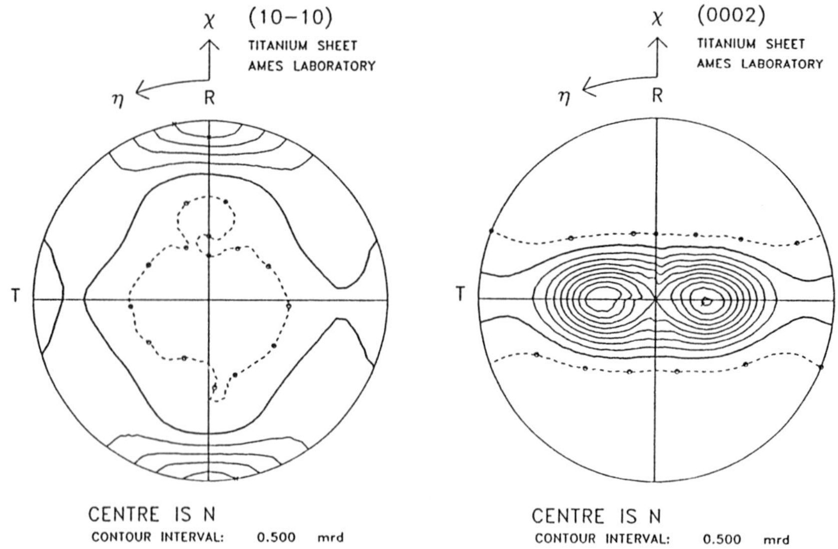

Fig. 1. Pole figures for the Ti samples

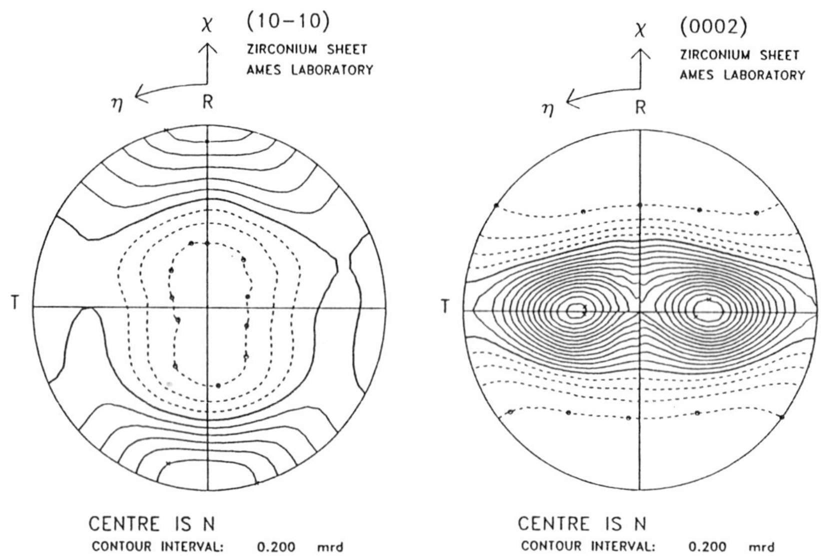

Fig. 2. Pole figures for the Zr sample

Table II. ODC's calculated from neutron pole figures

l	m	$W_{\ell m0}$ (error) -- Ti	$W_{\ell m0}$ (error) -- Zr
2	0	0.01709 (59)	0.01300 (40)
2	2	-0.00545 (17)	-0.00575 (34)
4	0	0.00574 (24)	0.00088 (18)
4	2	-0.00687 (20)	-0.00497 (06)
4	4	0.00217 (06)	0.00236 (16)

Table III. Measured ultrasonic velocities and wave numbers for the Ti

Measurement method		mode	(unit)	0°	45°	90°	135°
zero-crossing	Abs + Rel	S_{0lim}	(mm/μs)	5.3236	5.3778	5.4506	5.3830
		SH_0	(mm/μs)	2.9540	2.9582	2.9523	2.9611
Fourier-phase	Abs	S_{0lim}	(mm/μs)	5.3191	5.3791	5.4646	5.3846
		SH_0	(mm/μs)	2.9511	2.9575	2.9510	2.9572
		SH_1	(bk/π)	0.6367	0.6757	0.7011	0.6796
	Rel	S_{0lim}	(mm/μs)	0.0000	-0.0655	-0.1508	-0.0741
		SH_0	(mm/μs)	0.0000	-0.0070	0.0008	-0.0072
		SH_1	(bk/π)	0.0000	-0.0410	-0.0685	-0.0405

samples can be generated using the generalized Rayleigh-Lamb equations of anisotropic plates [8,9]. On the other hand, using Fourier phase absolute measurement method, experimental dispersion curves can be reconstructed. The comparison of the theoretical dispersion curves and the ultrasonically reconstructed dispersion curves are shown in Fig. 3 for the Ti sample for wave propagation in the rolling direction. In Fig. 3 the continuous curves are the theoretical curves and the curves with crosses are experimental dispersion curves. There are four experimental dispersion curves: low wave number S_0 mode, high wave number S_0 mode, SH_0 mode, and SH_1 mode. In this paper, the information pertaining to the high wave number S_0 mode will not be discussed.

Generally good agreement can be observed from Fig. 3 between the theoretical and experimental dispersion curves. There is, however, one acute problem associated with the Fourier phase method: an ambiguity of a multiple of 2π in determining the relative phase of the two signals being deconvolved. When using Fourier phase method, the phase computation involves an inverse tangent function. The returned value of the inversion is always within the $-\pi$ to π range while the true phase does not have such a limit. In other words, the true phase is the computed phase plus/minus $2n\pi$. Physically, this corresponds to an n cycle misalignment in time domain. To choose the correct value n, we have utilized an algorithm that will be discussed elsewhere because of the page limitation here.

To make sure that we have selected the right multiple, we have plotted the experimental dispersion curves with a positive and a negative offset of 2π phase in Fig. 4. The comparison between the theoretical and experimental curves shows beyond doubt that we have chosen the right multiple in our fitting. Fig. 4 also demonstrates that if we do not have the prior knowledge of neutron diffraction results, it can be very difficult to identify the correct multiple of 2π for relatively strongly dispersive waves.

Table IV. Measured ultrasonic velocities for the Zr

Measurement method		mode		0°	45°	90°	135°
zero-crossing	Abs + Rel	S_{0lim}	(mm/μs)	4.0941	4.0805	4.0987	4.0811
		SH_0	(mm/μs)	2.3309	2.3594	2.3309	2.3605
Fourier-phase	Abs	S_{0lim}	(mm/μs)	4.0862	4.0648	4.0901	4.0579
		SH_0	(mm/μs)	2.3286	2.3584	2.3239	2.3560
	Rel	S_{0lim}	(mm/μs)	0.0000	0.0143	-0.0073	0.0155
		SH_0	(mm/μs)	0.0000	-0.0309	0.0015	-0.0305

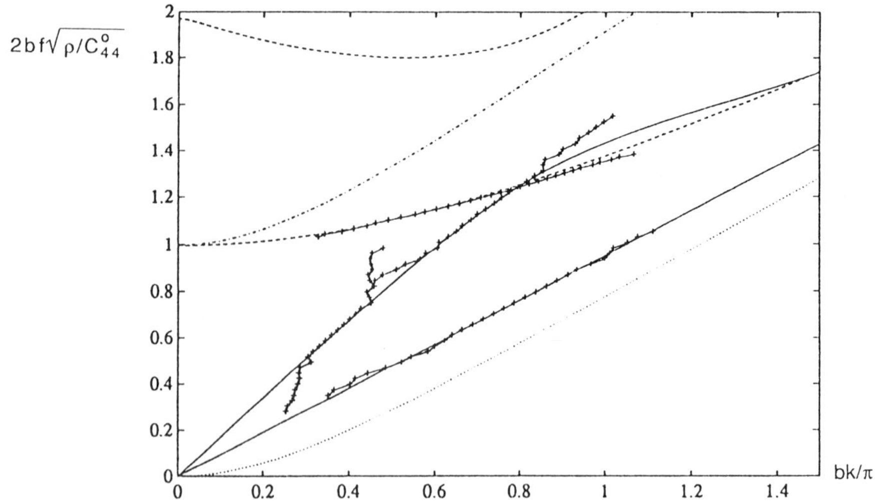

Fig. 3. Some dispersion curves for the Ti sample

Table V gives a data analysis and comparison example. In this case, the conventional S_{0lim} velocity data are used to obtain a linear combination of W_{220} and W_{420} using the fourth equation in Eqs. (1). The equation at the beginning of the table expresses a simplified relation between W_{220} and W_{420} and ultrasonic velocities. The coefficients $C^{S_0}_{220}$ and $C^{S_0}_{420}$ given at the bottom of the table are constants computed based on the fourth equation in Eqs. (1) and the elastic isotropy and anisotropy constants listed in Table I. The values of these coefficients are for the Hill averaging method while the bounds of the values correspond to the Voigt and the Reuss averaging methods. On the left hand side of Table IV, the values are computed based on neutron diffraction data and the errors correspond to the standard deviation in W_{220} and W_{420}. On the right hand side of Table IV, results calculated from ultrasonic velocity measurements are listed along with the standard deviations based on the three repetitions of the experiment for each of the different measurement methods.

From Table V, one can see that there is a good agreement between ultrasonic data and neutron diffraction results for the Ti sample, but the comparison for the Zr plate is not so good, even though consistent results are obtained from different ultrasonic techniques and measurement methods. We do not have a good explanation for this discrepancy at this stage.

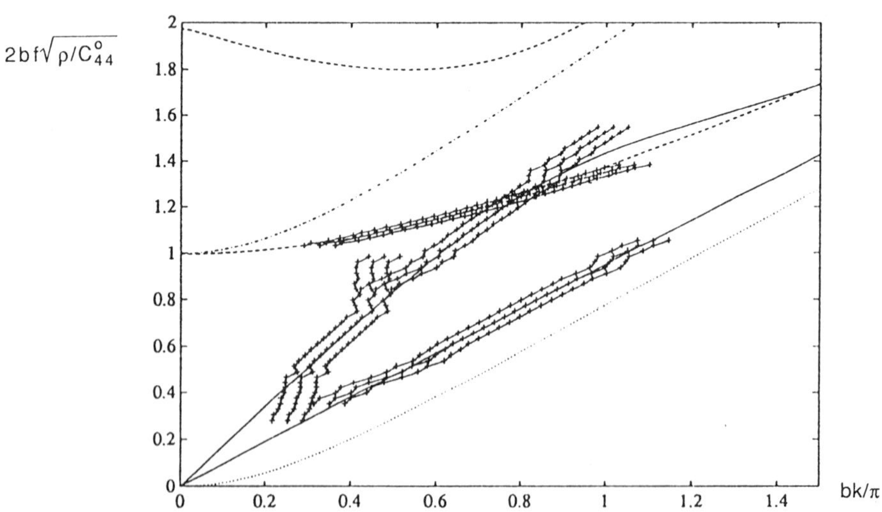

Fig. 4. Ambiguity of multiple of 2π

Table V. Data analysis for W_{220} and W_{420} using S_{0lim} modes

	$C_{220}^{S_0}$ W220 + $C_{420}^{S_0}$ W420	=	$\rho[V_{S_0}^2(90)$ - $V_{S_0}^2(0)]$	
Ti:	6.6375 ±0.2573		6.1712 ±0.1875	(zero-crossing)
			7.0763 ±0.2577	(Fourier phase absolute)
			7.3273 ±0.0863	(Fourier phase relative)
Zr:	0.8586 ±0.2070		0.2452 ±0.1191	(zero-crossing)
			0.2074 ±0.1846	(Fourier phase absolute)
			0.3871 ±0.3022	(Fourier phase relative)
Based on	neutron		ultrasound	(measurement method)
Ti:	$C_{220}^{S_0} = -1491.6^{+8.1}_{-8.2}$;		$C_{420}^{S_0} = 217.1^{+80.1}_{-80.2}$	
Zr:	$C_{220}^{S_0} = -601.7^{-78.5}_{+78.5}$;		$C_{420}^{S_0} = 523.4^{+44.3}_{-44.5}$	

Table VI is another data analysis example. In this example, information of the SH_1 mode measurements is used to obtain another linear combination of W_{220} and W_{420}. Here only the analysis for the Ti sample is presented since we did not get reliable waveforms for the Zr because of the nonuniform thickness of the plate. The analysis in this table is presented in the same fashion as in Table V, except that the squares of dimensionless wave numbers, rather than velocities, are used on the ultrasound side. The zero-crossing measurements are missing from this table because the SH_1 waves are strongly dispersive and the zero-crossing measurement method is unsuitable for such a situation. Again, a very good agreement between diffraction and ultrasonic data can be observed.

INVERSION AND CONCLUSIONS

We have calculated the predictions of $W_{\ell mn}$ based on ultrasonic data for different combinations of techniques and measurement methods discussed in this paper. Comparisons of neutron diffraction data to the best and the worst estimates by ultrasound are given in Table VII. For the Ti sample, except for W_{400}, the best estimates by ultrasound are in very good agreement with those by neutron diffraction. We currently do not have a good explanation on why the estimate of W_{400} by ultrasound is so much different from neutron data. The worst estimates for the Ti are sometimes considerably off from the neutron results except for W_{220} and W_{420} where the differences are relatively small. For the Zr sample, W_{220} and W_{420} remain coupled since reliable ultrasonic data were unavailable for the SH_1. For other ODC's, the best estimates are very close to the neutron estimates. On the other hand, the worst estimate for W_{400} is again far off from the neutron data, even the sign has changed.

Table VI. Data analysis for W_{220} and W_{420} using SH_1 modes

	$C_{220}^{SH_1}$ W220 + $C_{420}^{SH_1}$ W420	=	[K(90) - K(0)]	
Ti:	0.0922 ±0.0024		0.0861 ±0.0036	(Fourier phase absolute)
			0.0923 ±0.0243	(Fourier phase relative)
Based on	neutron		ultrasound	(measurement method)
Ti:	$C_{220}^{SH_1} = -13.66^{+0.02}_{-0.03}$;		$C_{420}^{SH_1} = -2.59^{-0.91}_{+0.95}$	

Table VII. Comparisons of neutron diffraction data to the best and the worst estimates by ultrasonic techniques

		W_{200}	W_{220}	W_{400}	W_{420}	W_{440}
	Neutron	0.01709	-0.00545	0.00574	-0.00687	0.00217
Ti	Best	0.01746	-0.00542	0.02787	-0.00646	0.00231
	Worst	0.00672	-0.00571	0.04263	-0.00415	-0.00696
	Neutron	0.01300	-0.00575	0.00088	-0.00497	0.00236
Zr	Best	0.01092	*	0.00060	*	0.00239
	Worst	0.01040	*	-0.00320	*	0.00410

In conclusion, we have shown that ultrasonic techniques can be used to characterize texture in plates of hexagonal polycrystallites. The distinguishing aspect of texture characterization in hexagonal materials is the coupling of W_{220} with W_{420} and W_{200} with W_{400}. Conventional ultrasonic techniques that have been used for texture determination of cubic polycrystals cannot along determine all five ODC's for hexagonal materials; therefore other ultrasonic technique or techniques must be used or developed. This paper addresses one of the techniques we are using: the SH_1 mode technique. Other techniques may also be used [10]. When using the SH_1 mode technique, along with the conventional techniques, all five $W_{\ell mn}$ for ℓ up to 4 can be determined. The results presented in this paper has shown a promising future of ultrasonic determination of texture, but more work needs to be done on the issue of ambiguity of multiple of 2π before transferring the technology to industry.

ACKNOWLEDGEMENT

Ames Laboratory is operated for the U. S. Department of Energy by the Iowa State University under contract No. W-7405-Eng-82. This work was supported by the Director for Energy Research, Office of Basic Energy Sciences.

REFERENCES

1. R. B. Thompson, S. S. Lee, and J. F. Smith, Ultrasonics **25**, 133 (1987).
2. R. B. Thompson, et al, Met. Trans. **20A**, 2431 (1989).
3. Y. Li, J. F. Smith, and R. B. Thompson, in Nondestructive Characterization of Materials 3, edited by P. Höller, V. Hauk, G. Dobmann, C. Ruud, and R. Green (Springer-Verlag, New York, 1989), p. 312.
4. E. P. Papadakis, et al, "An Automatic Instrument For The Ultrasonic Measurement Of Texture", These Proceedings.
5. R.-J. Roe, J. Appl. Phys. 36, 2024 (1965).
6. R.-J. Roe, J. Appl. Phys. 37, 2069 (1966).
7. Yan Li and R. Bruce Thompson, J. Appl. Phys. **67**, 2663 (1990).
8. Y. Li and R. B. Thompson, in Review of Progress in Quantitative NDE 8A, edited by D. O. Thompson and D. Chimenti (Plenum Press, New York, 1989), p. 189.
9. Yan Li and R. Bruce Thompson, J. Acoust. Soc. Am. **87** (5), 1911 (1990).
10. Y. Li, R. B. Thompson, and S. S. Lee, in Review of Progress in Quantitative NDE 9B, edited by D. O. Thompson and D. Chimenti (Plenum Press, New York, 1990), p.1781.

LASER GENERATION AND INTERFEROMETRIC DETECTION OF ULTRASOUND IN ANISOTROPIC MATERIALS

James B. Spicer, Andrew D. W. McKie, John B. Deaton, Jr., and James W. Wagner

Center for Nondestructive Evaluation
The Johns Hopkins University
Baltimore, Maryland 21218

INTRODUCTION

Current investigation into the influences of material anisotropy on the signature of laser generated waveforms in single crystals using interferometric detection has shown that the orientation of the lattice not only influences the speed of wave propagation but also the overall shape of the detected waveform. In few grained crystalline materials which exhibit strong anisotropy, the misorientation of grains relative to one another may be observed in the signature of detected waveforms. As a result of the reproducibility and the inherent directivity of laser generated ultrasound, the lattice orientation of a specimen relative to its boundaries may be inferred by observing both wave velocities and wave signatures.

THEORY

Laser ultrasonic investigations of anisotropic, crystalline materials differ from conventional ultrasonic testing of materials in several important aspects. In conventional ultrasonic testing, ultrasonic displacements in a material are produced directly by displacing the boundary of the material using a contacting transducer. Ultrasonic displacements produced in this manner primarily are sensitive to the elastic properties of a material. For laser generation of ultrasound in the thermoelastic regime, the electronic, thermal and elastic properties of the material being tested can be important. The strength and penetration of the laser ultrasonic source into the material directly depend on these material properties. First, the radiation incident on the material boundary is absorbed by the material. For metals the mechanisms of radiation absorption and thermalization depend primarily on the electronic properties of the material.[1] Absorption and thermalization variations with crystallographic direction necessarily vary the thermal source region within the material. However, for many metals, the absorption depth for optical radiation is on the order of nanometers and is not consequential in the interpretation of laser ultrasonic waveforms.[2] As a result, variations in the absorption and thermalization characteristics with crystallographic direction are assumed to have minimal effect on the laser ultrasonic waveforms detected in single-crystal metals. Secondly, the thermal transport properties of the material must be taken into account. From the theory of thermoelasticity, the parabolic heat conduction equation in the thermal stress approximation may be written as follows:

$$k_{ij}T_{,ij} - c_v\dot{T} = -Q \tag{1}$$

where k_{ij} is the thermal conductivity tensor, T is temperature increment relative to ambient, c_v is the thermal heat capacity, and Q is the thermal energy source.[3] Subscripts after a comma indicate differentiation with respect to the corresponding spatial coordinate, and the repeated subscript summation convention holds. The effect of the thermal conductivity anisotropy is important in laser ultrasonics as the temperature field resulting from laser heating serves as the source for elastic displacements. This is shown by the theory of thermoelasticity where the differential equation governing elastic displacements in an anisotropic material may be obtained from the Duhamel-Neumann relations and may be written as follows:

$$c_{ijkl}u_{k,lj} = \rho\ddot{u}_i + \beta_{ij}\,T_{,j} \tag{2}$$

with

$$\beta_{ij} = c_{ijkl}\alpha_{kl} \tag{3}$$

where c_{ijkl} are the second order elastic constants, u_i is the ith component of the elastic displacement vector, ρ is the material density, and α_{kl} are the coefficients of thermal expansion.[3] Specifically, equation (2) shows that the spatial derivative of temperature multiplied by the appropriate constants acts as the source for elastic waves in the material.

In general, for thermoelastic laser-generated ultrasound in metals, all of the anisotropic properties discussed above contribute to the character of the observed displacements. As a result, the interpretation of recorded waveforms can become quite involved especially when the normal to the sample surface is not in a high symmetry direction. However, crystal lattices with cubic symmetry behave in an isotropic fashion for thermal conduction and for thermal expansion (which is homogeneous deformation process).[4] This thermal behavior in cubic systems has been verified experimentally and has been used to separate thermal from elastic effects in thermoacoustic imaging systems.[5] Consequently, significant simplification to the theoretical problem of thermoelastic, laser generation of ultrasound occurs when the material studied has a cubic lattice. In cubic systems, only the elastic anisotropy of the material contributes to the observed effects of material anisotropy.

EXPERIMENT

The purpose of the experimental study presented here was to determine the presence of ectopically nucleated grains in an otherwise single crystal specimen. The material under investigation was a nickel-based superalloy with a cubic lattice. Two parallelopiped samples of the alloy were used in the experiment to study the effects on laser ultrasonic wave propagation by elastic wave interaction with a grain boundary and the effects of lattice orientation on the observed ultrasonic displacements. The effect of the lattice orientation on the observed ultrasonic displacements produced by laser generation served both to indicate the smallest variation in lattice misorientation between grains which could be detected and to indicate the orientation of the sample under study. Laser ultrasonic determination of lattice orientation provides a potentially noncontacting method, similar to X-rays, for crystallographic orientation analysis. For all of the studies shown here, laser generation was performed in the ablative regime. The ablative laser source can behave like a mechanical impulse[6,7] to the sample surface. In the general case for material anisotropy, such a source simplifies data interpretation compared to the analysis required for interpretation of data obtained using a thermoelastic source since the effects of elastic anisotropy would dominate the observed waveforms.

Laser ultrasonic wave propagation across a grain boundary was studied using the apparatus shown schematically in Figure 1. Laser ultrasonic displacements were generated using a Nd:YAG laser having a pulse energy of 18 mJ and a pulse width of 4 ns. The laser pulse was focussed to a 3 mm long line source on the sample surface. At a distance perpendicular to the line source a stabilized Michelson interferometer was used to transduce the resulting displacements.[8] The output of the interferometer was recorded with a transient-digitizing recorder. The superalloy sample used in this study was prepared by etching half the sample surface to reveal the location of grain boundaries and by optically polishing the other half to provide good reflectivity for the interferometer. Three locations of laser ultrasonic testing were performed for which the resulting waveforms are shown in Figures 2a, 2b and 2c. Figure 2a shows the waveform recorded for which the laser source and receiver were on the primary grain of the sample such that no interaction with any other grains occurred. Figure 2b shows the waveform obtained for transgranular propagation of the ultrasonic wave where generation was performed on a second grain in the sample and receiver was on the primary grain. Figure 2c shows the waveform recorded when both source and receiver were on the second grain in the sample. Note that the signature of the waveform changes dramatically as generation and detection are moved from one grain to the next. Such changes result directly from the differing crystallographic orientations of the two grains.

To determine the sensitivity of the laser ultrasonic method to lattice orientation, a series of experiments were performed in which the laser source and receiver were fixed in space relative to each other. The sample being tested was rotated such that the crystallographic direction from source to receiver was changed for each waveform recorded. The sample was rotated through 360^0 about the normal to the sample surface and waveforms were collected at 20^0 intervals. The laser pulse was focussed to a line source 3 mm long with the source-to-receiver distance at 7.4 mm. Significant variation in the signature of the waveforms occurred for lattice rotations of 20^0. From these waveforms, the times-of-flight for the quasi-surface-skimming longitudinal wave were recorded and were plotted on the polar plot shown in Figure 3. The radial coordinate

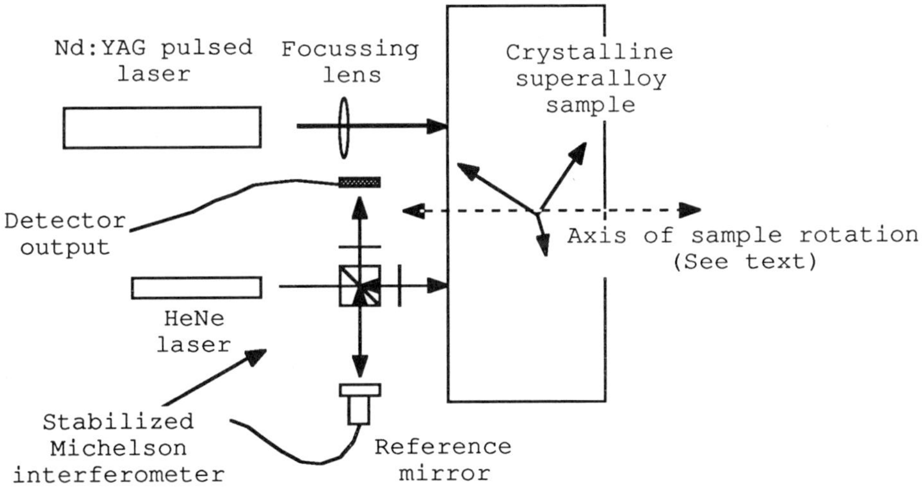

Figure 1. Experiment schematic for the study of crystalline anisotropy in a nickel-based superalloy using laser ultrasonics.

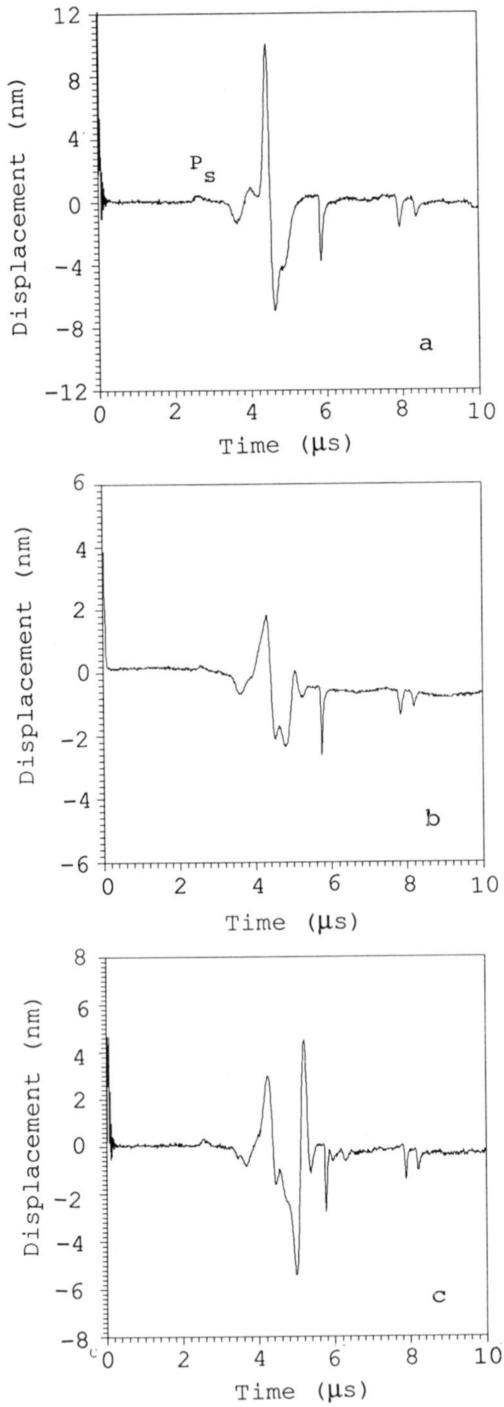

Figure 2. Effects of transgranular wave propagation on laser ultrasonic waveforms. (2a) Generation and detection on the primary grain in the superalloy sample. (2b) Generation on the second grain and detection on the primary grain in the sample. (2c) Generation and detection on the second grain in the sample.

(Note: P_s = quasi-surface-skimming longitudinal wave)

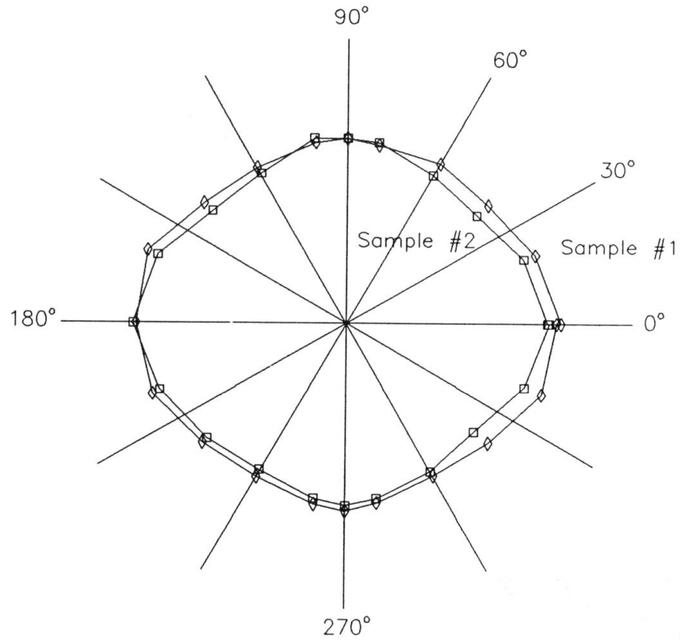

Figure 3. Variation of the time-of-flight for the quasi-surface-skimming longitudinal wave with the angular orientation of the superalloy sample.

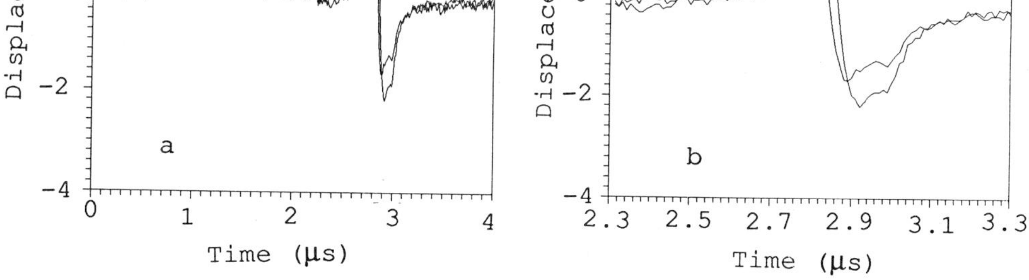

Figure 4. Variation of the laser ultrasonic waveform with angular orientation of the superalloy sample. (4a) Full waveform taken for the sample angular orientation of ϑ superposed on the waveform taken for sample orientation of $\vartheta + 2^0$. (4b) Figure 4a shown on an expanded time scale.

represents the time-of-flight and the polar angle represents the angular rotation of the sample. The clear two-fold symmetry of this plot indicates that for the samples tested, the surface normals are in the <110> direction. The deviation from two-fold symmetry by the time-of-flight plot for sample #2 indicates that the surface normal for this sample might deviate from the <110> direction. Next, to improve the angular resolution of the experiment, the source and the receiver were focussed to reduce the angle subtended by the source at the receiver and, conversely, by the receiver at the source. The results for rotating the sample through a 2^0 angle are shown in Figures 4a and 4b. In Figure 4a two waveforms are overlaid. One represents the displacement recorded for a given orientation of the sample; the second, the displacement recorded when the orientation is changed by 2^0 . Note that for such small rotations of the lattice, differences in wave time-of-flight are small. Figure 4b shows the portion of Figure 4a for which relatively large ultrasonic displacements occurred. With the expanded time scale, the difference in the times-of-flight for the displacements shown is apparent. Perhaps more significant is the marked difference in the signature of the waveform. Eventhough the time-of-flight difference approaches the resolution limit of the digitizer used for recording the waveforms, the change in the waveform signature clearly indicates a difference in the crystallographic direction of ultrasonic propagation.

CONCLUSIONS

It has been demonstrated that laser ultrasonics is a viable method for studying anisotropic, crystalline materials. For laser generation in the thermoelastic regime, the interpretation of waveforms may be complicated by the anisotropy of the thermal, the thermal-mechanical, and the elastic properties of the material. However, for cubic systems only the elastic properties contribute to the observed effects of material anisotropy on laser ultrasonic waveforms. Using laser ultrasonics, the orientation of a single crystal may be determined. Orientation determination was performed for a single crystallographic axis in two nickel-based superalloy samples by observing the symmetries in the times-of-flight for quasi-surface-skimming longitudinal waves. Changes in the signature of waveforms indicate that crystal lattice orientation may be determined to within 2^0 or, similarly, that misorientation of adjoining grains by 2^0 may be observed. Theoretical studies are required to interpret laser ultrasonic waveforms so that the orientation of the crystal lattice may be determined to the precision which is indicated by the experiment presented.

REFERENCES

1. G. Hass and L. Hadley, Optical Properties of Metals, in: "American Institute of Physics Handbook," D. E. Gray, ed., McGraw-Hill, New York (1972).
2. U. Schleichert, K. Langenberg, W. Arnold, and S. Fasbender, A Quantitative Theory of Laser-generated Ultrasound, in: "Review of Progress in Quantitative Nondestructive Evaluation," Vol. 8A, D. Thompson and D. Chimenti, eds., Plenum Press, New York (1989).
3. W. Nowacki, "Dynamic Problems of Thermoelasticity," Noordhoff, Leyden, The Netherlands (1975).
4. J. F. Nye, "Physical Properties of Crystals," Oxford University Press, London (1957).
5. J. W. Maclachlan and J. C. Murphy, Anisotropy in Thermoacoustic Imaging of Single Crystals, in: "Photoacoustic and Photothermal Phenomena," P. Hess and J. Pelzl, eds., Springer-Verlag, Berlin (1988).
6. C. B. Scruby, R. J. Dewhurst, D. A. Hutchins, and S. B. Palmer, Laser Generation of Ultrasound in Metals, in: "Research Techniques in Nondestructive Testing," Vol. V, R. S. Sharpe, ed., Academic Press, New York (1982).

7. J. D. Aussel, A. Le Brun, and J. C. Baboux, Generating acoustic waves by laser: theoretical and experimental study of the emission source, Ultrasonics 26:245 (1988).
8. R. J. Dewhurst, C. E. Edwards, A. D. W. McKie, and S. B. Palmer, Comparative study of wide-band ultrasonic transducers, Ultrasonics 25:315 (1987).

NONDESTRUCTIVE MEASUREMENT OF TEXTURE AND PLASTIC STRAIN RATIO OF STEEL SHEETS USING EMATs:COMPARISON OF TWO APPROXIMATIONS TO CALCULATE CODFs

Katsuhiro Kawashima

Electronics Laboratories
Nippon Steel Corporation
Tokyo, Japan

INTRODUCTION

All the lowest order independent CODF coefficients, W_{400}, W_{420} and W_{440} of a rolled metal sheet were satisfactorily measured using only EMATs by Kawashima(1990). In that method only the Voigt approximation was used to calculate CODF coefficients from measured ultrasonic velocities and known values of single crystal elastic constants. In this paper both the Voigt and Reuss approximations are used for the calculation, and the results are compared. Stress in a steel sheet was neglected because, in general, it has significant texture which gives the major contribution to elastic anisotropy. In the 1st section the basic equations for the CODF, elastic constants, Young's modulus, polefigure diagrams and ultrasonic waves in a sheet are reviewed. In the theory of the 2nd section, it is shown that W_{400}, W_{420} and W_{440} can be obtained as the solutions of three linear equations. Two sets of three linear equations are given by using the Voigt and Reuss approximation methods respectively. In the experiment, the through thickness resonance method is used with an EMAT that can simultaneously measure all three modes of ultrasonic waves propagating in the through thickness direction. A set of EMATs to measure precisely SH0(shear horizontal)plate wave velocity is also used. In the 3rd section, the CODF coeficients are calculated by the equations given in the 2nd section using the measured ultrasonic velocities. Further, X-ray polefigures and ultrasonic polefigures obtained using the calculated CODF coefficients are compared. All nine elastic constants of the steel sheets and Young's modulus in the sheet plane are also calculated and compared with plastic strain ratio. The paper is concluded in section 4.

THE BASIC EQUATIONS

In this section the basic equations used in this article are reviewed in order to make it self-contained.

Relations between CODF coefficients and elastic constants

A rolled steel sheet is a polycrystalline aggregate of cubic crystallites. Symmetry arguments for the sample with orthorhombic symmetry and for the cubic crystallites considerably reduce the number of independent CODF

Nondestructive Characterization of Materials IV
Edited by C.O. Ruud *et al.*, Plenum Press, New York, 1991

coefficients, W_{lmn}. The lowest order independent coefficients are $W_{000}=1/(4\sqrt{2}\pi^2)$ (normalized constant) and W_{400}, W_{420} and W_{440}. These relations have been discussed in detail by Roe (1965, 1966) and later have been given in a compact form by Sayers (1982). An orthorhombic continuum has, in general, nine different elastic rigidity constants. Following the Voigt procedure for averaging elastic constants of a polycrystal having such a CODF, Sayers has found that the nine polycrystalline elastic constants C_{ij} of an orthorhombic plate may be expressed in terms of only six independent constants, W_{400}, W_{420} and W_{440} and the single crystal elastic constants, C_{011}, C_{012} and C_{044}.

$$C_{11}= C_{011}-2C_0/5-2C_0X_1,\quad C_{44}= C_{044}+C_0/5+C_0X_4,\quad C_{23}= C_{012}+C_0/5+C_0X_4$$
$$C_{22}= C_{011}-2C_0/5-2C_0X_2,\quad C_{55}= C_{044}+C_0/5+C_0X_5,\quad C_{31}= C_{012}+C_0/5+C_0X_5 \qquad (1)$$
$$C_{33}= C_{011}-2C_0/5-2C_0X_3,\quad C_{66}= C_{044}+C_0/5+C_0X_6,\quad C_{12}= C_{012}+C_0/5+C_0X_6$$

where $C_0=C_{011}-C_{012}-2C_{044}$

and

$$X_1=-(6/35)\sqrt{2}\pi^2[W_{400}-(2/3)\sqrt{10}\,W_{420}+(1/3)\sqrt{70}\,W_{440}]$$
$$X_2=-(6/35)\sqrt{2}\pi^2[W_{400}+(2/3)\sqrt{10}\,W_{420}+(1/3)\sqrt{70}\,W_{440}]$$
$$X_3=-(16/35)\sqrt{2}\pi^2\,W_{400}$$
$$X_4=-(16/35)\sqrt{2}\pi^2[W_{400}+\sqrt{5/2}\,W_{420}]$$
$$X_5=-(16/35)\sqrt{2}\pi^2[W_{400}-\sqrt{5/2}\,W_{420}]$$
$$X_6=(4/35)\sqrt{2}\pi^2[W_{400}-\sqrt{70}\,W_{440}]$$

Sayers also obtained the equations for elastic compliances using the Reuss averaging procedure.

$$S_{11}= S_{011}-2S_0/5-2S_0X_1,\quad S_{44}= S_{044}+4S_0/5+4S_0X_4,\quad S_{23}=S_{012}+S_0/5+S_0X_4$$
$$S_{22}= S_{011}-2S_0/5-2S_0X_2,\quad S_{55}= S_{044}+4S_0/5+4S_0X_5,\quad S_{31}=S_{012}+S_0/5+S_0X_5 \qquad (2)$$
$$S_{33}= S_{011}-2S_0/5-2S_0X_3,\quad S_{66}= S_{044}+4S_0/5+4S_0X_6,\quad S_{12}=S_{012}+S_0/5+S_0X_6$$

where $S_0=S_{011}-S_{012}-S_{044}/2$

The equations for the Young's modulus $E(\alpha)$ in the xy plane, with angle α with x axis can be found elsewhere.

$$1/E(\alpha)=S_{22}\sin^4\alpha+S_{11}\cos^4\alpha+(S_{66}+2S_{12})\sin^2\alpha\cos^2\alpha \qquad (3)$$

where $S_{11}=(C_{22}C_{33}-C_{23}C_{23})S$, $S_{22}=S(C_{11}C_{33}-C_{13}C_{13})$

$S_{12}=(C_{13}C_{23}-C_{12}C_{33})S$, $S_{66}=1/C_{66}$

$S=1/(C_{11}C_{22}C_{33}+2C_{12}C_{23}C_{13}-C_{11}C_{23}^2-C_{22}C_{13}^2-C_{33}C_{12}^2)$

Polefigure diagram

Sayers and later Hirao and Aoki et al. (1987) have given an equation for a polefigure diagram $q(\zeta,\eta)$ in which only the terms up to l=4 are included:

$$4\pi q(\zeta,\eta)=1+4\pi P\{(3/8\sqrt{2})(35\zeta^4-30\zeta^2+3)W_{400} \qquad (4)$$
$$+(9/2)\sqrt{5}(1-\zeta^2)[1-(7/6)(1-\zeta^2)]W_{420}\cos 2\eta$$
$$+(3/8)\sqrt{35}(1-\zeta^2)^2\,W_{440}\cos 4\eta\}$$

where, for example, $P=-4\pi/3$ (111) polefigure
$P= 2\pi$ (100) polefigure
$P=-\pi/2$ (110) polefigure

Ultrasonic velocities in an anisotropic sheet material

CODF coefficients, W_{400}, W_{420} and W_{440}, can be related to the ultrasonic velocities in a rolled metal sheet through the elastic constants of the sheet. In this subsection, the equations for the ultrasonic velocities in an anisotropic sheet necessary to calculate the CODF coefficients using the new method given in the next section are reproduced. An equation for an ultrasonic velocity in a sheet material is, in general, different from that for an ultrasonic velocity propagating in the same direction in an infinite medium because of the stress-free boundary conditions at the sheet surfaces (see Fig.1). However, the equations for the velocities of selected modes of ultrasonic waves in a sheet material are the same as those for the velocities of the plane ultrasonic waves in an infinite medium. Three of those ultrasonic waves are the waves that propagate in the through thickness direction (Oz). These waves are the pure longitudinal waves of velocity V_{zz}, the pure shear waves of velocity V_{zy} polarized in the transverse direction (Oy) and the pure shear waves of velocity V_{zx} polarized in the rolling direction (Ox).

$$\rho V_{zz}^2 = C_{33}, \quad \rho V_{zy}^2 = C_{44}, \quad \rho V_{zx}^2 = C_{55} \tag{5}$$

where, ρ is density

The other two such waves are the shear horizontal plate waves of velocity $V_{xy}=V_{SH0}(0°)$ propagating in the rolling direction and polarized in the transverse direction, and the shear horizontal plate waves of velocity $V_{yx}=V_{SH0}(90°)$ propagating in the transverse direction and polarized in the rolling direction. The examination of the solutions for their plane shear wave counterparts in an infinite medium shows that these plane shear waves generate no dynamic stresses in an xy plane. Thus the shear horizontal plate waves propagating along the material symmetry axes are expressed by the same equation as that for the plane shear waves. The velocities of those waves are expressed by the following simple and exact equations.

$$\rho V_{xy}^2 = \rho V_{SH0}^2(0°) = C_{66}, \quad \rho V_{yx}^2 = \rho V_{SH0}^2(90°) = C_{66} \tag{6}$$

For all the waves expressed by the equations (5), (6), the velocity is independent of the frequency, and the skewing angle, the angle between the wave vector and the energy vector, is zero. These characteristics allow us to directly compare the measured velocities with theoretical velocities.

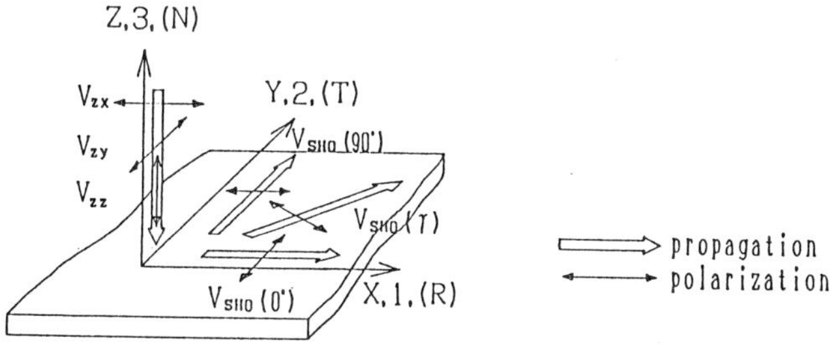

Fig.1. Propagation and polarization directions of ultrasonic waves in a rolled metal sheet (Ox,Oy,Oz:rolling, transverse and normal direction)

It is necessary to find one more ultrasonic velocity to calculate the CODF coefficients by using the method given in the next section. The shear horizontal plate waves of velocity $V_{SH0}(\gamma)$ propagating in the xy plane along the direction other than the material symmetry axes is no longer purely shear because the particle velocity vector is not perpendicular to the wave vector, although it lies in the xy plane. It is not an easy job to obtain the exact solution for the velocity $V_{SH0}(\gamma)$ because the nonlinear, anisotropic wave equations must be solved under the stress free boundary conditions on the plate surfaces. Thompson and Lee et al.(1987) have given an approximate solution to this problem under the condition of small thickness-to-wavelength ratio.

$$\rho V_{SH0}^2(\gamma) = C_{66}[1 + \frac{\beta}{4}(1-\cos 4\gamma)]^2 \tag{7}$$

where

$$\beta = \{\frac{1}{2}[\frac{C_{11}+C_{22}}{2} - C_{12} - \frac{(C_{13}-C_{23})^2}{2C_{33}}] - C_{66}\}/C_{66}$$

When γ equals 45° , the solution is simplified to

$$\rho V_{SH0}^2(45°) = C_{66}(1 + \frac{\beta}{2})^2 \tag{8}$$

Thus the velocity $V_{SH0}(45°)$ is independent of the frequency under the long wavelength condition. Further, Allen and Langman et al.(1985) and Thompson and Lee et al.(1986) have shown that the skewing angle for $V_{SH0}(45°)$ is zero to first order in anisotropy. These characteristics also allow us to compare the measured velocity with the theoretical velocity under such conditions.

NEW ULTRASONIC METHOD TO OBTAIN THE CODF COEFFICIENTS

Theory to calculate the CODF coefficients using the relative ultrasonic velocities

The method of velocity combinations of longitudinal and shear waves has been used by Allen and Sayers(1984) and Hirao and Hara et al.(1988), and that of shear horizontal plate waves Thompson and Lee et al.(1987). The method given here uses both the velocity combination methods in order to obtain all the lowest order independent CODF coefficients, W_{400}, W_{420}, and W_{440}. Firstly a method using the Voigt approximation is described. Combining Eqs.(1) and (5) gives

$$K_1^2 = \frac{\rho V_{zy}^2}{\rho V_{zz}^2} = \frac{C_{44}}{C_{33}} = \frac{C_{044} + C_0/5 + C_0X_4}{C_{011} - 2C_0/5 - 2C_0X_3} \tag{9}$$

Similarly

$$K_2^2 = \frac{\rho V_{zx}^2}{\rho V_{zz}^2} = \frac{C_{55}}{C_{33}} = \frac{C_{044} + C_0/5 + C_0X_5}{C_{011} - 2C_0/5 - 2C_0X_3} \tag{10}$$

Eqs.(9) and (10) are linear equations of W_{V400} and W_{V420} and the solutions are given by

$$W_{V400} = \frac{7}{16\sqrt{2}\pi^2}(\frac{5C_{044}+C_0}{C_0})[1-(\frac{5C_{011}+10C_{044}}{2C_0+10C_{044}})(\frac{K_1^2+K_2^2}{K_1^2+K_2^2+1})] \tag{11}$$

$$W_{V420} = -\frac{7}{16\sqrt{5}\pi^2}(\frac{5C_{044}+C_0}{C_0})(\frac{5C_{011}+10C_{044}}{2C_0+10C_{044}})(\frac{K_1^2-K_2^2}{K_1^2+K_2^2+1}) \tag{12}$$

Combining Eqs.(6) and (8) gives

$$K_3^2=\frac{\rho V_{SH0}^2(45^\circ)}{\rho V_{SH0}^2(0^\circ)}=\frac{C_{66}(1+\beta/2)^2}{C_{66}}=(1+\frac{\beta}{2})^2 \tag{13}$$

One can obtain a linear equation of W_{V440} through an elementary mathematical procedure after substituting into Eq.(13) the values of W_{V400} and W_{V420} calculated by Eqs.(11),(12) and neglecting the second order terms in X_i. Its solution is given by

$$W_{V440}=\frac{7}{8\sqrt{35}\pi^2}(\frac{5C_{044}+C_0}{C_0}+\frac{4\sqrt{2}}{7}\pi^2 W_{V400})(\frac{K_3-1}{K_3}) \tag{14}$$

Secondly,a method using the Reuss approximation is described. Combining Eqs.(1) and (5)

$$\frac{1}{K_1^2}=\frac{\rho V_{zz}^2}{\rho V_{zy}^2}=\frac{C_{33}}{C_{44}}=\frac{S_{44}(S_{11}S_{22}-S_{12}^2)}{D(S_{mn})} \tag{15}$$

Similarly

$$\frac{1}{K_2^2}=\frac{\rho V_{zz}^2}{\rho V_{zx}^2}=\frac{C_{33}}{C_{55}}=\frac{S_{55}(S_{11}S_{22}-S_{12}^2)}{D(S_{mn})} \tag{16}$$

where

$$D(S_{mn})=S_{11}S_{22}S_{33}+2S_{12}S_{23}S_{13}-S_{11}S_{23}^2-S_{22}S_{13}^2-S_{33}S_{12}^2$$

Substituting Eqs.(2) into Eqs.(15),(16) and neglecting the second order terms in X_i gives

$$\frac{1}{K_1^2}=\frac{S_{044}+4S_0/5+4S_0X_4}{S_a-S_bX_3} \tag{17}$$

$$\frac{1}{K_2^2}=\frac{S_{044}+4S_0/5+4S_0X_5}{S_a-S_bX_3} \tag{18}$$

where

$$S_a=\frac{(S_{011}+2S_{012})(S_{044}+4S_0/5)}{2(S_{011}+S_{012}-S_0/5)} \qquad S_b=\frac{2S_0(S_{011}+2S_{012})^2}{(S_{011}+S_{012}-S_0/5)2}$$

Eqs.(17) and (18) are linear equations of W_{R400} and W_{R420} and the solutions are given by

$$W_{R400}=-\frac{7}{16\sqrt{2}\pi^2}(\frac{5S_{044}+4S_0}{4S_0})[1-(\frac{10S_{011}+15S_{012}-S_0}{5S_{011}+10S_{012}})(\frac{K_1^2+K_2^2}{K_1^2+K_2^2+1})] \tag{19}$$

$$W_{R420}=\frac{7}{16\sqrt{5}\pi^2}(\frac{5S_{044}+4S_0}{4S_0})(\frac{10S_{011}+15S_{012}-S_0}{5S_{011}+10S_{012}})(\frac{K_1^2-K_2^2}{K_1^2+K_2^2+1})] \tag{20}$$

One can obtain a linear equation of W_{R440} through an elementary mathematical procedure after substituting into Eq.(13) the values of W_{R400} and W_{R420} calculated by Eqs.(19),(20) and neglecting the second order terms in X_i. Its solution is given by

$$W_{R440}=\frac{7}{8\sqrt{35}\pi^2}\left(-\frac{5S_{044}+4S_0}{4S_0}+\frac{4\sqrt{2}}{7}\pi^2 W_{R400}\right)\left(\frac{K_3-1}{K_3}\right) \quad (21)$$

Once W_{400}, W_{420} and W_{440} are found and the single crystal elastic constants are known, one can calculate all the nine elastic constants and Young's modulus of a sheet metal by using Eqs.(1),(2) and (3). Ultrasonic pole figure diagrams can also be calculated using Eq.(4).

Experiments

Cold rolled steel sheets(44 test pieces,size:50cm x 50cm ,thickness: 0.67~0.82mm) were used for the experiments. All the test pieces were collected from 7 different steel coils. Data of the 7 test pieces collected from the 7 coils are shown in Table I. Two types of specially designed EMATs were made and the ultrasonic wave velocities were measured quickly and accurately. One of the two types of EMAT was made to measure the longitudinal ultrasonic velocity V_{ZZ} and the shear ultrasonic velocities V_{ZY} and V_{ZX}. It works in the thickness resonance mode that is suited to measuring the ultrasonic velocities propagating in the through thickness direction in a metal sheet. The other type of EMAT having one transmitter and two receivers was made to measure SH0 plate wave velocity in a metal sheet. According to Eqs.(11),(12),(14),(19)-(21),it is only necessary to find the relative velocities and the relative values of single crystal elastic constants,not the absolute values,to calculate the CODF coefficients. The relative velocities K_1 and K_2 were obtained as the ratio of the resonance frequencies measured by the thickness resonance EMAT. The relative velocity K_3 was obtained as the reciprocal of the relative time of flight measured by the SH0 plate wave EMAT. The details of these EMATs,the importance of the relative velocity measurements and the advantage of using only EMATs were reported by Kawashima(1990).The all automatic measurement system was developed using this new measurement principle.

ANALYSIS OF EXPERIMENTAL RESULTS

CODF coefficients W_{V400}, W_{V420} and W_{V440} (Voigt approximation),and W_{R400}, W_{R420} and W_{R440}(Reuss approximation) were calculated by Eqs.(11), (12),(14),(17)-(19) using the measured relative velocities and the known values of the single crystal elastic constants of α-iron, C_{011}=237 GPa, C_{012}=141 GPa,C_{044}=116 GPa,given by Schmid and Boas(1935).The calculated results are shown in Table I. The (100) ultrasonic polefigures were calculated using the CODFs obtained as shown in Table I. Two of these polefigures are shown in Figs.2(b) and (d). The one shown in Fig.2(b) was drawn for No.1 test piece using $W_{V400}=-1.56\times10^{-2}$, $W_{V420}=-1.27\times10^{-3}$,$W_{V440}=7.18\times10^{-4}$ and the other shown in Fig.2(d) for No.2 test piece using $W_{V400}=-3.94\times10^{-3}$, $W_{V420}=-1.36\times10^{-3}$,$W_{V440}=1.83\times10^{-3}$. The polefigures measured by the X-ray method for thin samples (75μm) taken from No.1 and No.2 test pieces are also shown in Figs.2(a) and (c) for comparison. An overall correspondence including relative pole densities is seen between the ultrasonic and the X-ray polefigures,but the fine structure is missing in the ultrasonic polefigures. Another difference is found in the angles of the high pole density peaks along the rolling direction. The angles in the ultrasonic polefigures are larger than those in the x-ray polefigures. These differences are the result of only the lowest order CODF coefficients being used to make the ultrasonic polefigures.

Table I. Data of the test pieces(cold rolled steel sheets) ;CODF coefficients and elastic constants were calculated using the measured data and the known values of single crystalelastic constants, C_{011}=237GPa, C_{012}=141GPa and C_{044}=116GPa.

Test piece No.		1	2	3	4	5	6	7
d	(mm)	0.690	0.670	0.690	0.745	0.750	0.815	0.800
$\bar{r}$		2.10	1.24	1.74	1.98	1.61	1.25	0.86
V_{zx}	(m/s)	3050	3230	3090	3120	3110	3240	3290
V_{zy}	(m/s)	2980	3160	2990	2980	3060	3140	3170
V_{zz}	(m/s)	6200	5970	6120	6220	6180	6090	5960
$K_1=V_{zy}/V_{zz}$		0.4805	0.5292	0.4886	0.4791	0.4949	0.5154	0.5324
$K_2=V_{zx}/V_{zz}$		0.4918	0.5409	0.5309	0.5017	0.5034	0.5319	0.5510
V_{SH0}(0°)	(m/s)	3310	3310	3320	3280	3340	3300	3280
V_{SH0}(45°)	(m/s)	3270	3190	3230	3290	3190	3220	3190
K_3		0.9861	0.9643	0.9724	1.0021	0.9537	0.9745	0.9743
W_{V400}		-1.56×10^{-2}	-3.94×10^{-3}	-1.31×10^{-2}	-1.45×10^{-2}	-1.24×10^{-2}	-6.61×10^{-3}	-2.30×10^{-3}
W_{V420}		-1.27×10^{-3}	-1.36×10^{-3}	-1.75×10^{-3}	-2.56×10^{-3}	-9.67×10^{-4}	-1.90×10^{-3}	-2.23×10^{-3}
W_{V440}		7.18×10^{-4}	1.83×10^{-3}	1.43×10^{-3}	-1.03×10^{-4}	2.45×10^{-3}	1.30×10^{-3}	1.30×10^{-3}
W_{R400}		-7.78×10^{-3}	2.79×10^{-3}	-5.55×10^{-3}	-6.83×10^{-2}	-4.94×10^{-3}	3.64×10^{-4}	4.28×10^{-3}
W_{R420}		-1.15×10^{-3}	-1.23×10^{-3}	-1.59×10^{-3}	-2.33×10^{-3}	-8.79×10^{-4}	-1.73×10^{-3}	-2.03×10^{-3}
W_{R440}		5.99×10^{-4}	1.53×10^{-3}	1.19×10^{-3}	-8.56×10^{-5}	2.04×10^{-3}	1.08×10^{-3}	1.08×10^{-3}
$\bar{E}$(Voigt)	(GPa)	242.8	231.5	240.5	241.8	239.8	234.2	229.8
$\bar{E}$(Reuss)	(GPa)	203.4	191.5	200.8	202.3	200.2	194.1	189.9

All the nine elastic rigidity constants, C_{ij} of the steel sheets and Young's modulus in the sheet plane were also calculated from Eqs.(1),(2) using the CODFs, W_{V4m0} and W_{R4m0} obtained as shown in Table I. Fig.3 shows the relations between the average value of the Young's modulus, $\bar{E}(=[E(0°)+2E(45°)+E(90°)]/4)$ obtained here and the average value of the plastic strain ratio, $\bar{r}(=[r(0°)+2r(45°)+r(90°)]/4)$ measured with a tension test. The relation obtained by Stickels and Mould(1970) is also shown in the same figure. It is shown in Fig.3 that the Young's modulus values obtained here using the Voigt approximation procedure (large circles) are larger than those obtained using the Reuss approximation procedure (large squares) by about 20 %. This is probably due to the inherent difference in the basic assumption of the two models. There are also large differences between the results obtained here and the results obtained by Stickels and Mould (small circles). This is most likely caused by the different methods of obtaining Young's modulus because the plastic strain ratio was also measured by Stickels and Mould with a similar tension test. They measured Young's modulus by a magnetostrictive method in which a sheet specimen is cut to 4 inch long by 1/4 inch wide. The non-destructive ultrasonic method here is of course completely different from their method. However it is interesting to find that the relation obtained here by taking the average of both approximations and their relation are quite similar to each other. In fact the averaged Young's modulus obtained here is slightly smaller than their Young's modulus. However, it is not correct to conclude that the value of Young's modulus obtained here using the average of the Voigt and Reuss approximations is near to the true value, because the true value was not measured here. The present ultrasonic method is also affected by the chosen values of single crystal elastic constants used in the calculation. Young's modulus calculated using different sets of single crystal elastic constants obtained by Schmid and Boas(1935), and by Rayne(1961) are shown in Fig.4. Fig.4 shows that Young's modulus obtained using Schmid's values is smaller than that obtained using Rayne's values but both relations have similar gradient. Young's modulus calculated using sets of single crystal elastic constants obtained by other authors lie between them (not shown here). In spite of the differences between the relations in Fig.4, the both relations have similar gradient which makes the present method well adapted for the consistent non-destructive prediction of the average plastic strain ratio $\bar{r}$ provided the same values of C_{011}, C_{012}, and C_{044} and the same approximation method is always used.

CONCLUSION

Equations to calculate from ultrasonic velocities all the lowest order CODF coefficients, W_{400}, W_{420}, and W_{440} of an orthorhombic plate of cubic crystals have been obtained. The equations include only the relative values of the single crystal elastic constants and the relative ultrasonic velocities. Both Voigt and Reuss approximations were used in the calculation. The ultrasonic method using EMATs here can produce polefigures but can not fully reproduce the fine structure. Average Young's modulus, $\bar{E}$ calculated using the ultrasonic CODF coefficients were in good correlation with average plastic strain ratio $\bar{r}$, showing that the ultrasonic method here is well adapted for practical use since it is very quick and nondestructive. However, some other technical and theoretical background work remains to be carried out before this method is put in real use. This involves the influence of temperature, stress, and magnetic fields upon relative ultrasonic velocities.

ACKNOWLEDGEMENT

The author would like to acknowledge useful discussion with Dr.S. Nagashima. He also acknowledges advice of Mr.K.Kawasaki on measurement of

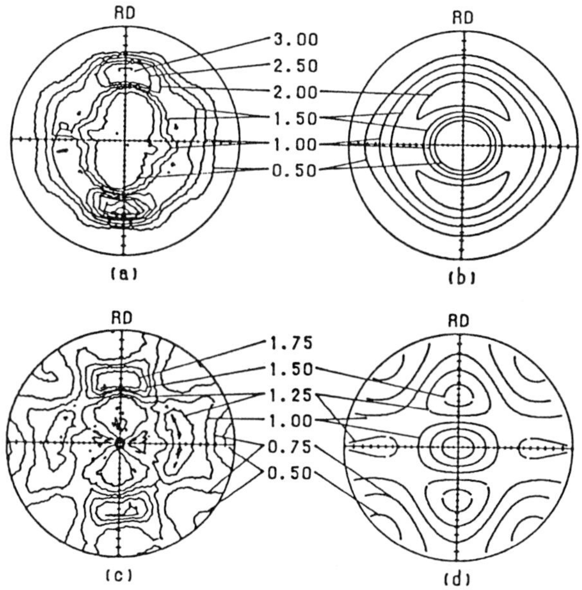

Fig.2 (100)polefigure diagrams:(a)x-ray polefigure on No.1 sample,(b)ultrasonic polefigure on No.1 sample,(c)x-ray polefigure on No.2 sample,and(d)ultrasonic polefigure on No.2 sample.

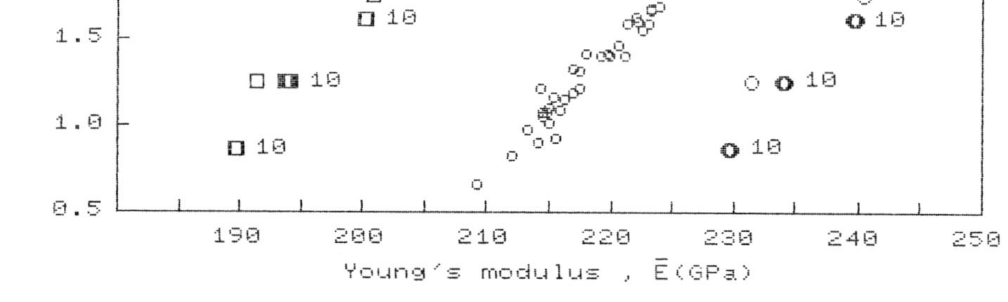

Fig.3 Relations between average Young's modulus $\bar{E}$ and average plastic strain ratio $\bar{r}$:large circles,measured here using the Voigt approximation; large squares,measured here using the Reuss approximation;small circles, measured by Stickels and Mould. The numbers refer to the number of different samples used for the overlapping data points indicated.

Fig.4 Relations between average Young's modulus $\bar{E}$ and average plastic strain ratio $\bar{r}$, measured here using the Voigt approxi mation; circles (Schmid), triangles (Rayne)

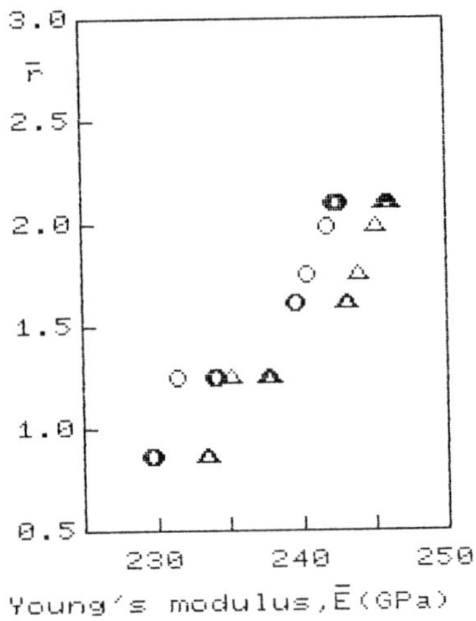

X-ray polefigures and he thanks Dr.O.B.Wright for his checking the equations. They are with R&D Labs.I and Electronics R&D Labs., Nippon Steel Corporation.

REFERENCES

Allen,D.R. and Sayers,C.M.(1984). The measurement of residual stress in textured steel using an ultrasonic velocity combinations technique. Ultrasonics,22,179-188.

Allen,D.R.,Langman,R. and Sayers,C.M.(1985). Ultrasonic SH wave velocity in textured aluminum plates. Ultrasonics,23,215-222.

Hirao,M.,Aoki,K. and Fukuoka,H.(1987). Texture of polycrystalline metals characterized by ultrasonic velocity measurements. J.Acoust.Soc.80(3),1434-1440.

Hirao,M.,Hara,N.,Fukuoka,H. and Fujisawa,K.(1988). Ultrasonic monitoring of texture in cold rolled stell sheets., J.Acoust.Soc.84(3),667-672.

Kawashima,K.(1990). Nondestructive characterization of texture and plastic strain ratio of metal sheets with electromagnetic acoustic transducers ,J.Acoust.Soc.Am.87(2),681-690.

Rayne,J.A. and Chandrasekhar,B.S.(1961). Elastic constants of iron from 4.2 to 300 K. Phys.Rev.,122,pp1714-1716.

Roe,R.J.(1965). Description of crystallite orientation in polycrystalline materials. J.Appl.Phys.36,2024-2031.

Roe,R.J.(1966). Inversion of pole figures for materials having cubic crystal symmetry. J.Appl.Phys.37,2069-2072.

Sayers,C.M.(1982). Ultrasonic velocities in anisotropic polycrystalline aggregates. J.Phys.D,15,2157-2167.

Schmid,E. and Boas,W.B.(1935). In Kristallplastizitat,pp21,Berlin:Springer.

Stickels,C.A. and Mould,P.R.(1970). The use of Young's modulus for predicting the plastic strain ratio of low carbon steel sheets. Metallurgical Transactions,1,1303-1312.

Thompson,R.B.,Lee,S.S. and Smith,J.F.(1986). Angular dependence of ultrasonic wave propagation in a stressed,orthorhombic continuum: Theory and application to the measurement of stress and texture. J.Acoust.Soc.Am.80(3),921-931.

Thompson.R.B.,Lee,S.S. and Smith,J.F.(1987) Relative anisotropy of plane waves and guided modes in thin orthorhombic plates:implication for texture characterization. Ultrasonics,25,133-137.

MATERIALS CHARACTERIZATION FROM ELASTIC WAVE ANISOTROPY IMAGES

A. G. Every [†], Wolfgang Sachse and M. O. Thompson [‡]

Department of Theoretical and Applied Mechanics
[‡] Department of Materials Science and Engineering
Cornell University, Ithaca, New York - 14853 U S A
[†] University of the Witwatersrand, Johannesburg, South Africa

SUMMARY

We describe in this paper the use of transient ultrasonic waves to image wave anisotropy effects in single crystal specimens of silicon. The measurements rely on the *point-source/point-receiver* (*PS/PR*) technique with a Nd:YAG laser as a scanned source and a point piezoelectric transducer as a detector. The image is based on the measurement of the signal arrival times and corresponding amplitudes of wave groups along various propagation directions in the single crystal specimens. It is found that along some directions there are strong focusing effects. An interpretation of the measured images based on an analysis of elastic waves in anisotropic solids is given. A potential application of this imaging procedure as a materials characterization tool is also shown.

INTRODUCTION

Anisotropic materials, including single crystals and certain composite materials, are playing an increasingly important role in structural applications and this has resulted in renewed efforts for the development of suitable non-destructive techniques for their characterization. Determination of the characteristics of the medium as well as their spatial variation, which are related to the composition and micro- and macrostructural features of the material, is the *inverse medium* problem, but a general solution of this is not yet available. Instead, measurements of the ultrasonic attenuation and wavespeeds as a function of frequency often form the basis of materials characterization procedures. The focus here is to explore the factors governing the anisotropy-related ultrasonic amplitudes in a signal and to consider their application for characterizing anisotropic materials and, in particular, single crystal specimens of silicon.

PS/PR TECHNIQUE

If a point source is used to generate ultrasonic signals in a material and if these are detected with a small aperture receiver, the so-called *point-source/point-receiver*

(PS/PR) technique results [1]-[3]. This technique has shown great promise in a number of materials inspection and characterization applications. The source may be a normal force obtained by the fracture of a capillary or a horizontal double force obtained by using a focused laser beam operating in the thermoelastic regime [4]. The latter is particularly useful when repetitive or scanned ultrasonic measurements are made. The receiver may be a small element piezoelectric transducer, a laser interferometer or a small aperture capacitive sensor [4]. Although the wave propagation in *PS/PR* measurements is more complex than in conventional plane-wave ultrasonic measurements, many of the problems and limitations of the latter are minimized or avoided. Since the theory of transient ultrasonic signals in an ideal, bounded isotropic medium has become well established, its application to interpret the signals propagating in a real material from a source to a receiver of known characteristics can be the basis of a powerful *quantitative* ultrasonic measurement technique for determining the characteristics of a material. As pointed out in previous publications, simultaneous longitudinal and shear wave measurements can be extracted from one waveform. Procedures have also been demonstrated for using *PS/PR* measurements to determine the frequency-dependent wavespeeds and attenuations of longitudinal and shear waves propagating in homogeneous, isotropic, viscoelastic materials [5,6]. Also of essential importance for the measurements to be described, the ultrasonic signals emitted by an ultrasonic point source are simultaneously propagated in a wide range of directions in a specimen. The orientation dependence of the wave amplitudes and speeds of propagation from each excitation of the source can be found by using an array of sensors or by scanning either source or receiver over the specimen.

The *PS/PR* technique shows particular promise for measurements in thick and highly absorptive composite materials since broadband pulses rich in low frequencies are typically generated. However, for this application it will be necessary that the propagation of *PS/PR* signals in anisotropic materials be well understood.

PS/PR MEASUREMENTS IN ANISOTROPIC MATERIALS

We consider *PS/PR* measurements made for a number of source and receiver configurations in an anisotropic material. We then consider the orientation dependence of the wave amplitudes and the focusing effects which can arise.

The elastic properties of an anisotropic material are specified in terms of a matrix of elastic constants, C_{ilkm}, which are often written in terms of the contracted notation, $C_{\alpha\beta}$. The propagation of ultrasonic waves through such a material is described by *Christoffel's* equation

$$(C_{ilkm} n_l n_m - \rho v^2 \delta_{ik}) U_k = 0 , \tag{1}$$

where the n_i's refer to the components of the wave normal $\hat{\mathbf{n}}$, v is the *phase* velocity, ρ is the density of the material, the U_k represent the components of the polarization vector of the wave and δ_{ik} is the Kronecker delta. The solution to the secular, or characteristic equation, given by

$$|C_{ilkm} n_l n_m - \rho v^2 \delta_{ik}| = 0 , \tag{2}$$

permits finding the phase velocities corresponding to the *quasi-longitudinal* and two *quasi-transverse* wave modes which can propagate in a particular direction in an

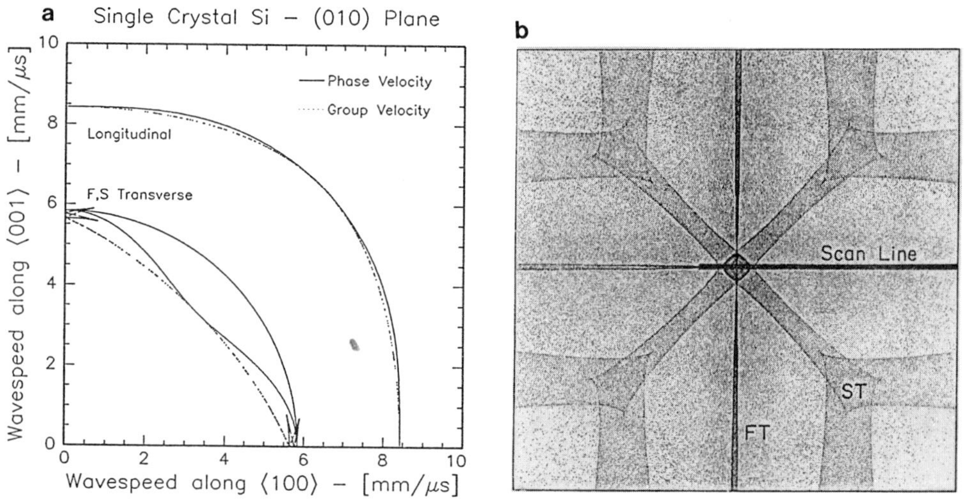

Figure 1. (a) Phase and group velocities as a function of direction in the (010) plane in single crystal Si; (b) Computed anisotropy image of ⟨100⟩ Si.

anisotropic material. The directional dependence of the phase velocities for Si are shown in Figure 1(a). Algorithms for determining the elastic constants of an anisotropic material from wavespeed measurements made along non-principal directions of a specimen has recently been demonstrated [7,8].

The peaks and other discontinuities in a *PS/PR* waveform correspond to the arrival of energy of a particular wave mode at the sensor and hence the wavespeeds related to the arrival times of features in a waveform are expected to correspond to the *group velocity* of the waves in the material. The group velocity $\mathbf{V}$ of a wave packet having center frequency ω is given in terms of the phase velocity v according to [9]

$$\mathbf{V} = \nabla_{\mathbf{k}}\,\omega(\mathbf{k}) = \{v(\hat{\mathbf{n}}) - \hat{\mathbf{n}}\cdot\nabla_{\hat{\mathbf{n}}}\,v(\hat{\mathbf{n}})\}\,\hat{\mathbf{n}} + \nabla_{\hat{\mathbf{n}}}\,v(\hat{\mathbf{n}}) \qquad (3)$$

where $\hat{\mathbf{n}}$ is the wave normal. It follows that for waves propagating in an anisotropic material, the group velocity, in general, does not equal the phase velocity in either its magnitude or direction. A comparison between the orientation dependence of the phase and group velocities of a single crystal of silicon is also shown in Figure 1(a). We note that in specific directions for each of the wave modes the two velocities are equal. The relationship between the two velocities is $\mathbf{V}\cdot\hat{\mathbf{n}} = v$. Hence, when the group velocity $\mathbf{V}$ is parallel to $\hat{\mathbf{n}}$, the group and phase velocities are equal. This is the case for the longitudinal and shear waves propagating in the ⟨100⟩ and ⟨110⟩ directions and for longitudinal waves propagating in the ⟨111⟩ direction of a material of cubic elastic symmetry. In fact, a measurement of the *PS/PR* waveforms and the corresponding group velocities along these directions of a specimen of Si therefore readily yield the values for the elastic constants of this material [10].

ANISOTROPY FOCUSING

The decay of wave amplitudes of *PS/PR* signals as they propagate through a material is a consequence of the geometric divergence of the waves as well as the material's attenuation. In an anisotropic solid, these are not the only factors affecting the amplitude of a *PS/PR* signal. Also important is an effect known as *anisotropy* or *phonon focusing* c.f. [11], by virtue of which the energy flux radiated from an ultrasonic point source in an anisotropic material is highly concentrated in particular directions. Past observations of this effect were made at liquid helium temperatures in non-metallic crystals with laser-generated, thermal phonons at THz frequencies which were detected in most experiments with bolometers. The spatial variation of detected intensities, comprising an image, display the expected *phonon focusing* effects [11].

It is possible to compute the influence of focusing on the intensities of the signals for various directions of a single crystal specimen by means of a *Monte Carlo* simulation. For this, one considers an incoherent source to radiate longitudinal as well as fast and slow shear waves with wave normals distributed in all possible directions in the specimen. The calculated ray vectors associated with these normals are then projected onto the viewing surface where the receiving transducer is located. Computing the integrated intensity at a number of receiver points over a surface, one can synthesize a *focusing pattern* [11,12]. The pattern computed for a single crystal specimen of Si of $\langle 100 \rangle$ orientation, is shown in Figure 1(b) [12]. Due to bunching or focusing of the ray vectors in certain directions, the intensity pattern is highly non-uniform. Indeed, there is a pattern of caustics where the predicted flux intensity diverges. It is also possible to compute the flux intensity as a function of receiver location and time.

An anisotropy image can be obtained by scanning either the source or the receiver in a systematic manner over the surface of a specimen. If the signals detected for a number of closely-spaced configurations of the source/receiver are stacked together, one can generate an image which facilitates the detection of patterns of wave arrivals and amplitudes arising in the propagation of waves in different directions in a specimen.

PS/PR WAVEFORMS IN Si

In order to illustrate the characteristics of transient waves propagating in anisotropic materials, we have measured the propagation of ultrasonic waves from a point source in the three symmetry directions of a single crystal specimen of Si [10]. The source was a normal force obtained from a capillary fracture on the surface of silicon single crystal disks, 7.62 cm in diameter and 1.0 cm thick, of $\langle 100 \rangle$, $\langle 110 \rangle$ and $\langle 111 \rangle$ orientation. The signals were detected with a capacitive displacement sensor of 1.0 mm aperture and bandwidth 10 kHz to 10 MHz. The displacement signals and their time derivatives, corresponding to the normal velocity signals, detected at epicenter for each of these specimens are shown in Figure 2.

These waveforms differ significantly from those measured in isotropic materials. In the displacement signal measured in the $\langle 100 \rangle$ oriented crystal, the arrival of the first longitudinal (L) mode appears as a sharp upward displacement of the surface which is followed by a plateau. This is followed by a dip at the predicted slow shear (ST) wave arrival time and then a very steep upward displacement. This anomalous behaviour is evidently associated with the complex folded shape of the slow shear sheet of the wave surface near the $\langle 100 \rangle$ axis which can be seen in Figure 1(a). The signal detected in the $\langle 101 \rangle$ oriented crystal displays sharp upward displacements corresponding to peaks in the surface velocity at the arrivals of the L and fast shear (FT) waves. The FT sheet of the wave surface has a narrow fold around the $\langle 101 \rangle$

direction, and this is evidently the reason why there is not simply a change in slope of the displacement signal at the *FT* arrival. There is no significant *ST* signal for this case. The *PS/PR* signal for the ⟨111⟩ oriented crystal contains no appreciable shear mode signal. The mode-converted signals *L2S* and *3L2S* are also absent. The analysis of *PS/PR* waveforms to recover the elastic constants of single crystals of Si has been described in considerable detail in Refs. [10,13].

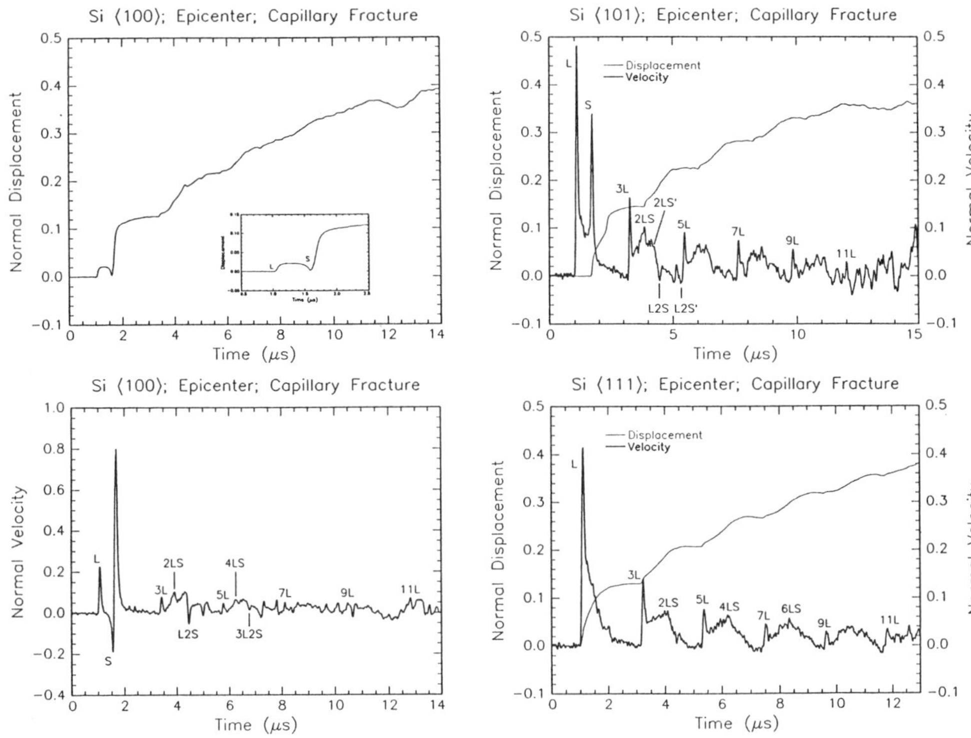

Figure 2. Epicentral displacement and velocity waveforms in Si crystal specimens of ⟨100⟩ , ⟨101⟩ and ⟨111⟩ orientations. Step excitation source.

ANISOTROPY FOCUSING IN Si

An anisotropy image, computed for a crystal of Si of ⟨100⟩ orientation, is shown in Figure 3(a) [12]. The horizontal axis refers to the position of the source point relative to the receiver (mounted on the opposite side of the plate). The vertical axis is the time axis of the *PS/PR* waveforms. The dark regions in this figure correspond to concentrations of energy flux. Visible are the arrivals of the longitudinal (*L*) and the slow shear (*T*) and head waves (*HW*). These arrivals are followed by the arrivals of the multiply-reflected and mode-converted signals.

If a laser point source is scanned along a line across a single crystal specimen of Si of ⟨100⟩ orientation and the generated signals are detected with a 1.3 mm aperture piezoelectric transducer attached to the back surface of the sample, the scan image shown in Figure 3(b) is obtained when all the detected waveforms are displayed together. Although the piezoelectric sensor gives rise to considerable ringing in the signals, the predicted and measured images are in reasonably close agreement. The

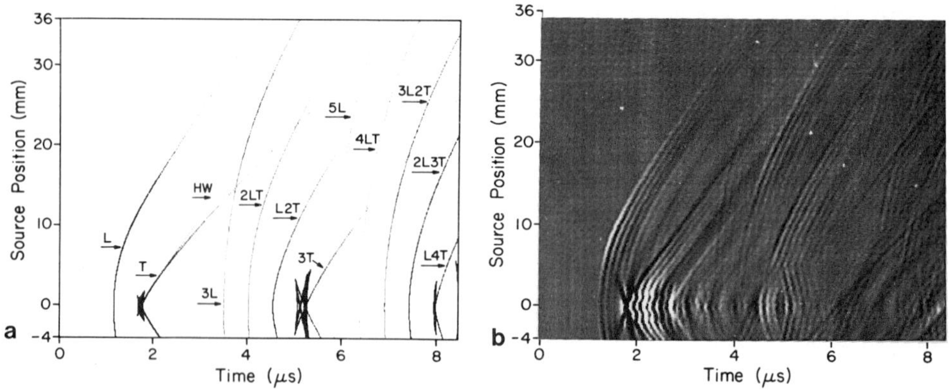

Figure 3. Anisotropy images in a Si crystal specimen of ⟨100⟩ orientation and ⟨010⟩ scan. (a) Computed; (b) Measured with pulsed laser source, piezoelectric detector.

orientation of the scan line with respect to the focusing pattern is shown in Figure 1(b). Of particular interest is the intense amplitude of the slow transverse wave near epicenter which is a consequence of the intense focusing of this mode evident at the center of Figure 1(a). Also striking is the concave structure of the $3T$ arrivals in the epicentral region. This is evidence of the complex folded shape of the goup velocity surface near the ⟨100⟩ direction of Si.

A confirmation of these focusing effects can also be observed in specimens of other orientations. For example, if measurements are made on a silicon crystal of ⟨111⟩ orientation, a large peak is observed at the arrival of the shear wave not at epicenter (i.e. ⟨111⟩ direction), but rather, at the off-epicentral position, again corresponding to the ⟨100⟩ direction. This is illustrated in Figure 4.

Evidence that the anisotropy images are sensitive indicators of the properties of the material was observed when scanning a Si specimen that had been damaged

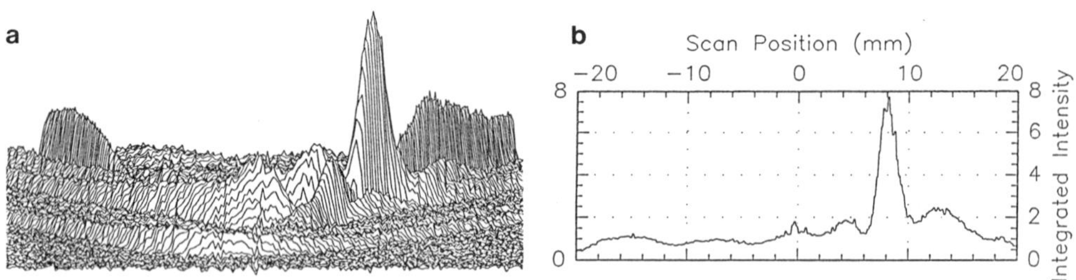

Figure 4. Anisotropy images in a Si crystal specimen of ⟨111⟩ orientation. (a) "Head-on" view of measured image; (b) Time-integrated intensity of the waveforms as a function of position.

by laser ablation at several points on its surface. The measured image is shown in Figure 5(a). Each of the damage points is visible because the amplitudes of the *PS/PR* waveforms at these points exhibit a behavior which differs markedly from the waveforms obtained in the undamaged regions of the specimen. A comparison of two waveforms illustrating this effect is in Figure 5(b). The damage resulting from ablation extends into the interior of the specimen allowing penetration of the laser excitation pulse. Consequently, the sound source is no longer localized at the surface, but rather extends into the interior of the material. From the arrivals of the signals detected over the damaged region of the specimen, we estimate that the source region extends up to 4 mm into the interior of the specimen.

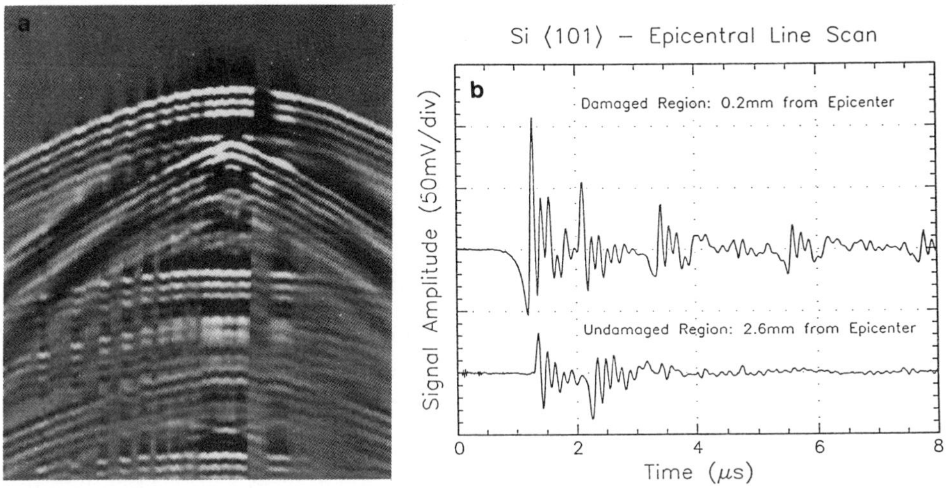

Figure 5. (a) Anisotropy image and, (b) waveforms of a surface-damaged Si crystal specimen of ⟨110⟩ orientation; Pulsed laser source, piezoelectric detector.

CONCLUSIONS

We have summarized in this paper measurements utilizing the *PS/PR* technique to image wave anisotropy effects in single crystal specimens of silicon. We have shown that the amplitudes of *PS/PR* signals, corresponding to the amplitudes of various ray arrivals in particular directions, exhibit strong focusing effects. An interpretation of the measured images based on an analysis of elastic waves in anisotropic solids has been given. A potential application of this imaging procedure as a materials characterization tool was also shown.

ACKNOWLEDGEMENTS

We thank Dr. K. Y. Kim for the results shown in Figure 2. Aspects of the work described here have been supported by the Office of Naval Research (Solid Mechanics and Physical Acoustics Programs). The work of A. G. E. has been supported by the National Science Foundation through the Materials Science Center at Cornell University and by the University of the Witwatersrand.

REFERENCES

[1] W. Sachse and K. Y. Kim, "Point-source/Point-receiver materials testing", in *Ultrasonic Materials Characterization II*, J. Boussière, J. P. Monchalin, C. O. Ruud and R. E. Green, Eds., Plenum Press, New York (1987), pp. 707-715.

[2] W. Sachse and K. Y. Kim, "Quantitative acoustic emission and failure mechanics of composite materials", *Ultrasonics*, **25**, 195-203 (1987).

[3] W. Sachse, B. Castagnede, I. Grabec, K. Y. Kim and R. L. Weaver, "Recent developments in quantitative ultrasonic NDE of composites", *Ultrasonics*, **28**, 97-104 (1990).

[4] W. Sachse, "Transducer considerations for point-source/point-receiver materials measurements", in *Ultrasonics International '87 - Conference Proceedings*, Butterworths, Guildford, UK (1987), pp. 495-501.

[5] R. L. Weaver and W. Sachse, "Asymptotic viscoelastic rays in a thick plate", T&AM Preprint, Cornell University, Ithaca, NY (June 1988). In Press: *Journal of Applied Mechanics.*

[6] R. L. Weaver, W. Sachse and L. Niu, "Transient ultrasonic waves in a viscoelastic plate; Part I. Theory; Part II. Applications to materials characterization", *J. Acoust. Soc. Am.*, **85**(6), 2255-2261; 2262-2267 (1989).

[7] B. Castagnede, W. Sachse and M. O. Thompson, "Determination of the elastic constants of anisotropic materials with laser-generated ultrasound", in *Ultrasonics International'89: Conference Proceedings*, Butterworth, Guildford, UK (1989), pp. 71-77.

[8] B. Castagnede, J. T. Jenkins, W. Sachse and J. S. Baste, "Optimal determination of the elastic constants of composite materials from ultrasonic wavespeed measurements", *J. Appl. Phys.*, **67**(6), 2753-2761 (1990).

[9] F. I. Federov, *Theory of Elastic Waves in Crystals*, Plenum Press, New York (1968).

[10] K. Y. Kim, W. Sachse and A. G. Every, "On the determination of sound speeds in cubic crystals and isotropic media using a broadband ultrasonic point-source/point-receiver method", Materials Science Center Report #6979, Cornell University, Ithaca, NY (June 1990). Submitted for publication.

[11] G. A. Northrop and J. P. Wolfe, "Phonon imaging: Theory and applications", in *Nonequilibrium Phonon Dynamics*, W. E. Bron, Ed., Plenum Press, New York (1985), pp. 165-242.

[12] A. G. Every, W. Sachse, K. Y. Kim and M. O. Thompson, "Phonon focusing in silicon at ultrasonic frequencies", Materials Science Center Report #6930, Cornell University, Ithaca, NY (May 1990). In Press: *Phys. Rev. Letts.*

[13] A. G. Every and W. Sachse, "Determination of the elastic constants of anisotropic solids from acoustic wave group velocity measurements", Materials Science Center Report #6960, Cornell University, Ithaca, NY (June 1990). In Press: *Physical Review.*

TEXTURE ANALYSIS WITH 3MA-TECHNIQUES

I. Altpeter

Fraunhofer Institut für zerstörungsfreie Prüfverfahren
D-6600 Saarbrücken, FRG

INTRODUCTION

Texture is generated by rolling and heat-treatment processes during the production of sheets. After such processes (hot and cold rolling, recrystallization), the crystal orientations are not equally distributed in the alloy. Consequently elastic and magnetic properties of the sheets are anisotropic. There is a correlation between elastic and plastic anisotropy. Accordingly, the usually applied techniques to determine the texture are different: for instance, differences in plastic deformation can be detected by the determination of the average Lankford plastic strain ratio r_m, elastic anisotropy by the determination of the Young's modulus and x-ray texture analysis. In addition to the x-ray measurements, ultrasonic and magnetic techniques offer the possibility for fast and relatively simple analysis of texture. The x-ray measurements, which can be applied under practical conditions as well, are the state of technology[1]. The measurement of ultrasonic velocities in different directions is an alternative approach to the x-ray measurement and is being developed for practical applications[2]. The elastic anisotropy will be determined by the crystal texture, whereas the plastic anisotropy by the microstructural parameters like dislocation density and grain boundaries. Since the same microstructural features restrict the mobility of Bloch-walls, magnetic techniques offer the possibility to analyze the plastic anisotropy.

PHYSICAL BACKGROUND

In the body-centered cubic iron single-crystal, the magnetization is directional - dependent. The reason for this is the crystal-anisotropy energy. By magnetization parallel to the crystal directions [100], [110] and [111], different magnetization curves can be obtained. The quasi-static magnetostriction shows a similar anisotropy (Fig. 1). In order to find nondestructive techniques, i.e. techniques which allow a direct access to the material surface with pick-up-transducers, other measuring processes than hysteresis or magnetostriction measurements are necessary. The reason is that hysteresis measurements can be performed only with circumferential coils into which the specimen is introduced. Therefore, only the specimens with special shapes (cylindrical) can be evaluated. Larger specimens must be destroyed and cut in spheres or ellipsoids.

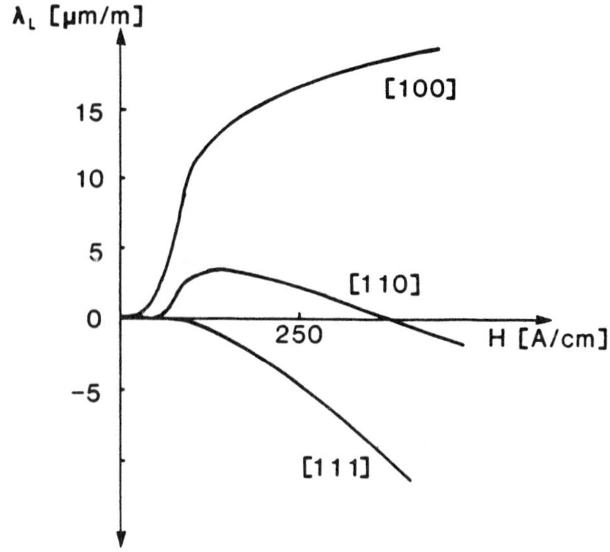

Fig. 1. Anisotropy of magnetostriction in a stress-free iron single crystal

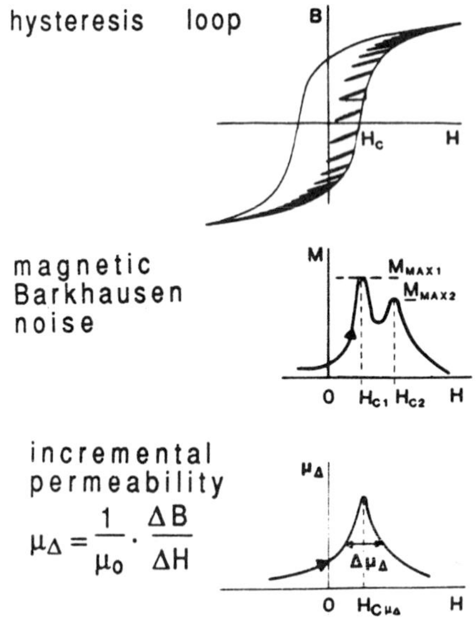

Fig. 2. Micromagnetic parameters (schematically)

Magnetostriction measurements are available by fixing strain gauges on the surface under inspection. Because the texture measurement principally uses the data acquisition in different directions, the technique is very expensive and cannot be applied on-line in a steel mill.

We use the 3MA-method (Micromagnetic-, Multiparameter, Microstructure- and Stress-Analysis)[3] as a nondestructive technique for texture evaluation. 3 MA-parameters are the magnetic Barkhausen noise, the incremental permeability and the dynamic magnetostriction.

Magnetic Barkhausen noise M[4]

The electrical voltage pulses induced in a magneto-inductive sensor (tape recorder head, air-coil) by irreversible Bloch-wall jumps during magnetic reversals, is named magnetic Barkhausen noise. Figure 2 shows a typical hysteresis curve, the magnetic Barkhausen noise amplitude and the incremental permeability as a function of the magnetic H-field. The Barkhausen noise is recorded in the frequency range between 0.5 kHz-5 MHz. After rectification and low-pass-filtering of the Barkhausen noise, signal profile curves can be detected versus the magnetic field strength. Depending on the microstructural state and the anisotropy of the material, multiple maxima can be observed. Measuring parameters are the amplitude of the maxima and the magnetic field values where the maxima occur. In textured material, the Barkhausen noise depends on the angle between the magnetization direction and the magnetically soft direction of the grains (for example [1 0 0] in iron). The magnetically soft direction is related to the texture and hence, texture influences the Barkhausen noise. When the [1 0 0] direction is parallel to the magnetization direction, 180° and 90° Bloch-wall motions take place. When [1 1 0] is parallel to the magnetization direction 180°, 90° Bloch-wall motion and rotation processes take place. When [1 1 1] direction is parallel to magnetization direction, 180° Bloch-wall motion and rotation processes take place[5]. The strength of the magnetization field needed to initiate the 180° and 90° Bloch-wall motions and rotation processes is different. Depending on the texture, which influences the type and intensity of Bloch-wall motions and rotation processes, the Barkhausen noise amplitude changes. The amplitude maxima of the Barkhausen noise curve (M_{MAX1}, M_{MAX2}) plotted as a function of the direction of magnetization, lead to the so-called magnetic pole figure of the Barkhausen noise.

Incremental permeability μ_Δ[4]

With this measuring parameter, reversible as well as irreversible magnetization processes can be measured. At smaller incremental magnetic ΔH-field amplitudes more reversible processes and at greater ΔH-fields irreversible magnetization processes are increasingly measured by μ_Δ-measuring parameters. The incremental permeability is a function of the dynamic magnetization following the hysteresis curve. The incremental permeability curve is shown in Fig. 2. From the incremental permeability, different measuring parameters can be derived, like for example $\Delta\mu_\Delta^{50\%}$ defined as the width of the μ_Δ(H)-curve at 50% amplitude (full width at half maximum, FWHM). Depending on the H-field range, crystal groups with different orientations are measured by these parameters similar to the magnetic Barkhausen noise. If one wishes to measure the [110]-texture for example, it is reasonable to use the quantity $\Delta\mu_\Delta$ as this parameter is mainly sensitive to magnetostrictively active 90° Bloch-wall motion. When the [110] lattice plane distribution increases in one direction, $\Delta\mu_\Delta$ has to increase as well, if the other directions are magnetically softer.

Dynamic Magnetostriction E_λ [6]

This nondestructive measuring parameter is influenced by the magnetostrictively active reversible and irreversible Bloch-wall motions and rotation processes. A quasi-static magnetic field is superimposed by a magnetic ΔH field with a frequency $\leq$ 3 MHz. In order to detect the generated ultrasonic waves, we use an electromagnetic ultrasonic sensor. The amplitude of the ultrasonic wave is used as a measuring parameter. The $E_\lambda(H)$-curves are represented in Fig. 3 for the several crystalline directions. The amplitude of E_λ at 280 A/cm (M_2) is used as a parameter for texture analysis.

Tangential field-strength H_t [7]

Further 3MA-quantities can be derived from the time signal of the tangential H-field strength by harmonic analysis for instance:
- the amplitudes and phase shifts of the higher harmonics
- the distortion factor K which is given as

$$K[\%] = \sqrt{\frac{|\tilde{A}_3|^2 + |\tilde{A}_5|^2 + |\tilde{A}_7|^2}{|\tilde{A}_1|^2}} \cdot 100$$

$\tilde{A}_1$ is the coefficient of the fundamental wave, $\tilde{A}_3$ - $\tilde{A}_7$ are the coefficients of the higher harmonics.

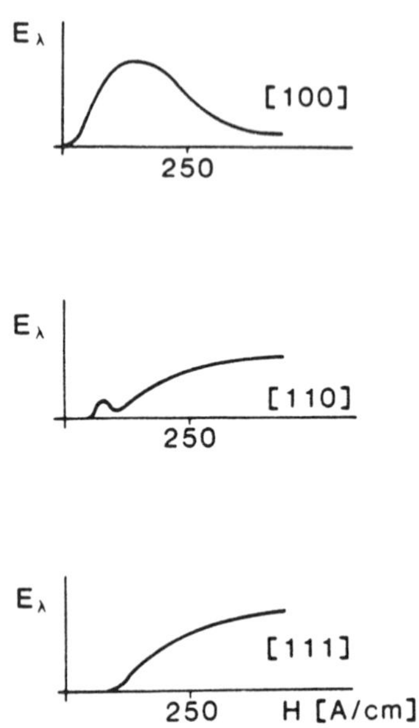

Fig. 3. Dynamic magnetostriction E_λ as a function of magnetic field strength for different crystallographic directions

Tab. 1. Chemical composition and thickness of the sheet qualities (Mass Fraction in %)

Steel	C	Si	Mn	S	Al	N	P	thickness [mm]
ST 14	0.06	0.01	0.28	0.035	0.06	0.004	0.018	0.88
ST 12	0.036 - 0.072	0.01 - 0.02	0.23 - 0.41	0.002 - 0.023	0.041 - 0.042	0.001	0.008 - 0.011	0.8 - 1

The generated higher harmonics and the distortion factor K depend on the shape of the hysteresis curve. In the body-centered cubic iron single-crystal, the magnetization is directional-dependent. Consequently the distortion factor K is directional-dependent. Therefore the distortion factor K can also be used as a texture sensitive measuring parameter.

EXPERIMENTAL PROCEDURE

The magnetic excitation of the specimen, in most cases, is performed by an u-shaped electromagnet. A function generator in connection with a bipolar power supply is used for continuous cyclic magnetization. In most cases the maximum magnetic field H_{max} is $\leq$ 100 A/cm. The magnetization frequency is usually in the range 0.1 Hz - 100 Hz. The tangential magnetic field strength is measured by using appropriate Hall-elements. For the direction-dependent magnetic Barkhausen noise measurements, air coils or coils with a ferrite core are used as magneto inductive sensor. Both are located in the central position between the pole-shoes of the electromagnet. The chemical composition and thickness of the cold-rolled strips is shown in Tab. 1.

RESULTS

Anisotropic magnetic measuring quantities were correlated with the crystal texture as well as with the plastic anisotropy. IzfP-prototype devices[3] and laboratory equipment were used. The magnetic measuring quantities were compared with x-ray texture measurements (pole figures) and destructively determined anisotropy values (r_m-values).

As mentioned earlier, by recording micromagnetic quantities in different magnetization directions, the so-called magnetic pole-figures are obtained. A comparison of these magnetic pole-figures with x-ray pole figures suggests a similarity concerning the information obtainable. The measuring direction is defined mainly by the magnetic field direction. In addition to this, the direction-dependency is increased with the sensors, by which the tangential or normal magnetic components are weighted, received or excited. Measurements on different cold rolled strip specimens have shown that the magnetic pole figures correlate with the x-ray pole figures. Fig. 4 shows the distribution of the prameter M_{MAX2}. Fig. 5 shows a [110]-lattice plane distribution of a cold-rolled strip of steel quality ST 14 (called strip B) obtained by x-ray analysis. A good correlation between magnetic and x-ray pole figures can be observed. Both pole figures indicate the rolling direction as preferred orientation.

The incremental permeability curves were also recorded in different directions. The curves differ in the curve broadening as measured by FWHM as well as in the coercivity $H_{C\mu}$. The higher the coercivity $H_{C\mu}$ and the FWHM, the stronger is the hinderance of Bloch-wall movements during the remagnetization process. Plotting $H_{C\mu}$ and $\Delta\mu_{\Delta}^{50\%}$ as a function of the direction of magnetization, we obtain the so-called incremental permeability pole figures. Fig. 6, for example, shows the pole figure of $H_{C\mu}$ for the cold-rolled strip specimen B. This pole figure also correlates well with the x-ray [110] pole figure shown in Fig. 5. In all the three parameters (incremental permeability, magnetic Barkhausen noise and dynamic magnetostriction) the influence of texture can be seen. Fig. 7, for example, shows the dynamic magnetostriction pole figure for a cold rolled strip B. The amplitude of the 2nd maximum of the $E_\lambda(H)$-curve[7] is plotted as a function of the angle of rotation. Fig. 8 shows the x-ray pole figures of the [110]- and [111]-lattice planes for a tilt angle of 37°. From this a preferred direction which is about 30°-45° to the rolling direction is found. Since the amplitude M_2 was recorded at ~ 280 A/cm, the contribution from the [100]-crystals to the amplitude M_2 compared with other crystal directions decreases.This is because $d\lambda_L/dH$ takes smaller values with increasing H-field strengths for the [100]-direction than for the relatively magnetically harder [110]- and [111]-directions (Fig. 3). This means that one has to correlate micromagnetic results in the high H-field range (> 200 A/cm) with [111]- and [110]-x-ray results. Since the relative intensity of the [111]-plane is much smaller than the intensity of the [110]-crystal contribution (Fig. 8) the M_2-pole figure is mainly obtained by the [110]-texture[6].

The combination of measuring quantities, which are derived from the magnetic Barkhausen noise and the tangential field strength are after first investigations, in linear correlation with the r_m-value which characterizes the plastic anisotropy. R_m is defined by:

$$r_m = \frac{1}{4} \cdot [r\ (0°) + 2\,r\ (45°) + r\ (90°)];$$ r is defined in ref. 2.

If one defines - taking analogy to the definition of the r_m-value - and plots the ratio H^*_{CM}/M^*_{MAX} against the objective quantity r_m (Fig. 9), one finds a linear correlation between H^*_{CM}/M^*_{MAX} and r_m for the examined steel quality St 12. After the application of a multi-regression algorithm the derived quantity H^*_{CM}/M^*_{MAX} shows a correlation to the r_m-values with a correlation coefficient of 0.86. The scattering of the measuring points can first be explained by the error in r_m and secondly by different residual stress states in the cold rolled-samples, as the 3MA-parameters are influenced by both the residual stress state and texture state. Fig. 10 shows the distortion factor of the tangential field strength

$$K^* = \frac{K(WR) + 2K(45°) + K(90°)}{4}$$ as a function of the r_m-values for ten cold rolled strips of steel quality ST 12.

Here, a correlation coefficient of 0.7 is obtained.

CONCLUSION

The 3MA-quantities M_{MAX2}, $H_{C\mu}$, M_2 were measured for different directions of the steel strips with respect to the magnetization direction and shown as "magnetic pole figures". The results were compared with X-ray pole figures. A good correlation was obtained between the magnetic and x-ray pole figures. This paper dealt with the experimental results obtained from specimens of low-alloy steel grades with small carbon content (deep drawing steels) and well-defined textures, characterized by the deep drawability values r_m. The measurements were performed using the direction dependence of the 3MA-quantities. At the present stage of development, Barkhausen noise- and incremental permeability profiles were recorded by means of a rotating yoke-transducer, i.e. by varying the magnetic direction. Laboratory equipment, the IzfP-3MA-prototype and automated probe manipulation systems were used. According to these first results, combina-

$$H^*_{CM} = \frac{1}{4} \cdot [H_{CM}(0°) + 2H_{CM}(45°) + H_{CM}(90°)]$$

and

$$M^*_{MAX} = \frac{1}{4} \cdot [M_{MAX}(0°) + 2M_{MAX}(45°) + M_{MAX}(90°)]$$

H_{CM} and M_{MAX} are defined in ref. 4.

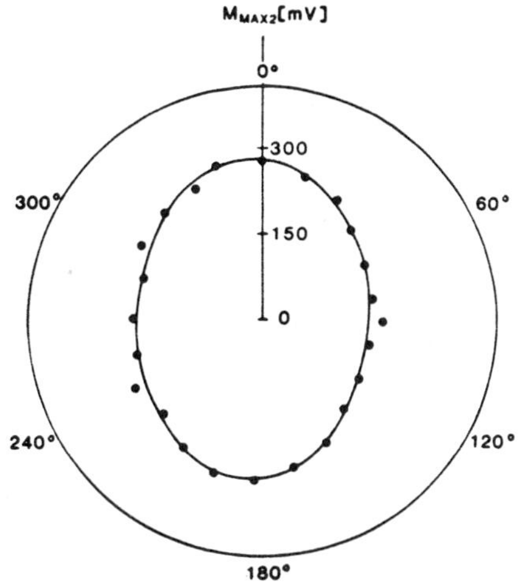

Fig. 4. "Magnetic pole figure" with Barkhausen noise amplitude M_{MAX2} (Specimen of cold rolled strip B)

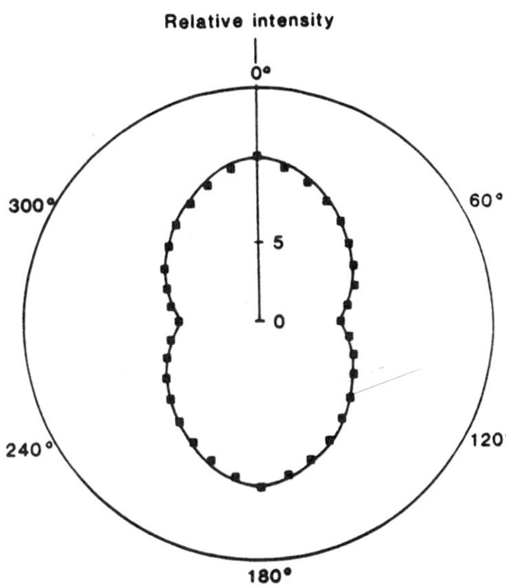

Fig. 5. (110)-lattice plane distribution in the sheet plane of cold rolled strip B as determined by x-ray

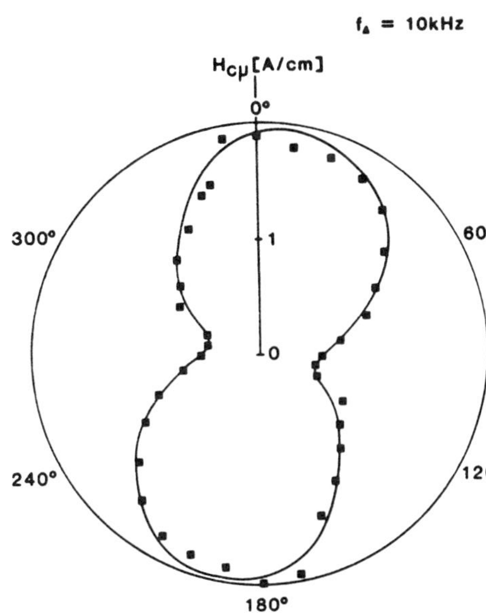

Fig. 6. "Magnetic pole figure" with coercivity $H_{C\mu}$ (Specimen of cold rolled strip B)

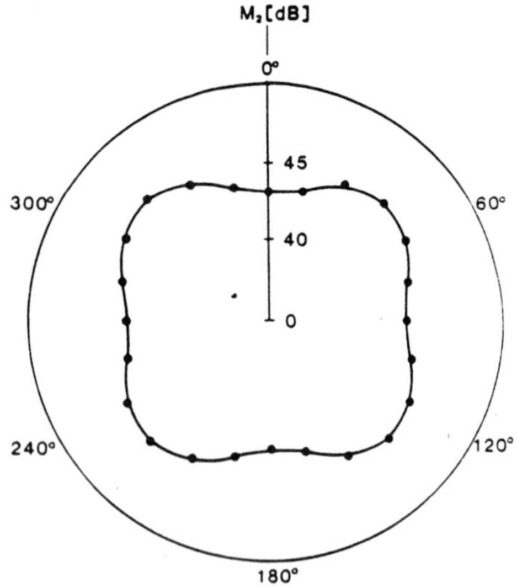

Fig. 7. Magnetic pole figure with dynamic magnetostriction (Specimen of cold rolled strip B)

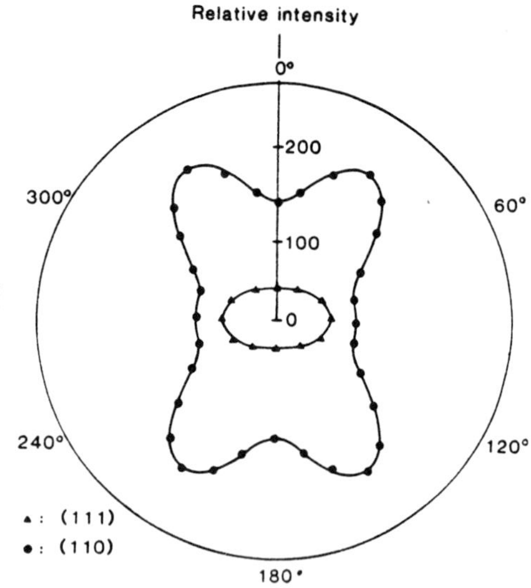

Fig. 8. Lattice plane distribution for a tilt angle of 37°, x-ray investigation (Specimen of cold rolled strip B)

$$M^*_{MAX} = \frac{M_{MAX}(0°) + 2M_{MAX}(45°) + M_{MAX}(90°)}{4}$$

$$H^*_{CM} = \frac{H_{CM}(0°) + 2H_{CM}(45°) + H_{CM}(90°)}{4}$$

$$K^* = \frac{K(0°) + 2K(45°) + K(90°)}{4}$$

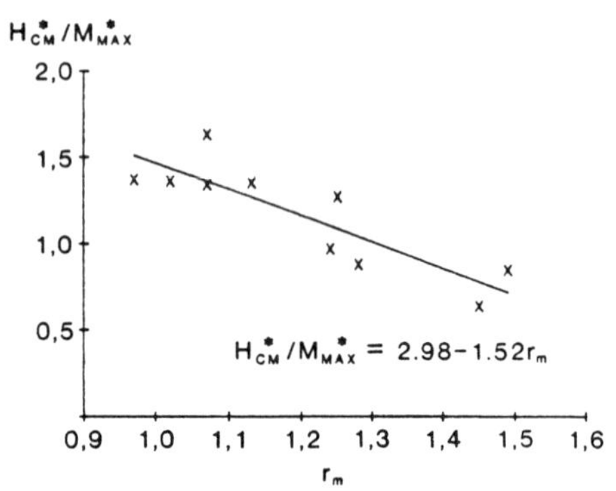

Fig. 9. Measuring quantity $H_{CM}{}^*/M_{MAX}{}^*$ as a function of r_m-value (deep drawing steel ST 12)

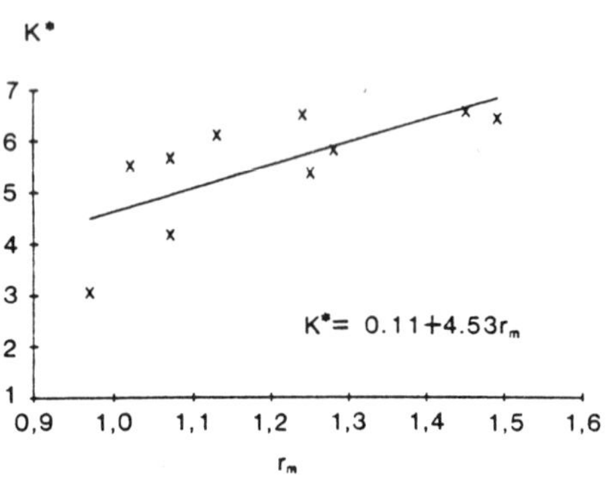

Fig.10. Measuring quantity K^* as a function of r_m-value (deep drawing steel ST 12)

tion of measuring quantities derived from Barkhausen noise-, incremental permeability- and tangential field strength-measurements were found to have linear correlation with the deep drawability parameter r_m suggesting that 3MA technique can be used for nondestructive characterization of deep drawing steels.

REFERENCES

1. H.J. Kopineck, International Conference on: Monitoring Surveillance and Predictive Maintenance of Plants and Structures Taormina , Italy, Proceedings 1: 55 - 66, (15th - 18th October 89)
2. M. Spies, E. Schneider, Nondestructive Characterization of Materials (Eds. P. Höller, V. Hauk, G. Dobmann, C.O. Ruud, R.E. Green, Jr.) Berlin, Springer Verlag: 296-302 (1989)
3. W.A. Theiner, B. Reimringer, H. Kopp, M. Gessner, ibid.2.: 699-706
4. W.A. Theiner, I. Altpeter, in: New Procedures in Nondestructive Testing (Proceedings). Editor: P. Höller, Springer-Verlag Berlin, Heidelberg, 575 (1983)
5. A. Seeger, "Moderne Probleme der Metallphysik". Berlin: Springer: 229 (1966)
6. R. Koch, P. Höller, ibid.2: 644-651
7. G. Dobmann, H. Pitsch, ibid.2.: 636-643

AUTHORS INDEX

SUBJECT INDEX